The Sciences in
Enlightened Europe

From Bernard de Fontenelle, *Entretiens sur la pluralité des mondes* (The Hague, 1728).

THE SCIENCES IN
Enlightened
Europe

EDITED BY

William Clark, Jan Golinski, and Simon Schaffer

THE UNIVERSITY OF CHICAGO PRESS
Chicago & London

WILLIAM CLARK is visiting lecturer in history and philosophy of science at Cambridge University.
JAN GOLINSKI is associate professor of history at the University of New Hampshire.
SIMON SCHAFFER is reader in history and philosophy of science and fellow
of Darwin College at Cambridge University.

The University of Chicago Press, Chicago 60637
The University of Chicago Press, Ltd., London
© 1999 by The University of Chicago
All rights reserved. Published 1999
08 07 06 05 04 03 02 01 00 99   5 4 3 2 1

ISBN (cloth): 0-226-10939-9
ISBN (paper): 0-226-10940-2

Library of Congress Cataloging-in-Publication Data

The sciences in enlightened Europe / edited by William Clark,
Jan Golinski, and Simon Schaffer.
       p.   cm.
     Includes bibliographical references and index.
   ISBN 0-226-10939-9 (alk. paper). — ISBN 0-226-10940-2 (pbk. : alk. paper)
   1. Science—Europe—History—18th century.   2. Philosophy, Medieval.
   3. Foucault, Michel—Influence.   I. Clark, William, 1953–   .
   II. Golinski, Jan.   III. Schaffer, Simon, 1955–   .
   Q127.E8S356   1999
   509.4'09033—dc21   98-51992
                                                                         CIP

It is more easy to account for the rise and progress of commerce in any kingdom than for that of learning; and a state, which should apply itself to the encouragement of one, would be more assured of success than one which should cultivate the other. . . . You will never want booksellers while there are buyers of books: but there may frequently be readers where there are no authors. . . . We may therefore conclude, that there is no subject in which we must proceed with more caution than in tracing the history of the arts and sciences, lest we assign causes which never existed, and reduce what is merely contingent to stable and universal principles.

David Hume, "Of the Rise and Progress of the Arts and Sciences" (1742)

Contents

PREFACE

Though frequently ambivalent about them, historians need names for the periods they study. "The Enlightenment" is sometimes no more than a name for the eighteenth century, in Europe and the parts of the world where its influence was felt. But for many intellectuals then and since, "the Enlightenment" has meant more than this; used to designate a more or less coherent movement with a certain set of values and beliefs, it has served both to rally people to the cause and to warn them of its dangers. The moral and political legacy traced to the eighteenth-century program of enlightenment has continued to be fiercely contested, not least in the twentieth century. The reputation of the "Age of Reason" has reflected that of reason itself, questioned many times in the turmoil of European thought and experience since that period.

Until quite recently, however, historians of the sciences have shown relatively little interest in the specific experiences of the eighteenth century, and they have encountered major problems in making precise the interactions between the Enlightenment and the natural and human philosophies of the period. For too long, the eighteenth century was overshadowed by its precursor and its successor, the epochs of "revolutions" in scientific thought and industrial production. In recent years, however, historical studies have begun finally to do justice to the age that many contemporaries thought of as witnessing the triumph of science and reason. It is this welcome development that has made the present volume possible.

The work included here derives from a number of recent conceptual developments leading to a reappraisal of the role of the sciences in enlightened Europe. These include the rejection of one of the legacies of eighteenth-century thought about science, namely positivism. A postpositivist understanding of science has sought to comprehend it as a feature of human culture, rather than as a uniquely objective account of empirical reality. This approach has intersected in productive ways with the research of those historians who have worked to situate the Enlightenment as a complex formation within European social and cultural history.

On a related front, the work of Michel Foucault has inspired a range of studies concerning the intersections of scientific knowledge with the formations of power. Attention has been directed at the techniques that produced knowledge of the human body and the regimes of discipline that surrounded it. Foucault also insisted that scholarship should distance itself from the humanism of the eighteenth century. This stipulation has been influential in leading scholars to scrutinize the category of "human nature" as it was deployed in the period. These concerns have also informed work on the sites where particular knowledges were made, and the linkages that allowed them to be distributed, shared, and challenged. Operations of knowledge making have then been mapped across the dimensions of physical space and in relation to human understandings of locality and distance. We now better appreciate how the Enlightenment witnessed the definition of a properly European sphere, in part through comparisons—frequently violent, dramatic, or ironic—with the other worlds that European traders, naturalists, astronomers, and navigators encountered and described. So we see the enlightened making a new map of knowledge in company with a new map of nature.

These and other issues will be found to be represented in the chapters gathered here under the headings "Bodies and Technologies," "Humans and Natures," and "Provinces and Peripheries." The introduction traces in greater detail the roots of this research in historiographical debates since the 1930s, which we take to be a decisive moment in the long debate on Enlightenment, science, and their relationship. Three other chapters also help to explicate the approaches taken here and their implications. Chapter 1, by Dorinda Outram, provides an alternative orientation toward the problem, by considering what it means for the twentieth century to have treated the eighteenth as its "contemporary." Chapter 14, by Nicholas Jardine, links the categories used by some influential twentieth-century historians back to the emergence of a new form of historical sensibility at the end of the Enlightenment itself. Chapter 15, by Lorraine Daston, provides an afterword to the volume, commenting upon the previous chapters and considering the reasons for the continuing pertinence of the "ethos" of enlightenment nearly two hundred years after the end of the century it conventionally names.

As editors, our most substantial debt is to our contributors. The high quality of their work has exemplified the very best scholarship in this field. They have responded to our (sometimes impertinent) requests with good humor and professionalism. They have also shown considerable patience over the years it has taken to bring this project to completion. The volume was launched at a workshop at Darwin College, Cambridge, in July 1995. The generous hospitality of the college and the willingness of our contributors to attend at their own expense made for a remarkably good-natured and productive meeting, perhaps the nearest we shall come to recapturing the Enlightenment ideal of knowledge as polite conversation. We are grateful also to guests at the workshop who did not contribute papers to the volume, but whose influence is felt in these pages, including Marie-Nouëlle Bourguet, Michael Bravo, and Patricia Fara. In its written form, the project owes a great debt to the editorial staff of the University of Chicago Press, especially to Susan Abrams, whose encouragement has sustained our spirits and whose gentle but firm guidance has been critically important.

BIBLIOGRAPHICAL NOTE

In the notes to each chapter, references are given in the form of short-title citations. Primary sources are listed at the end of the chapter in which they are used. Primary sources are defined, for our purposes, as texts written before 1850, regardless of their date of publication.

Secondary sources for all of the chapters have been compiled into a single comprehensive bibliography, which appears at the end of the volume. Our aim in preparing this has been to provide a convenient reference tool for other scholars engaged in the study of the sciences in enlightened Europe.

Unless otherwise noted, translations from both primary and secondary sources in any given chapter are by the contributor.

PART

1

Orientations

Introduction

This book aims to open up new paths for historical writing on the sciences in Europe during the "long eighteenth century." The collection does not aim for exhaustive coverage, although most of the major European nations and scientific disciplines are touched upon in the chapters that follow. We have been guided in choosing the contributions and structuring the volume by our perception of the historiographical issues that frame current attempts to understand the sciences in the century of the Enlightenment. "Enlightenment" names both a period of European history and a process continuing into the present and touching all parts of the world. We take this problem of nomenclature as a fundamental condition of attempts to understand the eighteenth century and its scientific accomplishments. Historical inquiry into the role of the sciences in that period has been profoundly shaped by the fact that the moral and political values associated with the Enlightenment have remained pertinent and contested.

The task of this introductory chapter is to show how this has been so and to draw out the implications for writing histories of the sciences in enlightened Europe. We shall indicate how the chapters in this volume rely upon, and critically engage with, the historiographical debates of recent decades. Recent inquiry has retrieved three particularly significant dimensions of the history of the sciences in eighteenth-century culture. First, the effects of scientific reason have been traced in the realms of material technology and bodily discipline. The domain of instrumental rationality has been at the center of arguments about the political and moral implications of enlightened reason. Second, eighteenth-century thinkers have been shown to have recognized the mediating role of human activity and character in the creation of scientific knowledge. Multiple connections have been found between Enlightenment natural knowledge and the notions of human nature shaped by changing social experiences at the time. And third, a new geographical consciousness of locality and distance has been seen to have emerged in the eighteenth century as an integral part of the experience of enlightenment itself. A sense of participation in the diffusion of new knowledge led both to consciousness of spatial hierarchies of origin and periphery and to reflection on the specific character of enlightenment in particular locations. These three dimensions provide the bases for the three substantive sections of this book, "Bodies and Technologies," "Humans and Natures," and "Provinces and Peripheries," each of which carries a separate introduction to the chapters gathered therein. In this way, we trace the route by which the three themes have arisen from the historiography of the past seven decades.

It may seem perverse to begin a discussion of eighteenth-century science by reference to the 1930s. Yet it is that decade that we find particularly salient for the production of Enlightenment historiography. It was then that leading philosophers and historians articulated a characterization of the Enlightenment as both a period and an epistemic and administrative enterprise. In this sense, the 1930s have continued to shape the ways in which enlightened science has been viewed. The emergence of Nazism and Stalinism in the 1930s seemed to many intellectuals to threaten a collapse of the ultimate values of European culture and secular rationality. Some saw this "crisis of civilization" as the liquidation of the heritage of the Enlightenment, others as the consequence of it.

As we shall show, it was at this point that intellectual historians such as Ernst Cassirer forged a characterization of the Enlightenment in terms of its "mind." Responding to what they feared was an impending collapse of the fundamental values of European civilization, they drew upon idealist notions of "culture" originally articulated in the late eighteenth and early nineteenth centuries. The vocabulary in which enlightened thinkers had reflected upon their own era, at the moment of its close and in its aftermath, was recycled in the 1930s as the terminology of an idealist historiography of the period. Set against this mode of validating the Enlightenment was the secular eschatology of Theodor Adorno and Max Horkheimer, which held enlightened rationality responsible for the rise of totalitarianism itself. Instrumental reason led, in their bleak dialectical vision, to the elimination of human freedom, to anti-Semitism and other forms of intolerance, and ultimately to the extermination of human beings.

However, as we shall also argue, the pioneers of professional history of science in the 1930s showed a surprising lack of interest in the eighteenth century. Positivistic historians of the sciences, such as George Sarton, who were looking to map steps along the path of accumulating factual knowledge, could discern few landmarks in this terrain. Writers influenced by historical materialism generally leapt from the seventeenth century to the aftermath of the Industrial Revolution. Some, such as Boris Hessen, supposed that seventeenth-century science relied on a technological imperative already in place before the Enlightenment. Others, such as J. D. Bernal, held that the application of the sciences to technological change had to await the complete industrialization of society during the nineteenth century. Those who developed the idealist program in the history of science, especially Alexandre Koyré, highlighted the "Scientific Revolution" of the seventeenth century as an intellectual transformation in a way that also drew attention away from the period of the Enlightenment. Notwithstanding the more general works of philosophers and intellectual historians, in which the scientific rationality of the era was hailed or condemned, the result was a lack of attention to the specific features of eighteenth-century science.

The Mind of the Enlightenment

The search for the Enlightenment's mind was a central theme of 1930s historiography. The contents of that mind, it was proposed, provided the reservoir for past thought and culture and the source of the revolutions to come at the end of the eighteenth century. The positivist outlook, which, in the nineteenth and early twentieth centuries, had tried to exclude all metaphysical and theological elements from philosophy and base it solely on empirical science, was called into question. During this decade of crisis, positivism became a topic for inquiry at least as much as an assumption of method. Positivisms of all kinds were under scrutiny—not least because successive historians found that behind the Enlightenment's vaunted liquidation of metaphysics lurked tacit metaphysical agendas to be excavated and documented.

"The mind of the Enlightenment" was, tellingly, the title of the first chapter of Ernst Cassirer's *Die Philosophie der Aufklärung*, published in 1932 but not widely read until its translation into English (as *The Philosophy of the Enlightenment*) in 1951. After the Nazi rise to power, Cassirer left Hamburg for Oxford in 1933, and then went to Göteborg in 1935 and to Yale in 1941, ending up at Columbia in New York City from 1944 till his death in 1945. On the third page of the preface, he termed his book "a phenomenology of philosophical spirit [*Geist*]." The following pages declared that the Enlightenment had begun in Britain and France, from which it progressively diffused into other parts of the European world. The process was to be grasped not in the specific contents of thought but rather in certain forms of thinking. In later chapters, these forms were specified as modes of "analysis" and "critique," along with the belief in "progress." As a purely intellectual phenomenon, the Enlightenment was traced through natural science, psychology, theology, historical writing, law, sociology, and aesthetics, with philosophy as leitmotif. The approach was fundamentally antipositivistic, exhibiting the unity of enlightened culture as a retort to the positivists' assertion that religion, aesthetics, and metaphysics were simply obstacles to rational progress. Idealist historical works like Cassirer's were concerned with "climates of thought and opinion," which usually meant the prevailing—and above all the strong—winds. Concepts such as these—climate, fashion, style, and so on—served as instruments against positivistic views of science, in which only rationality was supposed to hold sway.

The notion of "culture" underwrote Cassirer's project of finding an ideational unity in spheres of social action as diverse as natural science and aesthetics. In this respect, he drew upon terms of analysis proposed in late-eighteenth-century Germany, by Immanuel Kant among others. Kant's philosophy served as the goal of much of Cassirer's exposition of the Enlightenment mind, the movement that had begun in England and France being seen to reach its apotheosis in Germany. Kant's renowned essay of 1784, "What is Enlightenment?" with its culminating dictum that the eighteenth century was an age of enlightenment but not yet an enlightened age, sustained the urgency of Cassirer's inquiry into the fate of enlightened values in his own time. In this inquiry, the category of "culture," which Kant had elsewhere distinguished from the superficiality of Anglo-French "civilization," was a critical tool.[1] As Cassirer was later to explain in his American work, *An Essay on Man* (1944), he saw culture as a coherent project of "progressive self-liberation." Defense of culture as a single enterprise of human self-expression was, for Cassirer, central to "philosophical anthropology," a bulwark against dehumanization and the fragmentation of intellectual activity that accompanied it.[2]

In the English-speaking world, the projects of the History of Ideas Club, launched at Johns Hopkins University in 1923 by Arthur Lovejoy and his colleagues, shared much the same agenda. Members including Lovejoy and Marjorie Hope Nicolson devoted their study to the intellectual, especially literary, culture of Enlightenment sciences. In 1930 Lovejoy lectured at the Modern Language Association of America on rationalism, a "characteristic idea-complex" he judged indispensable for understanding the thought of the German Enlightenment, the *Aufklärung*, and which, though held in full by no one individual, nevertheless was "widely accepted as too self-evident to need, as a whole, formal exposition or defense." This literary-historical excavation of thought-styles dominated modes of exposition of the Enlightenment and its great chain of ideas, and, in turn, drew much of its own legitimacy from what Lovejoy called the "uniformitarianism" of that epoch.[3]

French explorations of the period were similarly motivated. One literary historian at the Collège de France, Paul Hazard, argued in his *Crise de la conscience européenne* (1935) that the Enlightenment, especially in its most scientistic aspects, was simply the idealized Renaissance "without a smile," a rather laggard and sober aspect of "the unending quest of the mind of Europe." Hazard would pursue this quest further in a study of eighteenth-century thought published at the end of World War II, indicating again that the men-

---

1. Compare, in this respect, the work of Norbert Elias, written in the same period, *Über den Prozess*, 1:8.
2. Cassirer, *Essay on Man*, 1–22, 222–28.
3. Lovejoy, *Essays*, 78.

tality of the enlightened spread outward and downward as the Revolution approached.[4] Daniel Mornet, a literary historian from the École des Hautes Études Sociales, in his *Les origines intellectuelles de la Révolution française* (1933), also asserted that ideas determined the Revolution. These ideas were just those held by the *philosophes* and diffused by them to the lower orders and the provincial periphery; they defined the Enlightenment.[5] Exemplary thinkers were thus to be treated as emblems of thought styles—Jean d'Alembert, for example, could well stand for the positivism and rationalism which were supposed best to represent the place of the sciences in the mind of the Enlightenment. D'Alembert's celebrated remarks in his *Éléments de philosophie* (1759) on the eighteenth-century "exaltation of ideas" and "general effervescence of minds" thus became an almost obligatory text for the Enlightenment's historical expositors down to our own day. The text appears prominently in an important collection of primary readings on Enlightenment attitudes to Newton and Locke. It opens the most recent monograph on science and the Enlightenment, whose author, Thomas Hankins, carefully argues that "the Enlightenment was not a fixed set of beliefs but a way of thinking."[6] In an earlier biography of d'Alembert, Hankins pointed out that d'Alembert's statement owes much of its status to its appearance at the very head of Cassirer's first chapter.[7]

The advent of Mornet's and Cassirer's books on the Enlightenment also witnessed the Cornell historian Carl Becker's *The Heavenly City of the Eighteenth-Century Philosophers,* based on lectures given at Yale in 1931. Unlike Cassirer's book, Becker's was widely read from the outset. The first chapter was called "Climates of Opinion," a phrase said to have been revived by A. N. Whitehead. The expression served as another means to obviate positivism, not so much in favor of idealism as for the sake of one of its manifestations, namely, "historicism"—the belief in the historical embeddedness of thought and action. On this view, much of what was thought and done in eras long gone, in the Middle Ages, for example, could be described by modern historians but no longer rendered fully meaningful. Thus, Becker argued, "The one thing we cannot do with the *Summa* of St. Thomas is to meet its arguments on their own ground. We can neither assent to them nor refute them. . . . [W]e instinctively feel that in the climate of opinion which sustains such arguments we could only gasp for breath. Its conclusions seem to us neither true nor false, but only irrelevant."

This thesis on medieval philosophy carried over implicitly to the sciences

4. Hazard, *European Mind,* 497–99.
5. Mornet, *Origines intellectuelles,* 2–3; Chartier, *Cultural Origins,* 3–7.
6. Buchdahl, *Image,* 61–63; Hankins, *Science,* 1–2.
7. Hankins, *D'Alembert,* 2–3.

of the eighteenth century, another climate of opinion. Only a few techniques had changed: "When philosophy became a matter of handling test tubes instead of dialectics everyone could be, in the measure of his intelligence and interest, a philosopher." The air gets even thinner here, and breathing more difficult, for Becker's aim was to demonstrate "that the underlying preconceptions of eighteenth century thought were still essentially the same as those of the thirteenth century." This estranges us from the Enlightenment, by pushing it back into the Middle Ages. Moreover, in the end Becker also pushed the Enlightenment into the 1930s. The last chapter of *The Heavenly City* traced the movement into the French Revolution and, in the final few pages, to the Russian. The trope of the "mind" allowed Becker to give the Enlightenment memories and anticipations, even if unintended ones, for which, like a person, an epoch might be held responsible.[8]

A year after Becker's work, his Cornell colleague Preserved Smith prefaced his comparably broad account of the Enlightenment's intellectualism, utilitarianism, optimism, and above all, scientism with a remark about mind's social place. He reckoned the Age of Reason was so because of the dominance of the intelligentsia and the exclusion of labor. Writing during the Great Depression, Smith held that for the plebs, "too strenuously engaged in the struggle for a livelihood to devote much time or strength to thought, the emotions rather than the intellect have always supplied the criterion of value." This allegedly eternal truth of the condition of the working classes helped Smith see how what he called the intellectual "graphocracy" came to power. All this chimes neatly with Becker's views. Smith contrasted the Enlightenment, which "brought to leadership just that literate public whose claims could be vindicated only by an appeal to reason," with "modern times, [which] since the end of the eighteenth century have been increasingly irrational partly because they have been increasingly democratic." The common theme is that the collective mind's fate was the right way to interpret enlightenment and the Enlightenment. The lesson of that interpretation is the fragile but admirable rule of mind over matter.[9]

The end of the 1930s saw the publication of Abraham Wolf's *A History of Science, Technology, and Philosophy in the Eighteenth Century* (1938). Wolf taught at the London School of Economics and had already produced a similar work on the sixteenth and seventeenth centuries. He expressed his belief in the notion of a "spirit of the age," which he specified as composed of secularism, rationalism, naturalism, and humanism, with the last as fundamental.[10] In this light, his project was not unlike the others we have been describing in

8. Becker, *Heavenly City*, 12, 58, 31, 121.
9. Smith, *The Enlightenment*, 34–36.
10. Wolf, *A History*, 1:27, 34–36.

this section. However, the notion of a unifying spirit or mind collapsed under the weight of the detail Wolf presented. Thus, although Hankins, for example, sees Wolf as his immediate predecessor, he notes that "Wolf's book emphasized technology and instrumentation, whereas mine emphasizes science and ideas."[11] Wolf's two volumes on the eighteenth century are indeed encyclopedic in scope, covering technology, natural science, medicine, and some social and human sciences, and concluding with two chapters on philosophy. While Cassirer's work on the Enlightenment falls into the mold of an idealist spirit of a culture, Wolf's seems a positivistic anatomy of scientific civilization in the eighteenth century, during which "the intellectual and moral forces of the age were harnessed to the chariot of human progress . . . [but] forces of darkness and oppression were too well entrenched to be easily dislodged."[12]

Peter Gay's *The Enlightenment: An Interpretation* (1967–69), though published more than thirty years after Cassirer's text, appears in many respects as the culmination of the idealist way of writing. Gay, born Peter Fröhlich in Germany in 1923, moved to the United States in 1938. He took his advanced degrees at Columbia University and joined the faculty of government in 1947, two years after Cassirer's death. Gay moved into the history department in 1956, then to Yale in 1969. On the first page of his preface, he situates his own work in the long historiographical tradition pro and contra the Enlightenment since Edmund Burke and the German Romantics: "And so scholars have turned to polemics. I have had my share in these polemics, especially against the Right, and I must confess that I have enjoyed them. But the time is ready and the demand urgent to move from polemics to synthesis."[13] Gay's text is nonetheless unapologetic in its defense of the Enlightenment, understood as a system of values—liberal, secular, rational—of continuing moral and political import. Whereas Cassirer's Enlightenment reached its telos in Kant's philosophy, Gay's achieved its climax in the birth of the American republic. Benjamin Franklin and Thomas Jefferson, in particular, represented the institutionalization of the idea of enlightenment in the foundation of Gay's adopted land.

Toward this end, Gay's Enlightenment unfolded via a dialectical process: a revival of classical antiquity, fused with a Christian inheritance, from which synthesis emerged a modern secularism. Though divided into two separately published volumes, each with its own subtitle, the whole work was actually apportioned into three continuously numbered "books": "The Appeal to Antiquity," "The Tension with Christianity," and "The Pursuit of Modernity." They traced a dialectic that mirrored almost to the point of parody Cassirer's

11. Hankins, *Science*, vii.
12. Wolf, *A History*, 1:27.
13. Gay, *The Enlightenment*, 1:xi.

designation of his own work on the Enlightenment as a phenomenology of philosophical spirit. Appropriately, Gay recorded, "My greatest debt is to the writings of Ernst Cassirer." Gay made only limited attempts to give his narrative a dimension of wider social experience: The small group of like-minded intellectuals, the "family of philosophes," strove "to assimilate two pasts they had inherited—Christian and pagan—to pit them against one another and thus to secure their independence." [14] As had been the case for previous exponents of the search for the Enlightenment's mind, so for Gay this group's dialectical struggle for autonomy constituted the "cultural climate" of the age.

Enlightenment Science Impoverished

While Cassirer resorted to the values of the Enlightenment to sustain his vision of humanism, another group of mid-twentieth-century intellectuals traced part of Europe's crisis to the positivistic turn taken by secularization. This assessment issued in a similar focus on the critical turning point of the eighteenth century, but produced an opposite appraisal of the moral worth of the Enlightenment enterprise. In the same decade that Cassirer set out the intellectual climatology of the culture which positivism ignored and betrayed, Max Horkheimer and his colleagues at the Frankfurt Institute for Social Research began telling their story about the Enlightenment's positivist wake. In his first book, *Anfänge der bürgerlichen Geschichtsphilosophie* (1930), Horkheimer had already damned positivists for treating the metaphysical assumption of nature's uniformity as though it were an empirical fact. He indicated that this metaphysics was part and parcel of bourgeois domination of nature and culture, and he then traced this secular and aggressive program to the seventeenth-century philosophy of Thomas Hobbes and the subsequent Enlightenment, "enlightenment arising from material motives which paves the way for revolution." [15] In wartime exile in New York, Horkheimer and Theodor Adorno would turn this history into a ferocious diagnosis of Enlightenment's self-liquidation, *Dialektik der Aufklärung* (1944, translated as *The Dialectic of Enlightenment*). The secularism and especially the iconoclasm contra metaphysics of scientific reason was said to have led straight to a culture in which "science itself was worshipped and applied down to the last details of the murder factories." Adorno went so far as to discern the interweaving of modern occultism and secular rationalism. The "postulate that positivism should become a kind of religion is fulfilled ironically—science is hypostatized as an ultimate, absolute, truth." [16]

14. Ibid., 1:423, xiii.
15. Horkheimer, *Between Philosophy and Social Science*, 353.
16. Horkheimer, "Reason against Itself," 364; Adorno, *Stars Come Down*, 116.

Horkheimer and Adorno's indictment provides forceful reasons why the eighteenth century, and the Enlightenment project associated with it, should remain especially salient for any discipline called history of science. Horkheimer traced to the Enlightenment the desire "to speak of . . . science in terms of a unified history of ideas that encompasses long stretches of time [and] that restricts itself to developmental trends that are purely intellectual." [17] The import of this denunciation is that eighteenth-century science should be brought under review from the vantage point of criticism of the positivist ways of thinking *about* science bequeathed to us by the Enlightenment. Writers of that period, including d'Alembert, Joseph Priestley, and the marquis de Condorcet, narrated the history of the sciences in a continuous sequence with their own achievements. Any postpositivist history of the sciences is thus likely to produce a very different view of the epoch from that prevailing in the eighteenth century itself.

It was, however, with the production of a narrative of acknowledged positivistic lineage that the history of science was associated at the time Adorno and Horkheimer wrote. Histories were generally written of specific disciplines, across extended periods of time. In this form, the history of science acquired institutional recognition, preeminently through the efforts of the Belgian historian George Sarton, an émigré to Harvard in the 1920s. The first volume of Sarton's *Introduction to the History of Science* appeared in 1927, explaining in its opening pages that the history of science is essential to comprehension of "our civilization." The introductory chapter echoed the views of the nineteenth-century founder of positivism, Auguste Comte, equating science with "systematized positive knowledge" and insisting that only science "is truly cumulative and progressive," thereby making our civilization "essentially different from earlier ones." [18] Insofar as positivism makes scientific action an autonomous sphere in relation to state and society, such histories facilitated a positivistic view by their periodization alone, as though, regardless of other human activities, science marches to its own beat. Moreover, as many critics of Mornet, Cassirer, and Gay point out, such narratives easily take the form of origin myths. [19] The historiography of the sciences has suffered especially from this atavistic mode. Thus, in his introductory essay to John Yolton's recent *Companion to the Enlightenment,* Lester Crocker, the eminent biographer of Rousseau and Diderot, notes the "unprecedented" scale of Enlightenment public interest in natural sciences, but then traces a one-way street from the sciences to intellectual culture and concludes that "science in the last analysis

17. Horkheimer, *Between Philosophy and Social Science,* 361.
18. Sarton, *Introduction,* vol. 1, part 1, 3–4.
19. Venturi, *Utopia and Reform,* 2–5; Chartier, *Cultural Origins,* 4–7.

depends on discoveries from which theories are induced, and the eighteenth century provided a small share of these." [20] The long impoverished state of the history of science of the Enlightenment can thus be traced to such things as a positivistic tendency of historians to suppress typical historical periodizations in the case of science.

An additional reason for this impoverishment can be found in historians' construction of the "Scientific Revolution," the label applied to a singular break in the secular intellect that has assumed institutional centrality in the historiography of early modern science. On this showing, the Enlightenment did not precede a revolution but followed one. As an example, consider Crocker's characteristic assumption that, because it was during the seventeenth century that the physical sciences had been securely established, the direct influence of those sciences on the subsequent Enlightenment was correspondingly modest. What Clifford Truesdell in 1960 called "a program toward rediscovering the rational mechanics of the Age of Reason" has only slowly and unevenly changed the popular view, among scientists not least, that scientific modernity started with Newton's completion of the Scientific Revolution. The term "Scientific Revolution" itself was coined in the 1930s and 1940s by just those historians of science, such as Alexandre Koyré, who in dialogue with Cassirer and in the face of totalitarianisms of various kinds, sought to locate the great change of European mind that had spawned the most typical (often virtuous) aspects of human thought. Just as it has recently been suggested that Cassirer's work be read not as a history of Enlightenment thought but as a response to Fascism, so it has been suggested that the idealist version of the Scientific Revolution was part of the struggle against mid-twentieth-century dictatorship. [21]

Koyré, initially an expert on the histories of Russian Hegelianism and German mystic alchemy, helped invent the "Scientific Revolution" during the 1930s, labeling it, "at the same time the source and result of a profound spiritual transformation which overturned not only the content but the very frames of our thought." His disciple and colleague Hélène Metzger used this periodization and approach to explore eighteenth-century natural theology, natural philosophy, and chemistry, understood as an aftermath—a story of how European intellectuals coped with the Newtonian synthesis. In his New York exile during the war, Koyré published his argument in the *Journal of the History of Ideas*, launching a dialogue in which Cassirer's climates of thought were assimilated to the History of Ideas Club's "idea-complexes." In a remark-

---

20. Crocker, "Introduction," 7–9.
21. Truesdell, *Essays*, 90–93, on Newton; Schmidt, "What is Enlightenment?" 43, on Cassirer; Porter, "Scientific Revolution," on Koyré.

able public lecture in the spring of 1945, just before returning to Paris, Koyré found himself compelled to defend the thesis on the Enlightenment intellect as mere aftermath against its "bad press." There was rather little he could offer: "[N]o doubt it did not produce great metaphysical systems, [or] great scientific theories (Newton), but it was a century which was and wanted to be reasonable."[22]

In the 1950s, in order to pursue his research to trace the legacy of the Scientific Revolution, Koyré initially sought the establishment of a chair in the history of scientific thought at the Collège de France. He won support from his long-time colleague Lucien Febvre, visionary editor of the journal *Annales,* who in his submission on Koyré's behalf made much of the centrality of the Scientific Revolution, "this great movement from the approximate *[l'à peu près]* to precision," both in Koyré's historiography and in the "history of the development of our modern mind." The bid to sanction such a program at the very center of French intellectual life only just failed: Koyré eventually divided his time between Princeton and his newly established section of the École Pratique des Hautes Études devoted to the study of history of the sciences. There his work helped the project, launched by Gaston Bachelard and pursued by Georges Canguilhem, to study the ways in which successive concepts and techniques became pure enough to join properly scientific inquiry. Infuriated by one consequence of this approach, an insistence that past sciences be studied on their own isolated terms, Truesdell in 1966 perversely attacked Koyré's work on the Newtonian aftermath in enlightened science as but another example of a modish model of science as "time-conditioned, social and institutional." In fact, the Koyré project saw the seventeenth and eighteenth centuries as replete with successive purification rituals, inaugurating scientific fields and leaving value-laden contexts well behind. Enlightened sciences were thus camp followers in the onward march of scientific emergence.[23]

A similar story can be told of institutional and intellectual developments in Britain, where in the 1940s and 1950s Herbert Butterfield and A. Rupert Hall, among others, promulgated the notion of a great discontinuity of revolutionary scope during the seventeenth century and set up pedagogies in the history of science centered entirely on this periodization. Butterfield's *The Origins of Modern Science* (1949), based on lectures delivered at Cambridge, extended its chronological range generously back into the medieval period, but had little to say about the eighteenth century. Notoriously, chemistry was shown as simply waiting around during the century of Enlightenment until it experienced its "delayed scientific revolution." The latest edition of Hall's text-

22. Koyré, *De la mystique,* 128, 61, 99.
23. Ibid., 132; Truesdell, *Essays,* 146.

book, *The Revolution in Science, 1500–1750* (1983, originally published in 1954), stretches into the eighteenth century, albeit only halfway and again under the heading of aftermath, the final chapter being entitled "The Legacy of Newton." Eighteenth-century sciences were also commonly understood as but extensions and interpretations of the Newtonian legacy in the postwar American historiographical tradition, inaugurated by I. Bernard Cohen in his 1941 edition of Franklin's electrical work—for which he received the second Ph.D. in history of science awarded in the United States. The approach was furthered in Cohen's own *Franklin and Newton* (1956), which was dedicated to Koyré, and in the schematics of later works such as *Mechanism and Materialism* (1969) by Cohen's student Robert E. Schofield. In these ways, the postwar institutional and historiographical construction "Scientific Revolution" has had ramifications on our understanding of enlightened science just as great as the issue of positivism. In particular, the Scientific Revolution, posited as the central event of early modern science, has tended to make the Enlightenment into a sort of middle ages between "classical" and "modern" eras, with little other to do than mopping up.[24]

The poverty of the historiography of enlightened science up until the 1960s can be gleaned from broad synthetic works not per se committed to the Scientific Revolution. We confine ourselves to two examples, one published in Princeton in dialogue with Koyré, the other in Paris in dialogue with Canguilhem: Charles Gillispie's *The Edge of Objectivity* (1960) and Michel Foucault's *Les mots et les choses* (1966, translated as *The Order of Things*).

Gillispie's *The Edge of Objectivity* concerned not so much a scientific revolution as the slow formation of "classical" sciences in various domains. Like many histories of science in the wake of Koyré, this work came less to praise and more to bury positivism. In astronomy and physics, classical science was said to have begun with Galileo and culminated in Newton. In chemistry, it centered on the work of Lavoisier; in biology Darwin was the key. The nineteenth-century physics of energy and fields brought classical science to an end, although an epilogue continued the story to the birth of modern science with Einstein. Gillispie traced the development of the classical natural sciences in disciplinary categories, and his periodization was consequent upon this. Chapter 5, "Science and the Enlightenment," is an anomaly within this structure and is weakened by the author's uncertainty as to what story to tell about the eighteenth century. At one point, he flirts with the notion that "the history of the influence of science in culture is bound to be the history of a misunderstanding," then swiftly outlines aspects of the *philosophes'* involvement with

24. Butterfield, *Origins;* Hall, *Revolution.* For the historiography of the Scientific Revolution, see Porter, "Scientific Revolution"; Shapin, *Scientific Revolution*, 1– 4, 168–70.

the sciences, before relaxing into a leisurely castigation of the influence of Romanticism, "the wrong view for science."[25] Gillispie's narrative of classical versus modern science gave only an uncertain and anomalous role to the Enlightenment.

On the surface, *Les mots et les choses* could hardly be more different from *The Edge of Objectivity*. But in the end Foucault and Gillispie were not so far apart in their views of the Enlightenment. While Gillispie wrote on natural sciences, Foucault treated of the emergence of the human sciences. He divided his story into four periods: the Renaissance, the "Classical Age," the nineteenth century, and the twentieth century (the last treated only implicitly). For Foucault, the "Classical Age" ran from Descartes to Kant, crosscutting the periodization of Baroque and Enlightenment. Though he interwove a number of disciplines in each period (for instance, grammar, natural history, and economics in the Classical Age), Foucault still effected a periodization with respect to these particular sciences and, moreover, essentially told a story whose narrative focus is the transition from classical to modern, with a prologue (the Renaissance) and an epilogue (the twentieth century). Although he was scolded by existentialist reviewers as a positivist, Foucault's account was rather closer, as Canguilhem soon noted, to that of the antipositivist Cassirer in its exploration of the preconditions of the critical philosophy and of Comtian positivism itself. But in this project, clearly, Foucault gave the Enlightenment per se little or no part to play.[26]

It seems safe to say that the Enlightenment, as a named entity and as distinct from the chronological eighteenth century, long lay distant from the center of surveys, general histories, and textbook narratives about science in Europe. Until recently, general histories of science have tended toward an impoverished estimation of this period. Recoiling against positivism, mid-twentieth-century historians of science showed little sympathy for the estimation of eighteenth-century writers themselves that they lived in an age distinguished by acceleration of the patterns of production and circulation of knowledge established during the Renaissance. D'Alembert had seen his own age as building ever more ambitiously upon the rebirth of learning after the "Dark Ages." Priestley had looked with apocalyptic expectations toward the climactic fulfillment of intellectual and social enlightenment in his own time. Both writers would surely have been mystified to be told that they lived in the aftermath of a much more significant intellectual revolution completed the century before. Insofar as they adopted this view, postpositivist historians of science left themselves without the means to address the period that, if any,

25. Gillispie, *Edge of Objectivity*, 10, 155, 201, 495–506, 521.
26. Foucault, *Les mots et les choses;* Canguilhem, "Death of Man?"

deserves to be called the "Age of Science." They deprived themselves further of influence in the political and philosophical debate over its moral legacy.

Enlightenment Science Anatomized

For many practitioners of the history of the sciences, withdrawal from explicit engagement with the grand narrative of Enlightenment and its legacy has seemed a virtue. A salient theme of recent historiography of the sciences has undoubtedly been its localism, its sense that the best analyses of past scientific work would study specific practices in their own, highly delimited, settings. For example, some historians have produced narratives of individual disciplines with little apparent reference to the wider culture of the time: eighteenth-century mathematics, chemistry, optics, and electricity have all been studied in this way. Others have charted the local interactions between scientific development and economic change in such areas as steam technology, agriculture, and the textile trade. Such histories engage in what might be called an anatomy of their subject. Reenacting the gestures by which eighteenth-century anatomists sought the bodily seat of the soul or the mind, historians have attempted to localize intellectual attributes and functions by ever finer dissections and more microscopic scrutiny. It might have been expected that such localism would question an overarching account of a singular Enlightenment mentality. But though these critical anatomies of enlightened cultures might superficially have seemed to challenge the intellectualism of Cassirer and his epigones, one feature, and not the least important, of that earlier historiography remained. The unifying principle was frequently still a reliance on the Enlightenment's *mind,* now assumed as the guarantee that somehow and somewhere these anatomical studies would find their synthesis.

The measure of convergence between intellectualist historiography and its successors in disciplinary anatomy was publicly raised in a telling conference exchange in Oxford in 1961 between Koyré and the Cornell historian of Enlightenment chemistry, Henry Guerlac. The moment was opportune. At that meeting, Joseph Needham, the great scholar of Chinese science, attacked the Eurocentrism of Gillispie's newly published *Edge of Objectivity,* which he said distorted "the salutary enlightenment of all men without distinction of race, colour, faith or homeland" achieved by "modern universal science." There, too, Thomas Kuhn set out an early version of his thesis on dogmatic "normal science," soon to assume a crucial role in his *Structure of Scientific Revolutions* (1962), finding his prime examples in the eighteenth century.[27] In the wake of the Enlightenment, as Guerlac then pointed out, the scientific disciplines were at last granted their own specialized histories. Late Enlightenment his-

27. Needham, "Poverties and Triumphs," 148–49; Kuhn, "Function of Dogma," 351–52.

torians, such as Priestley, Jean-Etienne Montucla, and J.-B.-J. Delambre, provided his examples, followed by Comte, who chronicled the general revolution in the sciences of early modern Europe and promoted history of science
as an academic discipline. Ever since positivism, according to Guerlac, specialization had brought in its train regrettable attempts to demarcate scientific disciplines from their social context. Guerlac himself had just finished a
highly detailed monograph on just one year of the chemical career of Antoine
Lavoisier. (Guerlac was also the doctoral adviser for Roger Hahn's self-
proclaimed "anatomy" of positivistic specialization at the eighteenth-century
Paris Academy of Sciences.) To speak of economic or technological factors in
such episodes would court the charge of being *marxisant,* yet to continue along
the idealist path meant historians of science would be unheard by other historians. "When we overlook the social connexions of scientific thought and action," Guerlac concluded, "we are in danger of distorting . . . our image of
science itself." In response, Koyré accepted that historians of science had to
run the risks of overspecialization, and accepted, too, the label "idealist." Scientific thought was not "the product of the will to power," though Koyré
seemed to acknowledge that, in some modern societies, the sciences might indeed have degenerated into just such willful schemes of domination. True scientific culture, Koyré insisted, hinged on the pursuit of understanding by the
leisured classes. Where leisure was available, understanding a goal, and this
attitude valued, then science would appear and flourish. Otherwise (as in ancient oriental or modern bureaucratic despotisms) it would not. For these reasons, Koyré believed, historians must focus on the thought of individual sciences whose development was obviously "inherent and autonomous." [28]

Guerlac and Koyré agreed that the sources of their work lay in the Enlightenment: "[I]t is the eighteenth century that gave birth to our historiography."
They were at one in hoping that from particular tales the general cultural account they both sought would eventually emerge. Like most protagonists in
the debates of the time on "internalism" and "externalism," they also agreed
that linkages with technology and society should not be used to denigrate
scientific enlightenment. Koyré thought so because such linkages were barely
present in properly scientific thought. In 1966, Guerlac echoed Koyré's eulogy
for enlightened sciences: "[T]he characteristic natural philosophy of the liberal
competitive era of early modern history" was said to be intimately linked with
"our deepest political beliefs," reasonableness, tolerance, individualism, representative democracy. But in the age of "giant corporations, great labor
unions, top heavy bureaucracies and heavily armed leviathan states," this ideal

28. Guerlac, "Some Historical Assumptions," 812; Guerlac, *Lavoisier;* Koyré, "Commentary,"
852–53, 856; Hahn, *Anatomy of a Scientific Institution.*

of the sciences was in danger of passing away, and historians must chronicle the transformation.[29]

This defense of enlightened scientific culture, as an autonomous and unified entity whose fate it was to succumb to external powers and internal fragmentation, defined itself in the wake of the onslaught launched by Horkheimer and Adorno. Guerlac's highly localized studies of eighteenth-century Newtonianism, for example, were undergirded by the faith, shared with Koyré, that the sciences of the Enlightenment could be characterized through loyalty to a great mind. Although they could be said to have rescued the Enlightenment from the Frankfurt school's denigration, both writers did so by reasserting a humanist vision of its cultural integrity. Anatomical studies of microhistorical topics were not to be allowed to put this at risk. Intellectual unity was still judged to be the principal feature of Enlightenment thought. This unity allowed the liberal efflorescence of local studies that did not significantly call it into question.

Similarly, studies of local cases of engineering, industry, and commerce have relied on a very generalized model of the Enlightenment intellect. Positivist historiography saw the business of the sciences as the production of fundamental theories, found no such theories in industrial enterprises of the eighteenth century, and thence deduced that only after major and autonomous industrialization could the sciences interact effectively with economic life. This was the argument of Hall's influential work on the science of ballistics, published in 1952, an apparently conservative insistence on the divorce of theory and practice in early modern Europe, which, interestingly, was cited with approval the following year in Bernal's Marxist analysis of science and industry. Bernal learned from Hall that, during most of the eighteenth century, "science had been an affair of courts, gentlemen and scholars, and . . . hardly affected ordinary life."[30]

So, insofar as historians were prepared to analyze the entanglement of eighteenth-century sciences and technical enterprises, they did so principally by using a very broadly defined scientific spirit that was supposed to characterize the Enlightenment. This was the position Gillispie adopted in an important essay on eighteenth-century French industry in 1957. All "eighteenth-century scientists [said] that science illuminates the arts, that it enlightens the artisans, and that this process honors the century."[31] The mind of the Enlightenment, it seemed, affected industry through its diffusion of the spirit of classification and analysis. This complemented earlier accounts of the modern economy's

29. Koyré, "Commentary," 849 n; Guerlac, *Essays*, 52–53; Shapin, "Discipline and Bounding."

30. Hall, *Ballistics*, 1; Bernal, *Science and Industry*, 8–9.

31. Gillispie, "Natural History of Industry," 403–4.

origins. Max Weber himself had notoriously seen the Enlightenment as but the "laughing heir" of the Protestant work ethic whence flowed the spirit of capitalism.[32] In her recent studies of the Enlightenment sciences and industry, Margaret Jacob, another of Guerlac's students, appeals to what she calls the "magisterial historiography" of Cassirer, Becker, Hazard, Mornet, and Gay for her "litmus test for assessing participation in the European Enlightenment, . . . namely the willingness to accept the new science." Jacob uses this feature of the Enlightenment to correct the weaknesses of Weberian accounts of Protestantism and capitalism, for it was "the moral economy of applied science," the essence of Enlightenment mentality, that provided eighteenth-century entrepreneurs with their sense of legitimacy. "The Enlightenment permitted them to imagine their industry had universal meaning." Intellectual unity therefore becomes, on this account, the principle feature of the Enlightenment's economic significance.[33]

One might say that, since Lovejoy, a certain intellectual uniformitarianism has been viewed as the key to Enlightenment culture. This has had substantial implications for historical accounts of eighteenth-century science. Uniformitarianism can be understood as the claim that a single mentality should be assumed as the deep structure of Enlightenment views of humanity and nature. The uniformity of such a mentality was treated as a precondition of human emancipation, since all social formations could be judged by the same moral and critical standards. Hence, humanism and liberalism have drawn encouragement from the vision of a unified Enlightenment mind. Furthermore, the single mentality was understood as confronting a uniform natural order. This uniformity became the precondition of nature's domination, since local principles would work and be made to work everywhere. No other explanation for the global spread of enlightened science and technology needed to be supplied. The worldwide reach of the Enlightenment led to spatial hierarchies of center and periphery: at the center, a zone whose inhabitants basked in the glow of illumination, while in the outer shadows lived people whose idiosyncratic mental habits deprived them of membership in the singular culture of rationality. Intellectual uniformitarianism has brought along with it a mental geography of metropolis surrounded by colonial hinterland.

The Sciences in Enlightened Europe

Since the 1960s, studies of the sciences in eighteenth-century Europe have flourished in the absence of any new large-scale historical synthesis. Some of

32. Weber, *Protestant Ethic*, 182.

33. Jacob, *Living the Enlightenment*, 218, and *Scientific Culture and the Industrial West*, 126–29.

these studies have remained in thrall to the dictates of positivism, cataloging discoveries and advances in the onward march of science, which is occasionally still viewed as occurring at the expense of religion and metaphysical philosophy. Some have opted for what seemed the safer course of mere anatomical dissection and microhistory, offering one more case study as a contribution to the putative eventual assembly of an integrated historical vision. This program, positivist in its own methodological suppositions, projects the reconstruction of the Enlightenment mind onto the millennial future "when all the facts are in."

There is also, however, a third course, in relation to which we situate our own project in this volume. Certain kinds of anatomical studies have subjected the uniformity of Enlightenment science to scrutiny. Rather than assuming its unity a priori, or simply fragmenting it into homogeneous parts, some contextual studies of the sciences in enlightened Europe have called the uniformitarian principle into question. These studies take as their topic one or another aspect of the construction of cultural unity in eighteenth-century Europe. By investigating the mechanisms of this construction, they provide us with a variegated geography of the Enlightenment and perhaps also permit us to recuperate a sense of the period as a whole. They show, not just that Enlightenment science was different in different places, but just how concrete forces of unification were operating in conjunction with those of local differentiation. Three guiding themes of these studies underlie the three sections of this book. First, historians seeking the means by which scientific knowledge was extended across time and space have focused upon structures of discipline, particularly those embodied in material technology and oriented toward human subjects. Second, the frameworks of humanity and nature have been brought under review as cultural constructs, intimately bound up with one another and shaped by new institutions of public conversation and judgment. Third, creation of geographical hierarchies and the formation of local or regional identities have been shown to have been closely associated with participation in enlightened culture. Local studies, particularly of the areas of the European domain traditionally regarded as "peripheral" to the Enlightenment, thus have the potential to reconfigure our sense of the geography of the movement and its unity.

At the wellspring of these studies, we can set the new kind of "anatomy" pursued by Foucault from his chair at the Collège de France after 1970. Foucault's chair in the "history of systems of thought" replaced one in the "history of philosophical thought." His initial lectures there, as preparation for *Surveiller et punir* (1975, translated as *Discipline and Punish*), introduced a triad of key techniques of "power/knowledge": measures, inquiries, and examinations. Each technique comprised a means of simultaneously producing knowledge and instituting control over its subjects. Foucault went on to outline a triple change in the physics of disciplinary power at the Enlightenment's end, constituted by a new optics, mechanics, and physiology. By following these themes, a "history of moral ideas" was to be replaced by a "history of bodies," with immense consequences for the reorientation from Enlightenment mentality to Enlightenment anatomy. In this way, the critique mounted by Horkheimer and Adorno found some empirical warrant: Foucault showed that the ferocity of the Enlightenment's disciplinary order was a central aspect of its own mechanisms for making knowledge, rather than a response to external pressures or a side effect of internal incoherence. Foucault's subsequent attention, in work on penology and biopolitics of 1973–76, to the productive schemes that engrossed living beings, social bodies, and entire populations stimulated new studies of enlightened regimes for knowing subjects and governing them. Jeremy Bentham's plan for an ideal prison, the "panopticon," became the model for systems of administration and knowledge in a variety of enlightened disciplines. Medicine, for example, was seen to be linked with tactical classification and bureaucratic surveillance of diseased subjects, mephitic sites, and neurotic individuals.[34]

Those studying Enlightenment disciplinary structures have remained in Foucault's debt, but other sources can also be found for these concerns. Kuhn's portrayal of paradigmatic dogmas might also find its location here, for example, with its suggestion that the putative intellectual liberation offered by the Enlightenment's sciences was rather to be understood as an aspect of systems of discipline and control. Kuhn's insistence, most clearly set out in a paper on mathematics and experiment first published in *Annales* in 1975, that during the Enlightenment the key to scientific change was the process of quantification, has prompted historical work on enumeration, instrumental measurement, and metrology—the "quantifying spirit." For example, historians of statistics and probability show how, in the German lands, at first under the aegis of Gottfried Wilhelm Leibniz, the practices of enumeration of population were identified by the servants of enlightened absolutism with the very essence of the nation-state. Distinctions into civil and military, sick and healthy, productive and sterile, were made and charted in this way. Empowerment of the intelligentsia went hand in hand with the effects of power and knowledge on disciplined subjects.[35]

SYSTEMS OF MEASURE AND DISCIPLINE

When anatomized, then, the principle of uniformity is seen to be embodied in, and produced by, calculated measures of discipline and accountability, which

---

34. Foucault, *Resumé des cours*, 20–21, 48–49; *Surveiller et punir*.
35. Kuhn, "Mathematical versus Experimental"; Hacking, *Taming of Chance*, 18–26; Buck, "People Who Counted"; Daston, *Classical Probability*.

were preconditions of Enlightenment judgments of equality and balance. Historians have indicated that enlightened management of the natural and human forces of production required the simultaneous control of both the physical and the social realms. Humans were not merely numbered; they were also mobilized as subjects of measured evaluation. The measure of work appears as an archetypal science of enlightened Europe. In Swedish mining and forestry industries, for example, as in French military schools and Prussian offices, links between institutional authority and technical competence helped sponsor cultures of meritocracy—the intercomparability of persons—as they promoted schemes for efficient technical dominance over nature and manpower.[36] Recent histories have reorganized the chronology of local Enlightenments to take account of this process. The first event of the Scottish Enlightenment, it has been plausibly suggested, was the foundation of the Edinburgh University medical school and its imposition of strict disciplinary order on the unruly sects of outdated physic and philosophy. Thence developed the disciplinary measures of physiology, pneumatics, and the systematic public medicine cultivated by enlightened physicians. If the key term of the Scottish version of enlightenment was "improvement," its culmination was surely the publication of John Sinclair's huge *Statistical Account of Scotland* (1791–99), which imported the term "statistics" from German to English.[37]

The British nation-state, to which improvers wished so actively to link Scotland, was home to the Newtonianism that intellectual historians use to unlock the key to Enlightenment mentality. Here, too, a history of the Enlightenment as a system of measure and disciplines has turned attention toward steam engines and stock markets as the proper concern of Newton's disciples. The universalism of the Newtonian cosmos underwrote, and was evidenced by, the sphere of operations of ingenious projectors and rational engineers. Lacking police science, *Policeywissenschaft* (the eighteenth-century German enterprise of rational governance), the British "fiscal-military state" nevertheless hosted practices of scrutiny and evaluation typical of the grand measures of enlightened absolutism and the mundane ones of mercantile bookkeeping. Recent arguments by political historians that even—or especially—in Britain the networks of excise and taxation were remarkably developed, thus helping make and reproduce the values of secular enlightenment, have buttressed this new account of measures and disciplines as essential to the sciences of enlightened Europe.[38]

36. Lindqvist, "Labs in the Woods," 313; Alder, *Engineering the Revolution;* Picon, *L'invention.*
37. Cunningham, "Medicine to Calm the Mind"; Golinski, *Science as Public Culture*, 11–49; Phillipson, "Scottish Enlightenment"; Hacking, *Taming of Chance*, 27.
38. Jacob, *Cultural Meaning;* Stewart, *Rise of Public Science;* Brewer, "Eighteenth-Century British State," 60.

NARRATIVES OF THE PUBLIC SPHERE

Viewed as a set of disciplinary structures, the uniform reason of the Enlightenment is seen to have been defined by processes of classification and exclusion. Therefore, recent studies attend to those barred from the realm of reason or subjected to its rule willy-nilly. The monstrous and the irrational are always threateningly present at the boundary; the smooth functioning of the Enlightenment public domain is shown to have been disturbed by the social friction between groups of different class and gender. Here narratives of the "public sphere" and its varying sociabilities have taken center stage. This move has been motivated partly by a reading of Norbert Elias's magisterial analysis, first published in 1939, of what he called the "civilizing process," a term that explicitly made the reformation of manners a crucial aspect of Enlightenment projects. According to Elias, the term "civilization" became a watchword for Enlightenment reformers in late-eighteenth-century France, located in salons and other extra-academic sites, in contradistinction to the term "culture," the defining notion of contemporary German self-reflection on the age. This argument pointed historians' attention to the forms of sociability that sustained intellectual life in the public realm. A further impetus to this reorientation was Jürgen Habermas's *Strukturwandel der Öffentlichkeit* of 1962 (translated as *The Structural Transformation of the Public Sphere* in 1989), which posited publicity as a category of bourgeois society. Habermas rooted this version of modernity in early-eighteenth-century Britain, providing good grounds for linking the worlds of coffeehouses, reading clubs, and print journalism to the enterprises of liberal critique and entrepreneurial markets. The numbers and schemes cascading across enlightened Europe are thus found to have been gathered by, and debated within, early forms of these public networks.[39]

Mapping these forms of enlightened knowledge, historians have linked Habermas's public sphere with the realm of publishing, an enterprise undergoing significant commercial growth in the eighteenth century. By reinterpreting the public sphere in terms of the communication circuits of print culture, Robert Darnton was an early advocate of this turn in the "social history of ideas." In a 1971 critique of Cassirer and Gay, he suggested that the "clouds of vaporous generalizations" in which the Enlightenment had become obscured could be dispelled by hard labor in dusty archives. Quantitative studies of literacy rates, publication runs, booksellers' catalogs, and the contents of institutional and personal libraries would disclose the extent of enlightenment at ground level. Darnton's work succeeded in charting the readership for different

39. Elias, *Über den Prozess;* Habermas, *Strukturwandel;* Chartier, *Cultural Origins,* 20–37; Calhoun, *Habermas.*

editions of the *Encyclopédie,* thereby taking the temperature of various cultural microclimates in the French provinces. But, perhaps symptomatically, Darnton's positivistic use of statistical methods was accompanied by an unwillingness to call into question the role of the sciences in enlightened culture. The identification of a (supposedly homogeneous) "science" with progressive, secular rationality tended to be taken for granted.[40]

Roger Chartier has pointed the study of Enlightenment publishing in a slightly different direction. He calls for the history of texts to break with the notions of universal taste and uniform human nature that were products of the eighteenth century itself. Rather than counting numbers of texts, under the assumption that they conveyed attitudes and values unambiguously, historians are enjoined to attempt to recover the multiple ways in which they were read. A model of diffusion and consumption is replaced by an emphasis on differentiated uses and plural appropriations. Texts can then be shown to have been read with variant meanings by readers who differed in class, religion, gender, or regional origin. Chartier insists that this does not simply lead to a more minute anatomy of local cultures. Unexpected patterns of circulation of texts and exchanges of meaning between groups of readers are brought to light. Even the gulf between literate and illiterate populations cannot be considered absolute in a society in which reading aloud brought written materials within reach even of the unlettered.[41]

As Chartier has noted, such an approach implicates in the history of publicity the private spaces in which texts were read, making the complex relations between readership and public sociability the key to enlightened culture. Enlightenment books and periodicals did not so much distribute a universal human reason as provide tools for multiple constructions of meaning by their readers. The marquis de Condorcet pointed out, in his *Sketch for a Historical Picture of the Progress of the Human Mind* (1793), that it was the printing revolution that had ultimately made possible this relationship between written works and the private exercise of the reader's reason. That reason was, in turn, displayed publicly as, "A tribunal . . . independent of all human power, from the penetration of which it is difficult to conceal any thing, from whose verdict there is no escape." [42] The illumination shed so universally by the tribunal of publicity was made possible by the independence and seclusion of the sovereign reader.

Thus, the enlightened reader was invited to witness scenes of labor in the experimental sciences, in natural history, and in the practical arts. The labora-

40. Darnton, "In Search of the Enlightenment"; *Literary Underground,* 179–81; *Business of Enlightenment.*
41. Chartier, *Forms and Meanings,* 1–5; *Cultural Uses of Print,* 3–12.
42. Condorcet quoted in Chartier, *Forms and Meanings,* 9–10.

tory and the workshop were opened to public gaze, and their openness celebrated in such works as the plates of Denis Diderot and d'Alembert's *Encyclopédie.* Publication of such texts mirrored the movement toward display in the
work of public lecturers and scientific showmen, who exploited the venues of
coffeehouse, assembly room, and exhibition hall for new forms of experimental demonstration. In terms of the settings in which they were read, these texts
also appealed to changes in the domestic sphere that accompanied and complemented the formation of a new public domain. Writers such as Benjamin
Martin, Luigi Algarotti, and John Newbery situated scientific discourse within
domestic scenes involving women and sometimes also children. The technique
served to validate the reproduction of knowledge in the public sphere by linking it to familial and feminine values, such as refinement, politeness, and piety.

Feminist scholars have insisted on the importance of women's role in this
process. In England, G. J. Barker-Benfield has traced the reconfiguration of
public discourse in the eighteenth century around the values of politeness and
conversability, a development that reflected a "feminization" of manners and
speech under the aegis of the "culture of sensibility."[43] For France, Dena
Goodman has asserted the centrality of the salons in enlightened sociability,
taking their lead from the role of the *salonnières,* who fostered politeness and
civility by governing men's conversation. The *salonnières* made use of domestic
gender relations to cultivate a playful style of discourse that was nonetheless
capable of bearing serious content. Beyond France, Benjamin Franklin and
David Hume were among those who paid tribute to the female "sovereigns of
the empire of conversation."[44]

Public knowledge showed itself dependent on the private exercise of reason, also, in the clubs and societies that were a ubiquitous feature of the enlightened public sphere. Many of these, ranging from formally constituted
academies to ad hoc groups of like-minded friends, retained a degree of privacy surrounding their activities. In elite academies, such as the Parisian
Academy of Sciences, the public sphere was defined in terms that were (perhaps necessarily) exclusive of much of the population. The Freemasons, recently discussed by Margaret Jacob, were obliged to maintain a degree of secrecy; in France they were subjected to police harassment and the strenuous
opposition of the church. The lodges nonetheless organized themselves as
miniature versions of the civic society they aspired to bring about: hierarchical
but meritocratic, secularist, fraternal, and bonded by an ideology of universal
human rights.[45] In a rather different way from the salons, the lodges mobilized
private relations in an orientation toward the public realm.

43. Barker-Benfield, *Culture of Sensibility.*
44. Goodman, *Republic of Letters,* 53–135.
45. Jacob, *Living the Enlightenment.*

In this light, the revolutionary convulsions of the end of the Enlightenment can be seen to have called into question the linked constitutions of the
public sphere and various private domains. Antirevolutionary paranoia, articulated by such writers as the abbé Barruel and John Robison, took aim at
the loose mixing of the sexes and the conspiratorial privacy of enlightened
sociability. The Freemasons found the terms of their self-representation reversed: a group that claimed to practice public virtue while perforce adopting
a cloak of secrecy was now manifested as a conspiratorial association that pretended allegiance to civic responsibility. Edmund Burke castigated the patterns
of interpersonal relations that had characterized debating clubs, coffeehouses,
and public meetings. "Clubability" was a dangerous solvent of social hierarchy,
Burke argued, which led to a general loosening of moral restraint and the
adoption of extremist positions: "association" was all too liable to bring on
fanaticism or "enthusiasm." Hoping to bury the already declining Enlightenment, Burke accurately identified modes of projecting the private domain into
the public sphere that had been fundamental to its character.[46]

ENLIGHTENMENT GEOGRAPHIES

Emerging from the microscopic or little enlightenments of the cafés, salons,
societies, and clubs, the Enlightenment as a macroscopic phenomenon is now
seen not as some mind or spirit, but rather as something projected, circulated,
and negotiated day by day by agencies such as the "Republic of Letters." A
good student of the Frankfurt school, Habermas tied this emerging public
sphere throughout to its economic double, the market: "The unfolding bourgeois public sphere rests on the fictive identity of the public called forth from
private individuals in their joint role as property owners and as humans
per se."[47] To be human no longer meant to be able to hear the word of God,
but rather to have something to trade. As articulated in the cafés and salons,
enlightened conversation—exchange often for its own sake—embodied con-
spicuous consumption par excellence, just as the subscription journal, the
material basis of the Republic of Letters, was the ethereal epitome of the joint-
stock company, whose capital was regularly consumed in the potlatch of its
dividends or "periodical issues." The free market, liberal politics, and enlightened science dovetail in "publicity," which thus replaces a universalist notion
of humanity as key to Enlightenment constructions of human nature and of
nature itself.

A similar reorientation has effected a change in our understanding of the
relations between the enlightened center and peripheral regions. The unifor-

46. Golinski, *Science as Public Culture,* 176–87.
47. Habermas, *Strukturwandel,* 121; Daston, "Ideal and Reality."

mitarian outlook had assumed that these relations were governed by a liberal and humanitarian "cosmopolitanism," which extended itself over a geographical space undistorted by inequalities of power. Recent studies, however, have pointed to forces of consumer appropriation and colonialism underlying the patterns of interdependence and local autonomy that are seen to have arisen in this period. Enlightenment cosmopolitanism is then exposed as a metropolitanism, in which experiences of cultural distance, centrality, and provinciality were produced conjointly with enlightened knowledge of nature.

When Napoleon led his troops into Egypt in 1798, as Elias pointed out, the process of civilization was viewed as complete in Europe, or at least in France. Napoleon saw himself "as the bearer of an existing or perfected civilization, . . . as the vanguard of civilization abroad."[48] A century earlier, in *Novissima Sinica* (1697), Leibniz could still hope that Europe might learn something from the civilization of China, which he held to be equal in merit. While, like the Jesuits, contemplating a Christianization of the Chinese through the propagation of European civilization, especially its sciences, Leibniz explicitly cast his interest in China in terms of his "cosmopolitanism." This meant exchange of goods both material and intellectual, as Leibniz hoped, via Russia, then well on its way to colonizing and civilizing Siberia. In Leibniz's China fantasy, the projection of cosmopolitanism outside Europe arose as an aspect of the Christianization of the world. And, though in the looking-glass world of enlightened satire, Swift's *Gulliver's Travels* (1726) and Voltaire's *Zadig* (1747) might contest the question of who was civilized and who was not (Europe or the "Others"), for Napoleon en route to Egypt the matter was settled as cosmopolitanism revealed its imperial guise.

Recently, historians have tended to move from charting cosmopolitanism to anatomizing its underpinnings. Enlightened cosmopolitanism is then made to reveal its ties to European imperialism as an ideology of consumerism and self-reflection. Taking their cue from social and economic histories of the eighteenth century, historians of science have begun anatomizing the tentacles of power and sinews of empire that bound colonies with the capitals of the Republic of Letters.[49] Natural history in all its branches has assumed a central place in topographies of enlightened science as a global enterprise. Studies of patronage, museums, and cabinets of curiosities have shown the importance of cycles of reciprocity in gift exchange and the growing culture of curiosities and scientific luxury goods in Renaissance and Baroque Europe. For some historians, it was just the displacement of the accumulative connoisseur by the philosophical naturalist that most distinguished the civilizing process as it was

48. Elias, *Über den Prozess*, 1:63.
49. Brewer, "Eighteenth-Century British State."

worked out in eighteenth-century museums and cabinets.[50] Translated into the emerging consumer society of eighteenth-century Europe, cycles of reciprocity and acquisition of luxuries acquired the force of immense global networks of scientific colonialism and extraction. In the new landscape of scientific consumerism, the voyages of Captain James Cook, with Joseph Banks and then Johann and Georg Forster in tow, emerge as an exemplary event of enlightened science, with the Pacific as a space for extended grand tours. We see better now how such "Romantic" scientific projects as that of Alexander von Humboldt grew from these enlightened ones.[51] In anatomizing the material bearers of Enlightenment, and the tours and raids it spawned, historians have found some of their best tools in Foucault's and related notions of circulating and diffused power. In Foucault's terms, power is no longer seen as vested in centralized institutions but as continuously reenacted in the relations of actors, at the periphery as much as at the center. Bentham's panopticon suggests itself as a better model for these colonial trading relations than Adam Smith's free market governed by the invisible hand. Invisible hands become most visible in the colonies, and, if we follow them back, we find them attached to such bodies as the London and Paris stock exchanges, as well as Kew Gardens and the Jardin du Roi.

Understanding of the global networks of science as part of consumer culture has gone hand in hand with a new awareness of Europe's eighteenth-century self-definition. Fernand Braudel's work on the Mediterranean radically historicized the notions of center and periphery in regard to Europe, arguing persuasively that the Mediterranean littoral remained central to European cultural self-consciousness well into the early modern period. Behind the cosmopolitan ideology of the Enlightenment lay the establishment of new intellectual and cultural centers in Europe, producing in turn the refashioning of Europe itself. The modern myth of "Europe" and its redefinition in the face of the "Orient" took shape in the wake of enlightened science. Renaissance humanism had embraced three ancient peoples as central to its myth of origins: Greeks, Romans, and Hebrews, with the Egyptians always somewhere in the background. The origin myth of enlightened Europe dispensed with the Hebrews, and in large part with the Egyptians too.[52] Enlightened European "civilization," which Napoleon would bring *to* Egypt, defined itself now as much in

50. Pomian, *Collectors;* Findlen, *Possessing Nature;* Daston, "Neugierde"; Smith, *Business of Alchemy.*

51. Gascoigne, *Joseph Banks,* 160–83; Miller and Reill, *Visions of Empire;* Dening, *Performances,* 107–10; Dettelbach, "Humboldtian Science."

52. Said, *Orientalism;* on Hebrews and Egyptians see Bernal, *Black Athena,* vol. 1, and the ripostes in Lefkowitz and Rogers, *Black Athena Revisited.*

terms of Greco-Roman roots as of Judeo-Christian ones, if not more. During the Enlightenment, the spiritual capitals of northern and western Europe shifted from Jerusalem and Christian Rome to Athens and classical Rome, with London and Paris as the Enlightenment's material centers. Europe's new capitals constituted the bases of its presumptions to embody humanity, enlightenment, and rationality, as well as the legitimation for its new imperialism and racism.

Conclusion

This selective review of recent historical research has given us the three main themes traced in this volume: the instrumental production of disciplinary order ("Bodies and Technologies"); the institutions of public taste and conversation in making a common conceptual framework for human and nonhuman natures ("Humans and Natures"); and the refashioning of cultural geographies and their associated anthropologies, including surveys of cultured and barbarian places and peoples ("Provinces and Peripheries"). Taken together, the chapters in the three principal sections of the book effect a series of reductions from the transcendental to the mundane: from the mind to the body, from humanity to society, and from the universal to the local. The book also contains, by way of orientation and departure, essays that canvass the possibility of recovering a sense of the Enlightenment as a whole. All of the chapters can be seen to historicize the forces of integration in the period, rather than hypostatizing them under the rubric of some climate, mind, or spirit of the age. Universalism thus emerges as a historical construction, set against powerful countervailing forces of the period that include the growth of individual expression through consumerism, the cultivation of domesticity, and a persistent (indeed resurgent) localism.

In conclusion, let us consider the balance between unity and diversity, between forces of consolidation and those of differentiation, as they emerge from consideration of each of these three themes. This enables us to give recent tales of biopolitics and modernization, of the Republic of Letters and the public sphere, of the expansion of Europe and the anthropological sleep, their proper historiographical import. It also points toward the discussion of individual chapters in the introductions to each section.

First, then, Foucault's work and others' gives us a set of tools for mapping the extension of material technologies of control and their effect in creating structures of discipline, particularly those invested in the human body. This draws attention to the circulation of disciplinary practices and artifacts that connected many different local sites. The integrative forces here included the administrative projects of states and "improvement" societies and the expansion of markets for industrially manufactured goods. But each case study poses the question as to how far the distribution of these practices and artifacts actually constrained individuals' actions in particular locales. The chapters in "Bodies and Technologies" show enlightened groups using disciplinary practices, such as measurements of population, for a variety of distinct purposes. Artifacts, whether barometers or automata, are seen to have been applied and interpreted in very diverse ways. Enterprises of administrative discipline are found to have served individual and group aspirations that were, in part, products of the same process of economic development. In view of this, disciplinary structures come to appear anything but hegemonic and universal in this period.

Second, the chapters in "Humans and Natures" explore the contexts in which notions of humanity and notions of the natural world were linked. As Roger Smith has recently noted, the concept "human nature" provided a powerful unifying framework for enlightened reflection on human characteristics.[53] The widely distributed mechanisms of social interaction grouped under the label "the public sphere" supported this ideological construct, since it was recognized that human nature was best expressed in society. But invocation of human nature in this kind of setting immediately raised the issue of what was truly universal and what was specific to certain classes, genders, and peoples. Interpretations of human nature in terms of sociability could not ignore the fact that different kinds of humans partook of different kinds of social institutions. The problem has been highlighted particularly by recent scholarship focused on the topic of gender. Many eighteenth-century writers were concerned with the physical and mental differences between the sexes. The elaboration of these descriptions of gendered attributes of mind and body challenged the putative universality of the Enlightenment's "science of man." Yet women could not simply be expelled from a monolithic masculine construct of human nature. Their presence in the institutions of enlightened sociability, where natural knowledge was produced, was marginal, perhaps, but crucial. The public sphere intersected with the semiprivate world of the salons and even the domestic world of the home. For this reason, gender was an unavoidable issue in enlightened reflection upon the nature of humanity.[54]

Third, attention to the manifestations of enlightenment at the peripheries of the European world recognizes that locality was significant both as a limitation on the universality of enlightened knowledge and as a category within it. The chapters in "Provinces and Peripheries" show how putatively universal knowledge was put to its local uses. We propose that this process is best seen

53. Smith, "Language of Human Nature."
54. Outram, *Enlightenment*, 80–95.

in what have been thought of as marginal locations, where cultural forces impinging from elsewhere are exhibited in their general and specific aspects. The large-scale formation of the Enlightenment is thus reflected in surveys of local populations and the construction of local identities. Too often, the Enlightenment has been seen as a purely mental construct, granted a geography and a temporality only insofar as it began in a certain place and diffused elsewhere. Like the free market, the purported cosmopolitanism of the Enlightenment has traced its origin principally to Britain and France, London and Paris. It was largely the Germans who actively opposed enlightened cosmopolitanism as the vehicle of Anglo-French hegemony or "civilization," inventing, if we are to believe Elias (who drew in turn upon Kant), the notion of "culture" to do so.[55] Since its coinage in the late eighteenth century, "culture" has proven a helpful term for the restoration of humanity and rationality to peasants and provincials, as well as for the illumination of marginal and peripheral peoples. This is surely a valuable way to reappropriate the term "culture" from its service to idealist universalism—not the least of the benefits of a historical reconsideration of the Enlightenment as a totality.

55. Daston, "Ideal and Reality," 372–75.

🙙 *1* 🙚

The Enlightenment Our Contemporary

DORINDA OUTRAM

No, I was no-one's contemporary, ever.
That would have been above my station.
How I loathe that other with my name
He certainly never was me.

Osip Mandelstam, poem 141 from *Selected Poems* (1924)

The Enlightenment has suffered a particular fate and a particular career. No other historical period has been defined with such intensity in relation to our own. No other has so insistently been viewed as the latency period of the twentieth century. Thus no other has been so heavily defined, since 1945 in particular, not as itself but as a way of elucidating the identity of the twentieth century. The Enlightenment has become our contemporary.

It is the purpose of this chapter to examine how this view has been fashioned, and with what consequences for our understanding of the Enlightenment, and of the place of science within it. It goes without saying that such a formulation of the Enlightenment also raises fundamental questions about our understanding of what it is that we do when we use one historical epoch to comment implicitly or explicitly upon another, in this case on what we are pleased to call, somewhat possessively, "our own."

The argument that in the Enlightenment can be found the origin of our own times received certainly its most powerful and enduring statement in a work now more than fifty years old. Max Horkheimer and Theodor Adorno, in their 1944 *Dialektik der Aufklärung,* or *Dialectic of Enlightenment,* argued that the Enlightenment had carried within it forms of thought and attitudes toward nature that were the necessary forerunners of the technological horrors of the twentieth century, the Holocaust in particular.[1] In defining the Enlightenment as our contemporary in this way, Horkheimer and Adorno also defined science and technology as fundamental cultural systems that cannot be readily distinguished from the power structures of society. That is, what has defined our

---

1. Horkheimer and Adorno, *Dialektik der Aufklärung* (1944); English translation, *Dialectic of Enlightenment* (1973). All references and quotations are from the 1973 translation and are cited in the text by page. Some early versions of the ideas of the *Dialectic* are to be found in Horkheimer, "End of Reason."

times as latent in the Enlightenment is the direct consequence of men's successful attempts to gain sovereignty over nature, which is no longer seen as the expression of a divine order and intention lying beyond and outside itself, but as reducible to a series of mathematical expressions. On this argument depend the other definitions of the Enlightenment given in their work, as "the destruction of gods and qualities," as "bourgeois society," as "ruled by equivalence and universalism," as "totalitarian," as "mythic fear turned radical," as a "nominalist movement," or as "wholesale deception of the masses." [2]

The situation in which their book appeared makes their interpretation entirely comprehensible. In 1944, both authors were refugees in America from Nazi persecution. They had started to confront the enormity of the Holocaust, realizing that an organized mass murder on such a scale would have been impossible to carry out without technology, and without an attitude that reduced thought to a calculus of means and ends, undercutting any independent moral ground for resistance. Insofar as the Enlightenment can be said to have laid the foundations for the development of both that technology and that attitude, it can be called "totalitarian." Both authors were also members of the Frankfurt school of social philosophers. Beginning before the Second World War, this group had aimed to use a revived Marxism to criticize industrialization and the growth of the bureaucratic state, both of which increasingly defined people as objects, and culture as a pure commodity. People and objects, as well as nature itself, were judged solely in terms of their exchange value, and not by any intrinsic worth or quality. The basis of this society was a technology that premised the complete domination of nature by man. The events of 1939–45, during which the Frankfurt school was forced to find refuge in California and New York, the onset of the Cold War, and the consequent division of Germany and Europe, caused some members of the group to abandon their earlier commitment to the revival of Marxism. Horkheimer and Adorno in particular drew back from former supporters, such as the philosopher Ernst Bloch, who remained committed to Marxist solutions. But this separation did not alter the group's determination to find a form of philosophy, a "critical theory," that would change the self-conscious actions of human beings and thus help to bring about social change. This commitment deepened with the return of Horkheimer and Adorno to Germany and their refoundation of the Frankfurt school. Because of their desire to assign meaning to the present, and working as they did at the intersection of fact and value, Horkheimer and Adorno made little distinction between philosophy and history.

The *Dialectic of Enlightenment* opened this program of engagement and criticism. The authors asked "why mankind, instead of entering into a truly

2. All these definitions appear in *Dialectic*, chap. 1.

human condition, is sinking into a new kind of barbarism" (xl). The answer to this question was a paradox: "[T]he Enlightenment has always aimed at liberating men from fear and establishing their sovereignty. Yet the fully enlightened earth radiates disaster triumphant. The program of the Enlightenment was the disenchantment of the world: the dissolution of myths and the substitution of knowledge for fancy" (3). Man, Horkheimer and Adorno argued, had gained sovereignty over nature and then over his fellows by "rationality," exemplified by the calculation of ends and means demanded by technology. This triumph of rationality involved a refusal any longer to see nature as the location of mysterious and inexplicable powers and forces: "Technology . . . does not work by concepts and images, by the fortunate insight, but refers to method, the exploitation of others' work and capital. What men want to learn from nature is how to use it in order wholly to dominate it and other men. . . . On the road to modern science, men renounce any claim to meaning" (3–5).

These insights link the *Dialectic* to ideas now commonplace in environmental thinking. They also engage with Enlightenment assumptions that once released from superstition and fear, once "disenchanted," humans could find purely rational solutions to problems, solutions therefore acceptable to all. Rationality would bring with it the end of politics. But of course, such agreement is never to be found, even in the calculation of means and ends. Since Enlightenment denies the validity of means, such as tradition or revelation, in solving problems, it is in the end impossible to resolve such conflicts without the use of force. This is why Enlightenment is called "totalitarian": it admits the validity of no other value system beside itself. It was not difficult to develop this point and allege that the Enlightenment had left no legacy that could be used to resist the general use of political terror, let alone the technologically ensured terror and mass death of the Holocaust.

The Enlightenment project of the "disenchantment of the world" made it difficult to focus aspirations on "the regions beyond." Rather, enlightened thought "transferred them as criteria to human organisation" (87). This transferral is what the *Dialectic* famously called "the administered life," a rational organization of men, nature, and knowledge itself for the achievement of objectives not oriented onto otherworldly or transcendent values. Rationality, defined as the calculation of ends and means, knew nothing outside itself. In support of this view, Horkheimer and Adorno quoted Immanuel Kant's dictum that "[r]eason has . . . for its object only the understanding and its purposive employment" (87).[3]

3. The quotation is from Immanuel Kant, *Kritik der reinen Vernunft* (1781).

The Enlightenment idea of rationality also had important cultural consequences. It turned knowledge into a commodity. Knowledge could become
part of systems of supply and demand, exchange and use, and its value could
be calculated in these terms. Knowledge therefore became disconnected from
what was beyond the economic sphere, that is, from wisdom and from ethics.
Knowledge thus became a means to an end, and existed purely as an exchange
value "without permeating the individuals who possessed it" (197). Relationships with knowledge became possessive rather than mimetic, thus losing their
capacity to act critically on society. Later, in his 1970 *Aesthetic Theory,* Adorno
was to argue that it is only aesthetic life that still, post-Enlightenment, allows
critical moments to occur, because aesthetics contains the possibility of mimesis, of creating reminders of the transcendent world.[4]

Such arguments gave the *Dialectic* iconic status for the revolutionary and
feminist movements of the 1960s. Along with works such as Herbert Marcuse's
One Dimensional Man, it seemed to offer an explanation for the evident in-
ability of Western society to produce an independent ground for moral action:
that society had become its own object of veneration.[5] Nor was the Frankfurt
school alone in raising such concerns. Simone Weil, in her posthumously published *L'enracinement* (1949), made the same case for different reasons, as did
Martin Heidegger's *Holzwege* (1950).[6] The rejection of science and the values it
seemed to foster took place all over Europe, and it is possible to see the crisis
of the late 1960s partly as the political culmination of an ideological rejection
of science, its associated value systems, and its social impact, which had been
growing ever since 1945.

The arguments of the Frankfurt school in general, and of the *Dialectic* in
particular, were also important because of their impact upon the rise of specifically feminist philosophies, politically and epistemologically critical, which
coincided with the revolts of the 1960s, and which also strongly attacked science. Particularly important was the thesis of the *Dialectic* that the rise of experimental science since Francis Bacon was based on the search for dominance
over, and possession of, a natural world essentially viewed as female. While
gender is far from being a major concern of the *Dialectic,* the long shadow cast
by such arguments is apparent, for example, in Evelyn Fox Keller's influential
Reflections on Gender and Science (1985). Keller examines the gender bias in the
language used by the "fathers" of modern science, which is seen as originating
with Bacon.

To point out the impact of the *Dialectic* is not, of course, to imply an

4. Adorno, *Ästhetische Theorie.*
5. Marcuse, *Eindimensionale Mensch.*
6. Heidegger's "Question Concerning Technology" is also apposite here.

argument for its originality. Criticism of Enlightenment thought began with the Enlightenment itself. The famous 1784 competition announced by the *Berlinische Monatschrift* for the best answer to the question "What is Enlightenment?" revealed the unease of contemporaries with the defining concept of their time. The now best known answer to this question, that of Kant, emphasized the idea that Enlightenment was not a completed process, but one still in progress. Kant's essay opened with his well-known definition of Enlightenment as "man's liberation from his self-incurred immaturity," by which he meant dependence on external authorities rather than the use of man's reason. But he also argued that reason should be used in an unrestricted way only in the private realm. In the public realm, the unrestricted use of reason was dangerous, and led to the subversion of the very monarchical authority that made at all possible the security of the private sphere. So Kant's definition of Enlightenment had a double legacy for the future. It defined Enlightenment as a process, not as completed stage of human understanding. And it also pointed out central contradictions in the very idea of Enlightenment. Most important, Kant's remarks about the different ways in which critical reason should or should not operate in the public and private spheres meant that it was difficult to use the notion of Enlightenment to ground any moral theory. Would such a theory apply to the use of reason in the public sphere or the private? Would it be grounded in the public or the private roles of the individual person? If science was part of critical reason, how could it be conducted in the public realm? If it was merely practical reason, the calculation of ends and means, then it was forever divorced from the sphere of value.[7]

To the problems inherent in Kant's definition of Enlightenment, Hegel added others. In the *Phänomenologie des Geistes,* he interpreted the Enlightenment as a perverted attempt to complete the movement of the Reformation toward the emancipation of reason, which substituted for the worship of the divine the worship of human reason as the foundation of all value. In former times, Hegel argued, men

> had a heaven adorned with a vast wealth of thoughts and imagery. The meaning of all that is, hung on the thread of light by which it was linked to that of heaven. Instead of dwelling in this world, presence, men looked beyond it, following the thread to an otherworldly presence, so to speak. The eye of the spirit had to be forcibly turned and held fast to the things of this world, and it has taken a long time before the lucidity which only heavenly beings used to have could penetrate the dullness and confusion

7. Kant, *Political Writings,* 54–60; Baehr, *Was ist Aufklärung?;* and Hinske, *Was ist Aufklärung?*

in which the sense of worldly things was enveloped, and so make attention to the here and now as such, attention to what has been called "experience," an interesting and valid enterprise. Now, we seem to heed just
the opposite: sense is so fast rooted in earthly things that it requires just
as much force to raise it. . . . The Spirit shows itself so impoverished that,
like a wanderer in the desert craving for a mouthful of water, it seems to
crave for its refreshment only the bare feelings of the divine in general.[8]

In the Enlightenment, human reason asserted the value of human reason
and, without an external reference point such as that provided by the transcendent or otherworldly, was unable to provide an independent ground for morals. These arguments were important influences on the *Dialectic*.

A third major influence, this time from the twentieth century, must also
be mentioned. It is to Max Weber that Horkheimer and Adorno were indebted
for the description of the post-Enlightenment period as the "disenchanted
world," a world from which "gods and qualities" had been banished, and in
which science had joined with bureaucracy to express all qualities as inter-
changeable mathematical expressions. It was also to Weber that the authors
were indebted for their concern with objectivity as the cultural value of science.
Weber had pointed in 1904 to objectivity as necessarily and rightly the central
value of the social sciences, as well as of the natural sciences. For him objectivity implied a willed distance between the opinions and personality of the
observer of society or nature and the object of his observations. Weber does
not argue that this definition of objectivity is necessarily separate from the
sphere of social values, which alone can give meaning and value to the results
of objective observation. But he gives little consideration to how meaning and
value might actually be constructed. Because of this, he provided no help
with the issue of how an inquiry based on objectivity might seek to ground its
value outside itself.[9] Weber's discussion of objectivity thus only reinforced
Horkheimer and Adorno's argument about the centrality of science to the restriction of the sphere of value in the post-Enlightenment world. This restriction, they argued, had in turn made possible the emergence of a purely instrumental form of reason, which saw knowledge as merely a tool in the calculation
of means and ends, not as a means of arriving at value. These were views that
enrolled their work in the quasi-Romantic rejection of positivistic science of
the 1960s.

Horkheimer and Adorno thus saw the Enlightenment negatively. Enlightenment thought, particularly Enlightenment science, was the precondition for

8. Hegel, *Phenomenology of Spirit*, 5.
9. Weber, "Die 'Objektivität' sozialwissenschaftlicher und sozialpolitischer Erkenntnis."

twentieth-century modernity, defined as the domination of technology and science, the restriction of the value sphere, and the consequent facilitation of technological mass murder. It is therefore interesting that their leading pupil, Jürgen Habermas, was to provide a far more positive characterization of the Enlightenment, equally influential in the historiography of science in this period. For Habermas, just as for his former teachers, the Enlightenment lay at the origins of modernity. But whereas for Horkheimer and Adorno, the central consequence of the Enlightenment was totalitarian mass murder, for Habermas, it lay in the creation of the public sphere as a location for critical reason. Ironically, however, Habermas developed this view in response to controversies over the Holocaust and the Nazi dictatorship. The German historian Ernst Nolte had argued that the concentration camp system, while undoubtedly criminal, was not unique in history, but had arisen in a reaction to the perceived threat of Bolshevism in the 1930s and 1940s. This attempt to portray Auschwitz as part of "normal history" aroused violent controversy, and provoked Habermas's intervention, culminating in the arguments of his *Structural Transformation of the Public Sphere* (1962).

Habermas emphasized Kant's insight that far from being closed, the Enlightenment even by the late twentieth century had still to be brought to completion. In his view, at least in Britain, the Netherlands, and France, the eighteenth century had seen the creation of a new "public realm" where critical public opinion was created. For all his divergences from the *Dialectic,* Habermas still agreed that the Enlightenment was characteristically bourgeois. And whereas Kant's essay had emphasized the limitations on the public sphere in states such as his native Prussia, Habermas argued for its efflorescence further to the west, a result of middle-class control and consumption of the flow of cultural materials, which became so important as commodities in this period. While Habermas's closing attack on modern mass culture differs little from that of Horkheimer and Adorno, he does nevertheless see the bourgeois public realm as a place of liberation, where critical reason can question traditional social forces. Habermas's public realm is a space where men can escape their role as subjects. There is no need to emphasize how important this view might be to a German historiography traditionally overshadowed by the creation of the state, and overwhelmed by the problem of the failure of German civil society and liberal humanist values to resist the rise of Nazism.[10]

In spite of the long gap between its first publication and its translation into English, and in spite of the fact that it deals explicitly only with France,

---

10. Habermas, *Structural Transformation of the Public Sphere;* Eley, "Nazism, Politics, and the Image of the Past."

Britain, and the Netherlands, Habermas's work has been influential for the history of science in Enlightenment Europe. In the historiography dependent on his work, science is often viewed as part of this revolution of media and consumption, part of the commodified flow of information that characterized the Enlightenment. This released historians from the dark view of science and technology put forward by Horkheimer and Adorno. It also released them from the internalist agendas of a previous generation of historians, and allowed them at least to pose the question of the relation between the history of science and the newly fashionable field of cultural history. However, the way in which historians such as Robert Darnton or Roy Porter have used Habermas's work has often made it difficult to assign a role to scientific ideas as such. Whereas Horkheimer and Adorno were sure that science and technology, for good or for ill, had supplied the central cultural structure of the Enlightenment, Habermas gave no specific place to science in the public realm. Historians who follow Habermas can find no way from within that theory to explain why, or whether, science could be assigned this centrality. The description of the Enlightenment as so strongly linked to the creation of middle-class self-identity through a relationship with print media also begs many questions. In the hands of the unwary, it can provide a license for seeing the Enlightenment as an innocent creation, the product of reading, writing, and conversation, of amateur science and genteel amusements: something that happened in the free time of the majority of its consumers, on their holidays from the realms of power and authority. The merit of Horkheimer and Adorno's approach is to make it impossible to divest consideration of the Enlightenment from the consideration of power. It is also important to note that Habermas's approach provides no solution to the problem of the relationship between the history of science and cultural history. It thus becomes difficult to justify the preference for portraying the Enlightenment as in some sense the origin of modern science.

What claim do we really make in such portrayals? Are we merely concurring with the insights of Horkheimer and Adorno into the relationship between the Enlightenment, modernity, and science, insights that seem to provide very little way of understanding science critically but positively, and thus enabling us to emerge from the current aggressive polarities of the science wars debates. How can we construct a history of science that is different?

A first step might be to ask ourselves what it is that we do when we describe another historical period as our own origin, in this case, when we describe the twentieth century as latent in the eighteenth. When we do this, we are identifying ourselves as contemporary to a certain time we think of as "ours," and we are simultaneously saying that the Enlightenment, connected to that time, is therefore also, to greater or less extent, "ours." Such thinking is highly possessive. It is also full of notions of identity, an idea the *Dialectic* repeatedly criticizes. By clinging too hard to such ideas as the justification of our interest in the Enlightenment, we run the risk of approaching it simply as a mirror to ourselves, of sacrificing its specificity to the need to find projections of ourselves in the past. In this sense, and with great irony, such assertions of contemporaneity have in fact a damaging effect on the possibility of writing the history of science as part of a critical cultural history, because they remove the possibility of the formation of external points of reference to validate that critical viewpoint. Maybe that is one of the reasons a great poet, who had good evidence of the wolflike nature of his century, should have refused in advance the attempt to make him the contemporary of anything.

The Enlightenment chose as its own projection field, not a historical era relatively close to it in time, but the societies of classical Greece and Rome. It was here that the era found its political and literary models. Other widely used projection spaces were described in the travel literature and utopian journeys of which the Enlightenment produced such great numbers. In other words, for the Enlightenment, very differently from our own period, origins were placed precisely through distance, not through some form of possessive identity. This is a perception we should consider before we designate any historical period as "ours"—and hence ourselves as "its"?

Primary Sources

Hegel, G. W. F. *The Phenomenology of Spirit.* Translated by A. V. Miller. Oxford: Clarendon Press, 1977. First German edition published 1807.

Kant, Immanuel. *Kant's Political Writings.* Edited by Hans Reiss. Cambridge: Cambridge University Press, 1970.

2

Bodies and Technologies

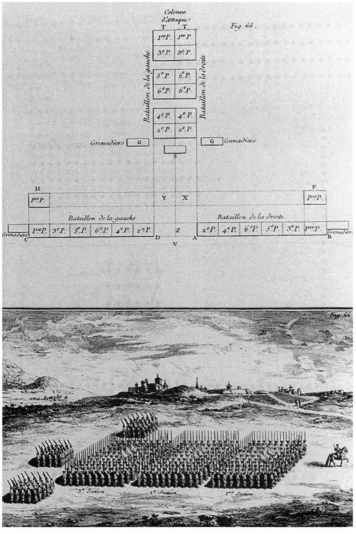

"Art Militaire: Evolutions." Plate 14 in Denis Diderot and Jean d'Alembert's *Encyclopédie: receuil des planches* (Paris, 1762), vol. 17.

In the name of liberty, the sciences of enlightened Europe promoted a remarkable range of new disciplines. Promoters of disciplinary schemes to arrange people and machines in healthy, effective, and moral spaces much celebrated their lucid accountability, in which arbitrary rule was to be displaced by enlightened reason. Techniques for producing knowledge of population, labor, climate, and mechanics sustained, and were made by, these systems of governance, while measures to promote these features of life depended on measurement. Individual bodies were combined as known and collective subjects, linked under a uniform therapy, climate, curriculum, or mechanism. Yet at the same time, local tactics challenged and resisted enlightened uniformities. Numbers supposed disinterestedly to settle dispute could be the topic of interested controversy. Mechanized and programmed citizens, soldiers, or laborers nevertheless possessed ineffable qualities of skill and ingenuity. Technologies designed to measure bodily changes could never quite command dogmatic assent and, instead, illustrated the dangers of overconfidence in the command of reason. Thus entangled with the enlightened program of discipline were rival programs of skeptical critique and resistance.

Andrea Rusnock introduces the key field of population arithmetic, in which the characteristic mentality of calculated judgment emerged. She cites as example the influential text of the baron de Montyon, whose pseudonymous work on the population of France (1778) argued that there could be no "well-ordered political machine" without such measures. In his account of the eighteenth-century era of "biopower," the integration of individual bodies under efficient disciplines and the measured supervision of variables of birth, health and death, Michel Foucault also cited de Montyon as one among several enlightened proponents of a calculated surveillance over national resources and inhabitants. The baron wanted to encourage a growing population, despite threats of atmospheric or agricultural exhaustion. Rusnock illuminates the technologies that proponents of counting heads invented, singling out the self-validating correspondence networks as milieus through which data were collected. She indicates, too, how new objects could thus be constituted through the taxonomies of the populace that different, often conflicting, categories of collection implied. These classification schemes, examined later in this book under the headings of natural history and the sciences of man, made new objects and also destabilized traditional orders of things. Rusnock shows that conflict was an ill for which political arithmeticians reckoned their tabulations offered a cure. It was through such data, allegedly, that the right conduct of married couples, tactics against smallpox, and choices of domicile could all be settled. But the numbered populace understood well that states had their own

motives for data collection, tax collection as often as the reform of grain provision during dearth. Moreover, the savants could use numbers for their own
secular purposes. Montesquieu influentially argued that the French population
was quickly declining, so condemned monasteries because of their rule of chastity; Voltaire quickly worked out that through warfare and persecution more
than nine million people had died because of the actions of Christianity.

Several intriguing qualifications of a general account of inevitable quantification emerge. Rusnock shows that the management of individual bodies
through these technologies was inseparable from the general technologies that
so taxed them. She also begins to indicate the limits to, and local queries of,
the apparent powers invested in enumeration. In his remarks on health politics in the eighteenth century, Foucault noted the many sites in "the social
body of health" where technologies and disciplines of salubrity and statistics
emerged. This was by no means a uniform process of incorporation by the
aggressive state. Foucault rightly pointed out the role of aerial spaces, of the
identification of morbid airs and mephitic places. These ambiguities are as
apparent in the technologies of meteorology that form the subject of Jan
Golinski's work. Knowledge of atmospheres was an integral part of disciplines
of population and salubrity. Rusnock describes the work of the Manchester
physician Thomas Percival, who blamed luxury and bad air for Manchester's
high mortality rates in the 1770s. Like his friends in the community of enlightened dissent, Percival also proposed highly regulated social systems to control
and direct dangerous and unhealthy classes. In his analysis of life expectancy,
the baron de Montyon had similar quantitatively based recommendations for
healthy places and healthy seasons on the basis of data from marshy and
mountainous parishes, from mortal and salubrious months. Golinski focuses
on the key instrument of such technologies, the barometer, topic of enlightened verse and satire, advertisement and enthusiasm. Just as Rusnock recalls
that Royal Society tables could not effortlessly sway populations in inoculation's favor, so Golinski shows the considerable ambiguities and ruses surrounding the medical politics of barometry.

No doubt the new science of quantitative meteorology set out to challenge
and displace what the enlightened judged the past superstitions of astrology
and plebeian weather-lore, outdated techniques highly dependent on the individual body as sympathetic to macrocosmic changes. Savants' technologies
were thus designed to calibrate and undermine mere bodily responses to airs
and waters. But through metaphorical slippage between glass instrument and
fleshly body, both subject to the vagaries of atmosphere and passion, the barometer could scarcely be imposed as a universal technology of weather and
health. The 1780s debates around mesmerism, described later in this book by

Simon Schaffer and by Mary Terrall, showed graphically the widespread instability of vulnerable bodies and the equally important need to regulate them. Such anxieties had their precedent in the culture of meteorology, which equally indicated subversive bodily sensitivity and the possibility of rationalizing it. Mobilization of public opinion was seen as indispensable and troublesome. Golinski indicates that the bourgeois public was recruited to see if barometry was reliable at all, but they were not invited to join in an already unchallengeable technological project. The dubiety and merely moral authority that inspection of the glass could yield taught proper humility and modesty against the dogmatism of the pedantic systematizer. Such technologies nicely embodied the knowledge of enlightened Europe, not because of a pervasive quantification, but rather because of the means through which they helped place faith in numbers at the center of public debate, interest, and critique.

The same careful strategy that steered between dogmatic systems and plebeian crafts helped define the new forms of knowledge and discipline of the military. Ken Alder's study of the eighteenth-century French artillery corps charts what he calls a "middle epistemology" designed quantitatively to investigate and optimize technological capacity without lapsing into the errors of artisan corporations or solitary savants. Encyclopedism, Diderot's joint attack on artisan secrecy and airy gentility, went to war. Especially important, Alder shows, was that this social and cognitive program by no means excluded graded hierarchy from professional meritocracy. Aristocratic distinctions were indeed to be dissolved by enlightened disciplines, but only so as to introduce potent measures of social control. A feature of Alder's tale is the multiple functions mathematical technique could play in the system of military engineering. Mathematical training helped form the bodies of the artillery in highly disciplined roles.

This is a familiar lesson from Foucault's account of the mechanical production of docile bodies on the parade grounds of enlightened Europe, the individualized, efficient, and uninterrupted surveillance that produced "an art of the human body." But Alder considerably extends this account. The codes of analytical mechanics and descriptive geometry embodied from the 1760s in the textbooks of Etienne Bézout and his collaborators promoted a new science of machines, which relied on and secured the high status of analysts, reckoned analysis the medium of thought itself, promoted a calculus of variation fitted for dynamic field tactics and trials, and in the end, allowed cost minimization and utility maximization. So, Alder concludes, professional military ideals of state service engineered in these disciplinary settings reorganized divisions between private and public power and knowledge. The enlightened engineers set out to render public and accountable what had previously been private and inarticulate. Yet just as public technologies of enumeration of population, wealth, and health all traded on the ambiguities of private interest and the ruses of privilege, so here the expert body of engineers claimed the right exclusively to select, manage, and validate their own members and skills.

Self-knowledge and self-discipline therefore provided a vital theme of the technologies of enlightenment. Rusnock attends to ways in which self-policing medical correspondents sought the reliable accumulation of numbers; Golinski indicates how the barometer might fit into polite codes of self-control; Alder concludes that self-government of engineers and their allies was an integral part of their public role as servants of the state. Simon Schaffer picks out one major enlightenment image of such forms of knowledge and discipline, the self-moving machine. Automata intersected with the concerns of a self-confident enlightened intelligentsia and the bodies of knowledge they sought to produce. In common with several chapters in this book, such as Michael Hagner's study of monsters and museology in the German lands, Emma Spary's exploration of provincial naturalists in France, and Lissa Roberts's tour of Dutch learned societies and philosophical theaters, Schaffer follows the contrasting themes of utility and entertainment, the popular and the elite, through a range of enlightened sites. Automata clearly featured as a principal means for representing the productive and subjected body. They did so for military theorists of drill, philosophical analyses of workshops, and medical naturalists of the nervous system and its passions. Philosophical discourse on necessity and free will, which so characterized the metaphysics of Paris and Birmingham, Berlin and Königsberg, often found fruitful matter in the automaton's deeds. So did the politics of enlightened absolutism.

In his genealogy of this absolutism's philosophies and politics, Foucault picked out two components of eighteenth-century biopower, individualized discipline and collective enumeration. He also distinguished the combination anatomy-metaphysics, which understood human bodies as machines, from the enterprise of technology-politics, which controlled and corrected bodies' operations. Savants often combined these different registers. Vaucanson designed showy automata and efficient silk mills; La Mettrie worked as military physician and materialist metaphysician; Diderot pursued philosophical materialism and the mechanical order of the workshops; Bentham designed ways of mechanizing penal and productive labor while materializing philosophy and jurisprudence. In so doing, Schaffer suggests, the enlightened could simultaneously define the social body as materially knowable and their own roles as privileged analysts of reason. The varying forms of the public sphere hosted this ingenious self-definition. Automata put on show in theaters and fairgrounds were entertaining models of cunning skill—Schaffer traces the career of a notorious machine, a chess-playing automaton within which a human virtuoso lay concealed first introduced in the German lands, Paris, and London in 1784. That device of mechanical skill and hidden virtuosity became an emblem of the public face of enlightened reason, which mapped humans as machines and machines as humane. Foucault reckoned automata were then small-scale models of power. Their odd career, as jokes and entertainments as well as forms of domination, production, and control, reminds us that in enlightened Europe the critique of discipline went hand in hand with its imposition.

Biopolitics: Political Arithmetic in the Enlightenment

Andrea A. Rusnock

The numbers of dead, categorized by location, age, or cause, became something of a preoccupation among Enlightenment authors. "A Faithful and minute register of mortality, and of the various diseases most fatal to mankind, at different ages, must evidently be of the most important consequence," proclaimed John Haygarth in 1774, "to the politician, the philosopher, and the physician, in their several endeavours to relieve the miseries, and promote the happiness of human nature."[1] Haygarth, a Cambridge- and Leiden-trained physician who practiced medicine in the town of Chester in northwest England, made this morbid assertion in an essay he sent to the Royal Society. Subsequently published in the *Philosophical Transactions* under the unassuming title "Observations on the Bills of Mortality, in Chester, for the Year 1772," Haygarth's essay contributed to a growing literature on political arithmetic, a field of inquiry established in the late seventeenth century by two Englishmen, John Graunt and William Petty.

Political arithmetic, defined in 1698 by Charles Davenant as "the art of reasoning by figures, upon things relating to government," flourished in the eighteenth century and can be considered a science of the Enlightenment, as Haygarth's statement fully reflects.[2] Haygarth's concluding phrase in the above quotation invoked the Enlightenment themes of relieving misery and promoting happiness through the application of reasoned knowledge, and while other factors may have motivated political arithmeticians, these familiar tropes nonetheless remain the stated ideal and goal of many eighteenth-century authors and should not be dismissed out of hand. A humanitarian ethic guided many of these arithmetical inquiries.

In the hands of political arithmeticians, mortality registers yielded numerical information about population, and the importance of population and its various attributes were, according to Haygarth, obvious to "the politician,

---

1. Haygarth, "Observations," 67.
2. Davenant, "Of the Use of Political Arithmetic," 128.

the philosopher, and the physician." The demand on the part of seventeenth-
and eighteenth-century governments to determine the number of their respective inhabitants coincided with the growth of the modern state.[3] Governments
began to assume more functions, and politicians became interested in assessing
the number of inhabitants for purposes of taxation and finance, trade policy,
and military strength. "There can be no well-ordered political machine, nor
enlightenment administration," wrote the French intendant Antoine Auget,
baron de Montyon, in 1778, "in a country where the state of population is
unknown."[4] For the Enlightenment *philosophe,* population became a measure
of the relative decline of Europe since antiquity, of the perverse effects of
luxury, and of the imbalanced relations between the sexes. And to the physician, the study of populations helped determine broad patterns in the incidence and course of diseases, based on such factors as climate, geography, and
living situation.

The significance of Haygarth's method of compiling "faithful and minute
registers" of the population and then forwarding them to a scientific society
was not lost on contemporaries either, nor on scholars of the Enlightenment.
Recent studies, in particular the work of Michel Foucault, have emphasized
the new methods of collecting and representing information—what Foucault
termed techniques of power. Under the rubric biopower, Foucault sketched
the use of these technologies usually, but not necessarily, wielded by the state
to discipline individuals.[5] Biopower works on two levels. At the individual level
(anatomo-politics) the body is the focus of techniques of power; this has perhaps best been shown in Foucault's analysis of the prison in *Discipline and
Punish.* The second level (biopolitics) concerns the aggregate, where the population becomes subject to control through the sciences of demography and
statistics. According to Foucault, "the disciplines of the body and the regulations of the population constituted the two poles around which the organization of power over life was deployed."[6] But he only briefly outlined this second
level of biopower in his essay on eighteenth-century health and later in *The
History of Sexuality.*[7] For Foucault, biopolitics assumed its power only in the
nineteenth century with the growth of institutions such as hospitals, schools,

3. Brewer, *Sinews of Power,* chap. 8; Brian, *La mesure de l'état,* part 2; Kreager, "Quand une
population est-elle une nation?"; and Revel, "Knowledge of the Territory."
4. Moheau [Montyon], *Recherches et considérations,* 20; cited in Pasquino, "Theatrum Politicum," 115.
5. For a critical discussion of Foucault's idea of biopower, see Donnelly, "On Foucault's Uses
of the Notion 'Biopower.'" Juri Mykkänen developed a related Foucauldian idea—that of governmentality—in his essay on William Petty (Mykkänen, "To Methodize and Regulate Them").
6. Foucault, *History of Sexuality,* 1:139.
7. Foucault, "Politics of Health"; Foucault, *History of Sexuality,* 1:139–45.

and prisons, as well as state bureaucracies, which recorded and monitored information about their populations. The origins of biopolitics, however, are to be found in the eighteenth century, and not in any particular institution, but rather as a diffuse cultural problem associated with the social and economic repercussions of population growth. In Foucault's words:

> The great eighteenth-century demographic upswing in Western Europe, the necessity for co-ordinating and integrating it into the apparatus of production and the urgency of controlling it with finer and more adequate power mechanisms cause "population," with its numerical variables of space and chronology, longevity and health, to emerge not only as a problem but as an object of surveillance, analysis, intervention, modification, etc.[8]

While Foucault's sketch of biopolitics is persuasive, the actual mechanisms and agents through which it developed remain obscure. He looked to broad demographic and economic changes to explain the rise of quantitative studies of population. I will examine more specific sources of biopolitics by focusing on the work of political arithmeticians and their place in Enlightenment culture.

To begin, it must be admitted that it is difficult to specify exactly who the political arithmeticians of the eighteenth century were. A diverse group of individuals (mathematicians, natural historians, ministers, physicians) contributed to the literature, and the social and political location of these authors varied from country to country as well as over time. Efforts to calculate the number of inhabitants as well as to account for changes in population were undertaken both by state officials and by private individuals, occasionally together, reflecting the absence of any clear demarcations between public and private in many eighteenth-century undertakings.[9] This fluidity has been discussed by Peter Buck, who has argued that a significant change occurred around the middle of the eighteenth century in Britain, when private individuals outside government became the leading political arithmeticians, thereby replacing the earlier calculators (such as William Petty) whose ties to court and state were much stronger. With this change in social location came a shift in the politics of enumeration: counting populations was recast as a republican activity, rather than a monarchical one.[10] And, as Ian Hacking has emphasized, "public amateurs" were much more willing to share their findings than "secret

---

8. Foucault, "Politics of Health," 171.

9. On the many meanings of private and public information in Britain during this period, see Brewer, *Sinews of Power*, 221–30.

10. Buck, "People Who Counted," 28–29; also Buck, "Seventeenth-Century Political Arithmetic." Margaret Jacob discussed the radicalism of British philosophical societies in the last half of the eighteenth century (Jacob, "Scientific Culture in the Early English Enlightenment," 154–57).

bureaucrats," who feared that public disclosure of population information would make the state vulnerable.[11] What can be said, then, with some confidence about political arithmeticians and the location of their practice is that the "public" aspects of their work were dependent upon specific cultural developments constitutive of the Enlightenment: public debate, a growing popular press and active Republic of Letters, a humanitarian ethic, and a widespread enthusiasm for natural philosophy.

Political arithmetic has generally been discussed within the context of the history of probability, statistics, demography, and to a lesser extent, public health.[12] The writings of prominent mathematicians who used vital statistics to develop the calculus of probabilities and to construct life tables, which were instrumental in the calculations of secure annuities, have received close attention in several recent studies on the history of probability.[13] Edmond Halley, Abraham de Moivre, Daniel Bernoulli, Jean d'Alembert, and Pierre-Simon de Laplace represent this tradition. In this chapter, I concentrate on the efforts of less mathematically sophisticated authors (and consequently less well-known scientists, at least to historians of science), who relied primarily on simple arithmetic and the calculation of ratios or proportions in devising quantitative approaches to the study of population. In this sense, they too shared in the "quantifying spirit" that infused scientists of many stripes during the second half of the eighteenth century.[14] As Theodore Porter has recently argued, quantification is especially appealing when communities are weak, when the bonds of personal knowledge and trust do not exist, and when individuals seek to produce universal rather than local and particular knowledge. While Porter's observations were specific to developments in the nineteenth and twentieth centuries, the move from private, local knowledge to public, universal knowledge, a widely professed ideal of the Enlightenment, made quantification an increasingly popular, if not always successful, choice of the eighteenth century.[15]

Traditionally John Graunt's *Natural and Political Observations made upon the Bills of Mortality,* published in 1662, has been considered the first political arithmetical work. In this pathbreaking book, Graunt applied his "shop arith-

11. Hacking, *Taming of Chance,* chap. 3.

12. For a recent historiographic review, see Kreager, "Histories of Demography."

13. Daston, *Classical Probability;* Hacking, *Emergence of Probability;* Hacking, *Taming of Chance;* Hald, *History of Probability;* Kreager, "Histories of Demography"; Pearson, *History of Statistics.*

14. Frängsmyr, Heilbron, and Rider, *Quantifying Spirit.*

15. Porter, *Trust in Numbers,* 46–47. In his earlier book, Porter distinguished political arithmetic from statistics by its close association with the monarchy and centralizing bureaucracy (Porter, *Rise of Statistical Thinking,* 18–26).

metic"—contemporary bookkeeping and accounting techniques—to numbers from the London bills of mortality and characterized various features of
the London population.[16] As he indicated in his title, his observations were
both natural and political: he calculated the total population of London and
the number of military-worthy men, ranked diseases in terms of gross mortality, and demonstrated numerically the high rate of infant mortality.[17] His
friend and sponsor, William Petty, one of the founding members of the Royal
Society, expanded this work and coined the term political arithmetic—a name
that stuck throughout the eighteenth century.[18]

Although Graunt, Petty, and other seventeenth-century arithmeticians addressed a broad spectrum of issues, political arithmeticians during the Enlightenment generally narrowed their focus to two topics: the determination of the
total number of inhabitants of politically defined geographies (a city, county,
or nation), and the construction of life tables that gave the number of deaths
per age group and were used to calculate annuities. A notable exception to
this generalization concerns eighteenth-century physicians, who investigated
a range of topics arithmetically including salubrity, fertility, and specific medical practices. In fact, this medical tradition became so well developed by the
last decades of the eighteenth century that the physician William Black was
able to recount a century-long history of it and create a new term—medical
arithmetic—to encapsulate this particular subset of issues within political
arithmetic.[19]

Political and medical arithmetic were primarily British sciences, and the
majority of examples presented here come from British writers. Efforts to
count and describe various populations, however, were not limited to Great
Britain, and the phrase political arithmetic occurs in continental writings of
the period. Although important differences exist among the national traditions—in Ian Hacking's phrase "Every state, happy or unhappy, was statistical in its own way"—the impulse and methods used to quantify population
were shared to some extent among European cultures.[20] The first section of
this chapter briefly addresses the significance of population to Enlightenment
thought in general, and in the second section some of the techniques developed by arithmeticians are analyzed. In the final section, I discuss several examples illustrating the impact of arithmetical arguments on Enlightenment

16. For a careful analysis of the historical sources of Graunt's method, see Kreager, "New
Light on Graunt."
17. James Cassedy emphasized the medical aspects of Graunt's works (Cassedy, "Medicine
and the Rise of Statistics," 294–97).
18. See, e.g., Petty, Political Arithmetick, 233–313.
19. Black, An Arithmetical and Medical Analysis, introduction.
20. Hacking, Taming of Chance, 16.

debates. Political and medical arithmeticians, I argue, contributed significantly first to contemporary conceptions of population through their construction of vital (or, more precisely, morbid) statistics, and second, on a more grand historical scale, to the development of biopolitics.

Population in Enlightenment Thought

Questions of population became central to numerous Enlightenment debates carried out in published works and in the Republic of Letters. One historian has noted that "pursuing this ubiquitous subject leads not only to the heart of the economic debates of the period, but to writings about somewhat less predictable topics such as toleration, slavery, primogeniture, climate, suicide, duelling, torture, prostitution, celibacy, monasticism, luxury, and the consequences of the development of the arts and sciences."[21] But most historians have treated population theory and quantitative studies of population in one of two distinct historiographic traditions: the history of political ideas or the history of demography.[22] A more integrated story needs to be told, and an essential ingredient to this revised account is the work of political and medical arithmeticians.

Beginning in the mid–seventeenth century and throughout the eighteenth, mercantilists, cameralists, and other political economists considered population an essential component of the wealth of nations. A growing population would increase trade and production and allow for more efficient manufacture through the division of labor. It would increase the base for taxation and the supply of able-bodied men. It would lead to progress through the decline of rural isolation and the growth of general commerce.[23] In short, there was widespread agreement among Enlightenment thinkers on the benefits of a large population.

Fueling this conviction was general concern among Europeans about depopulation: had the population of England, France, indeed the whole of Europe, declined since antiquity? Did the growth of luxury evident in most urban centers contribute to this depopulation? In his *Lettres persanes* (1721) and later in his *Ésprit des Lois* (1748), Charles de Secondat, baron de Montesquieu, argued these points persuasively, and other *philosophes* such as David Hume and Jean-Jacques Rousseau took up the issues.[24] While most of the contributors to

21. Tomaselli, "Moral Philosophy," 7.

22. Kreager made this point in "Early Modern Population Theory," 208–9. He argued that quantitative approaches to population were integral to the general development of ideas about population in political philosophy in early modern Europe, in particular on the relations between the individual and the state.

23. Stangeland, *Pre-Malthusian Doctrines,* chaps. 4 and 5.

24. Tomaselli, "Moral Philosophy," 9–15; Rusnock, "Quantification, Precision, and Accuracy," 23–25.

these public debates rehearsed the putative reasons for depopulation without providing any evidence that population had in fact declined, arithmeticians attempted to put these debates on a firm empirical foundation. In this way, larger cultural debates stimulated efforts to quantify population, and in turn the act of counting and calculating the number of inhabitants helped refine notions of population.

A similar process occurred in medicine. Population became interesting as well as conceptually relevant to eighteenth-century medical practitioners in part because of a marked revival in Hippocratic ideas that linked climate to the health of inhabitants (or populations) of specific areas.[25] This revival was stimulated by the awareness of new environments at home and abroad, which shifted medical interest from the individual to the aggregate. Many parts of Europe during the eighteenth century (especially in England) experienced a greater pace of urbanization and its attendant health problems. In addition, increasing exploration and colonization made Europeans more aware of the radical differences in climate around the globe and correspondingly of the exotic diseases and specific health practices of different peoples. The vogue for meteorology during the Enlightenment, reflected in the popularity of new instruments such as the barometer, further contributed to this renewed interest in climate, and remarks on the incidence of disease were frequently appended to meteorological observations.[26]

In most writings of this period, population was understood as the number of inhabitants in a specified geographical area at a given time. But even this apparently simple definition was not without its objections. The total number of inhabitants was frequently calculated from either the annual number of births or deaths, figures that could be tallied from parish registers. (This method assumed a stable ratio between the numbers of births or deaths and the total population.) But when individuals attempted to enumerate adults, the problem of categorization became immediately apparent. Should age, marital, economic, or social status, or some combination of these, be essential categories to count? Even births and deaths (or, more precisely, baptisms and burials) were not straightforward: how was one to account for religious minorities who were usually omitted from parish registers?

Quantification raised acute questions such as these, and who was counted and in what category had immediate bearing on the meanings of population. A telling example of the complexities political arithmeticians faced is found in a document William Petty titled "Fiant," where he outlined which categories of individuals should be enumerated.

25. Riley, *Eighteenth-Century Campaign;* Smith, *Hippocratic Tradition.*
26. Golinski, chap. 3, this volume; and Rusnock, *Correspondence of James Jurin,* 27–31.

1. Sex, age, marriage, widowhood
2. Births, burials, baptisms, marriages
3. Religion: Roman Catholics, Anglicans, Jews, Socinians, Quakers, Anabaptists, Presbyterians, Independent or Newters "to any of the seven said sorts abovementioned"
4. Titles: Princes, Dukes, Marqueses, Earls, Viscounts, Barons, Baronets, Knights, Esquires, Gentlemen, Yeomen, and "of what Degree in the University or Inns of Court."
5. Employment: Ecclesiastical, Civil or Military
6. Quality: Freeholders, Tenants, Freemen of corporations, Foreign Merchants, wholesale or retail sellers, artisans, handicraft men, mariners, fishermen, husbandmen, or laborers
7. Others: Prison, Workhouse, College or Free-school; impotent, beggar
8. "Who are absent at sea or in Foraine ports, or sick in their beds." [27]

The overlap among these categories is considerable; traditional, hierarchical demarcations of status are listed alongside the more leveling categories of burials and baptisms; institutional assignments vie with other locations such as "sick in their beds." As this list suggests, enumeration could be put to very diverse purposes. In the following section, I examine some of the methods political arithmeticians developed to quantify population and indicate the ways in which the measuring instrument—particularly the categories that were enumerated—defined population.

New Technologies of Inquiry: Registers of Mortality

Political arithmeticians relied heavily on registers, which displayed numerical information about births, deaths, marriages, diseases, and so on. One of the most important precedents for these registers, which constitute part of what Foucault labeled techniques of power, was set by John Graunt in his *Natural and Political Observations made upon the Bills of Mortality* of 1662. Graunt constructed registers in order to question and evaluate contemporary conceptions of population. "I had reduced into Tables" the weekly London bills of mortality, Graunt wrote, "so as to have a view of the whole together, in order to *the more ready comparing* of one Year, Season, Parish, or other Division of the City, with another, in respect of all the Burials, and Christenings, and of all the Diseases, and Casualties, happening in each of them respectively." [28] As Sergio Moravia has argued, comparison *(comparaison)* was a method central to the Enlightenment and to the development of the human sciences in particular. [29] In the sciences of political and medical arithmetic, *numerical* comparison as-

27. Petty, "Fiant," in *Petty Papers*, 258–59.
28. Graunt, *Natural and Political Observations*, 333–34. Emphasis added.
29. Moravia, "Enlightenment and the Sciences of Man," 249–50.

sumed paramount importance. The construction of tables containing numbers representing different facets of populations could yield new insights primarily through comparison, as Graunt himself recognized. "I did then begin not only to examine the Conceits, Opinions, and Conjectures, which upon view of a few scattered Bills I had taken up; but did also admit new ones, as I found reason, and occasion from my Tables." [30]

Graunt's source of numerical information, the London bills of mortality, did not provide a census of the living, but rather an account of the dead. They had been collected, published weekly, and distributed to subscribers by the Company of Parish Clerks on a regular basis since 1603. They contained information about the number of burials, the causes of death (especially the plague), and the number of christenings. Commissioned by the monarch, the bills were used originally as a public health guide—that is, as a warning against impending plague epidemics. Although there are scattered examples of bills of mortality from other European towns, a continuous series of bills was unique to London. In France, for example, there did not seem to be much interest in accounting for the dead. "The Bills of Mortality are something peculiar to the English," wrote a reviewer of Graunt's book in the *Journal des Sçavans,* who attributed them to English curiosity about the ebb and flow of the plague.[31]

The London bills of mortality, then, were unequaled both in their continuity and their form; without such a resource, arithmeticians outside of London and in other parts of Europe called for the collection of numerical information about living inhabitants. William Petty, for example, recognized this need and in 1662 advocated the gathering of information of the "Number, Trade, and Wealth of the people," but in his work relied upon figures taken from parish registers and tax records.[32] In France, at roughly the same time, Sebastien le Prestre de Vauban, adviser to Louis XIV, recommended an annual or biannual census to be carried out by the royal government, and he provided a model based on his survey of the province of Vézelay.[33] Both Petty and Vauban had close ties to the court, and their advocacy of a national census must be viewed in terms of strengthening royal power.[34] Despite the efforts of Petty, Vauban, and others, neither France nor Great Britain established a regular census until the beginning of the nineteenth century.[35]

The lack of a direct national census or continuous run of bills of mortality

30. Graunt, *Natural and Political Observations,* 334.
31. *Journal des Sçavans* 1 (2 August 1666): 585.
32. Petty, *Treatise on Taxes,* 34.
33. Vauban, "Description géographique de l'élection de Vézelay."
34. Buck, "Seventeenth-Century Political Arithmetic," 80.
35. In 1753 a proposal to establish a national census in Great Britain was soundly defeated in Parliament. For discussions of this proposal, see Buck, "People Who Counted," 32–35.

made securing accurate numerical information about population the foremost challenge to arithmeticians during the eighteenth century. Thus, a variety of indirect measuring techniques were developed, particularly in Britain. In the 1750s the clergyman William Brakenridge used tax records to evaluate whether the population of London and England had grown or declined. He relied on figures taken from records of the hearth tax and the malt tax, which allowed him to estimate the amount of wheat produced and hence the amount of bread consumed. From this he could calculate the population that could be supported by a given amount of food.[36] In the 1770s and 1780s, Richard Price calculated the English population from figures gathered from the Window Tax Office and the excise duty on liquor.[37]

Notwithstanding the creative use of tax records, parish registers remained the single most important source of numerical information about populations throughout the eighteenth century, not only in Great Britain but throughout Europe. Richard Forster, for example, in order to contest the estimates of Brakenridge used figures drawn from parish registers to calculate the population of London and England. Forster argued that his method was a more reliable—certainly a more direct—way of estimating the size of population.[38] In France, political arithmeticians relied almost exclusively on church records. Buffon, for one, constructed life expectancy tables based on mortality records taken from parish registers, and in the 1780s Laplace calculated the total population of France using figures for the annual number of births taken from parish registers.[39] As well in Germany, Johann Peter Süssmilch turned to parish registers to provide figures on the population in his monumental *Die göttliche Ordnung* (3d ed., 1765).[40]

Political arithmeticians were thus dependent on state and church records, although these institutions did not collect data for the purpose of enumerating the population. As contemporaries frequently noted, both parish registers and tax records had their faults when used as sources for numerical information about population. Parish clergy typically recorded christenings, marriages, and burials but often omitted the age of marriage and age of death, as well as entirely excluding individuals of different faiths. Taxation records were likewise incomplete and provided only indirect means for calculating population— techniques that often relied upon assumptions such as the number of persons

36. Brakenridge, "Concerning the Number of People in England."
37. Price, *Observations on Reversionary Payments;* Price, *An Essay on the Population of England.*
38. Forster, "An Extract of the Register of the parish of Great Shefford."
39. Buffon, *Histoire naturelle, générale et particulière,* vol. 2, chap. 11; Laplace, "Sur les Naissances."
40. Süssmilch, *Die göttliche Ordnung.*

per hearth, or the amount of bread consumed annually by an individual. In order to get around the limitations presented by parish registers and tax records, several arithmeticians developed a new method to collect information: correspondence networks.[41]

During the eighteenth century, correspondence networks were frequently organized by physicians and generally were designed to gather medical intelligence, for example, on the incidence and severity of diseases within different communities. These networks thrived because physicians shared interests and depended upon common professional standards, which ensured to some extent the accuracy and reliability of information imparted through correspondence. Physicians also can be considered active participants in the Republic of Letters. As Lorraine Daston has argued, the Republic of Letters promoted an ideal of disinterested criticism and evaluation of ideas, and clearly this ideal operated in the networks designed to collect numerical information about populations.[42]

One of the first and best examples of this method appeared in the work of the English physician and natural philosopher James Jurin. In the 1720s, Jurin became interested in smallpox inoculation and sought ways to evaluate the practice. Using his position as secretary to the Royal Society, Jurin solicited accounts of inoculation through an advertisement placed in the *Philosophical Transactions* and published separately in pamphlets. Specifically, Jurin wanted to document the risks of inoculation. In addition, he desired accounts of the mortality of natural smallpox. Many individuals throughout England responded to Jurin's advertisement and supplied him with accounts of inoculation, thereby enabling him first to construct tables to display mortality figures and second to calculate and compare the mortality rates for inoculated and natural smallpox. Jurin's method of collection and compilation explicitly demonstrated that numerical information about a group of patients could be used to quantify the benefits and risks of a medical practice.[43]

Medical societies established during the second half of the eighteenth century followed Jurin's method. The Society for the Improvement of Medical Knowledge in Edinburgh, for instance, turned to correspondents to document the use and success of different therapies in scattered communities throughout Britain. Accordingly, Alexander Monro *Primus,* secretary to the society, solicited accounts of the use of Peruvian bark to treat gangrene.[44] In midcentury

---

41. Bruno Latour drew attention to the role networks play in constructing scientific knowledge, and stressed the importance of "centers of calculation" (Latour, *Science in Action,* chap. 6).

42. Daston, "Ideal and Reality," 370–75.

43. For a detailed analysis of Jurin's project, see Rusnock, "Weight of Evidence."

44. Tröhler, "Quantification in British Medicine," 99.

France, Richard de Hautesierck created a correspondence network among physicians and surgeons employed in military hospitals. The plan was supported by the duke of Choiseul, the secretary of state for war, and resulted in the publication of two large volumes entitled *Recueil d'Observations de Médecine des Hôpitaux Militaires* (1766–82). The majority of observations included in these volumes took the form of case histories, but the section entitled "Topographical Memoirs" contained quantitative information about population.[45] Similarly, in the 1770s, Vicq d'Azyr promoted a correspondence network to supply the newly established French Royal Society of Medicine with observations and records of health and disease.[46] Printed forms were sent to physicians throughout France; one side of the form presented columns for meteorological information (daily temperature, barometric readings, monthly averages for the amount of rain, and so on); the other side was entitled "Nosology," where physicians were to make remarks about diseases. D'Azyr specifically wished to amass accounts of epidemics, which were to include both case histories and quantitative reports. "It is desired that if possible a register of the sick who recovered and who died be provided," urged the preface of the first volume of the memoirs of the Royal Society of Medicine.[47]

By the end of the eighteenth century, nonmedical inquiries about the population began to be conducted via correspondence. William Wales, master of the Royal Mathematical School at Christ's Hospital, designed a printed tabular form that he sent to parish clergy in order to gather information about the annual number of baptisms and burials for the years 1688 to 1780.[48] Likewise, the intendant La Michodière solicited and collected information on baptisms and burials from the clergy in various regions of France and presented his tabular summaries to the Royal Academy of Sciences.[49] Laplace then used figures drawn from these registers to refine his calculations of the total population of France.[50]

In all of these examples, the overarching success of the correspondence method rested on the solid reputation of a scientific or medical society. An established society encouraged, legitimated, and at times evaluated the exchange of information. It also provided an outlet, in the form of sponsored publication, for sharing the methods and results. Nonetheless, correspondence

45. Hautesierck, *Recueil*, 2:6–48.
46. Hannaway, "Medicine, Public Welfare, and the State"; Hannaway, "Vicq d'Azyr," 287–88.
47. *Histoire de la Société Royale de Médecine 1776* (Paris, 1779), xxxii. Jean-Pierre Peter, who has carefully examined this correspondence, noted that few physicians sent in quantitative information (Peter, "Disease and the Sick," 87, 93).
48. Wales, *An Inquiry*.
49. La Michodière, "Tables alphabétique."
50. Laplace, "Sur les Naissances."

networks had their limitations. They were dependent first of all on the dedication of the coordinator as well as the correspondents themselves. The networks were entirely voluntary organizations; no enforcement techniques existed to ensure a steady stream of contributions, a feature that distinguished eighteenth-century efforts from nineteenth-century bureaucracies. Second, the information gathered by this method varied considerably in spite of specific instructions (and occasionally preprinted forms) supplied by the coordinator.[51] And, as with other sources of numerical information (bills of mortality, parish registers, and tax records), correspondence networks required arithmetical skills on the part of the coordinator to reduce the information into useful tables and ratios.

Although the reality of the Republic of Letters did not always live up to its ideals, the correspondence method did enjoy some success, and part of the reason may lie in the very public nature of the network and the results of this technique of inquiry. Political and medical arithmeticians were usually eager to publish their work. Jurin, for instance, published the figures representing the comparative risks of natural and inoculated smallpox culled from his correspondence network in the *Philosophical Transactions* and in annual pamphlets. Later arithmeticians such as John Haygarth, Thomas Percival, and Richard Price also published in the *Philosophical Transactions*. In France, Vicq d'Azyr compiled the *Histoire et Mémoires de la Société Royale de Médecine*, an annual volume modeled on the *Histoire et Mémoires* of the Paris Royal Academy of Sciences. Moreover, publication encouraged potential correspondents to participate. Jurin included an advertisement at the end of his pamphlets soliciting accounts of inoculation. The preface to the annual volumes of the Medical Society of Edinburgh similarly entreated readers to contribute. And William Wales concluded his pamphlet by requesting "that such clergy as this little tract may fall into the hands of, will oblige the author with the annual number of baptisms, marriages and burials, in their respective parishes," and send them to him "in Christ's Hospital, London."[52] In short, what emerged, especially in Britain, was a decentralized model of knowledge production, dependent upon the cooperation of local institutions and the goodwill of scattered enthusiasts. Political and medical arithmeticians thus shared a commitment to public knowledge characteristic of the Enlightenment and more generally of natural philosophy since the Scientific Revolution.[53]

---

51. Marie-Noëlle Bourguet, in her book on Napoleonic statistics, clearly demonstrated that individual communities created their own categories, thus making quantitative comparisons difficult, if not impossible (Bourguet, *Déchiffrer la France;* see also Porter, *Trust in Numbers,* 35–37).

52. Wales, *An Inquiry,* 79.

53. Stewart, *Rise of Public Science.*

Political Arithmeticians and Public Debate

Political arithmeticians often became involved in Enlightenment debates concerning population because they perceived these public controversies to be marred by a high degree of acrimony and a regrettable lack of precision. "Many necessary and useful Things that have hitherto been made only Matter of Speculation and Dispute," argued the English physician Thomas Short in 1750, "but could never otherwise be truly determined, but by the Help of Registers."[54] Much like Haygarth, Short constructed registers to resolve otherwise interminable debates involving population. Yet despite such high-minded and rational motives, arithmeticians often simply confirmed existing opinions with the figures they generated, although occasionally they turned up something new. Quantification and the construction of registers served different purposes in public controversies, as illustrated by three specific debates to which arithmeticians contributed: polygamy, the effects of urban life, and smallpox inoculation.

The subject of polygamy was taken up by the earliest arithmeticians and continued to hold interest throughout the eighteenth century. John Graunt entitled one chapter of his book "Of the difference between the numbers of Males and Females," but only two paragraphs addressed "The Table of Males, and Females, for London," showing that more men than women were christened and buried in the city of London during the years 1644 through 1664. The remainder of the seven-page chapter discussed "some Inferences from this Conclusion," namely that "*Christian Religion,* prohibiting *Polygamy,* is more agreeable to the *Law of Nature,* that is, the *Law of God,* than *Mahumetism,* and others, that allow it."[55] According to Graunt, a surplus of men made polygamy a numerically unsatisfactory arrangement. Nor would polyandry do; however, no numerical argument was given to support this position. Graunt stated that women who had more than one sexual partner (in his word, "whores") did not procreate, thereby leading him to conclude that "it is no wonder why States, by encouraging Marriage, and hindering Licentiousness, advance their own Interest, as well as preserve the Laws of God from contempt and violation."[56]

In similar fashion, both John Arbuthnot and William Derham used the ratio of male to female christenings as evidence of God's design and to argue against polygamy in the early eighteenth century.[57] Thomas Short developed

54. Short, *New Observations,* xiv.
55. Graunt, *Natural and Political Observations,* 374.
56. Ibid., 377–78.
57. Arbuthnot, "An Argument for Divine Providence"; Derham, *Physico-Theology.* Ian Hacking analyzed Arbuthnot's argument in *Emergence of Probability,* chap. 18.

their arguments in midcentury by extending calculations of the ratio of males to females using mortality figures for different ages of each sex. "Here we may observe a remarkable Providence in the Production of a greater Number of Males than Females," he wrote, "for they run greater Hazard of Abortion between their Conception and Birth, are in more Peril at their Birth, seeing there are 10 still-born and chrysom Males, to 7 Females; they run greater Danger in Childhood, seeing 62 Boys die to 53 Girls; in greater Danger in Celibacy, for 12 Boys to 11 Girls, die; in more Peril in a Marriage State, seeing above 15 Married men die for 10 married Women: All which Dangers are increased by living in Cities or great Towns." With such an overwhelming catalog, Short could not but conclude: "Seeing the Dangers of Males, *in & extra uterum,* are so much greater than that of Females, then Polygamy is a most ridiculous, monstrous Custom."[58] Life, it seemed, was more deadly to males than females.

Debates about polygamy and other social arrangements between the sexes were certainly not confined to political and medical arithmeticians during the Enlightenment, yet birth and mortality rates could smoothly be incorporated into theological arguments supporting Christian monogamy. A somewhat different story can be told concerning salubrity, a topic that enjoyed renewed attention during the Enlightenment. Graunt and Petty had addressed the importance of geographical location in determining salubrity, and Petty suggested a measure of salubrity based on the incidence of acute and chronic diseases within a community. Others looked to contrasting mortality rates between communities. Gregory King, writing at the end of the seventeenth century, had developed yet another measure when he compared the number of children born to couples in London with those born to couples in rural areas. He discovered that rural marriages produced more children, and then calculated the fertility rates of city and country marriages. In this way, fertility became a phenomenon associated with particular geographies, rather than solely with individuals or families.

The implications of King's calculations were not quite as straightforward as those deduced by the antipolygamists. King speculated that the reasons for low fertility in the city were manifold: "1. From the more frequent Fornications and Adulteries. 2. From a Greater Luxury & Intemperance. 3. From a Greater Intensenesse or Businesse. 4. From the Unhealthfullnesse of the Coal Smoak. 5. From a greater Inequality of age Between the Husbands and Wives."[59] Clearly one way to improve fertility in urban areas was through individual moral reform; perhaps another would be to discourage persons from moving to the city in the first place.

58. Short, *New Observations,* 72–73.
59. King, "Natural and Political Observations," 28.

Some eighty years later, in 1774, the physician Thomas Percival calculated the mortality figures for parishes just outside of Manchester (1 in 56) and for those inside the city limits (1 in 28).[60] Just as King calculated comparative fertility rates, Percival quantified mortality rates of city and country dwellers. And like King, he took the opportunity to condemn the lack of morals among city inhabitants. "It must afford matter of astonishment even to the physician and philosopher," commented Percival, "when he reflects that the inhabitants of both live in the same climate, carry on the same manufactures, and are chiefly supplied with provisions from the same market. But his surprize will give place to concern and regret, when he observes the havoc produced in every large town by luxury, irregularity, and intemperance; the numbers that fall annual victims to the contagious distempers, which never cease to prevail; and the pernicious influence of confinement, uncleanliness, and foul air on the duration of life."[61]

As the works of King and Percival indicate, quantifying features of the population (in these cases, fertility and mortality) tended to confirm contemporary beliefs and prejudices—the wholesome country was more salubrious than the decadent city. And the majority of the justifications for these numerical disparities were equally ordinary or mundane: city life led to individual moral failing. But it is important to note that both King and Percival, although separated by a century, included environmental influences—King's "Coal Smoak" and Percival's "foul air"—that affected the population of an entire city and could not be rectified by the improved moral behavior of individuals. Percival in fact acted on this observation and formed a committee to improve the sanitation in Manchester's poorer neighborhoods.

Miasmatic views of disease espoused by King and Percival applied equally to rural as well as urban settings. Short, who provided much numerical evidence against polygamy, surveyed more than 160 country parishes throughout England and constructed twenty-five tables containing information on mortality, fertility, meteorology, and topography. Observations based on these tables led Short to recommend four specific actions to improve air quality and health in general: (1) drain marshes; (2) clear low, flat ground of woods and fences, which impeded the air circulation; (3) build houses with higher ceilings, larger lights, and clearer yards; and (4) urge farmers not to keep manure or plant orchards close to their houses, which were the common practices.[62] Evidence for these recommendations came from the Isle of Ely, a marshy area that had been drained in the seventeenth century. "[B]efore it was drained," Short

60. Percival, "Observations on the State of Population in Manchester . . . concluded," 323.
61. Ibid., 323–24.
62. Short, New Observations, 68–70.

concluded, "the Births were to the Burials as 61 to 70, now as 60 to 54. . . . This, with several others that are in the Table, is too dire a Proof of the mischevious Effects on human Bodies, of Marshy Ground, Standing Waters, and bad Air."[63] In sum, mortality and fertility patterns of different geographies and populations at different periods underscored the importance of climate and environment to health, and provided support for a variety of recommendations to improve both urban and rural settings.

The debates over smallpox inoculation provided an especially charged arena for the contributions of arithmeticians, from James Jurin in the early eighteenth century to those of William Black at the end of the century. From the accounts sent to him through his correspondence network, Jurin tallied the number of individuals who had been inoculated, and of those, how many had died. From these two figures, he constructed the ratio that roughly 1 in 90 individuals died from inoculation while 1 in 7 or 8 died from natural smallpox.[64] Jurin's motivation for constructing these ratios came from his desire to legitimate the practice of inoculation in as unobjectionable manner as possible. Since inoculation was contested on religious, ethical, and medical grounds, Jurin sought a new way to advocate the practice that would supersede these other types of arguments: he deployed an arithmetical argument to persuade individuals of the benefits of the procedure. "I have no Inclination to enter into this Controversy," he admitted, but "I heartily wish, that without Passion, Prejudice, or private Views, it may be fairly and maturely examin'd."[65] The results of Jurin's investigation were seemingly incontrovertible: the risks of dying from inoculation were far less than the risks of dying from natural smallpox.

The ratios Jurin calculated represented the mortality rates for smallpox inoculation and natural smallpox and encapsulated the risk of a procedure *within a population,* rather than to an individual.[66] A new meaning of population thus emerged through Jurin's work, namely, a group of individuals subjected to a particular medical practice, but not necessarily located within a specific geographical area. This novel concept of population was tied to the equally novel technique of collecting information—the correspondence network. Both the method and the approach emphasized the specific procedure over particular circumstances and represented a move away from traditional Galenic medicine, which focused on an individual's constitution and tailored

63. Ibid., 68–69.

64. Jurin, *A Letter to the Learned Caleb Cotesworth*, 17–18.

65. Ibid., 4.

66. As François Ewald has noted, "Risk only becomes something calculable when it is spread over a population" (Ewald, "Insurance and Risk," 203).

therapy to suit that constitution. Concern with populations of patients spread as physicians and surgeons began to evaluate numerically the success of other medical practices and therapies. Many of these efforts took place in hospitals, dispensaries, and other medical institutions, which grew in number and size during the second half of the eighteenth century.[67] Whether established as charities to serve the poor or financed by the state to serve the military, these institutions issued annual reports to their patrons that included accounts of the numbers of patients, the types of therapy administered, and the outcome. In this way, medical practice became represented in arithmetical accounts.

From these three examples, we see that political and medical arithmeticians contributed to public debates at several different levels. In the case of smallpox inoculation, the arithmetical approach provided a critical tool of persuasion. While some remained unconvinced of the applicability of arithmetic to medicine, contemporaries generally recognized Jurin's approach as highly influential; by midcentury, inoculation was firmly established in Great Britain in part because of numerical demonstrations of its benefits.[68] At another level, numerical representations of fertility, mortality, and salubrity confirmed contemporary beliefs about the ill effects of foul air and the unhealthiness of urban life and occasionally provided evidence for specific recommendations to improve the health of populations, not just of individuals. King's and Percival's work pointed to the ill effects of coal smoke and other foul airs, and in Percival's case, it led to direct action. Thomas Short was able to demonstrate numerically the salubrious effects of draining marshland by comparing birth to burial ratios in two different time periods. In all these instances, political arithmeticians deployed numerical evidence to inform policy, especially that concerning improvements—a leitmotif of the Enlightenment. Finally, the role of numerical arguments in debates over polygamy must be considered ideological. It remained almost inconceivable that polygamy would have been adopted in Christian Europe during the eighteenth century, even if the numbers came out in favor of such an arrangement. Nonetheless, arithmeticians from Graunt to Short constructed detailed numerical arguments to support monogamy. This approach, a form of natural theology, indicates how new types of arguments could be married to older traditions.

Conclusion

As a form of knowledge, political arithmetic comprised the virtues of quantification and the ideals of the Enlightenment. Protocols of sociability character-

---

67. Ulrich Tröhler discussed some of these initiatives in his "Quantification in British Medicine."

68. Rusnock, "Weight of Evidence."

istic of the period, especially those expressed in the Republic of Letters, were essential to its growth. Arithmeticians relied on correspondence networks to augment their individual findings, and British physicians, especially, were active in the development of this scientific method. Affiliation with local and national scientific and medical societies (which flourished during the Enlightenment) provided legitimation of their efforts and publication venues for their writings.

Political arithmeticians remained strongly convinced that quantification would bring clarity to contemporary debates concerning the health, wealth, and strength of nations. They gathered numerical information from parish registers, tax records, and bills of mortality about births, deaths, and diseases (among other things) to characterize and compare different features of populations. Quantifying phenomena such as fertility and mortality risks allowed political arithmeticians to take local observations and contribute to national (and in some instances international) discussions through comparison with similar numerical observations of other communities. Numerical representation facilitated comparison between geographical areas, chronological periods, and varieties of medical practice.

Comparison, classification, counting, and the construction of registers and ratios, all constitutive of the practice of political arithmetic, provided the methods for population to come under observation and analysis, developments that mark the origin of Foucault's biopolitics. Classification was central to this enterprise and the categories arithmeticians chose to quantify reinforced preexisting ideas of difference as often as they challenged dominant practices and prejudices. Advocating monogamy or limiting urbanization fit well with Foucault's emphasis on surveillance and control of the population. On the other hand, efforts to improve health and salubrity through the introduction of smallpox inoculation or the reduction of environmental hazards accentuated the more positive and humanitarian aspects of political arithmetic. From either perspective, obsession with the numbers of the dead reflected a deep concern for the living.

Acknowledgments

I would like to thank the contributors to this volume and Paul Lucier for their many thoughtful suggestions and criticisms.

Primary Sources

Arbuthnot, John. "An Argument for Divine Providence." *Philosophical Transactions* 27 (1710): 186–90.

Black, William. *An Arithmetical and Medical Analysis of the Diseases and Mortality of the Human Species.* 2d ed. London, 1789.

Brakenridge, William. "A Letter to George Lewis Scot, Esq; F.R.S. concerning the Number of People in England." *Philosophical Transactions* 49 (1756): 268–85.

Buffon, Georges Leclerc de. *Histoire naturelle, générale et particulière.* Vol. 2. Paris, 1749.

Davenant, Charles. "Of the Use of Political Arithmetic, in all considerations about the revenues and trade" (1698). From "Discourses on the Public Revenues and on Trade," in *The Political and Commercial Works of Dr. Charles D'Avenant,* vol. 1. London, 1771.

Derham, William. *Physico-Theology.* 1716. Reprint of 4th ed., New York: Arno Press, 1977.

Forster, Richard. "An Extract of the Register of the Parish of Great Shefford, near Lamborne, in Berkshire, for Ten Years: With Observations on the Same." *Philosophical Transactions* 50 (1757): 356–63.

Graunt, John. *Natural and Political Observations made upon the Bills of Mortality* (5th ed., 1676). In *The Economic Writings of Sir William Petty,* edited by Charles Henry Hull, 315–431. Cambridge: Cambridge University Press, 1899.

Hautesierck, Richard de. *Recueils d'Observations de Médecine des Hôpitaux Militaires.* 2 vols. Paris, 1766–82.

Haygarth, John. "Observations on the Bills of Mortality, in Chester, for the Year 1772." *Philosophical Transactions* 64 (1774): 67–78.

King, Gregory. "Natural and Political Observations and Conclusions Upon the State and Condition of England." 1696. In *Two Tracts by Gregory King,* edited by George E. Barnett. Baltimore: Johns Hopkins University Press, 1936.

Jurin, James. *A Letter to the Learned Caleb Cotesworth . . . Containing a Comparison between the Mortality of the Natural Small Pox, and that given by Inoculation.* London, 1723.

La Michodière. "Tables alphabétique des villes, bourgs, villages dans différentes régions de France: Moyenne des naissances, mariages, morts sur les années 1781, 1782, 1783." 2ème cartonnier, Académie des Sciences. Paris, 1785.

Laplace, Pierre-Simon de. "Sur les Naissances, les Mariages et les Morts à Paris, depuis 1771 jusqu'en 1784; et dans toute l'étendue de la France pendant les années 1781 et 1782" (1783). In *Mémoires de l'Académie Royale des Sciences de Paris,* 1786, 693–702.

Moheau, J. B. [baron de Montyon]. *Recherches et considérations sur la population de la France.* Paris, 1778.

Petty, William. *A Treatise of Taxes and Contributions* (1662). In *The Economic Writings of Sir William Petty,* edited by Charles Henry Hull, 1–97. Cambridge: Cambridge University Press, 1899.

———. *Political Arithmetick* (1676). In *The Economic Writings of Sir William Petty,* edited by Charles Henry Hull, 232–313. Cambridge: Cambridge University Press, 1899.

———. *The Petty Papers: Some Unpublished Writings of Sir William Petty.* Edited by the marquis of Lansdowne. 2 vols. London: Constable and Co., 1927.

Percival, Thomas. "Observations on the State of Population in Manchester, and other adjacent Places, concluded." *Philosophical Transactions* 65 (1775): 322–35.

Price, Richard. *Observations on Reversionary Payments; on Schemes for Providing Annuities for Widows, and for Persons in Old Age; on the Method of Calculating the Values of Assurances on Lives; and on the National Debt.* London, 1769.

———. *An Essay on the Population of England.* 2d ed. London, 1780.

Short, Thomas. *New Observations Natural, Moral, Civil, Political, and Medical, on City, Town, and Country Bills of Mortality.* London, 1750.

Süssmilch, Johann Peter. *Die göttliche Ordnung in den Veränderungen des menschlichen Geschlechts aus der Geburt, dem Tode und der Fortpflanzung desselben erwiesen.* 3d ed. 2 vols. Berlin, 1765.

Vauban, Sebastien le Prestre de. "Description géographique de l'élection de Vézelay" (1696). In *Projet d'une Dixme Royale suivi de deux écrits financiers,* edited by E. Coornaert. Paris, 1933.

Wales, William. *An Inquiry into the Present State of Population in England and Wales.* London, 1781.

3

Barometers of Change: Meteorological Instruments as Machines of Enlightenment

Jan Golinski

For the rising in the BAROMETER is not effected by pressure but by sympathy.
For it cannot be separated from the creature with which it is intimately & eternally connected.
For where it is stinted of air there it will adhere together & stretch on the reverse.
For it works by ballancing according to the hold of the spirit.
For QUICK-SILVER is spiritual and so is the AIR to all intents and purposes.

Christopher Smart, *Rejoice in the Lamb* (c. 1761)

WEATHER: Eternal topic of conversation. Universal cause of illness. Always complain about it.

Gustave Flaubert, *Dictionary of Received Ideas* (c. 1880)

Barometers might seem an unlikely topic for poetry, but during the eighteenth century, at least one writer was inspired to pen verses about them. Here are some lines by the (now justly forgotten) Augustan poet John Phelps, from his *The Human Barometer; or, Living Weather-Glass,* published in 1743:

The pois'd *Barometer* will sink or rise,
In Mode proportion'd to the changing Skies,
The Air serene th'inclosed Mercury shows;
And, as by Weight impell'd, it upward goes;
But when dilated Vapours crowd the Air,
Its sinking State will straitway make appear. . . .
The Air serene, from Clouds and Vapours clear,
Not burnt with Heat, nor chill'd with Cold severe;
Adjusts the Motion of the circling Blood,
The Pulse beats right, the Circulation's good;
Vapours and Storms aerial Weight abate,
Our Blood runs low, and languid is our State,

If Cold or Heat prevail to great Excess,
More than we ought, we then perspire or less,
Our passive Body Alterations finds,
And with our Bodies sympathize our Minds.[1]

Phelps's lines express commonplace ideas about the barometer in mid-eighteenth-century England. The mercury in the tube had been observed to move in response to changes in the air, apparently rising when the atmosphere was clear and serene, and sinking when it was charged with clouds and vapors. The instrument was also regarded as capable of predicting changes in the weather, though there was continuing debate about what kind of predictions could be made and how reliable they were. Its hermeneutics was admittedly problematic, but it was generally believed that the barometer, if interpreted in the light of other prognostic signs, could give some knowledge of future weather conditions.

Also commonplace was the connection Phelps made between the barometer's responses to atmospheric changes and those of the human body. More than two millennia previously, Aristotle's *Meteorologica* had articulated the notion, already developed by the Hippocratic writers, that the quality of the air was an important influence on bodily health. During the Renaissance, with the revival of the theory of humors and elaboration of the analogies between microcosm and macrocosm, belief in such a connection was strengthened. By the eighteenth century, the link was likely to be stated in terms of more recent physical and physiological discoveries—changes in air pressure, it was said, could affect the circulation of the blood—but the basic idea still held good. Indeed, the many projects to investigate the environmental causes of health and disease were premised, as the experimental philosopher Adam Walker explained in 1777, upon the conviction that it was proper for an age of "philosophy and enlarged sentiment" to locate in the air the causes of diseases that a previous age of "religious tyranny" had ascribed to divine retribution.[2]

Because it apparently mirrored physiological changes, the barometer offered itself as an instrumental means of assessing atmospheric effects on health. Phelps's poem expands upon the crucial metaphor of the body itself as a kind of barometer, undergoing physical changes in response to external conditions of which the mind is not directly aware. His verses build on the image of the human barometer to deliberate upon the mysterious connection between thought, emotions, and the physical environment. The poem thus takes its place in a cultural tradition Terry Castle has recently traced, in which meteoro-

1. Phelps, *Human Barometer*, 14–15.
2. Walker, *Unwholesome Air*, 29. See also Heninger, *Handbook of Renaissance Meteorology*; and Spitzer, "Milieu and Ambiance."

logical instruments have been imaged as devices to display the state of human, particularly female, emotions.[3]

This metaphorical application was a function of the increasingly pervasive presence of weather instruments themselves in eighteenth-century England, a presence that suggests barometers might provide a key to the unfolding of the Enlightenment. The devices emerged from the experimental philosophy of the late seventeenth century and were rapidly mass produced and distributed by the entrepreneurs of an expanding consumer market.[4] In a relatively brief period, they were transmuted from specialist instruments to household objects in many middle- and upper-class homes. In the 1660s, they had been "confined to the cabinets of the virtuosi," but by the 1710s they were described as common "in most houses of figure and distinction."[5] Their popularity parallels that of microscopes and telescopes, air pumps, globes, and orreries, which have been viewed as material agents for the transmission of scientific culture into the homes of the enlightened bourgeoisie. As these scientific artifacts were purchased by consumers and used in middle-class households, it has been proposed, they communicated the values and beliefs we associate with the Enlightenment.[6]

In this chapter, I shall argue for a more complex picture of the role of such artifacts in enlightened public science. By focusing on the barometer, in its transition from laboratory instrument to household furniture, I shall suggest that the device conveyed no single, prepackaged meaning. Far from transmitting specifically scientific ideas and aspirations, barometers were interpreted by their lay users in ambivalent, even contradictory, ways. While the continuing manufacture and sales of barometers in the course of the eighteenth century might suggest a picture of the progressive diffusion of objects and ideas, attention to the meanings that were spun from the devices recaptures some of the ambiguities and instabilities of enlightened culture. Barometers proved to have quite equivocal significance for their eighteenth-century users, in terms of their implications for scientific knowledge and human control of the natural world. No consensus was achieved as to how reliable a predictor of the weather the barometer could be, or the degree to which it was reasonable to trust it. The

3. Castle, "Female Thermometer."

4. For the connections between the English Enlightenment and the development of consumer culture, see Porter, "Enlightenment in England"; and Brewer and Porter, *Consumption and the World of Goods.*

5. Roger North, *Life of Francis North* (1742), quoted in Goodison, *English Barometers,* 31; Saul, *Historical and Philosophical Account of the Barometer,* 1.

6. Walters, "Tools of Enlightenment"; Butler, Nutthall, and Brown, *Social History of the Microscope;* Porter et al., *Science and Profit.* More generally, on the role of consumer goods in reshaping the domestic realm during the Enlightenment, see Goodman, "Public Sphere and Private Life"; Barker-Benfield, *Culture of Sensibility,* esp. chaps. 2, 4.

apparatus was simultaneously hailed as a triumph of contemporary science and as an object of superstitious faith.

This ambivalence can in fact be traced back to the experimental philosophers who pioneered the use of the device in the seventeenth century. The barometer did not emerge into the consumer marketplace as a "black-boxed" instrument with an agreed scientific rationale. Actually, as we shall see, one motive for experimenters who wanted to expand the number of users of the device was the hope of generating agreement about how it responded to weather phenomena and what its predictive potential could be. This goal could not have been achieved by experimental philosophers alone. Crucial to the transition from expert to lay realms was the work of commercial instrument makers (initially mostly clock makers), whose interests were not the same as the experimenters', but who saw opportunities in making and selling barometers to the public. They supplied the craftsmanship and the marketing skills that made it possible for the instruments to become popular commodities in a burgeoning consumer culture.

The success of barometers in finding purchasers did not convey a single or unified set of cultural values. On the one hand, the devices became tokens of virtuosity and philosophical prowess, of refinement and politeness; on the other hand, they continued to be associated with gullibility, vulgar superstition, and a feminine susceptibility to the lure of fashion. While they seemed to signify an understanding of weather and the climate as governed by natural laws, hopes that those laws could be known and their consequences predicted were continually frustrated. When barometers were first marketed, they were already accompanied by warnings not to take their predictions too seriously; and yet, at the end of the eighteenth century, promoters of the instrument were still arguing that accurate predictions would eventually be possible.

From Experimental Instrument to Philosophical Furniture

Standard histories trace the origins of the barometer to the Torricellian experiment of 1644, in which a column of mercury of a certain height was sustained in a sealed tube inverted over an open basin. But this is to give what is, in some respects, a misleading genealogy. As Jim Bennett has pointed out, the Torricellian tube shared a physical form with the later barometer, "but it fulfilled quite a different conceptual function."[7] Through the late 1640s and 1650s, the mercury column experiment served as the focus of arguments as to whether there was a vacuum at the top of the tube and (in principle a separate

---

7. Bennett, *Mathematical Science of Christopher Wren*, 83. Compare Middleton, *History of the Barometer;* Taylor, "Origin of the Thermometer"; and Daumas, *Scientific Instruments.*

issue) whether the mercury was supported by the weight of the air. Whether taken to the summit of the Puy-de-Dôme or to the bottom of a coal mine, or placed in the receiver of the air pump, the Torricellian apparatus served as a register of the weight or "spring" of the air. No attention was paid to the effect of the weather on the height of the column, except implicitly when it was noted that placid weather provided the best conditions for this kind of experiment.

It was only in the early 1660s that the Torricellian tube was converted by a kind of Gestalt switch into an instrument for measuring the small variation in the height of the mercury that would occur even if the apparatus was left in the same place, apparently independently of heat and cold. The coining of the terms "baroscope," by Robert Hooke in June 1664, and "barometer," by Robert Boyle in his *New Experiments and Observations Touching Cold* in 1665, signaled this reversal. It was asserted subsequently that the initial stimulus had come from Christopher Wren, who suggested that Torricelli's apparatus could be used to test the Cartesian theory of the cause of the tides.[8] If, as Descartes had claimed, the tides were caused by a mechanical pressure on the oceans, transmitted from the moon through the vortex surrounding the earth, then the mercury column also might move in response to the moon's motions. Boyle accordingly set up a tube in his house in Oxford and monitored the variation. No link with the moon's motions was found, but a correlation with the weather was quickly suspected. On 9 September 1663, Boyle asked at a meeting of the Royal Society for tubes of mercury to be kept at Gresham College and at Oxford, in order to observe how the variation in level related to the weather.

That the weather should immediately emerge as a candidate for correlation was unsurprising in view of the interest in meteorological phenomena long established among the society's leading members. Only the previous week, John Wilkins had suggested that they should revive plans to keep a comprehensive history of the weather, "in order to build thereupon an art of prognosticating the changes thereof."[9] In addition, many fellows already had experience of a meteorological instrument, namely, the common seventeenth-century device known as a "weatherglass." Originally, the weatherglass had taken the form of a glass bulb with a long tubelike neck. It was prepared by heating the bulb to dilate the enclosed air and inverting it over a basin of water. The water rose up the tube, trapping the air in the bulb as it cooled, and thereafter the level of water in the tube was taken as a sign of the state of the atmosphere. By the 1630s, London instrument makers were already advertising

---

8. Boyle, *New Experiments and Observations Touching Cold* (1665), in *Works,* 2:487; Birch, *History of the Royal Society,* 3:464 (meeting of 20 February 1679).

9. Birch, *History of the Royal Society,* 1:300–302

ready-made versions of this apparatus, and a convenient alternative design was being imported from Holland. With the recognition that atmospheric pressure varied independently of temperature, weatherglasses came to be seen as having confusingly responded to both phenomena conjointly. The construction of the barometer, its function clearly distinguished from that of the thermometer, was said by Boyle to have rendered the traditional weatherglass obsolete. By 1665, he was therefore describing weatherglasses as "these dead engines," and the following year Henry Oldenburg and John Beale wrote that, "the open Weather-glass is known to signifie nothing at certainty, having a double obedience to two Masters, sometimes to the *Weight of the Air,* sometimes to *Heat.*"[10]

So by the early 1660s, the barometer had been conceptually constructed as an instrument to measure the variation in pressure or weight of the air. Questions remained as to what caused this quantity to vary and what meteorological or other phenomena it might be related to. In October 1664, Boyle was recommending that records be made of barometer readings during eclipses and sunspots. In early 1666, he and John Wallis reported separately in the *Philosophical Transactions* that the barometer had dropped suddenly at the time of a recent earthquake in Oxford.[11] Hooke's "Method for Making a History of the Weather," published as part of Thomas Sprat's *History of the Royal Society* in 1667, stipulated that barometer readings should be recorded alongside temperature, humidity, wind strengths and directions, cloud formations, prevailing diseases, storms, and any extraordinary tides.[12]

Thus began an enterprise that was to grow over the succeeding decades, in which fellows of the Royal Society attempted systematically to gather meteorological data and to identify the correlates of barometric variation. Virtuosi in many places were enlisted in the program; some were even supplied with barometers by the society. In February 1665, John Beale had recommended this strategy to Boyle, noting that resistance to the new instruments could be overcome, "by recommending the sale of all sorts of weather-glasses, and baroscopes to some persons, that are fittest, and of best trust, in *London, Oxford,* and *Cambridge.* These are the liquid demonstrations. And some inventions may be better communicated in specie, than by books."[13] Beale himself was one of those who soon began to compile regular barometric observations, which he published in the *Philosophical Transactions* in 1666. Henry Power,

10. Boyle, *New Experiments and Observations Touching Cold,* 485; Oldenburg and Beale, "Observations upon the Barometer," 164.

11. Boyle to Oldenburg, October 1664, in Oldenburg, *Correspondence,* 1:254–55; Wallis, "Relation Concerning the Late Earthquake"; Boyle, "Confirmation of Former Account."

12. Hooke, "Method for Making a History of the Weather."

13. Beale to Boyle, 6 February 1665, in Boyle, *Works,* 6:393.

who had been working with the Torricellian experiment since the early 1650s, was another. Oldenburg enlisted the continental savants Adrien Auzout and Johannes Hevelius in the project, instructing them how to prepare their barometers and how to report their observations.[14] The ultimate aim, apparently, was to induce the general laws that would enable the barometer to fulfill its promise as a means of prediction.

Coinciding with the unfolding of this project was a continuing debate about what caused the mercury to rise and fall. At first, the predominant theory ascribed a rise in level to increased specific gravity of the air due to its being loaded with moisture and other vapors. Speculation, especially among those connected with the program of research on the physiology of respiration that had begun in Oxford in the 1650s, was that the presence of "nitrous vapors" or "aerial niter" was a crucial factor. As early as September 1663, Hooke somewhat prematurely declared that he had observed Boyle's barometer to rise in summer and fall in winter—a variation he ascribed to greater air density in the summer owing to increased exhalation of vapors from the earth.[15] This theory did seem to account for the descending mercury level frequently observed to coincide with the discharge of vapors in the form of rain, but it encountered an obstacle when observers uniformly reported that the mercury stood at a high level during winter frosts as well as during summer heat waves.

Modified versions of the exhalations theory nonetheless continued to hold sway. George Garden, from Aberdeen, and John Wallis, from Oxford, both articulated versions of it in papers in the *Philosophical Transactions* in 1685. Garden asserted that heavier air would have a greater capacity to support water vapor than lighter air. Although he could not offer a definitive explanation of why the air's weight varied, he canvassed a couple of possibilities, including changing concentrations in the air of some more subtle fluid medium or of "nitrous steams." Garden's point was that changes in the air's weight were the fundamental causes of meteorological phenomena, giving rise to rain when the air declined in weight and discharged its vapors, and causing winds to flow between places with unequal air density. Wallis broadly endorsed Garden's conclusions; he agreed that variation in air pressure was primarily due to the influence of dissolved exhalations, along with the effects of heat and cold.[16]

A year after these papers were published, however, an alternative to the exhalations theory emerged. It was proposed by the astronomer and navigator Edmond Halley, in a paper that promised "an attempt to discover the true reason of the rising and falling of the mercury, upon change of weather."

14. Oldenburg to Boyle, 13 March 1666, in Oldenburg, *Correspondence,* 3:57–60.
15. Birch, *History of the Royal Society,* 1:304 (meeting of 16 September 1663).
16. Garden, "Discourse Concerning Weather"; Wallis, "Discourse Concerning Air's Gravity."

Halley insisted that the main cause of variation in mercury level was the concentration or dilation of the air surrounding the barometer by winds. Converging winds would increase air density; diverging ones would lessen it. The loading of the air with vapors was almost entirely dependent on its density. A low density of air released vapors in the form of rain; high density absorbed them into the atmosphere and sustained them there. Far from being *caused by* variation in density, the winds were independent *causes of* that variation. The winds themselves were primarily caused by the physical forces set up by the earth's rotation, and they tended to follow regular paths across the earth's oceans, of which Halley supplied a map. The basic aim of Halley's account of the barometer was thus to remove it from the evidential context of Oxford physiology and the search for aerial niter, and to connect it with the realm of astronomy and navigation and with Halley's project of mapping the Trade Winds.[17]

Within the community of experimental philosophers, the barometer had attracted considerable interest by the late seventeenth century. There was, however, little agreement as to what caused the variation in the height of the mercury column or how useful it could be as a predictor of changes in the weather. A few correlations with weather patterns appeared clear: the mercury stood high during calm, fair conditions, but also during winter frosts; a fall in level generally presaged rain or a storm. But even these patterns were found not to be valid without exceptions, and little further could be established with any certainty. There was also, as we have seen, continuing debate about the cause of the variation. In 1679, fellows of the Royal Society were still discussing whether the air's density really changed or the atmosphere simply piled up to a greater height when the barometer column was high. In 1684, Martin Lister went so far as to doubt that the barometric variation really reflected a change in air pressure. He speculated instead that it revealed some peculiar aspect of the expansion and contraction properties of mercury in conditions of changing temperature.[18]

The philosophers' dispute about the causes of barometric variation was echoed, though sometimes in a rather garbled way, by the writers of texts for the nonexpert users of the barometer. The London clock maker John Smith, in his *Compleat Discourse of the . . . Quick-Silver Weather-Glass* (1688) followed Halley in ascribing fundamental importance to the winds as the causes of variation in the air's density; though by the time he brought out his *Horological Disquisitions* in 1694 he had reverted to the view that changes in the tempera-

---

17. Halley, "Discourse of the Barometer." See also Albury, "Halley and the Barometer."
18. Birch, *History of the Royal Society,* 3:460–63; Lister, "Discourse Concerning the Barometer."

ture and moisture content of the air were probably more important. Richard Neve, author of another layperson's guide, *Baroscopologia* (1708), while declining to enter into what he called "philosophical niceties," expressed the view that it was primarily moisture content that determined the air's weight.[19] Writers closer to the core of the experimental community, however, largely swung to Halley's side. John Harris, in his *Lexicon Technicum* (1704), quoted Halley's theory; as did J.T. Desaguliers in his *Lectures of Experimental Philosophy* in 1719. When Edward Saul, former Oxford don and rector of Harlaxton in Lincolnshire, produced his *Historical and Philosophical Account of the Barometer* in 1730, he dismissed Wallis's exhalations theory fairly summarily and declared for Halley's theory of the winds.[20]

Discussion of these issues in works directed at the literate public followed, of course, upon increased sales of barometers to consumers. Already in 1688 John Smith was noting that the instruments could be purchased from commercial instrument makers; by 1694 he recorded that many kinds of barometers were being sold in the London shops. Neve, in 1708, declined to give instructions for making a barometer because, he said, "any one may be furnished with it in *London,* at a cheaper rate than he can make it himself. . . . Few gentlemen [are] without one of them," he remarked, although, he went on, "few of them understand its right *Management* and use."[21] The virtuoso and former court lawyer Roger North gave an account of how commercial manufacture of barometers had started, in his "Essay of the Barometer," written in manuscript some time in the first two decades of the eighteenth century but never published. North ascribed a crucial role to his brother, Francis North, lord chancellor to Charles II and himself something of a virtuoso. Francis North, his brother reported, "thinking this instrument was too much confined among the curious, and that if the use of it was made more general, to which a shop exposition would tend, some catalogues of analogies might at length come forth," instructed his watch maker, Henry Jones of Inner Temple Lane, to make them and offer them for sale.[22] Jones soon lost interest, however, so North turned to Henry Wynne, a mathematical instrument maker in Chancery Lane, to continue the trade.

According to Roger North's account, then, the beginnings of commercial barometer sales did not follow the achievement of any kind of consensus as to their use among experimental philosophers. On the contrary, he suggests that

19. Smith, *Compleat Discourse of the Quick-Silver Weather-Glass,* 79–93; Smith, *Horological Disquisitions,* 73–77; Neve, *Baroscopologia,* 28–29.

20. Harris, *Lexicon Technicum,* s.v. "Barometer"; Desaguliers, *Lectures of Experimental Philosophy,* 135–36; Saul, *Historical and Philosophical Account of the Barometer,* 33–38.

21. Neve, *Baroscopologia,* 4–5.

22. North, "Essay of the Barometer," fols. 17r–v.

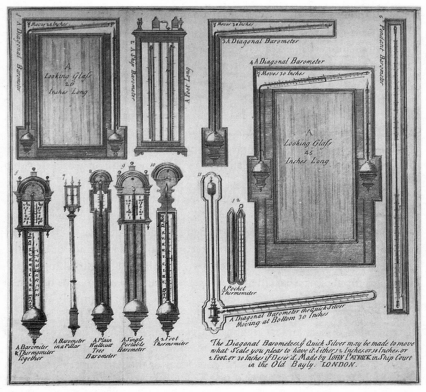

Fig. 3.1. Trade card of John Patrick, specialist barometer maker in the Old Bailey, c. 1710, showing two versions of the diagonal barometer mounted over looking glasses. Copyright British Museum.

the virtuosi might have sought wider public use of the apparatus in order to resolve uncertainty as to how it should be interpreted. To accept North's account as the whole story, however, would be to downplay the crucial role of the commercial instrument makers. Far from merely responding passively to instructions from philosophers, they seized opportunities to advance their own interests by designing, building, and marketing barometers. Even before Jones and Wynne, instrument makers who worked on commission for the leading experimental philosophers, such as Hooke's assistants Richard Shortgrave and Harry Hunt, had been selling barometers to the public. Thomas Tompion, probably the most renowned clock maker of his day, who also produced instruments for Hooke and Boyle, published the first known advertisement for a barometer in 1677. Other clock makers, including Thomas Tuttell and Daniel

Quare, followed the lead of Tompion and Wynne, and makers of navigational instruments and spectacles also moved into the market.[23]

Probably the first instrument maker to specialize in barometers was John Patrick, originally apprenticed as a cabinet maker, whose good connections with the scientific community were matched by excellent craftsmanship and advertising skills. Patrick offered a straight-tube, open-cistern instrument for two guineas. Most of the cost probably went into the wooden case: oak veneered with walnut was typical, with elaborate carved scrolls as ornamentation or even statuettes mounted on the top. Patrick also advertised an angled-tube apparatus at fifteen guineas, in which the barometer was bent into an inverted L, mounted around a mirror. The motion of the mercury was magnified along the length of the top segment of the tube, which was inclined at an angle just above the horizontal. Patrick's advertisement explained that the mirror was included so that, "Gentlemen and Ladies at the same time they dress, may accommodate their habit to the weather.—An invention not only curious, but also Profitable and Pleasant."[24] Patrick's conjoining of barometer with mirror indicated, as Castle has noted, the links between the two artifacts as reflections of the self. The barometer registered changes that were also experienced in a different way by the individual; it supplied an external reflection of internal feelings. Both mirror and barometer were therefore appropriately sited at the point where the individual comported herself or himself prior to emergence into the shared space of the public sphere. It was presumably with reference to this kind of device that George Sinclair reported in 1688 that, "Ladies and Gentlewomen at *London* do Apparel themselves in the Morning by the Weather-Glass."[25]

Following Patrick's example, several other instrument makers made a specialization of the manufacture of barometers. More commonly, the devices became part of the regular inventory of clock makers, opticians, and mathematical instrument makers. London manufacturers who included barometers on their trade cards included John Marshall, an optical instrument maker in Ludgate Street in the 1700s and 1710s; Stephen Davenport, who advertised mathematical and philosophical instruments in Holborn in the 1720s and 1730s; Edward Scarlett, who made and sold optical instruments in Soho for four decades after 1700; and Thomas Blunt, operating in Cornhill for about four decades from the 1770s. These and many other London manufacturers supplied a growing provincial demand; their products were retailed by

23. Goodison, *English Barometers,* 30–41; Taylor, *Mathematical Practitioners of Tudor and Stuart England.*

24. Patrick, *New Improvement of the Quicksilver Barometer,* 1.

25. Sinclair, *Principles of Astronomy and Navigation,* 48; Castle, "Female Thermometer."

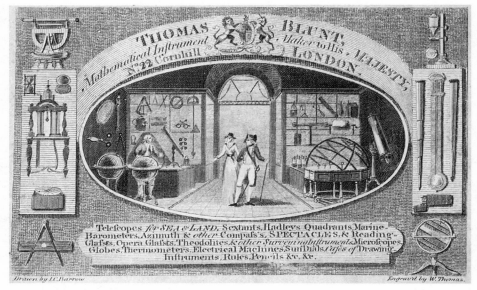

Fig. 3.2.   Trade card of Thomas Blunt, mathematical and optical instrument maker in Cornhill, c. 1790, advertising barometers alongside other philosophical apparatus. Copyright British Museum.

apothecaries, spectacle makers, and clock makers throughout the country. They were also joined by dozens of provincial barometer makers in the course of the eighteenth century, including opticians, jewelers, and gunsmiths, along with many clock makers.[26]

Whatever the reservations of the experts, there is little doubt that barometers were sold to the public on the expectation that they would predict the weather. As Roger North explained, the simple observation that the mercury rose before fair weather and fell before rain, though, "since found *egregiously, in many conjunctures,* to fail, having been for some time accounted almost certain, procured so great a credit to the instrument, that it was highly esteemed and courted. . . . For, . . . of all human kind who is he that lives, and is not extremely concerned to obtain (if possible) a certain presage of weather." Almost a century later, the instrument maker George Adams acknowledged it was still the case that "the far greater part of those who purchase meteorologi-

26. For information on barometer makers, see Goodison, *English Barometers;* Banfield, *Barometer Makers and Retailers;* Crawforth, "Evidence from Trade Cards"; Walters, "Tools of Enlightenment."

cal instruments, buy them, not so much to know the actual state of the elements, as to foresee the changes thereof."[27]

To gratify this demand, makers of the very first commercial barometers
began the practice of engraving the still familiar weather indicators that correspond to different heights of the mercury. In descending order, they are: "very
dry," "settled fair," "fair," "changeable," "rain," "much rain," and "stormy."
On some later models, separate scales for use in summer and winter were
mounted on either side of the tube. While writers regularly stressed that the
indicators were far from infallible (North described them as "like a cheating
oracle that . . . pretends to tell fortunes"), they admitted that most users did
place faith in them as trustworthy predictors. The usual advice, however, was
to rely more upon the movements of the mercury level than upon where it
stood against the scale. "Most *Gentlemen* that have *Weather-Glasses,*" wrote
Neve, "not knowing . . . this rule, . . . deceive themselves and others."[28]

The instruction to pay attention to the rising or falling of the mercury,
rather than simply to its position, was but the simplest of a series of rules for
weather prediction using the barometer. Almost all writers on the instrument
offered a version of these, or some compilation of the maxims of previous
authors. Halley gave eight rules, some of which required the observer to take
account of wind direction, season, or temperature, to interpret the device's
behavior. Patrick published ten rules, and Neve fourteen. Harris's *Lexicon
Technicum* indiscriminately listed Halley's rules and Patrick's, while Philip
Miller's *Gardener's Dictionary* (1752) continued the trend by adding eight rules
by John Pointer and sixteen by "another author."[29] Some of the maxims referred only to the barometer, but many required the observer to read its signs
in the context of other meteorological indicators: the appearances of the
sun, moon, and clouds; the winds and the season; even the behavior of animals. The Reverend John Laurence, for example, in his *Fruit-Garden Kalendar*
(1718), offered his readers a series of rules in which fluctuations of the mercury
column were related to other, more traditionally recognized, signs of the
weather. The same approach was followed by the agricultural writer John Mills
in his *Essay on the Weather* (1770). Mills remarked cautiously, "When the character of the season is once ascertained, the return of rain, or fair weather, may
be judged of with some degree of certainty in some years, though but scarcely
to be guessed at in others, by means of the barometer."[30]

27. North, "Essay of the Barometer," fol. 14v; Adams, *Short Dissertation on the Barometer,*
27–28.
    28. North, "Essay of the Barometer," fol. 18v; Neve, *Baroscopologia,* 33.
    29. Miller, *Gardener's Dictionary,* s.v. "Barometer."
    30. Laurence, *Fruit-Garden Kalendar,* 133–49; Mills, *Essay on Weather,* 74.

By presenting the barometer in the context of rules of interpretation that took into account other signs of the weather, the promoters of the device inserted it within traditional practices of forecasting or "weather-wising." "To *Prognosticate*, or foretel the alteration of the Weather," noted Thomas Willsford in 1665, "there hath been (in all Ages) diligent observers of *Nature*, who have prescribed rules and prenotations of the *Airs* mutability, grounded on judicial signs." Traditional maxims for predicting the weather abounded, and were being codified in such collections as *The Shepherd of Banbury's Rules*, edited by John Claridge in 1670 and substantially expanded in 1744. Claridge himself gave only a guarded welcome to the barometer; it was a "wonderful invention," he said, "curious" and "ingenious," but it had yet to prove itself as a means of long-range forecasting. Other authors more readily accommodated the readings of meteorological instruments within catalogs of traditional prognostic signs. John Smith, for example, recommended that the movements of the mercury should be watched along with the appearance of the sun, cloud formations, the phases of the moon, the wind's strength and direction, and a clutch of indicators from animal behavior: hooting owls, flying bats and gnats, cawing crows, and pigs "crying in an unusual manner."[31]

Barometers and the Climate of Enlightenment

The rise of the barometer as a means of predicting the weather coincided with the decline of astrology, which had previously claimed that role. Initially, the barometer was welcomed by the leading authors of astrological almanacs, John Gadbury and John Goad. One of the first announcements of the instrument written for a nonexpert audience appeared in the *Royal Almanack* of 1676. Gadbury and other compilers of astrological almanacs followed with their endorsements of the invention. Robert Plot, an Oxford virtuoso with a strong interest in astrology, saw the Royal Society's program of research on the barometer as quite compatible with the efforts of reform-minded astrologers like Goad.[32]

In the first year of the eighteenth century, however, a direct assault on astrology on behalf of the new instrument was launched by Gustavus Parker. Parker published a series of tracts in 1700–1701, in which he used a newly invented portable barometer to offer forecasts of the weather for up to one month ahead. He made it clear that he was aiming to replace the astrological

31. Willsford, *Nature's Secrets*, 72; Claridge, *Shepherd of Banbury's Rules*, iv–v; Smith, *Horological Disquisitions*, 78–88.
32. Gadbury, *Nauticum Astrologicum* and *Stars and Planets the Best Barometers*; Goad, *Astro-Meteorologia Sana*; Capp, *Astrology and the Popular Press*, 185; Plot, "Letter to Dr. Martin Lister," 930.

almanacs, which traditionally included long-range weather forecasts.[33] Parker enticed Gadbury into an exchange of pamphlets, with the astrologer insisting that "stars and planets [were] the best barometers." Gadbury pointed out that the rival instrument, although it could be useful for short-term predictions, such as warning of the approach of storms at sea, lacked the long-range accuracy claimed for astrological forecasts: "The Plenty, or Scarcity of Fruit, Grain, Hay, &c. depend not upon what Weather a *Day*, but what a whole *Season* produceth; and such Philosophical foresight belongs not to the *Barometer-Men*, as such, but to the *Astrologers*."[34]

Notwithstanding Gadbury's rearguard action, the intellectual stock of astrology was descending precipitately by the end of the seventeenth century. The barometer shared in the triumph of the new experimental philosophy; but it also assumed some of the expectations previously invested in astrology. Most early writers on the instrument stressed its status as a new invention, a "wonder," or a "curiosity of art." In 1666 Beale thought the instrument "one of the most wonderful that ever was in the world"; exactly a century later Benjamin Martin described it as, "both in Regard of Curiosity and Utility, . . . the first in Dignity among the modern Philosophical Inventions."[35] Early purchasers read that the device had been investigated by "many Ingenious and exquisite Searchers into Nature." The names of Torricelli, Boyle, and Hooke were routinely invoked in the promotional literature, and in 1718 John Laurence told his readers that, "the philosophical Reason of the Mercury's rising and falling . . . is made intelligible enough by the new Philosophy of *Gravitation* upon the Principles of the GREAT Sir *Isaac Newton*."[36]

Thus, the barometer came to be seen as a material token of the achievements of the new science; those who bought it were buying into the experimental sensibility and acquiring a symbol of intellectual refinement. The instrument was labeled a "Philosophical, or Ornamental Branch of Furniture." It joined a small class of devices, including telescopes, microscopes, and orreries, that were both consumer goods and vehicles for the dissemination of Enlightenment science. It was bought to adorn the domestic interiors of aristocratic or bourgeois households, but it was more than simply a typical piece of Georgian furniture. Rather, the barometer was a "conversation piece" in an

33. Parker, *Account of Portable Barometer; New Account of Alterations of Wind and Weather; New Baroscopical Account . . . for July 1701; New Baroscopical Account . . . for September 1701; New Baroscopical Account . . . for October 1701.*

34. Gadbury, *Stars and Planets,* preface (unnumbered pages).

35. [Oldenburg and Beale], "Relation of Mercurial Observations," 155; Martin, *Description of the . . . Simple Barometer,* 3.

36. Neve, *Baroscopologia,* 3; Laurence, *Fruit-Garden Kalendar,* 135.

age of enlightened conversation. Edward Saul wrote that his contemporaries, "having given themselves and their Parlors an Air of Philosophy, by the purchase of a Barometer, [were] . . . desirous of exerting now and then a superiority of understanding, by talking clearly and intelligibly upon it."[37]

The weather, as Samuel Johnson wearily complained, was a perennial topic of conversation in a land where its unpredictability was proverbial.[38] The barometer gave these conversations a new inflection, coloring them with an awareness of experimental philosophy, but also signaling a degree of ambivalence about its value. Benjamin Martin dramatized such a dialogue in his *Young Gentleman and Lady's Philosophy* (1772). The university-educated young man, Cleonicus, informs his curious sister, Euphrosyne (whose education has of course been confined to the home), that the barometer shows changes in the air that affect individuals' health and fortunes. When she demurs, he insists, "I can assure you, my *Euphrosyne*, that it is a Matter of more than mere Pleasantry." In travel, trade, and agriculture, Cleonicus goes on to say, the knowledge of imminent weather conditions can be of significant economic value. Euphrosyne is readily convinced and expresses astonishment that not everyone possesses such a useful instrument. "Who would [not], for the Sake of such a small Sum of Money, want a general Index for Life, Health, and good Fortune?" she asks.[39]

Euphrosyne's capitulation suggests that barometers might have been used to reinforce the scientific authority of men in the bourgeois household. The instrument, which Martin noted was particularly suitable for "a Gentleman's Parlour, or Study," provided opportunities for displays of sophistication to less scientifically literate women. But the barometer was an equivocal tool for this purpose. Aside from its mercurial unreliability, which was frequently described in feminine terms, the device was an object of fashionable consumer desire, and to that extent tainted with female folly. Martin explained that many purchasers were seduced by expensive and needlessly elaborate models, because "many Gentlemen, as well as Ladies, affect to have Things very fine and showy, and such as shall make a grand Appearance."[40]

The aura of feminine folly also surrounded the barometer because of its continuing links with superstition. Parker, who had used the instrument to challenge the claims of astrologers, was himself taken to task by Patrick for his pretensions to predict the weather for a whole month ahead.[41] Even after such

37. Saul, *Historical and Philosophical Account of the Barometer*, 1, 2.
38. Johnson, *Idler*, no. 11 (24 June 1758), in Johnson, *Works*, 2:36–39.
39. Martin, *Young Gentleman and Lady's Philosophy*, 1:323–24.
40. Ibid., 324–5, 328.
41. [Parker and Patrick], *New Account of . . . the Wind and Weather*.

exaggerated claims were scaled back, promoters of the device were likely to sound as if they were describing a kind of Enlightenment magic: a machine that could gratify the demands for prediction previously served by astrology but was endorsed by the leading experimental philosophers. To this degree, the barometer could be placed alongside the philosophical shows and spectacles, the automata and gadgets, displayed by entrepreneurial craftsmen and show-men.[42] Lofty moralists declined to distinguish this kind of science from vulgar superstition. Johnson eagerly drew such a conclusion in the *Idler* of 5 August 1758: "The rainy weather which has continued the last month, is said to have given great disturbance to the inspectors of barometers. The oraculous glasses have deceived their votaries; shower has succeeded shower, though they predicted sunshine and dry skies; and by fatal confidence in these fallacious promises, many coats have lost their gloss, and many curls been moistened to flaccidity." For Johnson, the message was clear: experimental attempts to second-guess divine providence, particularly for the sake of luxury and vanity, were to be condemned. It was a lesson he drew out at greater length in the issue of 2 December 1758, with the "Journal of a Senior Fellow"—extracts from the supposed diary of a Cambridge don, whose repetitive and unproductive life is regulated by a neurotic degree of attention to his barometer.[43]

Johnson was obviously more disdainful of the barometer than those who promoted the instrument, but even advocates routinely criticized users who naively invested too much trust in it, for example, by relying only on the weather indicators engraved on the scale. Neve's condemnation of those who deceived themselves and others in this way was quite typical. Disavowing naive credulity or an exaggerated appreciation of the barometer's capabilities, most promoters of the apparatus suggested that it could be relied upon only for short-term forecasts, with a reasonable degree of probability, provided other weather signs were also taken into account. The anonymous author of *The Meteorologist's Assistant* (1793) summed up the consensus view at the end of the eighteenth century, when he remarked that the barometer, "is not a certain and unerring prognosticator of the ensuing weather, but when its motions are carefully attended to, and the rules which experience has formed judiciously combined therewith, we may without doubt be enabled to form a very probable guess at the weather to be expected."[44]

It was within these very circumscribed limits that the barometer played its part in encouraging the adoption of a more naturalistic understanding of the

42. Stafford, *Artful Science;* Altick, *Shows of London.*
43. Johnson, *Idler,* no. 17 (5 August 1758), in Johnson, *Works,* 2:53–54; no. 33 (2 December 1758), in ibid., 101–6.
44. *Meteorologist's Assistant,* 2.

weather, which Peter Eisenstadt and others have identified as characteristic of the Enlightenment.[45] In the late seventeenth and early eighteenth centuries, it has been claimed, the weather came to be viewed less as a vehicle of providential intervention through singular events and more as a natural process accessible to human knowledge. Hence, observers relinquished the emphasis of a previous age on extreme and unusual meteorological events and the attempt to decode them as prodigies or divine punishments; instead, they devoted themselves to the relentless collection of routine observations. The aim (to cite the title of a 1723 book by John Pointer, chaplain of Merton College, Oxford) was a "rational account of the weather." In the work to achieve such an account, the barometer had a pivotal role. Pointer explained that meteorological instruments could be used to forecast the weather because they revealed how, "Natural Causes do Naturally (i.e. according to the settled Order and Nature of things) produce Natural Effects."[46]

Notwithstanding Pointer's enthusiastic naturalism, the barometer proved a disappointment as a key to unlock the weather's laws. It did nonetheless assume an important place in the eighteenth-century projects for systematic recording of meteorological observations. The significance of these activities may lie more in the routinization of observational practices than in the achievement of a naturalistic understanding of the weather. Individuals might indeed have participated in the discipline of collecting observations because they expected that laws would eventually be found, but the gratification of that expectation was indefinitely deferred. Presumably, then, the daily or twice-daily drill of tapping the barometer and taking a reading yielded some more immediate satisfaction, perhaps as a means of structuring experienced time by laying down the benchmarks of a reassuring ritual.

Inspired ultimately by Francis Bacon's "History of the Winds" and promoted by Hooke, Boyle, Robert Plot, Martin Lister, John Locke, and other members of the early Royal Society, this enterprise accrued participants and expanded its geographical coverage throughout the eighteenth century. Plot seems to have been the first to submit to the *Philosophical Transactions* a comprehensive weather diary, giving daily readings for the whole of the year 1684. Toward the end of the 1690s, William Derham took up the self-imposed duty, which he continued for several years. He was followed in the 1710s–30s by Henry Beighton, engineer and editor of the *Ladies' Diary*, who compiled weather records, including regular barometrical readings, for more than twenty years. In the 1720s, James Jurin became involved in correspondence

---

45. Eisenstadt, "Weather and Weather Forecasting in Colonial America." See also Reed, *Romantic Weather*, chap. 1.
46. Pointer, *Rational Account of the Weather*, vii.

with several weather observers, including John Horsley, a schoolmaster in Northumberland; John Huxham, a doctor in Plymouth; and Thomas Nettleton in Halifax. Jurin published a Latin invitation to European meteorologists to participate in reporting weather records using standard procedures. He laid particular emphasis on the barometer, which was to be read at least once a day.[47]

These meticulous and self-disciplined observers played their parts in an enterprise that assumed significant scale in the eighteenth century, as historians of meteorology have recognized.[48] Hundreds of individuals were involved, throughout Europe and the European colonies. In the course of the century, the *Philosophical Transactions* introduced an annual register of the weather in London, and published records from Massachusetts, New York, Pennsylvania, South Carolina, China, and Siberia. Many other societies, from the Oxford Philosophical Society in the 1680s to the American Philosophical Society in the 1760s, added their contributions. Nor was the effort confined to the scientific elite. Purchase of a barometer, in effect, implied acceptance of an invitation to participate in an experimental community. The instrument demanded some kind of use, and thereby exerted some degree of discipline upon its prospective users. Those who bought the early versions were expected to fill the tube with mercury themselves, taking precautions to clean the glass and avoid getting air bubbles in the column. In 1747, the *Universal Magazine* reiterated the standard advice on this procedure, and went on to give instructions about how to replenish the mercury regularly. John Warner, an instrument maker active in London in the first two decades of the eighteenth century, published a trade card incorporating a blank grid, on which daily barometer readings and other weather observations could be recorded.[49] The *Gentleman's Magazine* was one of the more popular periodicals that introduced a regular weather diary, including daily thermometer and barometer measurements and comments on prevailing diseases.

In conjoining weather records with notes on diseases, the *Gentleman's Magazine* reflected a widespread trend. Remarks on the illnesses of humans and animals were frequently added to records of instrument readings in weather journals. Jurin's correspondent John Horsley sent a sample of his notes for April 1723, in which he had recorded (at the beginning of the month) "The Ague Epidemical," and (toward its end) "The Ague has become more Epidemical." John Mills, an enthusiastic advocate of the barometer, quoted with

---

47. [Jurin], *Invitatio ad Observationes Meteorologicas.* Andrea Rusnock kindly supplied me with copies of letters collected for Rusnock, *Correspondence of James Jurin.*
48. Feldman, "Late Enlightenment Meteorology."
49. "To the Authors of the *Universal Magazine*"; Walters, "Tools of Enlightenment," 120–21.

approval the observations of the Oeconomical Society of Berne for the year 1766. In February, it was noted, "The barometer varied greatly, being sometimes very high, and sometimes very low"; in June, "Frequent aches were felt every where, and some dogs ran mad." [50]

The problem highlighted by these entries was that of linking states of health with states of the weather in a systematic way. Whereas the barometer might vary hourly and tended to be read at least once a day, diseases followed a slower cycle and were usually recorded on a weekly or monthly basis. John Huxham, who published his records of more than ten years' observations at Plymouth in 1739, solved the problem by noting the highest, lowest, and mean barometer readings every month, with some remarks on its changes; he then added comments on the prevailing diseases in that month. His entry for January 1728 reads:

> The Wind was for the most Part southerly, or from the West, and that often stormy: Near the End of the Month however a south-east Wind, and sometimes a north-east, rose the Barometer, and scattered the Clouds, now and then Frost and Snow intervening.—From the 26th to the 28th much Rain fell, though the Baroscope was at 30.0 [inches] and higher, a south-east Wind blowing: An Argument that the Atmosphere was still loaded with Vapours. . . . From the Beginning of the Month Coughs and Catarrhs were very frequent, oftentimes attended with a troublesome Tumor of the Fauces, and a slight Fever commonly. Rheumatisms and Squinzies up and down; great Lowness of Spirits, and frequent hysteric Paroxysms every-where.[51]

Following this kind of pattern, it became common for remarks on diseases to take the form of heterogeneous and irregularly recorded annotations in the last column or the footnotes of monthly weather tables. Many observers remained convinced of the Hippocratic connection between the quality of the air and the state of bodily health, despite the difficulties in quantifying and correlating the variables. It was not unusual, however, for a degree of skepticism to be expressed about the value of the barometer itself as an indicator of the conditions affecting health. Thomas Short of Sheffield, commenting on twelve years' records by Thomas Barker of Rutland (and no doubt also in the light of his own twenty-two years of observations), allowed that variation in air pressure could induce various diseases, but claimed that its effects were likely to be overwhelmed by changes in the air's moisture and temperature. City

---

50. Horsley to Jurin, 2 April 1723, Royal Society Early Letters, H.3.100; Mills, *Essay on Weather*, xii–xxix.

51. Huxham, *Observations on the Air and Epidemic Diseases*, 6–7.

dwellers who were inclined to place trust in the barometer, and to reject the weather lore of country people as superstitious nonsense, were therefore taken to task by Short as guilty of superstition themselves. They should recollect, he pointed out, that living beings possessed their own internal barometers, so that their responses to changing air conditions were likely to be more sensitive registers than the movements of the mercury column.[52]

Short's comments indicate the ambivalence that often attended upon use of the barometer to collect statistics of weather and health. Although observers such as Gilbert White proposed the systematic compilation of statistics as a means of correcting vulgar errors and superstitions "too gross for this enlightened age," the awkward fact could not be avoided that the barometer was an equivocal emblem of scientific knowledge, itself susceptible to vulgar faith. It was in part to distance the instrument from these connotations that the enterprise of meteorological record keeping went on, oblivious to the failure to produce any general correlations between the motions of the mercury and states of bodily health. Researchers were obviously aware that the facts they were collecting were not reducible to theory, but they did not consider that this invalidated their project. George Adams, in his *Short Dissertation on the Barometer* (1790), wrote as if systematic observations had only just begun. "We can scarce expect them in sufficient number in our own age, to deduce from them a general and perfect theory," he remarked. Philosophers of his day, he noted, were "continually labouring to accumulate unconnected facts," while carefully abstaining from premature attempts to reduce them to hypotheses.[53]

Even though no actual laws of weather emerged, then, the use of the barometer in this kind of enterprise indicated its centrality to a particular structure of expectations as to how the weather *would* (eventually) be understood. Indeed, the instrument served many writers as an emblem of the way human knowledge was almost always limited to a certain degree of probability. Pointer wrote that meteorological indicators were not to be taken as "always sure and certain, but only as, for the most part, probable and conjectural." Benjamin Martin and J. H. Magellan both remarked that, although the barometer could only yield probable predictions of the future, most human decisions had to be made on the basis of knowledge that was only probable, though none the less valuable for that.[54] John Laurence put the point well:

52. Short, *New Observations . . . on Bills of Mortality,* 334–55, 454–56.

53. White, *Natural History and Antiquities of Selbourne,* 203; Adams, *Short Dissertation on the Barometer,* 2–3.

54. Pointer, *Rational Account of the Weather,* xxiv; Martin, *Description of Simple Barometer,* i–ii; Magellan, *Description et Usage des Nouveaux Barometres,* 140–41.

I will not, *I dare not* say the Barometer is *always* a certain and *infallible* Guide; but I *venture* to say, the *better* we understand its Motions, and the more nice Observations we make of the Variation of the Wind and Weather thereupon, the *less* liable we shall be to be deceived on all Occasions.

It is undoubtedly the best *Guide* we have; and a *Guide* ought, and will not fail to be treated with Respect, so long as *Modesty* is preserved; so long as *absolute* Power and *uncontroulable* Dominion are not pretended to.[55]

This understanding of its limited predictive power was of key importance for the assimilation of the barometer in Enlightenment England. Some early promoters had made grandiose claims for the predictive accuracy of the apparatus, partly relying upon public awareness of weatherglasses, which had already been advertised and sold as prognostic devices. The initial boosterism, exemplified by Parker's month-at-a-time predictions, proved to have been overwrought. Roger North suggested that many of "the vulgar" "sacrificed the [instrument] to the displeasure conceived at the many disappointments it gave them."[56] But the barometer survived, and indeed flourished. As it did so, it became commonplace to criticize the vulgar expectations of prognostic accuracy, and to insist that the barometer's meaning was somewhat more recondite than this. To users who were encouraged to identify themselves as polite and enlightened, the barometer suggested that human knowledge could only be gained by meticulous and extended observation, and that predictions of the future could never be more than probable.

Richard Kirwan summed up the polite consensus in "An Essay on the Variations of the Barometer," published in the *Transactions of the Royal Irish Academy* in 1788. He introduced the instrument as a prime achievement of the experimental philosophy and a symbol of the ascendancy of enlightened Europe over other peoples. It was, he explained, the Torricellian experiment that first demonstrated "the advantages of experimental investigation, and thus excited the more civilized nations to those pursuits, which have since rendered them as superior both in arts and arms to the more ignorant, as men are to brutes." He then immediately went on to say, however, that, "Great . . . as has been the success . . . of the barometer, the bulk of mankind have always expected still more from it: Nothing less than a prognostic of the weather." These vulgar expectations were doomed to disappointment, since "its variations af-

55. Laurence, *Fruit-Garden Kalendar,* 141.
56. North, "Essay of the Barometer," fol. 3r.

ford no indications absolutely certain, though with certain restrictions they afford some ground for probable conjecture." [57]

This, then, was the message the barometer conveyed. To expect the instrument to serve as a totally reliable or long-range weather predictor was a vulgar error. Understanding the barometer correctly required attention to many other weather indicators. Users were invited to embark on their own programs of observation, recording precipitation, wind strengths and directions, temperature, and so on, alongside regular barometric readings. Used in this way, the device served as a concrete emblem of the experimental philosophy and its particular view as to how the natural world might be known. The world was to be understood by routine, meticulous, repeated, quantitative observation. Meanwhile, increasing the volume of facts collected would increase the probability with which the future could be predicted. And increased probability would have to suffice, for certainty was unattainable. Among other things, the barometer was symbolic of the limits of human knowledge and the inescapability of the path of painstaking empirical research.

In this and other respects, it was the flexibility and uncertainty of the barometer that undergirded its popularity. Its readings called for qualitative interpretation and judgment; they offered the opportunity to compare numerical results with subjective impressions and feelings, and to derive satisfaction from the agreement or the disagreement. Rather than displacing reliance on subjective interpretation with quantitative accuracy, the barometer continued to straddle the two realms. Similarly, the marketing of the instrument relied upon, but did not entirely stabilize, distinctions between polite and vulgar, urban and rural, male and female users. The tendency to superstitious, feminine, vulgar reliance upon the predictive value of the apparatus was regularly castigated, but also tacitly encouraged. It was primarily in these senses— as a token of scientific knowledge that also trumpeted its limitations, as an instrument of polite experimentation that was also liable to be mistaken for magic, and as a popular measuring device that also demanded qualitative interpretation—that barometers were indeed symptomatic of enlightened science.

Acknowledgments

The research project on which this paper draws was begun during my tenure of a visiting fellowship at the William Andrews Clark Memorial Library, University of California, Los Angeles, in the fall of 1989. I have subsequently

57. Kirwan, "Essay on Variations of the Barometer," 44–45.

received assistance from the Wheeler Fund of the Department of History, University of New Hampshire. For their advice, I thank the audiences who heard preliminary versions of the paper at All Soul's College, Oxford, in January 1993, and at the University of New Hampshire in April 1993. In addition to those involved with this volume, Patricia Fara supplied valuable information and welcome criticism.

Primary Sources

Adams, George. *A Short Dissertation on the Barometer, Thermometer, and other Meteorological Instruments.* London: R. Hindmarsh, 1790.

Birch, Thomas. *The History of the Royal Society of London for Improving of Natural Knowledge.* 4 vols. London: A. Millar, 1756–57.

Boyle, Robert. "A Confirmation of the Former Account [by John Wallis]." *Philosophical Transactions* 1 (1666): 179–81.

———. *The Works of the Honourable Robert Boyle.* 2d ed. 6 vols. Edited by Thomas Birch. London: J. and F. Rivington, 1772.

Claridge, John. *The Shepherd of Banbury's Rules to Judge of the Changes of the Weather.* London: W. Bickerton, 1744.

Desaguliers, John Theophilus. *Lectures of Experimental Philosophy.* 2d ed. London: W. Mears, B. Creake, and J. Sackfield, 1719.

Gadbury, John. *Nauticum Astrologicum; or, The Astrological Seaman.* London: Matthew Street, 1691.

[———]. *Stars and Planets the Best Barometers and Truest Interpreters of All Airy Vicissitudes.* London: John Nutt, 1701.

Garden, George. "A Discourse Concerning Weather." *Philosophical Transactions* 15 (1685): 991–1001.

Goad, John. *Astro-Meteorologia Sana.* London: Samuel Smith, 1690.

Halley, Edmond. "A Discourse of the Rule of the Decrease of the Height of the Mercury in the Barometer." *Philosophical Transactions* 16 (1686): 104–16.

Harris, John. *Lexicon Technicum; or, An Universal Dictionary of Arts and Sciences.* London: Daniel Brown et al., 1704.

Hooke, Robert. "Method for making a History of the Weather." In *The History of the Royal Society of London,* edited by Thomas Sprat, 173–79. London: by T. R. for J. Martyn, 1667.

Huxham, John. *Observations on the Air and Epidemic Diseases from the Year MDCCXXVIII to MDCCXXXVII Inclusive.* London: J. Hinton, 1759.

Johnson, Samuel. *The Yale Edition of the Works of Samuel Johnson.* 16 vols. Edited by W. J. Bate, John M. Bullitt, and L. F. Powell. New Haven: Yale University Press, 1958–.

[Jurin, James]. *Invitatio ad Observationes Meteorologicas Communi Consilio Instituendas.* London: William and John Innys, 1724.

Kirwan, Richard. "An Essay on the Variations of the Barometer." *Transactions of the Royal Irish Academy* 2 (1788): 43–72.

Laurence, John. *The Fruit-Garden Kalendar . . . To Which is Added, An Appendix of the Usefulness of the Barometer.* London: Bernard Lintot, 1718.

Lister, Martin. "A Discourse Concerning the Rising and Falling of the Quicksilver in the Barometer." *Philosophical Transactions* 14 (1684): 790–94.

Magellan, John Hyacinthe. *Description et Usage des Nouveaux Barometres.* London: W. Richardson, 1779.

Martin, Benjamin. *A Description of the Nature, Construction, and Use of the Torricellian, or Simple Barometer.* London: for the author, 1766.

———. *The Young Gentleman and Lady's Philosophy.* 3d ed. 3 vols. London: W. Owen, 1781.

The Meteorologist's Assistant in Keeping a Diary of the Weather. London: for the author, 1793.

Miller, Philip. *The Gardener's Dictionary.* 2 vols. London: J. and J. Rivington, 1752.

Mills, John. *An Essay on the Weather.* London: S. Hooper, 1770.

Neve, Richard. *Baroscopologia; or, Discourse of the Baroscope, or Quicksilver Weather-Glass.* London: W. Keble, 1708.

North, Roger. "Essay of the Barometer." British Library, Add. MSS. 32541. N.d.

Oldenburg, Henry. *The Correspondence of Henry Oldenburg.* 13 vols. Edited by A. R. Hall and Marie Boas Hall. Madison: University of Wisconsin Press; London: Mansell and Taylor and Francis, 1965–86.

[Oldenburg, Henry, and John Beale]. "A Relation of Some Mercurial Observations and their Results." *Philosophical Transactions* 1 (1666): 153–59.

———. "Observations continued upon the Barometer, or rather Ballance of the Air." *Philosophical Transactions* 1 (1666): 163–66.

Parker, Gustavus. *An Account of a Portable Barometer, with Reasons and Rules for the Use of It.* London: William Haws, 1700.

[———]. *A New Account of the Alterations of Wind and Weather, by the Discoveries of the Portable Barometer.* London: W. Hawes, 1700.

[———].*A New Baroscopical Account of the Daily Alterations of the Wind and Weather for July 1701.* London: J. Nutt, 1701.

[———]. *A New Baroscopical Account of the Daily Alterations of the Wind and Weather for September 1701.* London: J. Nutt, 1701.

[———]. *A New Baroscopical Account of the Daily Alterations of the Wind and Weather for October 1701.* London: J. Nutt, 1701.

[Parker, Gustavus, and John Patrick]. *A New Account of the Alterations of the Wind and Weather, by the Discoveries of the Portable Barometer . . . [with] A Journal of the Wind and Weather . . . as it was observ'd by J. Patrick, Barometer Maker.* London: J. Nutt, [1700].

Patrick, John. *A new Improvement of the Quicksilver Barometer.* London: Richard Newcomb, [1710?].

Phelps, John. *The Human Barometer; or, Living Weather-Glass: A Philosophick Poem.* London: M. Cooper, 1743.

Plot, Robert, "Letter . . . to Dr. Martin Lister." *Philosophical Transactions* 15 (1685): 930–43.

Pointer, John. *A Rational Account of the Weather.* 2d ed. London: Aaron Ward, 1730.

Saul, Edward. *An Historical and Philosophical Account of the Barometer, or Weather-Glass.* London: A. Bettesworth, 1730.

Short, Thomas. *New Observations, Natural, Moral, Civil, Political, and Medical, on City, Town, and Country Bills of Mortality.* London: T. Longman and A. Millar, 1750.

Sinclair, George. *The Principles of Astronomy and Navigation . . . To Which is Added, A Discovery of the Secrets of Nature, Which are Found in the Mercurial Weather-Glass.* Edinburgh: the heir of Andrew Anderson, 1688.

Smith, John. *A Compleat Discourse of the Nature, Use, and Right Managing of that Wonderful Instrument, the Baroscope, or Quick-Silver Weather-Glass.* London: Joseph Watts, 1688.

———. *Horological Disquisitions Concerning the Nature of Time . . . To which is added, The best Rules for the Ordering and Use both of the Quick-Silver and Spirit Weather-Glasses.* London: Richard Cumberland, 1694.

"To the Authors of the *Universal Magazine.*" *Universal Magazine of Knowledge and Pleasure,* no. 1 (June 1747): 18–20; no. 2 (July 1747): 57–59.

Walker, Adam. *A Philosophical Estimate of the Causes, Effects and Cure of Unwholesome Air in Large Cities.* [London]: for the author, 1777.

Wallis, John. "A Relation Concerning the Late Earthquake near Oxford." *Philosophical Transactions* 1(1666): 166–71.

———. "A Discourse Concerning the Air's Gravity." *Philosophical Transactions* 15 (1685): 1002–14.

White, Gilbert. *Natural History and Antiquities of Selbourne, in the County of Southampton.* New ed. London: White, Cochrane, 1813.

Willsford, Thomas. *Nature's Secrets; or, The Admirable and Wonderful History of the Generation of Meteors and Blazing-Stars . . . with Direction for Observing of a Weather-Glass.* London: N. Brooke, 1665.

French Engineers Become Professionals; or, How Meritocracy Made Knowledge Objective

KEN ALDER

Since Merit but a dunghill is,
I mount the rostrum unafraid.

W. H. Auden, "On the Circuit"

Among the most intoxicating slogans that the Enlightenment bequeathed to the French Revolution was the cry for a "career open to talent." At the same time, the Enlightenment laid the foundation for a radical new program for how talent might itself be nurtured, measured, and directed. This program operated under the sign of a new empiricist knowledge that eschewed both speculative natural philosophy and the routine of artisanal practice. Instead, it promoted a new form of hybrid, flexible, and "objective" knowledge, which we today associate with the modern professions. Steven Shapin and Simon Schaffer have memorably noted that "the solution to the problem of knowledge is the solution to the problem of the social order." This paper argues that the rules by which Enlightenment engineers sought to obtain successful results in the material world were inextricably tied to the rules by which engineers were themselves to be judged. That is, this paper offers an account of the origins of modern engineering that points to its interlocking epistemological and social bases. It is the story of how a group of "new men" were molded into a (relatively) cohesive social class—and how they developed new ways of knowing to do so. In the process, it attempts to understand the making of modern expertise, not simply as an institutional arrangement, nor even as an epistemological pose, but as a code for living. The engineers of the French Enlightenment sought to give the lie to Pascal's quip that "[m]en are not in the habit of inculcating merit, but only in compensating it where they find it ready-made." To this end, they subjected young men to a rigorous mathematical education. Their purpose was to create a *self*-disciplined individual, a being whose institutional habitat has been described by Max Weber, whose anatomy has been dissected by Michel Foucault, and whose proudest possession was a new kind of objective knowledge. The modern professions inhabit that peculiar panopticon called meritocracy.[1]

Meritocracy is today said to be the system by which persons are ushered into their proper station in life. We rise as high as we deserve—or if you prefer, as high as we merit. It is hardly an exaggeration to say that meritocracy is one of the key organizing ideals of modern society, a principle widely lionized as the only just and efficient way to award society's favors. These favors, whether they take the form of wealth, high office, or social prestige, are supposed to be the fitting reward of proven worth. For those who are satisfied with appearances, then, meritocracy thrives on a disarming circularity: we are to recognize the meritorious by the fruits they enjoy. And for those who investigate more closely, meritocracy is said to rest on real results: the title of merit, we are told, is conferred on the basis of actual performance as measured against objective standards. To the extent that such claims are believed, meritocracy would appear to be a system that transcends politics, a close approximation to a natural social order, which deposits each individual in his or her proper place even as it produces outcomes that are presumably "the best." Hence, those who question this system will necessarily appear to be pressing their particular interests in order to act politically—and out of sour grapes besides.

But the relationship between *who* succeeds and *what* succeeds depends on who defines success and how. In our modern conception of meritocracy, the standards that define the meritorious are at a minimum supposed to be objective in the sense of being "impersonal." That is, rewards are supposed to be granted, not by the secret pull of private connections, but by the judgment of those with no interest in the outcome. Ideally, that impersonality is guaranteed by means of standards that are open to public scrutiny or by the operation of an "invisible hand." But to assert that impersonal forces select society's winners—aggregate demand in the marketplace, computer-graded GRE scores, or the (double-blind) recognition of one's peers—is to ignore how social institutions structure the ways in which the title of merit is conferred. The design of these institutions, moreover, is a matter of considerable contention. It is one thing to say that impersonal public standards should be used to select the meritorious, and another to specify which standards should be used, how they will be implemented, and who decides upon them. The history of merit is the history of those institutions and the ongoing negotiations over how its rules

---

1. Shapin and Schaffer, *Leviathan and the Air-Pump;* Pascal, *Pensées,* 1:496; Weber, "Science as a Vocation," and "Politics as a Vocation," in *From Max Weber;* Foucault, *Discipline and Punish.*

should operate. In the professions, the professionals attempt to keep the writing of those rules in their own hands.

In the face of this circularity, the defenders of the professions also assert that the standards that designate the meritorious are also said to be objective in the grander sense of being "true"—or at least they are said to correlate in some straightforward way with real achievement. That is, a meritocracy is said to be structured around the supposition that laurels are parceled out to individuals, not according to the luck of high birth or by chance, but on the basis of proven success. Yet even this functionalist claim is a matter of contention. Even setting aside the difficult task of defining what counts as "success," the measure of merit remains notoriously vexed. Where the task is collective, who gets the credit for a particular success? What qualities enabled him or her to pull off their coup? And how can those qualities be measured if the relevant institution must base its decision to hire or promote on insufficient information (as it invariably must), or if it must predict performance in advance (as it is often called upon to do)? None of these questions can ever be easily or definitively answered. Most problematic, those standards that are amenable to public scrutiny (objectivity as impersonality) may be quite far removed from the sorts of standards that measure "real" abilities (objectivity as the "right" answer).

Of course, almost every society claims to operate in some fashion on the tautology that "the best" have accomplished "the most." Yet it was during the Enlightenment that the radical claim was first heard that the answers to these two questions (who succeeds and what succeeds) should be made to coincide—and that the way to do so was to bring into alignment standards that were both impersonal and indicative of effectiveness. No longer was the accident of high birth to be the basis for great reward; henceforth, the ability to achieve results was to be all that counted. This slogan—trumpeted by the *philosophes* themselves—has been widely reported. Seldom acknowledged is the fact that this shift in the social mechanisms for allocating rewards also depended on a shift in the kind of results considered worthy of reward, the type of methods considered proper for achieving them, and the sort of persons who were entitled to validate their success. Steven Shapin has recently documented how a gentlemanly code of honor shaped the truth-telling character of seventeenth-century natural philosophers in England. This paper examines the eighteenth-century repudiation of this aristocratic notion that a person's "degree" should somehow bolster his claims to speak about nature—or about practical problems, for that matter. It thereby closes the circle left open by those historians—notably Lorraine Daston and Theodore Porter—who have noted the rise of a new kind of "objective" knowledge making at the end of the eighteenth century. It suggests how the legitimation of this new kind of objective knowledge depended crucially on a new definition of individual merit, which was itself henceforth to be defined objectively.[2] Knowers were henceforth to be judged by the same sort of process they had begun to apply to their objects of study.

But at the same time, this paper does *not* accept the premise that this Enlightenment move to meritocracy succeeded in divorcing the realm of technological knowledge from the realm of social values. On the contrary, it uses the example of the French artillery engineers to show that the mathematical way of knowing developed in this period did not just guide proper design, it also guided proper social conduct. The character of this social epistemology was profoundly shaped by the conditions under which these artillerists labored, notably as servants of the military apparatus of both the absolutist and the revolutionary state.

To be sure, French engineers were not alone in traveling this route to meritocracy. Savants, doctors, lawyers, civil servants, and army officers all legitimated their own expertise by reference to the social mechanisms that governed their promotion. This is why the eighteenth century witnessed the rise of a new form of social organization, one that today goes by the name "professional."[3] The military engineers of the French artillery service offer an especially profound and striking example of this transition, one made particularly challenging by the aura of success that has clung to their individual and collective achievements. The cannon and muskets they designed, produced, and deployed spearheaded the conquest of Europe under Napoleon Bonaparte. After all, Napoleon, that most famous propagandist of a "career open to talent," was himself an artillery engineer.

Engineering the Hierarchy

To untangle this conflation of epistemological pose, practical success, and social mechanism of preferment, we must first be clear about the sort of career the Revolution actually made possible. The profession of engineering in postrevolutionary France—a pattern widely imitated abroad—was legitimated by an elaborately overdetermined system of anointing the meritorious. French

2. Shapin, *Social History of Truth;* Daston, "Objectivity"; Porter, *Trust in Numbers.* It should be obvious that the foregoing discussion has been inspired by the ongoing debate over affirmative action and standardized testing in the U.S.

3. The standard history of French elites does not examine the mechanisms that marked the new notability as meritorious; Chaussinand-Nogaret et al., *Histoire des élites.* For the example of lawyers, see Maza, *Private Lives,* 86–97; Bell, *Lawyers and Citizens.* On the relationship between the rise of professionalism and the social transformations that undergirded the French Revolution, see Alder, *Engineering the Revolution,* and Jones, "Bourgeois Revolution." For the distinction between the new professions of the eighteenth century and the older corporate model of the guilds, see Haskell, "New Aristocracy."

engineers in the nineteenth century, as in the twentieth, were ranked in a well-defined professional hierarchy undergirded by a cognitive hierarchy, itself reinforced by a well-defined hierarchy of schools, which in turn recruited from a corresponding hierarchy of social classes. At the top of the pyramid stood the graduates of the École Polytechnique who went on to attend the Grandes Écoles, and who were themselves graded in an ascending ladder of prestige: Artillerie, Génie, Géographes, Ponts et Chaussées, and Mines. Recruited largely from sons of the upper reaches of the civil service and the rentier class, the *polytechniciens* were trained in the most abstract mathematics and sciences and bound for state service at the highest level, often in the military. Below them in status stood the graduates of engineering schools such as the École Centrale des Arts et Manufactures, where the sons of midlevel functionaries and the industrial bourgeoisie learned an extensive range of applied sciences and mathematics, as well as practical skills in technical drawing and industrial processes. They were generally employed in private industry, or if in state service, as the minions of the *polytechniciens*. Below these, stood the industrial foremen, the *gadzarts,* trained in schools such as the Écoles des Arts et Métiers. There, the sons of artisans received basic instruction in mathematics and technical drawing, and extensive shop-floor experience in a specific trade. These interlocking hierarchies were accepted by contemporaries as the "natural" landscape of nineteenth-century France.[4]

Recently, the schools that validated and reinforced these strata have been extensively studied, and their underlying cognitive hierarchy—based principally on mathematical knowledge—has been connected to a particular style of French engineering. But scholars have continued to locate the origins of this system in the educational reforms of the Revolution and early nineteenth century.[5] In doing so, historians have implied that this meritocratic system emerged spontaneously once the French Revolution had ended the legal restrictions that had reserved high government posts for those of high birth. The unspoken assumption here is that meritocracy was a natural order given free rein by the new liberal polity. But let us examine this postrevolutionary hierarchy more closely. First, the system was in principle based on the assumption that entry was available to all qualified (male) citizens. The most visible qualification—a high degree of mathematical proficiency—was held to be democratic because it was a publicly verifiable standard that did not depend on the judgment of the examiners (objectivity as impersonality). But in practice, steep barriers in financial and cultural capital—tuition fees, the *baccalauréat,* pre-

---

4. Shinn, *Savoir technique;* Weiss, *Technological Man;* Day, *Education for the Industrial World.*
5. Fox and Weisz, *Science and Technology in France,* 1–28; Kranakis, "Social determinants," and *Constructing a Bridge.*

paratory tutoring, plus an additional requirement to know Latin—meant that these schools still recruited from a narrow social base. Despite exceptions, "birth" still determined social destiny, except that now birth correlated more often with wealth than with an august lineage.[6] Second, this meritocratic scheme implied that an individual's ability in mathematics measured (or predicted) something like real performance in tasks deemed important to the state (objectivity as real ability). Yet as we will see, this assumption was by no means self-evident, even to contemporaries. Leaders in all the engineering schools realized that practical training continued to be essential to producing effective engineers.[7] And finally, there is the most basic assumption that this hierarchical system was in fact meritocratic. But that can hardly have been the case where practitioners were permanently locked into the stratum they had first entered (once a *gadzart,* always a *gadzart*). In short, the postrevolutionary social and cognitive hierarchy was not a frictionless natural order, but a historical construct governed by institutional norms. This structure operated under the umbrella of the state and, at the same time, constituted the structure of the state.

Nobility and Merit

The "constructedness" of this social order is quite apparent from the point of view of the ancien régime. Why were certain forms of knowledge—particularly the mathematical sciences—selected as the marker of elite knowledge? How did the various social strata come to be associated with strata in the hierarchy of knowledge when such social strata could hardly be said to have existed in 1750? And most fundamentally, how did "merit" and "birth" come to be seen as intrinsically antagonistic principles?

As Jay Michael Smith has recently pointed out, the aristocrats of the ancien régime had also justified their preeminent status by reference to their "merits." Building on the work of David Bien, he shows how this older sense of merit was compatible with a system where the reward of high office was doled out according to high birth. Indeed, in a polity where the absolutist monarch personally dispensed all favors, a family history of loyal service would "merit" a nobleman his just reward. Yet from the reign of Louis XIV onward, the numbers of young men seeking office grew rapidly as the army expanded and the monarchy sold ennobling offices. Historians have characterized the eighteenth century as a time of manic competition for state office. In this period, proximity to the king's person could no longer distinguish the claim of any but the handful of nobles at Versailles. These court nobles retained their lock on the

6. Weiss, "Bridges and Barriers."
7. Picon, *Ingénieur moderne.*

top positions in the state. But for the vast majority of nobles, the monarchy was obliged to award office through intermediary institutions, such as the War Office. These institutions themselves, however, were not ostensibly to exercise discretion about who to hire or promote, a privilege that remained the prerogative of the absolutist monarch. Hence, they increasingly judged and promoted on the basis of uniform standards thought to measure worthy performance. In the army, these standards included seniority, the number of battles fought, and even the number of wounds sustained. As Smith has noted, the theoreticians of absolutism justified these impersonal measures as an extension of the sovereign's "universal gaze."[8]

In the long run, however, this shift toward an impersonal treatment of nobles exacerbated the contradiction between the nobility's allegiance to the state and the more traditional pattern of patronage and personal fidelity that had long been the primary mechanisms for awarding favor in the ancien régime. Indeed, personal connections continued to play a key role in advancement in the late eighteenth century (as they always do). Court nobles commanded regiments, while provincial nobles served as captains. But reformers increasingly condemned these distinctions in the name of the efficient operation of the state. Minister of War Argenson put it this way at midcentury: "Were everyone the child of his achievements and merits, then justice would be done and the state would be better served." And in the 1770s, Minister of War Saint-Germain noted that people close to the king often urged him to promote individuals who actually worked against the letter of his own law. The court nobility, he noted, "had everything . . . without meriting it," while the provincial nobility, "had nothing . . . even when they merit [better]." Merit, for these French reformers, was not a sign of disembodied talent; it indicated that one had the capacity and dedication to serve the central state.[9]

Revanchism and State Service

This contradiction in how "merit" was defined was both spawned and exacerbated by geopolitical pressures bearing down on eighteenth-century France.[10] And it is to the French state's response to these pressures that this discussion now turns. After all, to understand the engineers' new program, we must first understand the arena in which they labored—and where they met with their most daunting challenge. This is because a genuine analysis of mechanisms by

8. Smith, *Culture of Merit;* Bien, "Réaction aristocratique"; Lucas, "Nobles, Bourgeois, and the Origins of Revolution."

9. Argenson in Chaussinand-Nogaret et al., *Histoire des élites,* 232–33; Saint-Germain, *Mémoires,* 39–40.

10. Stone, *Genesis of the French Revolution.*

which the meritorious are selected cannot shy away from an analysis of the material tasks at which the meritorious are supposed to excel. An eighteenth-century French engineer was invariably a military engineer. In particular, the French engineers were the acknowledged masters of the siege battles that had dominated warfare between the 1660s and 1740s.

At the end of the seventeenth century, Sébastien Le Prestre de Vauban had put the finishing touches on a form of warfare that he hoped would ensure that Europe would never again suffer the uncontrolled carnage and social upheaval of the Thirty Years' War. Vauban developed a "science" for the building and taking of fortresses. His five-sided towns were geometrically calculated to aid in defense. And his famous *Traité des sièges* listed the twelve stages in the prosecution of a siege. To bring his tactics to this level of certainty, Vauban also brought the artillery under his authority and coordinated their bombardments with his schedule. Vauban's "science" of siege warfare brought the power of the absolutist state to bear on points of resistance, while minimizing the destruction and uncertainty entailed in the transfer of valuable capital.[11]

The goal of Vauban's reforms was to bestow on France a perimeter of double fortifications, transforming the entire kingdom into a defensible fortress. But in the latter part of his reign, Louis XIV used these defenses to project his dynastic ambitions across the continent, only to be hurled back in costly defeat. And Vauban himself contributed more to the art of taking fortresses than to their defense, so that by the eighteenth century, foreign imitators had broken his "impregnable" perimeter of fortresses in several places. Intelligent military theorists, like the *littérateur* Choderlos de Laclos, realized that this new superiority of offensive over defensive war-making constituted an unacknowledged instability in the security of the nation.[12]

The defeats of the Seven Years' War made this instability painfully clear. Prussia's rapid rise to great-power status was unprecedented in European memory. Frederick the Great's radical new mobile battle tactics stunned Europe, and his militarist constitution provoked universal admiration. Whereas Louis XIV had famously identified the state with his dynastic ambitions, Frederick proudly called himself the first servant of the state. And his battlefield successes seemed to flow directly from a concomitant subordination of his officers and men, a discipline that enabled them to achieve troop movements of devastating rapidity. Frederick had devised his new tactics because his small kingdom lacked the financial and human resources to sustain a lengthy war. A group of well-placed French reformers now concluded that even a great nation could not

11. Vauban, *Traité des sièges*. On siege warfare, see Lynn, "*Trace italienne* and the Growth of Armies"; Nef, *War and Human Progress*; Duffy, *Fortress*, 10, 29–31, 71–97.
12. Laclos, *Vauban*, in *Oeuvres*, 569–93.

sustain indecisive siege campaigns and must try to achieve rapid victories in field battle. Jacques-Antoine-Hippolyte de Guibert and other French reformers condemned their nation's effort in the Seven Years' War as arrogant and apathetic. The officer corps was full of court nobles angling for promotion and unwilling to subordinate themselves to the official chain of command. Their squabbles paralyzed the army. Guibert recognized that this radical new mobile warfare required a corresponding transformation in French military and political institutions. After 1763, a revanchist spirit in France brought the officer class under intense scrutiny. In particular, it underscored the need for capable— read meritorious—officers.[13]

In the years after 1763, reform-minded ministers, such as Choiseul and Saint-Germain, phased out the venality of military office and took sole charge of recruiting soldiers, two practices by which colonels and captains still claimed proprietary rights to their units. This reformulation of the duties of unit commanders was meant to permit new tactical arrangements, such as the combat division, by which a general staff could pursue field warfare with speed and flexibility. Inevitably, the reform of institutions so central to the ancien régime provoked considerable resistance. The divisional structure was abandoned after regimental commanders objected to surrendering control over promotion. Similarly, there were bitter protests against the decision to allow officers to fill vacancies as they opened, regardless of regiment, because this too subverted the ability of commanders to reward their own clients.[14] The mechanisms of promotion became so highly contentious within the military as to be potentially destabilizing to the state. After all, the core activity of the absolutist state was the prosecution of war, and military service was still the nobility's principal justification for its social preeminence.[15]

The bitter dispute that split the artillery corps in this period was part of this larger struggle over the definition of the nation (was France a defensive fortress or an aggressive combatant?), and the related question of which cognitive program and methods of promotion would best serve the nation. Ostensibly, the notorious polemic between the followers of Joseph-Florent de Vallière and the followers of Jean-Baptiste de Gribeauval turned on the proper design of cannon. The guns of the Vallière system—the five standardized calibers that had been in use since the 1730s—were designed for the slow pace and heavy bombardments of siege warfare. French four-pounders, for instance, weighed twice as much as their Prussian equivalents. But at Rossbach (1761),

---

13. Guibert, *Essai général tactique;* Kennett, *French Armies in the Seven Years' War,* 117–18.
14. Saint-Germain, *Mémoires,* 39–47, 59–61, 234–35.
15. Tilly, *Coercion.* Indeed, I have argued that this conflict did much to fuel the aristocratic revolt of 1788–89. Alder, "Stepson of the Enlightenment."

Frederick had won a stunning victory in a field battle, partially by repositioning his artillery to great effect. Between campaigns a frustrated Marshal Broglie took his Vallièrist cannon to Strasbourg to have them re-bored to accept larger projectiles. These irregular light pieces were manufactured without the participation of the artillery service. The Gribeauvalist system of field cannon introduced in 1763 were an officially sanctioned and carefully standardized version of this new sort of artillery.[16]

Gribeauval's cannon, which served throughout the revolutionary and Napoleonic wars, were far more mobile than the guns they replaced. But as the Vallièrists never tired of pointing out, they were also less powerful and accurate. As Donald MacKenzie has noted, neither accuracy, precision, nor any other objective is a "natural" technological goal. These objectives are in dispute just as much as the means to achieve them.[17] Gribeauval's cannon was only one element in an elaborate technosocial system intended to increase the *mobility* of firepower. The goal was to break out of the bounded siege wars of the past and initiate the unlimited field wars of an (unspecified) future.

The Gribeauvalists understood that this new technology, to work effectively, required a broader reformulation of the duties and authority of artillery engineers. Much like the rest of the officer corps, the military engineers of the French Enlightenment sought to marry technical competence to elite status. For engineers, however, the solution to this ancien régime contradiction lay in a new social and cognitive program—one that was to become the basis for a new professional ethos.

Neither a Savant, nor an Artisan Be

From its position as a second-class and vaguely dishonorable branch of the military, the eighteenth-century artillery corps developed an ethos of self-discipline and hierarchy that would ultimately prove emblematic of the new social order. Unlike officers of the line army, artillery engineers had never commanded by dint of their privileged birth alone. For them, authority depended on a claim to have mastered the material world. The push to create a new class of trained professionals begun by Vallière in the 1720s was intensified by Gribeauval in the 1760s. Drawing on those social strata that historians have come to associate with the "new men," these artillery engineers carved out a place for themselves in the ancien régime by a set of artful positionings. On the one hand, they positioned themselves as officer-experts between military patrons and cannoneer subordinates. Here, their role was to translate the state's

16. Alder, *Engineering the Revolution,* 23–55.
17. MacKenzie, *Inventing Accuracy.*

investment in weaponry into coercive power on the battlefield. And on the other hand, as technical experts, they interposed themselves between the state and private arms makers. Here their role was to use the power and purse of the state to secure plentiful supplies of battlefield weapons.[18]

Central to both these mediating roles—and to matching these roles to a program of marking individuals for preferment—was the "middle episte-mology" the engineers staked out, combining theory and practice in the pur-suit of technological novelty. Instead of studying the hidden mechanisms of nature, they turned directly to the investigation of human-built machinery. Theirs was a science of artifice, the systematic investigation of human con-structs. Their method—deceptively straightforward—was to describe quanti-tatively the relationships among measurable quantities, and then use these descriptions to seek out a region of optimal gain (as they defined it). In return, a vast array of social and political problems were expected to yield up easy solutions.

In a sense, these engineers were simply heeding the concurrent Enlight-enment call for a reformulation of knowledge making along empiricist lines. This was a two-pronged critique. The French *philosophes* derided the abstrac-tions of the Cartesians as a hubristic search for nature's hidden mechanisms and a priori cosmological "systems." And they castigated the routine practices of contemporary craftworkers as following from the secrecy, collusion, and venality of their artisanal corporations (or guilds). In contrast, the *philosophes* promoted a rule-governed *pratique* that was general, flexible, and effective.[19] Diderot's famous article "Art," for instance, echoed Francis Bacon's plea that the savant and craftworker offer one another mutual aid. Theoretical training was counterproductive unless combined with a practical knowledge of basic physical properties. At the same time, however, only an open technical discus-sion—conducted by means of rigorous analysis—would generate new discov-eries. Artisans needed to forsake their secret ways if they wished to throw off their slavish acquiescence to routine. Against the corporations, which re-strained producers and stifled free exchange, Diderot called for innovative public technological knowledge.[20] This agenda provided the epistemological and social opening into which engineers inserted themselves.

In the seventeenth century, the old art of artillery had been passed on through a practical apprenticeship, much like the art of the trade corporations. Formal science offered little assistance. Natural philosophers familiar with the new rational mechanics of Galileo and Newton may have touted the promise

18. For the engineers' role in production, see Alder, *Engineering the Revolution*, 127–251.

19. D'Alembert, "Discours préliminaire," in Diderot and d'Alembert, *Encyclopédie*, 1: i–vii, xxiv.

20. Diderot, "Art," in *Encyclopédie*, 1:716.

of a scientific ballistics in their scholarly writings, but experienced practitioners knew the accuracy of these models was derisory. As long as these models ignored the problem of air resistance, their parabolic trajectory represented, at best, an upper limit on the distance of a cannonball's flight. Instead, these craft-oriented gunners emphasized the practical recipes by which they minimized the uncertainties attendant on the firing of a cannon. At the core of their art was the unspecifiable, tacit skill encapsulated in the phrase "a gunner's eye" *(un coup d'oeil)*. Gunners acquired this "eye" on the practice polygon by repeatedly drilling themselves to work in concert with the company of cannoneers, who helped them position and aim the big guns.[21]

According to Enlightenment engineers, however, the art of artillery required formal study in the classroom. In 1720, the artillery founded the first schools in Europe dedicated to providing students with a scientific education, the type of training that today defines professional engineering around the world.[22] Though never as famous as the École du Génie at Mézières, the contemporary artillery schools were its equal in providing the most up-to-date technical education—and they turned out far more engineers. Nearly a thousand artillery engineers served at any one time between 1763 and 1789, as opposed to three hundred in the Génie. The state considered this training of great importance; by the end of the ancien régime, education expenses came to one-eighth of the artillery's total budget.[23]

School training included both practice and theory. Weather permitting, officers spent three days a week in the *école de pratique,* which included maneuvers, siegecraft, and gunnery. Other times, officers attended *école de théorie,* taught by a mathematics professor. Instruction included the science of fortification and sieges, plus various branches of mathematics, such as algebra, geometry, trigonometry, planimetry, and mechanics. All officers of captain's rank and below attended class. Every six months, the professor of mathematics tested students in the presence of superior officers.[24]

The purpose of this education was not simply to teach particular cognitive

21. For apprenticeship training, see Le Pelletier, *Famille d'artilleurs,* 27, 38–39. For a savant who advocated Galilean ballistics, see Blondel, *L'art de jetter les bombes;* and for the irrelevance of Galilean mechanics, see Hall, *Ballistics.* On the rules of practice and the gunner's "eye," see Saint-Rémy, *Mémoires,* 2:250–51; 3:154–55.

22. Earlier temporary schools had been established in all the services, including the artillery (1679); see Le Puillon de Boblaye, *Écoles d'artillerie,* 35–36; Hahn, "L'enseignement scientifique." The schools of the Ponts et Chaussées (1747) and the Génie (1748) soon followed; see Picon, *Ingénieur moderne;* Blanchard, *Ingénieurs du "roy."* For the parallel case of the Piedmontese artillery schools, see Ferrone, "Élites de la Maison de Savoie." Apprenticeship training of engineers persisted in the French civilian sector throughout the nineteenth century and in Anglo-American lands until the twentieth century; see Lundgreen, "Engineering Education."

23. "Dépenses annuelles de l'artillerie," [1788], MR1739, Service Historique de l'Armée de Terre, Vincennes, France (hereafter, Service Historique).

24. Saint-Rémy, *Mémoires,* 1:54–72; Le Puillon de Boblaye, *Écoles d'artillerie,* 53, 56–61.

skills, but to inculcate a particular social role. The artillery service asserted that this dual track of practical training and theoretical knowledge was central to the identity of the engineer. The engineer was enjoined to avoid the behavior of both artisan and savant. Not only was each mode of understanding considered inadequate, each was condemned as the sign of a defective social type.

> There are officers who devote themselves entirely to mechanical details; others regard such details as beneath their notice. Both types are deficient. The latter must be made to realize that [knowledge of] mechanics is an absolute necessity; the officer should know the language of the workman so as to make the workman understand, and on occasion to instruct him. On the other hand, those absorbed only in mechanical details must know that a knowledge of them alone, without a wider view, does not raise them above the level of a cannon founder, a powder-maker or a workman. . . . The artillery officer who knows his profession must not be ignorant of details, but he must know them, as an architect must know more than a mere stone mason.[25]

In an influential series of engineering texts published in the 1720s and 1730s and used throughout the century, artillery professor Bernard Forest de Bélidor steered young engineers through this social epistemology. It might be true, he admitted, that only practical experience brought one in contact with new phenomena, but without theory to discipline one's attention, one would not appreciate their novelty. In this regard, he noted how artisans blindly refused to adopt new ways of doing things. The degrading aspect of manual labor was less implicated here than the collusion of the artisanal corporations as they pursued their venal interests.[26] Rather than collude, engineers would vie in meritocratic competition. Vallière insisted that he would promote only those officers possessing ability consummate with their post. This requirement was emphatically underlined by Gribeauval. In the margin of a 1762 report, he wrote: "[S]eniority must be reduced to next to nothing, favoritism abolished, and superior talents given every reward, and these initiated before the age when the body begins to lose [its force] and the mind ceases to acquire [knowledge]."[27]

This competition among engineers would encourage innovation, even as it molded them to fit into narrow functional roles. The edict founding the new schools in 1720 assured young engineers that competition for personal preferment would enhance the corporate mission of the service.

25. *Ordonnance* of 1720, in Artz, *Technical Education*, 96–97.
26. Bélidor, *Architecture hydraulique,* 1:ii; Bélidor, *Science des ingénieurs,* 1:2.
27. Vallière, *Ordonnance,* 5 July 1729; marginal note by Gribeauval on Dubois, "Mémoire sur l'artillerie," [1762], 1a2/2, Service Historique.

Those who have ambition (and all should have it) will not content themselves with what they have seen and heard in school, but will study at home and in private lessons. And by deep thought and application, they will surpass the instruction given. The progress of their studies will every day encourage them, every day they will acquire new insights and they will thereby achieve the highest merit of their profession, which must be the unique goal of every officer.[28]

At the same time, however, Bélidor insisted that engineers must not engage in just any sort of study. He insisted that they avoid meaningless speculation, like those snobbish savants who despised practical problems and piled on endless calculations to no purpose. Once again, this cognitive inadequacy was the sign of a corresponding social failing. Artillery commanders condemned those solitary "metaphysicians" who studied science in the silence of their *cabinet*. The problem here was not simply that the savant could not master the tangible tools of the material world, but that he was asocial.[29] Unlike the savant, the engineer would acquire a detailed knowledge of tools and trades, even if he would never have to perform any physical manipulations himself. Indeed, an engineer needed to understand the theory of an artisan's craft "better" than the artisan himself in order to command him and keep an eye on his avaricious dealings. That way, despite their practical engagement with the world of production, engineers would eschew their personal interests and work cooperatively to advance the best interests of their employer, the state. This was what Bélidor had in mind when he said in 1737: "Whatever class one is born into; one must serve to the best of one's ability; that is what it means to be a good citizen."[30]

In short, the social epistemology of the Enlightenment engineers attempted to match individual achievements (merit) to a corporate ethos (state service) so as to realize that elusive oxymoron: institutionalized innovation. But what exactly was this marriage of theory and practice?

Mathematics as an Impersonal Standard

French artillery engineers received a broad technical training. But to outside eyes, it was the heavy mathematical education, more than anything else, that distinguished the school engineer from the craftworker or apprenticeship-trained mechanic. Skeptics (both then and since) have implied that for those

28. "Instruction que son Altesse Royale," 23 June 1720, in Saint-Rémy, *Mémoires d'artillerie*, 1:71–72.

29. Bélidor, *Science des ingénieurs*, 1:3–4; Bélidor, *Oeuvres*, xx; Bélidor, *Architecture hydraulique*, iv; [Du Puget], "Travail de l'officier," 1765, 2b5/1, Service Historique.

30. Bélidor, *Architecture hydraulique*, ii. For a similar point of view, see Frézier, *Coupe des pierres*, 1:i–vii.

engaged in practical tasks, mathematics served as mere ornament, a learned flourish to confer status on its practitioners. For instance, we should hardly be surprised to learn that, starting in the eighteenth century, mathematics acted as a marker of merit among gentleman scholars at Cambridge University.[31] But French engineers were not gentlemen scholars; they were men of action. Of what practical use was mathematics to them? David Bien has argued that the mathematics taught to young cadets at the École Militaire was intended as a form of discipline, not as useful knowledge. The traditional subjects that had provided a rigorous education since the Renaissance—Latin and rhetoric— had been denounced since the seventeenth century for inculcating skill in equivocation and sophistry, the very values that the new commanders ab-horred. Bien suggests that Enlightenment military leaders saw in mathematical rigor a tool to instill the rather different martial value of subordination.[32]

Mathematics served this function for the artillerists as well. Since Vallière's time in the 1720s, the artillery had used mathematics to provide a uniform course of study. This regimentation was meant to provide a standard education for aspirants. Artillery students, as we will see, came from a variety of social backgrounds. Upon graduation they were rotated through garrisons scattered across the kingdom. Only a uniform curriculum could provide a numerous and far-flung corps with a "common language" to discuss their work. Even small deviations from the curriculum were frowned upon. The artillery leader-ship chastised faculty who taught idiosyncratic methods of problem solving. The intention was to impose a uniformity of habit and thought, instilling a solidarity that was the technicians' equivalent of esprit de corps. Mathematics was particularly well suited for this role because it impressed on students the virtues of uniformity and precision. These were the virtues that made agree-ment matter.[33]

In a familiar paradox, this education also offered the artillery leadership a ready and impersonal scale for selecting and ranking pupils. Large numbers of nobles were seeking offices in the army. Mathematical testing offered a mecha-nism to admit some, without making invidious distinctions. In this sense, mathematics served to "democratize" the artillery service because it reduced the visible play of patronage in admission. And once the cadets were in school, gauging their achievement against a uniform curriculum fostered emulation and zeal. Recalling his arrival as a pupil, one artillerist evoked the almost comi-cal competition among the first entering class, as pupils covered the school's

31. Gascoigne, "Mathematics and Meritocracy," 547–84.
32. Bien, "Military Education"; Skinner, *Reason and Rhetoric.*
33. "École d'artillerie de Metz," October 1767, and letter to Gribeauval, 20 September 1788, 2a59, Service Historique.

doors, shutter, and tin plates with mathematical propositions.[34] Over the course of the eighteenth century, evaluation of students increasingly depended on academic prowess. In the first half of the century, examiners had still taken account of the cadet's moral qualities as well as whether he pursued his studies with "distinction," "did his best," or "did little." For instance, the mathematics examiner Camus passed Freard du Castel "with indulgence" even though his mathematical abilities were pitiful because his commanders had lauded his exemplary conduct. After the Gribeauvalists took control of the artillery, however, evaluations were increasingly tied to performance on exams. A 1761 law expressly prevented the military director of the school from influencing the mathematics examiner to pass weaker students. In the late 1760s, the new examiner Etienne Bézout began grouping graduates according to mathematical ability. Revealingly, a tie between two young men for first rank in the class of 1768 — Tardy, a noble-born officer, and Gassendi, a commoner—was broken in favor of Gassendi on the basis of "birth" because he was a distant descendant of the famous natural philosopher. By the 1780s, the examiner Pierre-Simon Laplace simply ranked the graduating class in numerical order from top to bottom. So that while a standardized education dissolved old social distinctions and bound young artillery engineers together, it simultaneously made them subject to new instruments of differentiation, and hence, social control.[35]

Michel Foucault, in *Discipline and Punish*, writes about those institutions of social control founded during the eighteenth century to mold the bodies and minds of "inmates." Foucault cites the Gobelins tapestry school as an example of how the pedagogues of the late ancien régime regimented the unruly children of artisans with regular drawing lessons. From the formal arrangements of the Gobelins school, Foucault deduces the discipline built into the organization of space (desk position), time (school periods), and supervision (the logbook). While studying under the schoolmaster's gaze, the students were to internalize the markers of merit and so acquire self-discipline. But a school is not simply an assemblage of ranked desks and methods of coercion, a mere conduit for miscellany; a school is also a curriculum. We also need to pay attention to the disciplinary power of particular subjects if we wish to understand the links that institutions forged between knowledge and social

34. "Inspection de 1721," Xd260, Service Historique; Le Pelletier, *Famille d'artilleurs*, 27, 38–39. Whatever else it was, mathematics was not, in this period, a covert stand-in for a high financial barrier; cadets were paid to study, housed at government expense, and the entry-level geometry could be acquired in any decent *collège*.

35. *Ordonnance*, 21 December 1761, in Le Puillon de Boblaye, *Écoles d'artillerie*, 56–57. For evaluations at various periods, see the following documents in the archives of the Service Historique: "Examen des officiers, Besançon," January 1748, 2a60; Camus, "Mémoire," 3 March 1762, Xd248; Bézout, "État des élèves," June 1768, Xd249; and Laplace to [?], 18 August 1789, Xd249.

role. For instance, both elite engineering students and artisanal pupils sat in ranked desks and imbibed a uniform curriculum. But the two groups studied quite different subjects. While artisans learned a variety of freehand drawing skills and practical procedures of construction, engineers learned a mathematized form of mechanical drawing and to calculate the forces of machines. The one was clearly being prepared to command the other. Schools do not simply provide a discipline *quelconque;* they discipline students for particular roles.

Mathematics, after all, did not simply regulate the internal affairs of the artillery engineers, it also mediated their relationship with the outside world—both with other social groups and with the material objects they were expected to master. The artillery leadership felt that the study of mathematics insulated its young officers from the corrupting world of dilettantism they associated with the study of rhetoric and belles lettres in favor at court.[36] This education also equipped artillery officers with an authority they could wield over subordinates. Uniformity in the execution of orders was seen as a prerequisite for military effectiveness. As noted above, there was a prevailing disgust among army reformers with the way junior officers of high noble status quibbled with the commands of their military superiors. One artillery general justified a mathematical education for officers principally as a means of maintaining a clear and effective chain of command. In other words, authority in a hierarchical social organization flows from clear and consistent rules for members at all levels. For subordinates to be properly disciplined, commanders must learn self-discipline, which they acquire by adherence to a common regime of knowledge.[37]

Mathematics as a Practical Tool for Gunners

Clearly, mathematics proved an alluring measure of merit because it allowed judgments about students to be made without discrimination (objectivity as impersonality). But one must still ask whether this mathematical education actually enabled artillery engineers to perform their tasks any better (objectivity as functionality). Bien is no doubt right in asserting that the theorems of their school days were of little practical use to officers who had trained in the École Militaire and who then went on to a career in the regular line army. Can the same be said of the far heavier dose of mathematics served up to officers in such technical services as the artillery? Did mathematics produce better gun-

---

36. Guibert, "Nouveau cours," 25 November 1761, 2b5/1, Service Historique. See Bien, "Military Education."

37. "Instruction du Corps de l'Artillerie," 19 thermidor, year IX [7 August 1801], 2a59, Service Historique.

ners? Or better designers of guns?[38] The artillery leadership knew perfectly well that a math exam did not test all the complex skills and virtues an officer needed to carry out his duties. Indeed, they warned that a narrow focus on mathematics could be dangerous for students. "Mathematics," one noted, "are all too alluring for those who have passed the first hurdles; so we must see that officers destined to fill these very important functions do not acquire too much of a taste for this science." Even the mathematics examiners admitted that memorizing formulae was less important than teaching students to use mathematics to appreciate the relative importance of facts. Such mathematics as was taught them was therefore intended to train the judgment of young officers and orient their approach to practical problem solving.[39]

The issue for engineers, then, was not mathematics per se, but what manner of mathematics. In particular, the geometry lessons that had been central to the Vallièrist curriculum were now derided as ostentatious and useless. In a broad critique of artillery education at the end of the Seven Years' War, Guibert complained that even after poring over Camus's official four-volume course, young artillery officers were still unable to shoot a cannon with accuracy. Not having been commissioned by the artillery, the text made no reference to artillery problems (unlike the old text of Bélidor). Now that Camus had promised additional volumes of his textbook, Guibert quipped, perhaps by volume thirteen he would have made astronomers of them. There was a real danger in submitting cadets (some of them adolescents) to this sterile study; it risked turning them into "metaphysicians" who would "despise the thousand necessary tasks which require less wisdom than *pratique*."[40]

Gribeauval agreed with his assessment. He strengthened the practical side of the curriculum. But Gribeauval did not toss out the mathematics professors. Instead he reformed the theoretical portion of the course. For the first time, the advanced group studied the infinitesimal calculus, as well as rational mechanics, hydraulics, the principles of machines, plus some physics and chemistry. The goal was a curriculum that replaced the "pure ostentation" of geometry with "a theory that applied to practice." After heated discussion, a four-volume textbook by Bézout was adapted for the artillery schools in the early 1770s. Prominent artillerists supplied ballistics questions, and the result-

38. "Rightly or wrongly," Bien notes with agnosticism, "it was assumed that aiming and firing cannon on land, as well as at sea, required a knowledge of geometry" (Bien, "Military Education," 53).

39. "Circulaire à MM. de Mouy, de Gribeauval," November 1765, 2b5/1, and Gomer, "Marche de l'instruction," 28 February 1783, 2a60/4, Service Historique.

40. Guibert, "Nouveau cours," 25 November 1761, also Bron, "Mémoires sur l'instruction," May 1766, 2b5/1, Service Historique.

ing mix of theoretical discussion and practical problem sets has since become the hallmark of the engineering textbook. Bézout's text was used at the École Polytechnique well into the nineteenth century.[41]

This engineering science—generally referred to as "mixed mathematics" in the eighteenth century—was based on a new *analytic* rational mechanics increasingly conveyed in algebraic notation. Even the engineers' new methods of technical drawing based on the descriptive geometry were tied to this type of analysis.[42] According to its advocates, mixed mathematics offered a method of research (for the design of new technology), a guide to practice in the field (for the operation of technology), and a program of pedagogy (for the shaping of technologists). Let us see whether mixed mathematics afforded a superior method of designing cannon, aiming cannon, and training cannoneers.

There were several (interrelated) reasons for the artillerists' turn toward mixed mathematics. First, there was the preeminent status of analysis among those savants perched atop the Enlightenment pyramid of learning. The Continental savants of the eighteenth century (the Bernoullis, d'Alembert, Euler, Lagrange, Laplace) created a new "foundational" rational mechanics. The relationship between these analysts' academic clout and the artillery engineers was symbiotic. Gribeauval appointed Laplace to the post of artillery examiner, his first important step up the ladder of state patronage. And Laplace, in his turn, used his position to launch his career as a scientific patron in his own right, which included using his position as gatekeeper to the artillery service to impose a curriculum based on analytic mathematics on the École Polytechnique.[43]

Second, analysis was a linguistic mode of mathematics in a period that took language as constitutive of thought. For Condillac, the identification was entire: to write an equation was to think rationally. The Gribeauvalists admitted that geometric synthesis had the advantage of presenting new subjects to pupils in a "palpable manner" *(manière sensible),* but analysis was superior for those "in the situation of needing to operate [mathematically]" because it "eased all problems of calculation" and "quickened and facilitated the discovery of the unknown." And for the engineer Lazare Carnot, the method did more than clarify thought, it acted *upon* thought, leading "infallibly to new truths." Analysis—as Antoine Picon has recently stressed—was a language of

41. "Circulaire à MM. de Mouy, Gribeauval," [Du Puget], "Travail de l'officier," 1765, 2b5/1, Service Historique; Bézout, *Cours de mathématiques.*

42. Monge, *Géometrie;* Daston, "Physicalist Tradition."

43. Brockliss, *French Higher Education,* 337–90; Duveen and Hahn, "Laplace's Succession," 423; Langins, "Sur l'enseignement et les examens."

technical innovators. And this was just the sort of mental orientation the Gri-beauvalists wished to foster.[44]

Third—and most substantive—analysis was well-suited for the representation of variation. This was particularly the case where there was no good "first principle" physical model of the phenomena in question. Analysis thus became a valuable tool in the engineers' new method of empirical investigations, a method of engineering research that Walter Vincenti has termed "variation of the parameters."[45]

What was this method? Historians have usually ascribed the eighteenth-century assault on technological problems to a recognition that Newtonian mechanics offered a new research program. But one may agree that Newton's tools prompted later investigators to tackle technological questions, and still note how often success eluded them. Time and again, engineers found that Newtonian mechanics did not lead directly to improved guns, waterwheels, boat hulls, or any of the other lumpy artifacts of material life. This is not to say that mathematics was without value to engineers. On the contrary, the new analytical mathematics helped the engineers model those technologies whose behavior could not be derived from first principles. For instance, the French engineers who created the science of machines (including waterwheels) in the years 1765–1830 deployed mathematics as a form of "descriptionism," a way to quantify the relationship between certain measurable parameters and other measurable parameters.[46] This was also the case for the new engineering science of ballistics.

In the seventeenth century, Galileo had been able to derive his parabolic trajectory for the cannonball from first principles by assuming that its motion took place in a vacuum. But as the Englishman Robins pointed out in the 1740s, in most cases air resistance significantly deflected the ball from that trajectory. Indeed, the effect of air resistance was so significant that any tests that compared the performance of different gun designs by measuring their relative range could be subject to widely different interpretations. As Robins explained,

44. On Condillac, see Rider, "Measure of Ideas," 113–40. Quotes from "Circulaire à MM. de Mouy, Gribeauval"; Carnot, "Calcul infinitésimal," 1797, in *Révolution et mathématique*, 2:457–58; Picon, *Ingénieur moderne*, 302–7, 622. See also Bos, "Mathematics and Rational Mechanics," 327–55.

45. Vincenti, *What Engineers Know*, 138–39, 160–62.

46. More properly, the rational mechanics labeled "Newtonian" was a creation of the eighteenth-century Continental analysts (Greenberg, "Mathematical Physics"; Truesdell, *History of Mechanics*, 85–183). In a sense, engineering descriptionism was in line with Newton's own insistence that scientific claims be empirically verifiable at every level of formulation (d'Alembert, *Fluides*, x, xxviii–xxix, xli–xlii; Heilbron, *Weighing Imponderables*, 29–31, 141–46). On waterwheels, see Reynolds, *Stronger than a Thousand Men*.

that was why the leaders of the various European artillery services used such tests: it allowed them to justify whatever design they chose.[47]

In France, tests of cannon had also been conducted in front of royal and aristocratic military patrons—and judged according to the cannon's range. These impressive displays were intended to reinforce the witnesses' sense of the cannon's raw power. In fact, they permitted diverse interpretations that were thereby subject to extratechnical (and military) authority. Guibert picked up on Robins's critique of such tests when he noted how often ballistics experiments were conducted in such a way that "the result was known in advance." This occurred because the hierarchy of rank or the interests of "party" could not be set aside. The attachments of military society undermined the impartiality necessary to be a good witness, "either because the authority of the presiding officers carries [the day] and covers up all [contrary] opinions, or because each observer comes with his own prejudice rather than his judgment, and the opinion that he wishes to maintain, rather than the impartiality that waits to see before judging." For this reason, the interpretation of tests on the practice ground always "conformed to the dominant opinion."[48]

To counter prejudicial opinion, Robins had proposed that a new device, a ballistics pendulum, be used to measure a parameter of the gun that did not depend on the air resistance: the cannonball's initial velocity. When given analytical form by Euler, this new engineering ballistics cleared the way for a more objective evaluation of a gun's capability.[49] That is, it promised to produce knowledge about variations in gun design that could not be so diversely interpreted—at least when interpreted by expert engineer-witnesses versed in impact theory and mixed mathematics.[50] Now, witnesses could proceed to make judgments about good cannon design without regard to social rank.

Most important, the objective knowledge produced by this new form of analysis made it possible for the engineers to assert their commitment to the goal of optimization that lay at the heart of their new professional ethos. Already in the 1730s, Bélidor was warning that geometry might teach one how to raise solid buildings, but one would needlessly waste materials until one learned the language of algebra and rational mechanics. With calculus, engineers could now subject their equations to "realistic" constraints, and designs could, in principle, be examined for moments of greatest strength or weakness. After 1770, Bézout's textbook taught students how to use Robins's methods to

---

47. Robins, "Practical Maxims," in *Mathematical Tracts*, 1:269–71.
48. Guibert, *Essai général tactique*, 1:447–49.
49. Euler, "Recherches sure la véritable courbe."
50. On the successes and limits of the new ballistics engineering, see Alder, *Engineering the Revolution*, 87–125; Steele, "Musket and Pendulum."

calculate such things as the maximum range of a gun and its greatest muzzle velocity. In principle, this allowed the engineers to assert that, if they were given the right to make their own technical judgments, they would be able to best serve the state by designing superior cannon.[51]

In sum, the new analytic mixed mathematics did more than provide a ready tool to rank and discipline candidates or socially distinguish artillery engineers from rude cannoneers. Any kind of mathematics—geometry, say—would have done that. But the geometric methods of the seventeenth century had expressed the limits of theory, and hence the limits of destruction. Geometry was associated with the static and bounded forms of siege warfare advocated by Vauban, and with the idealized Galilean description of the trajectory of a cannonball through a vacuum (literally, the upper limit on a cannonball's flight). By contrast, the new analytical mixed mathematics was open-ended and explosive. It was associated with the dynamic field tactics advocated by reformers, and with the maximizing methods of the new engineering ballistics. A mastery of this form of mathematical analysis associated the engineer with research, innovation, and a dynamic mode of thought. At the same time, it firmly tied the creators of this technological novelty to their duty to their employer: the state.

But putting the matter this way might seem to suggest that the Gribeauvalists' use of mixed mathematics led unambiguously to demonstrably superior results, and that their program was in that practical, functional sense meritorious. The situation, however, is not nearly so simple. Of course, there is no gainsaying the success of the revolutionary and Napoleonic armies—from Valmy to Austerlitz—nor that the mobility of the Gribeauvalist cannon contributed, in part, to these victories. And certainly by the middle of the nineteenth century, the engineering approach to ballistics would help artillerists design superior weapons and fire them with much greater accuracy (once advances in metallurgy had also made rifling and breech-loading practicable).[52] But it is important not to exaggerate the contribution of the new mathematical sciences to the reforms of the eighteenth-century artillery. In the 1760s, the Gribeauvalists still found it necessary to validate the advantages of their new mobile cannon by demonstrating their range in front of illustrious patrons. And even after the French adopted Robins's and Euler's model of ballistics in

51. Bélidor, *Science des ingénieurs*, 1:3–4. On optimization in the theory of machines, see Séris, *Machine et communication*, 283–341, 456–57; Bézout, *Cours de mathématics*, vol. 4.

52. For a corrective to the exaggerated claims regarding the impact of the artillery in the revolutionary and Napoleonic wars, see the most comprehensive work on the subject, Lauerma, *Artillerie de campagne française*. Rifling was known to improve the accuracy of guns long before Robins explained why it worked; see Hall, *Ballistics*.

the 1780s, they did not conduct a scientific investigation of cannon design using a ballistics pendulum. Such a program would have been extremely costly, and would have disrupted the ongoing production of guns and the training of gunners. The French state did not assemble a ballistic pendulum until the Restoration. True, in the late 1780s, artillery professor Jean-Louis Lombard (Napoleon's favorite professor) drew up new shooting tables partially based on Robins's methods, and these tables were used for the next twenty years by revolutionary and Napoleonic gunners. But empirical studies make clear that the accuracy of cannon fire did not noticeably increase during this period. The problem was that gunners in the field were still obliged to use the range of a shot to assess their aim because the initial velocity was not a parameter available on the battlefield. Gunners would then look in Lombard's table and adjust Gribeauval's new tangent sight and elevating screw. These two devices enabled the artillery officer to direct the vertical aim of the gun quantitatively and do so without any assistance from cannoneers. Hence, rather than possessing a *coup d'oeil* for the battlefield, expert judgment consisted of what Lombard called a *coup d'oeil* for the mathematics. But it did not improve the accuracy of fire, as Lombard himself acknowledged when he noted that even if the new tables could not improve a gunner's aim, "at least he will have a more exact notion regarding the most important object of his service."[53]

The functional contribution of mixed mathematics to the practical success of the late-eighteenth-century artillery engineers, therefore, was largely illusory. Its real contribution was social and pedagogical. No longer was an artillery officer to conceive of himself as an artisan working in concert with his cannoneer assistants, aiming his cannon with a practiced eye. Instead, his mastery of engineering science encouraged the artillerist to conceive of himself as an expert whose knowledge—knowledge that could only be corroborated by his peers—allowed him to manage the menial tasks of his cannoneer subordinates.

In other words, the artillerists' new sort of mathematical education highlighted the corps' status as the expert designers and managers of battlefield firepower. This education served primarily a social function, one located in the sphere of values, albeit a set of values dependent on the content of that curriculum. The engineers' cognitive training—the particular way that mixed mathematics merged theory and practice—carried particular moral lessons, positioning students for their managerial role, and marking them as worthy of

---

53. On the inaccuracy of Napoleonic guns, see Hughes, *Firepower*, 126, 167–68. On Robins in France, see Lombard, *Nouveaux principes,* and *Tables,* 30–31, vi. On the tangent sight and elevating screw, see Du Coudray, *Artillerie nouvelle,* 45–50.

autonomy in the eyes of both subordinates and patrons. Not only did their mastery of this sort of knowledge ensure that engineers would appear meritorious to one another and to outsiders (because mathematics could be judged by impersonal criteria), their mathematical skills also ensured that their judgments about meritorious engineering would be more objective (in the sense of commanding greater agreement among knowledgeable experts irrespective of their prior personal views).

But, as the engineers themselves recognized, judgments about engineering ability could not be reduced to judgments about mathematical ability, any more than judgments about how to best design or aim a cannon could be reduced to a mathematical calculus. Hence, the engineers arrogated the sole right to pass judgment on good cannon design, even as they arrogated the sole right to pass judgment on the abilities of their fellow engineers. In this sense, the kinds of rules with which engineers treated the material world of cannon were inextricably bound to the rules by which engineers treated one another.

Filling Out the Ranks

By adopting this cognitive and pedagogical program, the engineers challenged some of the most basic tenets of the ancien régime. In particular, the way they found an institutional framework to reconcile their marriage of elite status and technical competence set the pattern for the postrevolutionary social order.

This is the context in which we must understand one of the most seemingly radical innovations of the Gribeauvalist party: their method of promoting officers on the basis of a kind of election called "co-optation." The procedure worked as follows: When a vacancy appeared in one of the upper levels of the corps (lieutenant colonel, colonel, etc.), the members of that rank got together to nominate three candidates from the rank below "exclusively on the basis of merit." The final choice was then made by the minister of war on the recommendation of the first inspector general. For the rank of captain and below, Gribeauval instituted a hybrid system of seniority and co-optation, with three-fifths of first captains elected and two-fifths advanced by seniority.[54] In certain respects, this scheme resembles the hybrid system of promotion adopted by the revolutionary armies of the year II (1793–94) to balance seniority with talent, democracy with subordination. Historians of the Revolution have universally ascribed these innovations to the new ideals of 1789; in fact, certain elements can be found in the ancien régime artillery. But unlike the

---

54. *Ordonnance de l'Artillerie*, 3 October 1774, 15. This amplified the law of 1763.

revolutionary armies of the year II, the artillery never countenanced a bottom-up system in which soldiers elected their leaders. Even the army had only allowed such democratic practices in limited circumstances, and only temporarily.[55] Rather, the artillery pioneered a professional structure in which officers co-opted subordinates worthy to join their ranks. The artillery corps thereby sought to insulate themselves against powerful outside interests, particularly against those without sufficient knowledge to discriminate among the meritorious. Such ignorant outsiders included both those who hoped to impose patronage appointments from above and those who advocated a populist leveling of expertise from below. In the end, merit in the professions was a title conferred by fellow professionals. That is what made the professional meritocracy a peculiar kind of panopticon.[56]

More generally, the artillery did not emulate the usual pattern of officer recruitment in the ancien régime. One noteworthy feature of the service was the almost complete absence of court nobles. These *grands* engendered great resentment in the line army because of their monopoly on the highest commands. Instead, the artillery recruited its cadets from well-to-do commoners, the sons of its own corps, and recent *anoblis*. Even though only 14 percent of artillerists entering after 1763 were commoners, this was three times as many as in the regular army. The next largest group (23 percent) came from families with a history of artillery service. This intensive "autorecruitment" might seem incompatible with the service's ethic of meritocracy. But this policy was legitimated by the sensationalist psychology of the period, which held that one's upbringing determined one's abilities, and hence that only the sons of military families could acquire martial virtues. This enabled the artillery service to reconcile the two long-standing faces of aristocratic identity—its sense of kinship and its ideal of service—under the Enlightenment sign of talent. The largest group of cadets (63 percent) came from noble families. Their fathers did not belong to the traditional *noblesse d'épée*, however, but to the state's financial and juridical apparatus. Between 1763 and 1781, nearly 7 percent of artillerists had a father employed on the civilian side of military procurements, and 39 percent had a father who worked in a judicial, financial, or municipal office. These are the main groups of "parvenus" against which the notorious Ségur law was directed, yet even after 1781 they together accounted for 34 percent of

---

55. Bertaud, *Army of the French Revolution*, 88–89, 171–75.

56. In several respects, this system resembles the tenure and promotion methods still used in academia. This system is not exclusively a feature of "modern" organizations. Insider co-optation also enabled the artisanal corporations of the ancien régime to preserve standards and control access to their trade (Durkheim, *Professional Ethics*). However, the professions have a distinctly competitive mode of social promotion (Haskell, "New Aristocracy").

new artillery lieutenants. Hence, in the years between the Seven Years' War and the Revolution, the artillery recruited more than half its officers from the stratum of society Bergeron and Chaussinand-Nogaret identify as the backbone of the nineteenth-century notability.[57]

Not only did the artillery tap a wider base of recruitment, the Gribeauvalists also extended their social hierarchy both further up and further down the "chain of service," tying "each grade of officer to their talents and their accomplishments."[58] This process can be seen in the establishment of two controversial new ranks. The first was the *garçon-major*, an officer given the rank of third lieutenant, filled by worthy NCOs. While non-noble "soldiers of fortune" had occasionally served as officers in the general army, the fact that the artillery had now formally institutionalized their presence shocked contemporaries. The Vallièrists objected bitterly to admitting into the officer corps these men "without birth, education, talent, knowledge of mathematics, and often without principles of moral and [good] conduct." Worst of all, complained one Vallièrist, these parvenus would believe that they had risen on the basis of merit! As he put it, "a sergeant will always be more docile and obedient than this new officer, who, believing himself promoted for his merits, will have [all the more] pretensions." In short, severing the connection between birth and rank threatened to disrupt the social order that guaranteed effective command.[59]

The Gribeauvalists scolded the Vallièrists for this sort of caste snobbery. Shouldn't access to the officer corps be open to all estates, and merit alone guide promotion? This innovation, they affirmed, was justified on several grounds. First, it inspired NCOs with zeal for the service. Second, it introduced more battlefield experience into the lower ranks. And third, it freed the higher ranks to devote their time to the theory of their profession. In other words, this reform represented a further shift to a more carefully graded chain of command, based on function and education. Prospects for *garçon-majors* remained limited, and a formidable barrier still lay between them and the officer cadets.[60]

This carefully constructed stratification can be seen in the sort of training

57. *Registres,* 1763–89, Yb668–70, Service Historique. In these registers, the scribe listed the occupations of half the fathers. Only 4 percent of artillerists, as opposed to 14 percent in the line army, had been presented at court or had knightly antecedents (Bodinier, *Officiers combattants,* 97–99, 104). Even after the Ségur law of 1781 ostensibly ended the ability of commoners and recent nobles to enter the army, the number of commoners entering the artillery remained at 5 percent because the law still permitted entry to anyone with kin in the corps. For the line army, see Bien, "Réaction aristocratique," 30–35; Bergeron and Chaussinand-Nogaret, *Masses de granit,* 33–36, 63–64. An analogous pattern of recruitment can be found in the Corps du Génie; see Blanchard, *Ingénieurs du "roy,"* 229–46.

58. Gomer, "Constitution du Corps Royal," 1761, 2a60/4, Service Historique.

59. Saint-Auban, *Systèmes d'artillerie,* 101–2.

60. *Ordonnance de l'Artillerie,* 3 October 1774; Scheel, *Mémoires d'artillerie,* 439.

given to these *garçon-majors*. Granting formal instruction to NCOs was another shocking innovation of the Gribeauvalists. But the Gribeauvalists were
careful to give their sergeants only as much "theoretical principles" as was
consonant with their function: "But we must guard against pushing them too
far in this direction, or we risk making them into mathematicians, which would
be a great pity because they would then despise their jobs *(functions)* as sergeants. We must therefore limit [their education] to that which is strictly
necessary for them."[61] In their schools, NCOs learned basic writing, arithmetic, and "practical geometry" so that they could "translate" the plans of
their superiors to their subordinates. They did not learn mixed mathematics.
Their cognitive abilities were to be pegged to their social role. Matching the
table of ranks to a hierarchy of knowledge meant supplying education appropriate to their station.

This same concern with rank, role, and knowledge can be seen at the high
end of the hierarchy, where the Gribeauvalists created the position of *chef de
brigade,* an elite officer, chosen exclusively on the basis of merit and given
command of an autonomous artillery unit that could follow an army division
through all its maneuvers. In the person of the *chef de brigade,* the peacetime
and wartime operations of the artillery were finally fused, making possible the
sorts of mobile tactics with which the French revolutionaries went to war.[62]
The artillery hierarchy now stretched from the lowest cannoneer to the first
inspector general. It would ultimately extend to the chief artillerist of them all,
Napoleon Bonaparte.

Promoted on the basis of criteria judged by their immediate superiors and
attached to their common tasks by years of difficult study, the artillerists were
far removed from royal largesse. Their loyalty was therefore directed toward
state institutions, rather than toward the person of the king. As a result, they
proved far more likely than their peers in the infantry and cavalry to remain in
French service through the upheaval of the Revolution. Even during the Terror,
artillery officers were four times as likely to remain in state service (20 percent
versus 5 percent). Moreover, the majority of those who did remain were officers of noble birth and the *garçon-majors* they had promoted to the officer
corps before 1789. It was not simply a rump bourgeois artillery composed of
men born commoners. In short, the ancien régime artillery was already a professional artillery.[63]

61. "Circulaire à MM. Mouy, Gribeauval," 2b5/1, Service Historique.
62. Gribeauval, "Chef de brigade," [1765], 1a2/2, Service Historique.
63. *Registres,* [1795], AF II 293c, and *État par ordre d'ancienneté,* nivôse, year V [Decem-
ber 1796], AF III 153, Archives Nationales, Paris. For full statistics, see Alder, *Engineering the Revolution,* 83–85.

Rank and Revolution

"Hierarchy" has often been employed by social historians as a synonym for social stratification of any kind.[64] But the "orders" of the ancien régime were defined by legal entitlement, which might be acquired by birth or by purchase, but which, in either case, assured their holder of particular privileges and locked him (at least for the time being) into that estate. By contrast, in describing the engineering corps as a "hierarchy," I am using a word that ancien régime reformers refashioned to describe a stratified social order in which movement between rungs was theoretically possible. For centuries, the French word *hiérarchie* had been reserved exclusively for the rungs of celestial angels, and by extension, for the hieratic order of ecclesiastics. In the 1780s, however, an artillery captain privately used the word to describe the career ascent of the *garçon-major* up a ladder of ranks. "In the promotion of the sergeants and sergeant majors there exists as sort of hierarchy *(hiérarchie)*, a progression which at all time feeds their zeal, and [acts as] a lure to ambition. Before becoming a *garçon-major,* one must pass through all the grades."[65]

The sort of hierarchy referred to here implied a very different social order predicated upon a well-defined ladder in which advancement depended on individual achievement as recognized by one's superiors. This new use of the term only attracted widespread attention on the eve of the Revolution when Guibert entitled his 1788 law restructuring the army: "Ordinance Regulating the Duties of the Hierarchy of All Military Personnel." Guibert saw this new social structure as the solution to the populist ferment then brewing in France. Parisian wits apparently found his application of this term to a secular organization like the army quite amusing. The term, however, stuck. In 1791, the former minister Necker wrote of the *hiérarchie politique* that had been established by the National Assembly. Here, the new ladder of merit was explicitly coupled to the new sovereignty in which citizens were ruled by representatives that they themselves had elevated to high office.[66]

The juridical revolution of 1789 cleared the way for this engineering hierarchy to extend itself without obstruction. It also underlined the need for some "nonarbitrary" mechanism for selecting those worthy of receiving society's favors. In 1789, J.-G. Lacuée addressed this problem in the *Encyclopédie métho-*

---

64. For a classic use of the term, see Mousnier, *Hiérarchies sociales.*

65. "Remarques critiques," 1 February 1784, 2a60/4, Service Historique. The dictionaries all locate the first military usage in the post-Napoleonic era; see "Hiérarchie," *Le Robert.*

66. [Conseil de Guerre], *Ordonnance du roi portant règlement sur la hiérarchie de tous les emplois militaires,* 17 March 1788. On the mockery of the salons, see [Servan], "Hiérarchie." On Necker's usage, see "Hiérarchie," *Le Robert.*

dique. Lacuée was a career officer, a one-time aspirant to the artillery, and a future representative in the national legislature. In an article entitled "Exam," Lacuée asked the Assembly to "consecrate . . . this principle of the constitution: *All citizens are equally admissible to public dignities, positions and posts, without any other distinction than those of talent and virtues.*" But how were these "equally admissible" citizens to be chosen? Lacuée opposed training the new *citoyens-officiers* in public schools; such schools would inevitably monopolize admission to state offices. But military science was difficult, and its principles could not be easily acquired. Hence, Lacuée advocated exams, like those used by the artillery and fortification engineers, to choose capable men. He admitted he saw "no other way which was genuinely constitutional" or that "could eliminate the arbitrary in the selection of officers." [67]

Lacuée eventually came around to the idea of state-sponsored education. Indeed, as director of the École Polytechnique under Napoleon, he played a critical role in forging the foundational postrevolutionary links between a hierarchy of engineering knowledge and professional status. An 1802 regulation set out in excruciating detail both the theoretical knowledge and practical skills expected of each member of the artillery corps, from common soldier to lieutenant general. This regulation was a monument to the social structure of the new regime. Schooling was made appropriate for every level. Each rank was expected to master the theoretical knowledge—though not necessarily the practical skills—of all subordinate ranks. That way the activities of each level were subsumed as a particular application of the knowledge of the rank above. This allowed for specialization of tasks, while providing clear markers for a hierarchy of merit. [68]

Mathematics served to define this hierarchy, first because its public character made the choice of candidates appear fair, and second because the particular forms of mathematics the engineering schools adopted appeared to produce worthy results. But the former claim (merit as impersonality) depended on equal access to education, something generally available only to those with family money. This explains Napoleon's remark to Lacuée, "It is dangerous, where people who are not wealthy are concerned, to give them too great a knowledge of mathematics." [69] And on closer examination, the latter claim (merit as real ability) proved to be more of a promissory note than to be based

67. Lacuée, "Examen," *Art militaire,* 4:315–20.

68. Berthier (minister of war), "Règlement sur l'instruction des troupes d'artillerie," *Extrait des registres,* 3 thermidor, year XI [22 July 1803] (Paris: Imprimerie de la République, year XI [1803]), 3, 10, 2a59, Service Historique.

69. Napoleon to Lacuée, 2 germinal, year XIII [23 March 1805], in Bradley, "Scientific Education," 446.

on proven results. In the meantime, mathematical education oriented the thinking of young men so they would better serve the state. One artillery captain, looking back on his ten-year career in the revolutionary and Napoleonic wars, commented wistfully that his mathematical education had never helped him in his daily tasks. Yet he proudly asserted that the self-discipline he had acquired while mastering the equations had made him "a man," and had helped him to *un état*.[70]

Conclusion

The ideology of merit is so pervasive today that we find it almost impossible to disassociate a meritocracy that operates upon individuals from the superior merits of that society. Tocqueville implied as much: the egalitarian impetus that began to operate in the late eighteenth century eroded the restrictions based on birth, and the Revolution unleashed the talents that refurbished the central state. The proof is supposedly before our eyes: a more effective administration, a more potent army, and ultimately, a more dynamic economy.[71]

This aura of success has long clung to the exploits of engineers, and especially to the artillery engineers who nurtured Napoleon Bonaparte. But it would be a mistake to suppose that the social and technical victories of these ancien régime artillerists were preordained, or that the knowledge and technology they produced served them in the manner they claimed. These professional engineers lived in an in-between world of quasi-public markers of merit and quasi-public knowledge-making. On the one hand, they claimed to be judged by transparent and impersonal standards (mathematics exams), yet they indulged in a process of self-selection that was shielded from both patrons and the public. On the other hand, they produced knowledge that was publicly, even internationally, exchanged (ballistic science)—knowledge they asserted was superior in practice to older artisanal forms—yet they argued that this knowledge could only be validated by them. Furthermore, the empirical historical evidence suggests that this new ballistic knowledge had little practical impact in its own time. In short, the engineers' claim to have produced a new kind of objective knowledge depended far more on the credibility of their methods of promoting meritorious practitioners (meritocracy based on impersonal judgments) than it did on a practical evaluation of achieved results (meritocracy based on functional success). Hence, this paper has argued first

70. Pion des Loches, *Mes campagnes*, 31–34.
71. Tocqueville, *Old Regime*. This is also the implicit argument made by Thomas Ertman in *Birth of the Leviathan*, his study of military pressures, political regimes, and state structures prior to the Revolution.

that the new forms of objective scientific judgment that scholars have identified as having emerged in the late eighteenth and early nineteenth century received crucial validation from the concurrent application of those same objective criteria of judgment to the practitioners themselves. Meritocracy, then, made knowledge (appear) objective because it treated knowers according to the same logic with which they treated their objects of study. And this paper has argued, second, that the origins of meritocracy lie more in the need to assure skeptical outsiders and fractious insiders that practitioners had been selected, promoted, and given power on the basis of "fair" procedures (based on criteria divorced from birth or personal connections) than in any careful weighing of empirical evidence that this mode of selection picked experts (or expert knowledge) that were functionally superior—which, however, is not to say that the new criteria of selection did not transform the self-image and competencies of those practitioners over the long run.

Meritocracy is often described as a social system that transcends politics. Yet even the modern notion of merit as the title awarded to those individuals who have, by common acclamation, achieved real results presupposes agreement on what counts as success, what abilities go into achieving that success, and how to go about measuring those abilities. Agreements on all these matters must necessarily be mediated by social institutions. In the case of the professions, these matters have been relegated to the professionals' own judgment. I am not suggesting that the individual and collective achievements of professionals are illusory. What I am arguing is that the rise of the professional class was predicated on the creation of new forms of knowledge and that this knowledge was dedicated to the service of a new kind of master. France has long been held up as the nation that first developed an educational apparatus to produce and reproduce a stable professional elite. Yet this "planned meritocracy"—as Harold Perkin has called it—was a deliberate political construct, a social space in which a new kind of achievement was rewarded and directed toward new ends. Engineers were designed to serve.[72]

Primary Sources

Bélidor, Bernard Forest de. *La science des ingénieurs dans la conduite des travaux de fortification et d'architecture civile.* Paris: Jombert, 1729.
———. *Architecture hydraulique.* Paris: Jombert, 1737–53.
——— *Oeuvres diverses.* Amsterdam: Jombert, 1764.
Bézout, Etienne. *Cours de mathématiques à l'usage du Corps Royal de l'Artillerie.* Paris: Pierres, 1781.
Blondel, François. *L'art de jetter les bombes.* La Haye: Arnout, 1685.

72. Perkin, *Third Revolution;* Suleiman, *Elites.*

Carnot, Lazare. *Révolution et mathématique.* Edited by Jean-Paul Charnay. Paris: L'Herne, 1985.

D'Alembert, Jean Le Rond. *Nouvelles expériences sur la résistance des fluides.* Paris: David, 1777.

Diderot, Denis, and Jean Le Rond d'Alembert, eds. *L'Encyclopédie, ou Dictionnaire raisonné des Sciences, des Arts et des Métiers.* Paris: Briasson, 1751–72.

Du Coudray, Philippe-Charles-Jean-Baptiste Tronson. *L'artillerie nouvelle, ou examen des changements faits dans l'artillerie française depuis 1765.* Amsterdam, 1773.

Euler, Leonard. "Recherches sur la véritable courbe que décrivent les corps jetés dans l'air." *Mémoires de l'Académie de Berlin,* 1753, 321–55.

Frézier, Amédée-François. *La théorie et la pratique de la coupe des pierres et des bois . . . ou traité de stéréotomie.* Strasbourg: Doulsseker, 1737–39.

Guibert, Jacques-Antoine-Hippolyte de. *Essai général de tactique.* Vols. 1 and 2 in *Oeuvres militaires.* Paris: Magimel, year XII [1803]. Original edition published 1772.

Laclos, Pierre-Amboise-François Choderlos de. *Oeuvres complètes.* Edited by Laurent Versini. Paris: Gallimard, 1979.

[Lacuée, Jean-Gérard, comte de Cessac]. *Art militaire.* Vol. 4, supplement. Compiled 1789. Paris: Panckoucke, year V [1796–97].

Le Pelletier, Louis-August. *Une famille d'artilleurs, Mémoires de Louis-Auguste Le Pelletier, seigneur de Glatigny, Lieutenant-général des armées du roi, 1696–1769.* Paris: Hachette, 1896.

Lombard, Jean-Louis, ed. *Nouveaux principes d'artillerie, commentés par Léonard Euler.* Dijon: Frantin, 1783.

———. *Tables du tir des canons et des obusiers.* Auxonne, 1787.

Monge, Gaspard. *Géométrie descriptive, Leçons de l'an III.* Paris, year VII [1799].

Pascal, Blaise. *Pensées.* Edited by Louis Lafuma. Paris: Editions du Luxembourg, 1787.

Pion des Loches, Antoine-Augustin-Flavien. *Mes campagnes, 1792–1815: Notes et correspondance du colonel d'artillerie.* Edited by Maurice Chipon and Léonce Pingaud. Paris: Firmin-Didot, 1889.

Robins, Benjamin. *Mathematical Tracts.* Edited by James Wilson. London: Nourse, 1761.

Saint-Auban, Jacques Antoine Baratier de. *Mémoires sur les nouveaux systèmes d'artillerie.* 1775.

Saint-Germain, Claude-Louis. *Mémoires de M. le comte de Saint-Germain.* Switzerland: Libraires Associés, 1779.

Saint-Rémy, Pierre Surirey de, ed. *Mémoires d'artillerie, recueillis.* 3d ed. Paris: Rollin, 1745.

Scheel, Heinrich Othon von. *Mémoires d'artillerie contenant l'artillerie nouvelle, ou les changements faits dans l'artillerie française depuis 1765.* Copenhagen: Philibert, 1777.

[Servan, Joseph]. "Hiérarchie." In *Art militaire,* vol. 4, supplement. Paris: Panckoucke, year V [1796–97].

Vauban, Jacques-Anne-Joseph Sébastien Le Prestre de. *Traité des sièges et de l'attaque des places.* Edited by Augoyat. Paris, 1829. Original edition published 1704.

5

Enlightened Automata

SIMON SCHAFFER

> While justly respecting great geniuses for their enlightenment, society
> ought not to degrade the hands by which it is served.
>
> Jean d'Alembert, *Preliminary Discourse
> to the Encyclopaedia of Diderot* (1751)

Automata figure in the sciences of the Enlightenment as machines in the form
of humans and as humans who perform like machines. Some of these sciences
proposed the organization of productive bodies in disciplined settings, then
understood production in terms of the workings of automata. In 1757, the
Royal Academy of Sciences in Paris was presented with representatives of just
such enterprises during a contest for its post of associate mechanician. Backed
by court patronage and celebrated for his designs of automata and his work
in the silk mills, the engineer Jacques Vaucanson was appointed to the post.
Vaucanson's automata, first put on show in Paris and London, provided matter
for catchpenny theatrics, materialist theorizing, and industrial management.
The loser in 1757 was the editor of the *Encyclopédie,* Denis Diderot. His accounts and images of a vast range of machines and labor processes have long
been taken as a key to the Enlightenment's attitude to mechanical work. Diderot's handling of the mechanical arts was no less embroiled in the puzzles of
automatism. As Ken Alder points out in this book, the *Encyclopédie* repeatedly
announced its aim to free the mechanical arts from the condescension and
ignorance of the noble and the literate, yet did so in the name of an ideal of
rationalized labor processes under the guidance of enlightened managers. Jean
Ehrard has sagely observed that "the Encyclopedist apologia for labour is an
apologia for capital." [1] Laborers were there judged too secretive, incapable of
fully expressing the true principles on which labor processes relied. By making
techniques perfectly visible, they could apparently be reproduced anywhere
and everywhere. In general, as William Sewell has argued, "the workers of the

---

1. Doyon and Liaigre, *Vaucanson,* 308; Ehrard, "La main du travailleur," 53.

Encyclopédie are docile automatons who carry out their scientifically determined tasks with the efficiency and joylessness of machines."[2] The figure of the automaton thus had major epistemic and economic consequences. Because Diderot and his collaborators reckoned that active gestures were a fundamental source of knowledge, it was clear to them that valuable knowledge was locked up in the operations of mechanical workers. That knowledge could only be freed, reformed, and rendered efficient by the gaze of the enlightened, whose faith held that automatic machinery could displace these gestures. This belief legitimated the new devices representing working human bodies.[3]

The focus of this chapter is the relation that some Enlightenment savants developed between machinery viewed as human and humans managed as machines. Its immediate inspiration is a celebrated and tragic image employed by Walter Benjamin in his very final notes on "the problem of memory and forgetting," penned as a refugee from Fascism in 1940. The previous decade had understandably witnessed an efflorescence of studies, from Ernst Cassirer not least, of the Enlightenment. Meditating on the relation between the Enlightenment's two great Germanic legacies, idealist metaphysics and historical materialism, Benjamin invoked "an automaton constructed in such a way that it could play a winning game of chess." He found the automaton described in an essay by one of his favorite authors, Edgar Allan Poe, who had in turn plagiarized accounts of a late-eighteenth-century machine, fashioned in the form of a Turk, that seemed to possess unrivaled skill at chess though worked by a hidden (and human) virtuoso. Benjamin imagined "a philosophical counterpart" to this "puppet in Turkish attire" and to its "little hunchback who was an expert chess player seated inside." He wrote that "the winner, if it is up to me, will be the Turkish puppet, for whom the philosopher's name is materialism. It can easily be a match for anyone if assured of the services of theology, which today, however, is small and ugly and nowhere to be seen." The ugly theology to which Benjamin referred relied on the Romantic metaphysics of early-nineteenth-century Germany; the puppet of materialism embodied Benjamin's attitude to Marxism. Since Jürgen Habermas, at least, commentators have sought to define the relative power of the dwarf and the puppet, of theology and materialism. Responding to conservative attacks on the Enlightenment's legacy, Habermas has recently insisted that "it is of the very nature of the Enlightenment to enlighten itself about itself, and about the harm that it

2. For Diderot and workmen as automata, see Koepp, "Alphabetical Order," 243, 251, 255, and Sewell, "Visions of Labor," 277.
3. Puymèges, "Les machines dans l'*Encyclopédie*"; Proust, "L'image du peuple"; Picon, "Gestes ouvriers."

does." Some forms of automatic machinery—especially the amazing Turk—can be used to explore this kind of self-enlightenment and its ills.[4]

The automatic chess player was first designed in the 1760s by an ingenious Habsburg courtier to match his own virtues against the vulgar and the merely skillful. In the opening section of this chapter, it is argued that the status of enlightened philosophers was often supposed to depend on their capacity to comprehend the mechanical principles of nature and society alike. Once shown at court, the Turk became a major attraction of the fairs and theaters. The second section of the chapter therefore describes how such automata became captivating commodities, their meanings established in the market and their value assigned through commerce. Automata were never merely metaphors of social order. As commodity fetishes they played a significant role in the manufacturing economy and the mercantile system. The third section of this chapter connects major works of Enlightenment materialism with actively interventionist projects of Enlightenment engineering to indicate how such devices changed the management of industry and the workforce. Examples include Vaucanson's early theatrical automata, which then helped him establish systems of manufacture in the silk industry in the midcentury, and Lavoisier's chemical analysis of the animal economy, which let him evaluate the mechanical effects of both intellectual and manual labor. By the apparent mechanization of rational analysis, the show of the Turkish automaton then broached the issues of determinism and free will, obsessions of Enlightenment philosophy. The chess player appeared in the German lands at the same time as Kant's celebrated essay on Enlightenment. Late-eighteenth-century accounts of political order such as those of Kant, or Bentham, explicitly used automata as apt emblems of subjection and government. The fourth section of this chapter thus traces these arguments to emphasize how governmentality was worked out through the philosophy of automatic machinery. It at last becomes apparent why Benjamin was right to see in this exotic device a rich motif of Enlightenment philosophy and its legacies. In his survey of these legacies, Jean Starobinski has associated the end of the Enlightenment with "a technical expansion of the human will." Humanity's new knowledge of nature's mechanically determined laws enabled large-scale intervention in and command over the world defined by those laws. He picks out two aspects of this technical expansion—anthropogeny (the artificial production of humans) and utopianism (projects for ideal societies). This chapter thus examines con-

4. Benjamin, "Theses on the Philosophy of History," in *Illuminations*, 253; Nietzhammer, *Posthistoire*, 104–8 (citing Benjamin's revision of the Turkish image); Habermas, *New Conservatism*, 201. For Benjamin's theology see Roberts, *Walter Benjamin*, 210, 216.

nections between the Enlightenment construction of automata and of its idealized social order.[5]

Visible Technicians and Productive Labor

Some historians still deny that enlightened natural philosophies "fed the fires of the industrial revolution." Others more convincingly indicate the intimate connection between the machinery of natural philosophers' concerns and that of the new entrepreneurs and projectors. The lettered savants who plied their trade in a culture dominated by interests in economic improvement and civic sensibility were in fact noteworthy analysts of and contributors to mechanization and its consequences.[6] Enlightened philosophers tried to build for themselves a position from which they could describe the mechanisms that governed nature and humanity. We can see this process in exemplary sites of enlightened sciences such as Lowland Scotland and absolutist France. Diderot invited his bourgeois readers to admire the workings of an automated silk loom, in which "a machine makes hundreds of stitches at once . . . and all without the worker who moves the machine understanding anything, knowing anything or even dreaming of it." Workers resembled the very machine they managed. This is how the Scottish philosophic historian Adam Ferguson saw things in 1767:

> Many mechanical arts require no capacity. They succeed best under a total suppression of sentiment and reason, and ignorance is the mother of industry as well as of superstition. Reflection and fancy are subject to err, but a habit of moving the hand, or the foot, is independent of either. Manufactures, accordingly, prosper most, where the mind is least consulted, and where the workshop may, without any great effort of imagination, be considered as an engine, the parts of which are men.[7]

From the elevated viewpoint of the Edinburgh chair of moral philosophy, Ferguson here managed "without any great effort of imagination" to identify laborers with the machines they used. This identity, the disciplined productive body as automaton, played a salient role in the formation of enlightened culture. Automata were supposedly produced by rational design and modeled both the celestial system and the animal economy. Their history had taken them from religious and courtly ceremonies to eighteenth-century markets,

5. Starobinski, *Invention of Liberty,* 207–8.
6. Compare Jacob, "Scientific Culture," 136, and Stewart, *Rise of Public Science,* 29–30, with the response in Sutton, *Science for a Polite Society,* 211–12.
7. *Encyclopédie,* 2:98 ("Bas"); Ferguson, *An Essay,* 182–83.

squares, and theaters, a transition quite comparable with that of the enlightened themselves. In Ferguson's conjectural history, the division of labor purchased social progress at a price that limited the scope of laborers' interests to their immediate concerns. Thence arose "in this age of separations" a hierarchy defined in terms of the scope of individual attention: sympathy sustained the solidarity of intellectual elites; the laboring class was barely capable of refined feeling. According to Ferguson, "even in manufacture, the genius of the master, perhaps, is cultivated, while that of the inferior workman lies waste. The statesman may have a wide comprehension of human affairs, while the tools he employs are ignorant of the system in which they are themselves combined." This systematic overview defined the philosopher's place too: "thinking may become a peculiar craft."[8] Son of a Calvinist minister and grandson of a Perthshire artisan, Ferguson's own craft of thinking required careful specification. The efforts of laborers might benefit from progressive division and mechanization, but the social hierarchy did not. Discipline should be based on moral order, not on the mechanical relations found in manufactures. The effects of separation, which bred surplus value and leisured classes, would compensate for those of subordination. The workshops' inmates could be seen as automata, but the denizens of the enlightened academy must forge a crucial alliance of "the citizen and the statesman," which could compensate for the effects of "the subdivision of arts and professions." These hybrid citizen-statesmen were to be the privileged bearers of rational philosophy.[9]

In the same decade, Ferguson's opposite number at Glasgow, Adam Smith, drafted similar arguments for his planned treatise on political economy. "It was the division of labour," he wrote, "which probably gave occasion to the invention of the greater part of those machines by which labour is so much facilitated and abridged." But according to Smith, the key technical innovations could not be credited to confined workers but to those possessed of a much wider view. Water- and windmills were due to "no work man of any kind, but a philosopher or meer man of speculation; one of those people whose trade it is not to do any thing but to observe every thing, and who are upon that account capable of combining together the powers of the most opposite and distant objects." For the enlightened professors, spokesmen of a self-confident elite of improving landlords and civic humanists, philosophers' role was to exploit their relative leisure to coordinate powers sundered by the divi-

---

8. Ferguson, *An Essay,* 181–84. For Ferguson's politics, see Burchell, "Peculiar Interests," 119–50. For elitist sympathy, see Lawrence, "The Nervous System," 29.

9. Ferguson, *An Essay,* 230; on the militia and discipline, see Sher, *Church and University,* 220.

sion of labor. The development of steam engines and mill wheels, "the appli-cation of new powers which are altogether unknown and which have never before been applied to any similar purpose, belongs to those only who have a greater range of thought and more extensive views of things." Smith's Glasgow colleague the eminent chemist William Cullen voiced the same view about the subordination of arts and the superintendence of the enlightened philosopher. For Smith, philosophy itself became "a particular business, which is carried on by very few people who furnish the public with all the thought and reason possessed by the vast multitudes that labour." [10]

Philosophers' "particular business" was hard to define in this scheme of productive labor. In the 1760s, Smith met the protagonists of enlightened phi-losophy, notably physiocratic economists such as Turgot and Quesnay, during his visit to their capital, Paris. While he sought to demolish the physiocrats' claim that manufacturers and merchants were necessarily unproductive, a "humiliating appelation," Smith nevertheless assigned significant social groups to the "the barren or unproductive class." Magistrates and the military, plus "churchmen, lawyers, physicians, men of letters of all kinds; players, buffoons, musicians, opera-singers, opera-dancers etc." were all "unproductive of any value." [11] The most elevated and mercenary were among those social agents whose enterprises might be represented mechanically. Performance worked through entertaining artifice, armies behaved like well-ordered mechanisms, clerks were depicted as cogs in the machinery of state, the government itself was supposed to emulate the order of the world machine. But it was puzzling for Smith and his allies to specify how intellectual labor might be understood in the mechanical system. This was why Marx recorded the complaint that "Adam Smith invented the category of unproductive labourers out of pure malice, so that he could put the Protestant parsons in it." Marx observed that "Milton who wrote *Paradise Lost* was an unproductive worker. On the other hand, a writer who turns out work for his publisher in factory style is a pro-ductive worker." Smith's epoch was just when this style of literary production emerged and, it has been argued, came to dominate the business of enlighten-ment. Writers looked like automata. As Roger Chartier tells us, the commercial mechanization of print culture accompanied the ideology of "the urgency and absolute freedom of creative power," and this perverse relation between

10. Smith, "Early draft of part of *The Wealth of Nations*," in *Lectures on Jurisprudence*, 569–72. For Cullen, see Christie, "Ether," 95, and for Smith's comparable remarks on the chemistry laboratory, see Golinski, *Science as Public Culture*, 30–31; for Scottish improvement and enlight-enment, see Christie, "Scottish Scientific Community"; Phillipson, "Scottish Enlightenment."
11. Smith, *Wealth of Nations*, 1:330–31, 2:664.

subjection to the market and intellectual liberty was characteristic of the predicament of Enlightenment philosophers, Diderot and Smith alike.[12]

Better to understand the valorization of intellectuals' literary labor both as inspiration and merchandise, these philosophers could establish their power through their view of the machine. Smith's favored example of the division of labor, the pin factory, was taken from an article on a Normandy workshop he read in the *Encyclopédie*, the first volumes of which he bought for Glasgow University Library. The plates there illustrating the pin factory were made by the civil engineer Jean-Rodolphe Perronet, an expert on the disaggregation of the labor process keen to regiment piecework so as to subordinate production to the management of the enlightened engineer. The article on pin making was immediately followed by Diderot's editorial reflection on the virtues of enlightened authors. He there claimed the entry on pin making "will prove that a good mind can sometimes, with the same success, both raise itself to the highest contemplations of philosophy and descend to the minutest details of mechanics." Encyclopedists could "pass without disdain from a search for the general laws of nature to the least important use of its products."[13] Equable philosophy depended on the careful construction of the laws that governed the cosmos and primitive superstition on ignorance of their mechanics. However different their perspectives on civic society and rural virtue, Ferguson and Smith both shared this view. Ferguson attributed the "ignorance and mystery" of irrational superstition to subjects' perplexity when faced with "strange and uncommon situations." Truly enlightened social order hinged on "the study of nature by which we are led to substitute a wise providence operating by physical causes in the place of phantoms that terrify or amuse the ignorant." Smith argued similarly that "the lowest and most pusillanimous superstition" was spawned by "every object of nature . . . whose operations are not perfectly regular," and that were therefore "supposed to act by the direction of some invisible and designing power." Struck by David Hume's critique of innate causal powers, Scottish philosophers claimed that the progress of machine philosophy calmed these fears and wonders by imaginatively connecting phenomena under lawlike regularities. In a posthumously published essay edited by his literary executor, the chemist Joseph Black, Smith explained that "a ma-

---

12. Marx, *Capital*, 1:768n. 6 (on Smith) and 1044 ("Results of the Immediate Process of Production," 1863, on Milton). For literary production, see Darnton, *Business of Enlightenment;* Kernan, *Printing Technology;* and Chartier, *Order of Books,* 37.

13. Picon, "Gestes ouvriers," 135, on Perronet; Vidler, *Writing of the Walls,* 26, on workshops' depiction in the *Encyclopédie;* on the division of labor, see *Encyclopédie,* 1:717 ("Art"), and 5:807 ("Epingle"). See Smith, *Wealth of Nations,* 15n. 3 and 17n. 10, for his use of this source; Scott, *Adam Smith,* 179, for his purchase of it; and Alder, *Engineering the Revolution,* for the engineering concept of skilled work.

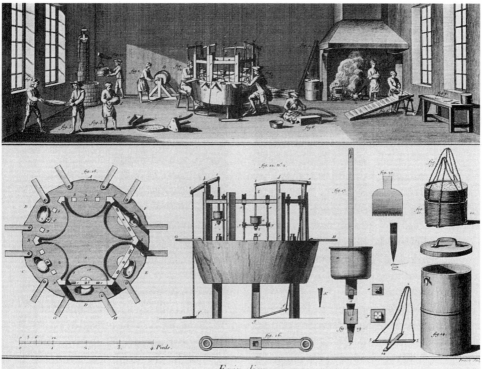

Epinglier.

Fig. 5.1.   The division of labor in a Normandy pin factory as Adam Smith saw it represented in the *Encyclopédie* in 1765. From Denis Diderot and Jean d'Alembert, eds., *Encyclopédie: Receuil des planches* (1765), vol. 21, plate 3, "Epinglier."

chine is a little system created to perform as well as to connect together in reality those different movements and effects which the artist has occasion for. A system is an imaginary machine invented to connect together in fancy those different movements and effects which are already in reality performed." And in the same work, he stipulated that the audience of any superior artwork was moved by "a very high intellectual pleasure not unlike that which it derives from the contemplation of a great system in any other science"—spectators of the *theatrum mundi* were moved by the same systematic pleasures as those of any other playhouse. In the mechanisms to which Diderot and Smith appealed, these systems switched the sources of power from invisible and capricious agents to the regular operations of predictable constructs.[14]

14. Ferguson, *Essay*, 90–91; Smith, *Essays*, 48–50, 66, 205.

The sciences of labor were integrated with laboratory techniques. As Lissa Roberts has argued, enlightened modernity viewed the body as a machine—and in chemistry as in industrial production surrounded it with ever more complex systems of human-built technologies. These technologies, she points out, then helped change the polity. After managing military works in the French West Indies in the 1760s, the eminent engineer and academician Charles Coulomb tried to evaluate the maximum effect extractable from labor. Coulomb's political economy taught him that work that commanded the same wage must involve the same amount of fatigue. Field trials allowed him to measure the effects of this labor.[15] During the early 1790s, Coulomb's close colleague, the chemist and economist Antoine Lavoisier, was involved in studies of the balance of profit and loss in the national agrarian economy and of means to reform the French political machine. At just the same time, he managed to develop a technique for the precise evaluation of the mechanical worth of intellectual labor. He reported at once on these techniques to his eminent chemical colleague Black, another authority on the animal economy and its thermal mechanics.[16] Lavoisier's laboratory methods treated all humans as so many machines absorbing vital air and nutriment. By evaluating pulse rate and air consumption, the prudent academician and his collaborators reckoned they could determine "how many pounds weight correspond to the efforts of a man who recites a speech, a musician who plays an instrument. Whatever is mechanical can similarly be evaluated in the work of the philosopher who reflects, the man of letters who writes, the musician who composes. These effects, considered as purely moral, have something physical and material which allows them, through this relationship, to be compared with those which a labourer performs." Lavoisier's chemical technology understood all humans as automata laboring in closed exchange systems. "It is therefore not without justice that the French tongue mixes together under the common name 'work' [travail] the efforts of the mind as well as those of the body, the work of the study and the work of the hireling." In the very months during which Lavoisier was preoccupied with evaluating agricultural productivity for the National Assembly, he here explained how the enlightened natural philosopher could "organize and regenerate" the careful balance of social, human, and natural machines. "In the silence of his laboratory and his study, the natural philosopher [physicien] can also exercise patriotic functions. . . . He can also aspire to the glorious title of benefactor of humanity." By constructing, displaying, and imagining such self-governing machine systems, the enlightened supposed

15. Roberts, "Death of the Sensuous Chemist," 507–8, 521; Coulomb, "Mémoire sur la force des hommes," 260. See Gillmor, Coulomb, 23–24, and Vatin, Le travail, 42–43.
16. Holmes, Lavoisier, 444–46, for Lavoisier and Black.

they could make their own social order and secure a powerful place within it. So automata had a political function in "the technologies of rationalism." [17]

Automata and the Parliament of Monsters

Histories of automata, drawing on the classical texts of Hero and Vitruvius, chronicled their hieratic role in ancient temples, where they could be used to impress the faithful, then traced the development of ingenuity in cathedral timekeepers and courtly entertainments. Such clockwork and hydraulic devices linked the workings of the cosmos, the state, and the human body. Renaissance horology allowed the construction of purely mechanical devices, and these typically involved shows that mixed celestial with mundane figures in planetaria or dances. Thus the automata installed by Tomaso Francini for the French monarch at Saint-Germain-en-Laye around 1600 included a Grotto of Perseus, in which alongside a set of smiths, weavers, and other artisans working steadily at their trades the eponymous hero slew his dragon. Francini's devices and their imitators at Fontainebleau were a plausible source of notorious Cartesian accounts of the animal machine. Baroque princes commissioned a host of similar devices, which mechanized the deeds of gods and heroes or else the labors of servants and workmen.[18] Automata were apt images of the newly disciplined bodies of the military systems of early modern Europe. In Golden Age Netherlands and absolutist Prussia, drill masters worked out systems to turn soldiers into machines. Automata as models of the well-regulated workshop also proliferated. Dan Christensen has recently described a remarkable eighteenth-century automaton built to represent the workings of the Kongsberg silver mine in Norway and its royal Danish administrators—there, for example, mercantilist management saw mineworkers as so many mechanical elements of a centrally driven automatic system. These projects ingeniously connected a culture that viewed laborers as machines with one that saw machines as sources of power. Access to the inwards of these machines was at least as vital as the display of their marvelous performance. Automata were both arguments and entertainments, designed seductively to place craft skill within the setting of power, and to allow the selective entry by that power to the inner workings of

17. Lavoisier and Seguin, "Première mémoire sur la respiration des animaux" (1790), in Lavoisier, *Oeuvres*, 2:697. See Holmes, *Lavoisier*, 454; Bensaude Vincent, *Lavoisier*, 220; Lavoisier, *De la Richesse Territoriale*, 10–11. For "technologies of rationalism" and Lavoisier's balances, see Wise, "Mediations," 220.

18. Useful histories are Brewster, *Letters on Natural Magic*; Chapuis and Droz, *Automata*; Price, "Automata in History"; Bedini, "Automata in the History of Technology"; Heckmann, *Die andere Schöpfung*, 91–106, 170–81; Beaune, "Classical Age of Automata." For Francini, see Chapuis and Droz, *Automata*, 43–47; for Cartesianism, see Jaynes, "Problem of Animate Motion," 224, and Dear, "Mechanical Microcosm," 58–60.

art and nature. Such devices thus played roles in early Enlightenment debates involving both clerics and court philosophers on the puzzles of good government—of the world by the deity, of the state by the prince, of the workshop by the master, and of body by spirit.[19]

In early modern cities, these machines moved from palaces and churches to fairgrounds, showrooms, and salons. The very term "automaton" had already been introduced into French in Rabelais's carnivalesque epics. As Peter Dear has perceptively indicated, the formalization of civility in early modern bourgeois societies helped make Cartesian accounts of mechanization current. By the eighteenth century, automata were not merely common in the shows but occupants of the linked worlds of court, marketplace, and theater. There, it was argued, "artificial men" wound up by internal and occult mechanisms played out their games of passion and interest, privy schemes were brought before the public in the name of commerce and spectacle.[20] Machines commissioned by Friedrich II from the émigré clockmaker Abraham-Louis Hugenin, or by Maria Theresa, for whom the inspector of the imperial physics cabinet Friedrich von Knaus built an automaton that could write its own messages during the 1750s, ended up on tour round European showrooms. In Paris, the automaton market boomed after Vaucanson's celebrated works of the late 1730s, a mechanical drummer, a flute-player, and a duck that could digest and even defecate. The young Grenoble engineer's machines were on show at salons and the great Paris fairs. Pleasure seekers paid 24 sous to see his mechanical flautist give concerts in the Tuileries in 1738. "At first many people could not believe that the sounds were produced by the flute which the automaton was holding. . . . The spectators were permitted to see even the innermost springs and to follow their movements." Automata soon began dominating the so-called *cabinets de physique* at Saint-Laurent and Saint-Germain. When a consortium of Lyonnais businessmen bought the three automata in 1742, the machines were taken to London, put on show at the Haymarket Theatre, and publicized with a translation of Vaucanson's pamphlets produced by the doyen of London engineers and demonstrators, John Desaguliers.[21]

19. For military discipline, see McNeill, *Pursuit of Power*, 125; Flesher, "Repetitive Order," 468–70; for philosophical automata, see Sutter, *Göttliche Maschinen*, 94–98; Robinet, "Leibniz, l'automate, et la pensée"; Gabbey, "Cudworth, More, and the Mechanical Analogy"; Giglioni, "Automata Compared." For Kongsberg, see Christensen, *Det modern Projekt*, 74.

20. Heckmann, *Die andere Schöpfung*, 208–10; Dear, "Mechanical Microcosm," 64–65. See also Beaune, "Classical Age of Automata," 431; Agnew, *Worlds Apart*, 162–69; Stallybrass and White, *Poetics and Politics*, 106–11.

21. Chapuis and Droz, *Automata*, 96–97 (Hugenin), 273–77 (Vaucanson), 289–92 (Knaus). For Knaus, see also Heckmann, *Die andere Schöpfung*, 235–38, and for the aftermath of Vaucanson's automata, see 228–32. For Paris automata, see Isherwood, *Farce and Fantasy*, 48–49, and Benhamou, "From *Curiosité* to *Utilité*." For the Lyons consortium, see Doyon and Liaigre, *Vaucanson*, 87–89.

Fig. 5.2.   Vaucanson's three automata: a drummer, a duck, and a flute player. From Jacques de Vaucanson, *An Account of the Mechanism of an Automaton,* translated by Jean Desaguliers (London, 1742).

London, metropolis of enlightened consumption, was peculiarly susceptible to the automata shows. The reputation there of John Merlin, who ran a mechanical museum in the West End, rivaled even that of Vaucanson. In his reminiscences of these London fairs, William Wordsworth punningly evoked the "Clock-work, all the marvellous craft / of modern Merlins. . . . All jumbled up together, to compose / a Parliament of Monsters." After winning prestigious finance from the backers of Boulton and Watt's new steam engines, Merlin kept a stock of figures "in brass and clockwork, so as to perform almost every motion and inclination of the human body, viz. the head, the breasts, the neck, the arms, the fingers, the legs &c. even to the motion of the eyelids, and the lifting up of the hands and fingers to the face." The Cambridge mathematician Charles Babbage, who later saw one of Merlin's mechanical dancers, remembered that "she used an eye-glass occasionally and bowed frequently as if recognizing her acquaintances." [22] Such machines were fine resources for contemporary debates on the art of acting and the "physics" of emotions. The fashionable Horace Walpole learned from one Parisian hostess, the marquise du Deffand, that among "the numerous company at my place yesterday evening, men and women seemed to me to be spring-driven machines . . . each played their role by habit." David Garrick's contribution, written in 1744 and indebted both to contemporary automata and to his reading of French materialist physiology, criticized "automaton Players who are literally such mere Machines that they require winding up almost every time before they act, to put them in action and make them able to afford any pleasure to an audience." But neither Garrick nor his more enlightened admirers, notably Diderot, severed the link between actor's bodies and the machines. On the contrary, automata showed that artificial organization could in principle display phenomena that were more powerful, and more dramatically effective, than mere clockwork. [23]

There were suggestive links between discussions of Garrick's powers and those of one of the most famous automata of the eighteenth century, a "Musical Lady," originally brought to London in 1776 by the great Swiss horologist Jaquet-Droz. The accomplished lady's eyes really moved, her breast heaved. "She is apparently agitated," a contemporary remarked, "with an anxiety and diffidence not always felt in real life." Such shows often turned to titillating effect modish materialist philosophies that, like enlightened theories of sensibility and mesmeric strategies for restoring health, sought to mechanize the passions. Terry Castle has shown how such an apparently innocent commodity as the weatherglass could be taken as an embodiment of the automatic movements of sexual desire, the *femme-machine*, while Jan Golinski's account of the marketing of the barometer in chapter 3 of this book demonstrates how consumer goods could carry Enlightenment values of curiosity, discipline, and patrician wit. [24] The hydraulic pornography of John Cleland's fictions and the erotic devices of James Graham's celestial bed were both much indebted to the philosophical materialism found in such texts as Julien Offray de La Mettrie's

22. French, *John Joseph Merlin*; Altick, *Shows of London*, 72–76; Babbage, *Passages*, 17. For Wordsworth, see Stallybrass and White, *Poetics and Politics*, 120.

23. Deffand to Walpole, 1766, cited in Benhamou, "From *Curiosité* to *Utilité*," 93; Garrick, 1744, cited in Roach, *Player's Passion*, 91.

24. Altick, *Shows of London*, 66; Castle, *Female Thermometer*, 31.

Man a Machine (1747). As Roland Barthes has argued, the late eighteenth century's notions of "the total machine," in which entire human groups were "conceived and constructed as a machine," reached their fullest elaboration in the libertine fantasies of Sade.[25] This was at least one way, as Joan Landes and Dena Goodman point out, in which the public sphere of the Enlightenment acquired its pronounced ambivalence toward the feminine. Erotic watches (the best were Genevan) were marketed carrying cunningly concealed automata of debauched monks and nuns. More visible automata adverts put love on sale:

> If the Poet speaks truth that says Music has charms
> Who can view this Fair Object without Love's alarms
> Yet beware ye fond Youths vain the Transports ye feel
> Those Smiles but deceive you, her Heart's made of steel
> For tho' pure as a Vestal her price may be found
> And who will may have her for Five Thousand Pounds.[26]

Since the 1760s, Merlin and his erstwhile employer James Cox had built extraordinary automata for the East India Company's China trade, opened shops in Canton where mandarins could acquire mechanical clocks, mobile elephants, and automatic tigers, and thus oiled the wheels of the tea trade. When this lucrative eastern commerce languished, Cox's firm went broke, and his successor, Thomas Weeks, never quite managed to revive it.[27] According to Weeks's advertisements, "these magnificent specimens which constitute almost all the labor of a long life, and were all executed by one individual, were originally intended as presents for the east, they have, indeed, all the gorgeous splendour, so admired there, and we can fancy the absorbing admiration they would create in the harems of eastern monarchs, where their indolent hours must be agreeably relieved by these splendid baubles, which however are so constructed as to combine in almost every instance some object of utility." [28] Caricatures of languor mixed with images of commerce and helped define the roles of automata in the mechanisms of Enlightenment culture.

Technico-Politics

According to a French dictionary of 1727, the word "machine" referred "in general to automata, and to all those things which move by themselves whether by

25. For Cleland, see Braudy, "*Fanny Hill* and Materialism"; for Graham, see Porter, "Sexual Politics of James Graham"; for Sade, see Barthes, *Sade, Fourier, Loyola*, 156–7.

26. For the feminine and the public sphere, see Landes, *Women and the Public Sphere;* Baker, "Defining the Public Sphere"; Goodman, *Republic of Letters;* Sutton, *Science for a Polite Society,* 341–48. The verse is cited from Ord-Hume, *Clockwork Music,* 42; for erotic watches, see Landes, *Revolution in Time,* 269.

27. Altick, *Shows of London,* 350–51; Chapuis and Droz, *Automata,* 107–110. For the long history of clockwork, trade, and orientalism, see Landes, *Revolution in Time,* 37–52, 401 n. 29.

28. Ord-Hume, *Clockwork Music,* 45.

art or naturally." Enlightened natural philosophies were sometimes associated with versions of materialism, including the claim that all living beings, and even humans, could be seen as machines. The new industrial and political systems of the classical age have comparably been interpreted as apparatuses of mechanical discipline, in cameralist administration and military tactics. In his genealogy of the prison, for example, Michel Foucault claimed that in the eighteenth century "the great book of Man-the-Machine was written simultaneously on two registers," that of "anatomico-metaphysics," pursued in roughly Cartesian terms by physicians and philosophers, and that of "technico-politics," manifest in military, classroom, and hospital management. Foucault wanted to distinguish these two registers—"there was a useful body and an intelligible body." He also wanted to link them, especially by appeal to La Mettrie's *Man a Machine,* which he read as a text both of materialism and of bodily manipulation, and to the regime of Friedrich II, "the meticulous king of small machines," who provided La Mettrie with asylum at Potsdam after Calvinist attacks on his work in 1748.[29] No doubt the technico-politics of the Age of Reason was especially apparent in military training. Prussian military regulations, highly influential in the British army too, called for at least six years' basic training in what was called "material exercise." Where soldiers were officers' property and often seen as an unreliable mob, mechanical repetition was understood as crucial to subordinate individual willfulness. Standard paces, musical cadence, and the decomposition of actions such as loading and firing were all parts of this automatic system. French programs such as that of Jean-Charles de Follard in the 1730s treated line soldiers as automata. French critics of the Prussian system adopted there after the 1760s noted that in drill depots "the wretched soldiers have all been put into a variety of artificial and enforced postures." Parade ground automatism was a model, barely realized on the battlefield. Foucault thus judged that "the celebrated automata," Friedrich's playthings and La Mettrie's exemplars, "were not only a way of illustrating an organism, they were also political puppets, small-scale models of power."[30]

Foucault was not alone in linking the mechanization of the body with the disciplinary apparatus of the old regime. As Ken Alder points out above, comparisons of the military-political strategies of Louis XIV and Friedrich II were and remain rich sources for an understanding of this apparatus. In his analysis of the Grand Monarch's theatrics, Jean Apostolides has insisted that by 1700

29. Thomson, *Materialism and Society,* 41, for the definition of machine; Foucault, *Discipline and Punish,* 136; Mattelart, *l'invention de la communication,* 38, on Foucault, Vaucanson, and La Mettrie.
30. Foucault, *Discipline and Punish,* 136; Houlding, *Fit for Service,* 259, 267–68; Childs, *Armies,* 67, 105; Duffy, *Military Experience,* 54, 105 (citation from Guibert, 1772); Alder, *Engineering the Revolution,* 116.

the old regime had displaced the powers of territorial reach by those of permanent, machinelike, organization: "the glorious body" of the king, "which functioned like clockwork, brought with it an extremely mechanized court ceremonial. At a moment when the actions of the first labourers working on industrial machines were decomposed and analyzed to improve performance, the body of the king-machine found itself being laid out in a multitude of mechanical actions." But in this intriguing analysis, the suggestion of some linkage between absolutist automata and the new industrial formation of the early eighteenth century remains coincidental.[31] Analyses of political philosophy, such as that of Reinhart Koselleck, have connected Enlightenment clubland with Hobbesian mechanization, asserting that Leviathan, "the automaton, the great machine," forced the philosophical construction of a secluded space of alienated critique where a politically impotent intelligentsia could assert its judgmental superiority. Unconcerned with Hobbesian natural philosophy or with its progeny in Enlightenment materialism, Koselleck, however, also ignored the issue of military and political discipline. He understood Friedrich II's predicament in terms of the dualism of morality and politics rather than the monism of his subjects' bodies.[32] Histories of technology, such as that of Otto Mayr, have associated clockwork models of physiological order with images of the absolutist state, and later eighteenth-century homeostatic machinery with the liberal critique of tyranny. Though he indicated the intimate connection between Frederician absolutism and the mechanical philosophy and noted the Prussian king's patronage of La Mettrie, Mayr was in the end unwilling to suggest more than an interesting coincidence between fashions for clockwork and for absolutism, and between self-regulating machines and liberalism.[33]

La Mettrie's *oeuvre* helps show that there was more than a coincidental connection between the mechanization of the animal economy and the handbooks of discipline and training. Brought up as a Jansenist in Paris, he may well have absorbed mechanist theories common among several of the sect's physicians. After training at Leiden in 1733–34, and practicing and publishing the version of Boerhaave's medicine he learned there for almost a decade back in Brittany, La Mettrie's first preferment was in Paris and on the battlefield as medical officer in the French guards. It was in this period, in the milieu of supreme barrack discipline, that he penned his first materialist treatises. In his eulogy, Friedrich II noted that La Mettrie's *Natural History of the Soul* (1745) was composed after experiences of camp fever at the siege of Freiburg. *Man a*

31. Apostolides, *Le roi-machine*, 156.
32. Koselleck, *Critique and Crisis*, 117. For a critique of Koselleck, see Jacob, *Living the Enlightenment*, 14–15.
33. Mayr, *Authority, Liberty, and Automatic Machinery*, 107–9.

Machine was written during 1746–47 when La Mettrie was in charge of the French military hospitals in the theater of war in the Low Countries. Violent controversies both with churchmen and with the medical establishment drove the military physician to the Netherlands, and thence, after his most notorious work had been burned by the public hangman, to Berlin.[34] On the basis of a sketchy summary of the active matter whose organization formed the human body, *Man a Machine* argued that "the human body is a machine which winds up its own springs. It is the living image of perpetual motion." According to the exiled physician, it was this fact that dispossessed priests of their authority. Physicians, who "alone have laid bare to us those hidden springs beneath the coverings which conceal so many wonders from our eyes," had the sole "right to speak here."[35]

From these principles, especially in his remarkable *Discourse on Happiness* (1748), a commentary on Stoicism that appeared soon after his arrival in Berlin, La Mettrie insisted on the linkage between determinism and medical management, for the education of individuals dominated their character. Adam Smith, for one, was impressed by La Mettrie's use of medical principles to construct an entire social therapeutics. According to the Breton physician, moral sentiment flowed from social discipline: "ideas of generosity, greatness and humanity have been tied to important actions of human commerce; respect, honour and glory to those actions which serve the country; and by these goads a great many animals with a human shape have become heroes." Thus political power acted on the social body the way the medical-military manager directed patients. In particular, because of the fragility of this form of socialization, and the tendency of the body to revert to its primitive state, it was crucial that training be maintained. The exercise of political power represented a constant struggle to bend the subject to public action and, as he put it, to exterminate the "mad dogs" who disrupted good order. La Mettrie's man machine was not an iatromechanical reverie of clockwork and pulleys, but a natural body subject to exercise and training, imbued with innate vitality and therefore simultaneously capable of, and requiring, external discipline.[36]

In 1747, La Mettrie assumed his readers would take the point that the human body was composed of self-moving matter at the highest stage of organization by citing the example of "Vaucanson, who needed more skill for making his flute player than for making his duck, and would have needed still more to

---

34. Vartanian, *La Mettrie's L'Homme Machine*, 4–7.

35. La Mettrie, *Man a Machine*, 93, 89. For mechanism and Jansenism, see Brockliss, "The Case of Philippe Hecquet," 212–13.

36. Wellman, *La Mettrie*, 226–28; Sutter, *Göttliche Maschinen*, 121–50. For the displacement of the mechanical body by the disciplinary body, see Foucault, *Discipline and Punish*, 155. For Smith on La Mettrie, see "History of Astronomy," 47.

make a talking man, a mechanism no longer to be regarded as impossible." By summer 1740, Friedrich had already contacted his agent in Paris to recruit Vaucanson to the Berlin court. The engineer was offered a pension of twelve thousand pounds to take up the Prussian offer; in the event, he refused in favor of better chances with the French government.[37] That government cared about these automata because of its interest in medical and economic reform. Parisian designers were experts in making "moving anatomies" of the human body. In the 1730s, several surgeons argued that by building hydraulic machines they could demonstrate physiological facts that corporate physicians were too prejudiced to accept. The young surgeon François Quesnay proposed "a human automaton in which will be seen the performance of the principal functions of the animal economy, by means of which the mechanical effects of bloodletting can be determined and several interesting phenomena which do not seem susceptible will be submitted to the balance of experiment." La Mettrie, a medical student in Paris in 1727–31, was a notable protagonist of the surgeons' cause from 1737.[38] There was a connection between these medical debates and economic theory because the economy itself was figured as a perpetually moving automaton. Quesnay began composing powerful physiocratic treatises on the hydraulic model of active fluid flow as an image of the right form of government. Here Adam Smith found his most important topic for political economic argument, and Lavoisier incentives for his important studies of agronomy and physiology. Meanwhile, Vaucanson had proposed the construction under royal commission of a dramatic automaton, "to perform experiments on animal functions, and thence to gather inductions to know the different states of health of men so as to remedy their ills. This ingenious machine," Vaucanson boasted in 1741, "which will represent a human body, could in the end serve to perform demonstrations in an anatomy lecture." The context of state regulation and medical controversy was important for Vaucanson's projects. His celebrated automata posed the problem of the working of the human machine and of the means by which it could be understood and governed.[39]

Mimesis and ingenuity remained nice matters. Vaucanson explained that he had left the workings of his mechanical duck "exposed to view" because he wished "rather to demonstrate the manner of the actions than to shew a

37. Doyon and Liaigre, *Vaucanson*, 133–36; La Mettrie, *Man a Machine*, 140.

38. Wellman, *La Mettrie*, 20–33; Doyon and Liaigre, *Vaucanson*, 120–24; Benhamou, "From *Curiosité* to *Utilité*," 101–2; Gelfand, "Empiricism and Eighteenth Century French Surgery"; Sutter, *Göttliche Maschinen*, 114–20.

39. Doyon and Liaigre, *Vaucanson*, 148; Puymèges, "Les anatomies mouvantes." Compare Fryer and Marshall, "Motives of Jacques de Vaucanson"; Stafford, *Artful Science*, 191–95. For Quesnay and perpetual motion, see Wise, "Mediations," 226.

machine. . . . I would not be thought to impose upon the Spectators by any conceal'd or juggling contrivance."[40] Transparency and cunning were connected. Nowhere was this more obvious than in Vaucanson's new initiatives of the 1740s. His project to build medical automata stalled, the government hired Vaucanson as inspector of the silk trade, sending him in 1741 to Lyons and to Piedmont to examine best practices in the silk mills. Back at his workshop in Paris, Vaucanson staged trials of reformed manufacture processes, and proposed a massive new state-regulated transformation of work practices, in particular, a system of surveillance centered in a royal factory based in Lyons. Vaucanson's regulations were fiercely resisted by the silk masters and artisans; he was forced to flee the turbulent city disguised as a monk, and the new state regulation collapsed. Instead, from 1744, Vaucanson began designing automatic silk-weaving machinery. Just as his mechanical flute player had relied on a carefully controlled rotating barrel, whose rate was fixed by calibration against an expert human player, so his silk machines used barrels timed to manage the weaving rates. He reckoned his automatic silk machine was a device "with which a horse, an ox or an ass can make cloth more beautiful and much more perfect than the most able silkworkers. . . . Each machine makes each day as much material as the best worker, when he is not wasting time." In 1751–54 Vaucanson tried installing these machines at a new-style silk mill in the Ardèche, but he soon found that lack of skilled workmen and local resistance again frustrated his plans.[41]

Diderot reprinted Vaucanson's accounts of his automata in the opening volume of the *Encyclopédie,* and in 1776 Vaucanson himself chose to publish in the proceedings of the Royal Academy of Sciences a series of plans of the idealized factories he had constructed: airy, light, disciplined, efficient, if ultimately bankrupt. The following year an authoritative dictionary echoed the view that adept artisans were those who lacked "any principle or any rule of movement," and this was what made them lowly. By 1804, Vaucanson's admirer, Jean Jacquard, had rebuilt the looms in Paris and was thus prompted to develop a new and decisive system of weaving automata, which in turn provided Babbage with one model of an intelligent calculating engine.[42] Real connections were forged between these endeavors to produce a disciplined workforce, an idealized workspace, and an automatic man. Vaucanson's London opposite number, Desaguliers, was as committed to these links between the

40. Vaucanson, *Automaton,* 23.
41. Gillispie, *Science and Polity,* 413–21; Doyon and Liaigre, *Vaucanson,* 197–203, 210.
42. Giedion, *Mechanization Takes Command,* 35–36. For Babbage and Jacquard, see *Passages,* 116–17. For the 1777 dictionary, see Doyon and Liaigre, *Vaucanson,* 39 n. For Vaucanson and Diderot, see *Encyclopédie,* 1:451 ("Androide"); 896 ("Automate").

mechanical modeling of human capacity and the labor discipline of the artisan trades.[43] Two decades before it hosted Vaucanson's automata, the Haymarket Theatre had staged shows of German strongmen. After collaborating with Alexander Stuart, a Leiden-trained expert on muscular anatomy, Desaguliers lectured on the means through which these theatrical feats could be replicated with an elementary knowledge of mechanics. Mechanical calculation over-mastered brute strength.[44] He applied the same public calculus to the new silk industry. In 1718 a Derby entrepreneur, John Lombe, introduced Italian silk-throwing machines after industrial espionage in Tuscany. Lombe collaborated with the hydraulic engineer George Sorocold to build a profitable silk mill. As the historian Thomas Markus points out, in Lombe's system the spatial distribution of social order, flowing from owner through supervisors to workers, did not yet match the flow of mechanical power, since workers sharing the same space did not use machines sharing the same power source. New automatic systems were required to make the distribution of regulated workers and of mechanical power correspond to each other. Desaguliers devoted some pages to these works in his lectures on natural philosophy and mechanics, explaining to his audience that despite the demands of industrial secrecy he could reveal that the "power of a hand" might be multiplied by Lombe's machines.[45]

The multiplication and intensification of power was Desaguliers's key theme. He explained that genteel "men of theory" could be deluded by their cunning workforce and by utopian optimism. "There is a Combination among most Workmen to make a Mystery of their Arts and they look upon him as a False Brother who lets gentlemen into their Manner of Working and the Knowledge of the Price of all Materials." To gain power over mechanical schemes, it was therefore necessary to employ engineers to master the workforce and direct the market. Speculators attributed power to the machines themselves, breeding false hopes of endless profit and perpetual motion: "the Vulgar commonly speak of a Machine as they do of an Animal, and attribute that effect to the Machine which is the Effect of the Power by means of the Machine." Wrong models of the animal-machine relationship spawned commercial and political disaster. The powers of machinery were the property of enlightened mechanics, not of the workforce nor the machines. These errors bred loss through the resistance of the workforce, or of friction.[46] Enlightened

43. Stewart, *Rise of Public Science*, 119–30, 213–54; Morton, "Concepts of Power," 67–70.

44. Stewart, *Rise of Public Science*, 126; Poni, "The Craftsman and the Good Engineer," 224; Stafford, *Artful Science*, 178.

45. Desaguliers, *Course of Experimental Philosophy*, 1:69–70; Markus, *Buildings and Power*, 263–6.

46. Desaguliers, *Course of Experimental Philosophy*, 2:415, 1:127. For perpetual motion and loss, see Schaffer, "Show That Never Ends."

Fig. 5.3. Henry Bridges's "Microcosm," built for the duke of Chandos in the 1730s. The clocks show solar time, lunar phases, and the Copernican system; above, a moving landscape of Parnassus and Orpheus; below, the automatic shipyards and carpenters' shops.

lecturers and engineers, who claimed to represent nature's laws, aimed for a striking shift of power, away from artisans toward the masters of the powers on which machines and projects depended. Automata were good signs of this shift. At the palace of Desaguliers's principal patron, the vicious speculator, entrepreneur, and oligarch the duke of Chandos, the engineer Henry Bridges built and installed a vast automaton clock. The "Microcosm," driven by more than one thousand wheels and pinions, displayed all the phenomena of the heavens, picturesque scenery, and an efficient wood yard. In 1741, the Microcosm went on show at a Charing Cross coffeehouse. A print of the device carried portraits of its designer, and of Isaac Newton, plus a fulsome dedication to Chandos. These were the owners and makers of the world machine. The Microcosm was a typical piece of showmanship whose concealed works and pompous exterior emblematized the relation between cunning engineering and outward show.[47]

But automatic machinery was never merely emblematic in the world of enlightened engineering. Alongside orreries and planetary machines, as the historian Alan Morton points out, Desaguliers showed paying audiences a machine that could measure the maximum effect of a laborer's power to raise a load, a device comparable to the new machines being introduced into the London coal trade to mechanize unloading at the docksides. Endemic struggles between the coal heavers and the managers culminated in the 1758 Coal Act and the introduction of a system of "whipping" on board ship.[48] Desaguliers's estimates of the maximum work extractable from labor remained a standard reference for later eighteenth-century engineers and industrialists keen to scotch the shows of dubious engines by projectors "who put them in motion for some minutes, by vigorous people, who make a momentary effort." These estimates helped establish later eighteenth-century maxims on the utmost work to be extracted from laborers: they were then cited, for example, by the Paris Academy of Science's preeminent mathematical chronicler Jean Montucla in a 1778 edition of a work on recreational mathematics. From the same year, Coulomb began to argue that the slight difference between theoretical and observed maxima of available human work established in French military projects and in the laboratory was due to "the instinct natural to all men" that "with a given burden" prompted them to adopt "the speed which most economizes their force." These academicians' accounts of the productive human machine then let their fellow Lavoisier establish during the later 1780s universal laws of labor,

47. Desaguliers, *Course of Experimental Philosophy,* 2:viii. For the "Microcosm," see Britten, *Old Clocks,* 430–31, and Chapuis and Droz, *Automata,* 128–31. For Chandos and Desaguliers, see Stewart, *Rise of Public Science,* 214–15.
48. Morton, "Concepts of Power," 75; Linebaugh, *London Hanged,* 311–12.

provided "the person submitted to experiments doesn't carry his efforts too near to the limit of his forces, since otherwise he would be in a state of suffering and leave the natural state." Lavoisier's respiration trials on these states demonstrated "that there exists for each person an unfailing law" governing body condition and work rate.[49]

Lavoisier's assays of the worth of philosophers and musicians alongside artisans and farmers moved well beyond the realm of metaphor. In summer 1791, he explained how his laboratory work on the animal economy would place the levers of power in the hands of the savant. Automatic systems of nature and society were susceptible to this careful management. "Like the physical order, the moral order has its regulators, and if it were otherwise human societies would long ago have ceased to exist, or rather they never would have existed." As Norton Wise points out in his study of the technologies of late-eighteenth-century French rationalism, strategies used to produce the sphere of enlightened action secured that culture to the extent that they disappeared from view, so that the nature of the late Enlightenment seemed an unmediated reality. Soldiers, engineers, prisoners, and workers were subjected to rather literal forms of enlightened discipline. The social order engineered their mechanization, and thus those who did this engineering could represent this condition as natural.[50] Vaucanson and Desaguliers, like Coulomb and Lavoisier, demonstrated machines well-matched to their accounts of the working human body and the principles that should govern disciplined labor. Peter Linebaugh, historian of the eighteenth-century London working class, connects the mechanization of coal heaving and silk weaving with measures against artisan resistance and customary labor processes. In his comparison of the Lyons initiatives of Vaucanson and comparable measures in the silk trade in Spitalfields, Linebaugh argues that strong legal codes were insufficient to impose new work systems. "Mechanical, organizational and geographical strategies were necessary accompaniments to strictly legal prohibitions." This is why the automata were vital for the materialization and evaluation of the laboring body.[51]

Machines under the Similitude of Men

It was not obvious to all eighteenth-century observers that social order could be secured by automatism. In the early-eighteenth-century exchanges involv-

---

49. Lindqvist, "Labs in the Woods," 307–10. Desaguliers's estimate of maximum work is cited in Ozanam, *Recreations*, 2:100. Coulomb's results are in his "Mémoire sur la force des hommes," 292, discussed in Gillmor, *Coulomb*, 78, and Vatin, *Le travail*, 41–51. Lavoisier's law is stated in Lavoisier and Seguin, "Mémoire sur la respiration," in Lavoisier, *Oeuvres*, 2:696.

50. Lavoisier and Seguin, "Premier mémoire sur la transpiration des animaux," *Oeuvres*, 2:704–14, 713 (read June 1791); Wise, "Mediations," 244–47; Bensaude-Vincent, *Lavoisier*, 221, 228.

51. Linebaugh, *London Hanged*, 270. See Stewart and Weindling, "Philosophical Threads."

ing Samuel Clarke, Leibniz, and the deists, as in responses to Joseph Priestley's materialism and determinism in the 1770s, it was a common theme that the identification of the human and the machine would spawn libertinism, atheism, and insurrection.[52] When issues of materialism and its morality were salient, connections with the automata shows and with bodily discipline were pervasive and threatening. In the immediate wake of Joseph Priestley's texts on philosophical materialism, for example, the Catholic priest Joseph Berington suggested that the notorious materialist should team up with James Cox, who would be able to stuff his Mechanical Museum with "two or three men machines of his own construction, that might really operate in a human manner, might gradually advance to the summit of knowledge in all the arts and sciences, and perhaps present the public with their several discoveries in religion, philosophy and politics." Among Priestley's contemporary supporters, the London journalist William Kenrick filled pages of his widely read *London Review* with ribald jokes and vicious attacks on materialism's enemies. He reckoned that a priest like Samuel Clarke "was confessedly so merely a reasoning machine that he would almost tempt one to think matter might think and that he himself was a living proof of it."[53]

These jokes depended on the familiarity of the metropolitan audience with the capacities of automata. Berington's squib set a precedent for connections between attacks on enlightened progressivism and on the capacities of such machines, prefiguring the satires of the *Anti-Jacobin* in the 1790s against Erasmus Darwin's similarly materialist doctrines.[54] Within the communities of enlightened philosophers themselves, furthermore, the discourse of self-moving machines was increasingly fashionable. Kenrick and his Grub Street friends, denizens of the lowlife of the Enlightenment, used automata to represent the behavior of the theater and the workshop. Debates on the mechanical passions prompted a duel between Kenrick and Garrick in the 1760s. Soon after, Kenrick published a series of essays on perpetual motion and automatism that defended materialism against the critique of pious divines, insisted on the indistinguishability of humans and automata, and concluded that it was feasible to produce self-moving machines. "A discovery of this nature," he announced, "would be of the utmost advantage to the commercial world."[55] In that world, of which Priestley and the rational dissenters were members, the automatism of humans was a crucial issue. Priestley's enlightened patron Josiah Wedgwood set out to "make machines of men as cannot err." His works were run according to a system of unprecedented time discipline, rendering

---

52. Vartanian, *Diderot and Descartes;* Yolton, *Thinking Matter;* Porter, "Barely Touching."
53. Yolton, *Thinking Matter,* 117–19.
54. Edmonds, *Poetry of the Anti-Jacobin,* 147 (April 1798).
55. Kenrick, *Lecture on the Perpetual Motion,* 9, 41, and *Account of the Automaton,* 24.

employees clockwork machines in an all-but-literal sense. Simultaneously, the educational programs touted by reformers like Wedgwood and his Midlands allies supposed that the human mind could be programmed if suitably planned and mechanically ordered.[56]

Thus the relation between the automaton, the rational human, and the social geography of industry, training, and display was fundamental for these late Enlightenment projects. In 1782, Wedgwood counseled his colleague James Watt on the right way of self-management: Watt's own body should be handled like "any other machine under your direction." "Production utopias," Thomas Markus's term for the visionary workshops of the late Enlightenment, often included provision for the mechanical training of their inmates. The ambitious lawyer Jeremy Bentham notoriously presented himself as this kind of self-regulated machine, seeking to contest plebeian cultures of the body with an exemplary public postmortem and the proposal that the corpses of the enlightened be used in "dialogues of the dead." With such inanimate bodies, "a commemorative festival might be exhibited after the manner of a comedy at one or more of the theatres," starring Bentham himself in conversation with Aristotle or Bacon.[57] This visionary self-image was developed alongside the palpable schemes of Benthamite panopticism, first designed by Samuel Bentham, Jeremy's brother, to accelerate production in tsarist wood yards in the 1780s. This was a scheme its protagonists reckoned would allow "the construction of a set of machines under the similitude of men." After 1795, Samuel Bentham and his collaborators, the engineers Marc Brunel and Henry Maudsley, used the resources of the Royal Navy to overhaul shipyards under a regime that connected public shows with automatic labor. The self-acting tools of the new precision workshops began to intrude on, and subvert, customary labor relations. Maudsley's lathes and blocks for the dockyard production line depended on, and in turn helped secure, reliable artisans. As in Vaucanson's silk mills, visibility was an invaluable aspect of this industrial reformation. Accountancy mechanized scrutiny while automation disciplined labor. "On entering the block mills, the spectator is struck with the multiplicity of its movements and the rapidity of its operations."[58] So the mills became impersonal, their inmates engines the parts of which were men. The automatism of the machines and of the workforce sustained—and was regulated by—the enlightened supervision of military and industrial production.

56. Thompson, *Customs in Common*, 385–86, 403; McKendrick, "Josiah Wedgwood," 34.

57. For Wedgwood, see Ignatieff, *Just Measure of Pain*, 68; for Bentham, see Richardson and Hurwitz,"Bentham's Self Image"; Schaffer, "States of Mind," 287–89. For Bentham and "production utopias," see Markus, *Buildings and Power*, 123–27, 286.

58. For "machines under the similitude of men," see Bentham, *Panopticon*, 127–28. For Portsmouth, see Cooper, "Portsmouth System," 213–14; Linebaugh, *London Hanged*, 399–401; Ashworth, "System of Terror."

In the work in which he described the panopticon as the culmination of enlightened political anatomy, Foucault argued that the knowledge of and power over "those who are stuck at a machine and supervised for the rest of their lives" accompanied the production of the modern soul as "the prison of the body." [59] To argue that the mechanics of social order produce the individual, who could scarcely possess a subjectivity without becoming a cog in this machine, is one version of the theme of the human automaton traced here. Foucault's final interrogation of the philosophy of the classical age involved a series of commentaries on Kant's essay "Answer to the Question, What is Enlightenment?" This paper, sent to the *Berliner Monatsschrift* at the end of September 1784, has been treated as a key to the Enlightenment's meanings. It has provided the occasion for a major exchange between Foucault's genealogy of the present and Habermas's theory of the public sphere. Scholars have rightly insisted on the immediate political context in which the question was posed in the reading clubs of Frederician Berlin, debates about the limits of civil marriage and of the rights of the state to deceive its subjects. They have also noted the transformations of that context implied by Kant's lapidary answer: "Enlightenment is humanity's emergence from its self-incurred immaturity." The theme soon became banal or, worse, risible. By 1790, under a more conservative Prussian regime, one pamphlet asked, "[I]f this is Enlightenment, what is nonsense?" while two years later appeared a similarly minded proposal "to abolish the fashionable term Enlightenment." [60]

Kant's text was linked with the discourse of automata that Foucault himself saw as a distinctive characteristic of the milieu that produced it. "Dogmas and formulae, those mechanical instruments for rational use (or rather misuse) of his natural endowments, are the ball and chain of his permanent immaturity. If anyone did throw them off, he would still be uncertain about jumping over even the narrowest ditch, because he would be unaccustomed to free movement of this kind." Priests, doctors, and tutors were all figured as parts of an oppressive apparatus, their immature victims turned into automata. Kant conceded the necessary restriction on private reason, since "we require a certain mechanism whereby some members of the commonwealth must behave purely passively so that they may. . . . be employed by the government." Subjects so employed, especially in its military, fiscal, and ecclesiastical functions, were seen as acting "as part of the machine." In his positive argument for the free use of public reason, Kant lauded Friedrich II's regime, argued that the

---

59. Foucault, *Discipline and Punish*, 29–30.
60. Kant, "What is Enlightenment?" The Habermas-Foucault exchange is documented in Kelly, *Critique and Power;* the background to Kant's text is described in Bahr, "Kant, Mendelssohn, and the Problem of Enlightenment from Above"; Birtsch, "Die Berliner Mittwochsgesellschaft"; Schmidt, "Question of Enlightenment," 269–91. The anti-enlightenment satires are described in Schneiders, *Die wahre Aufklärung*, 224, 233.

development of free thought would gradually increase subjects' capacity to act freely, and finally predicted that the state would find it could "profit by treating man, who is *more than a machine,* in a manner appropriate to his dignity." Kant's essay thus ended with an explicit rejection of La Mettrie, the burden of his essay insisting that the sphere of free reason lay outside the machine.[61]

As an academic philosopher and Prussian subject, Kant's argument that such mechanization was a limit on freedom could be contrasted with that of his most eminent predecessor, Christian Wolff. Appointed mathematics professor at Halle in 1706, Wolff thereafter attained a central role in the Republic of Letters. As principal writer for the Leipzig *Acta eruditorum,* Wolff was in a good position to judge and explicate the variety of machines touted round the European courts, and to link these new devices with Leibnizian doctrines of living force. As William Clark shows in detail in his contribution to this book, Wolff's expulsion from Brandenburg and his chair by Friedrich Wilhelm I followed a controversial lecture in which, by lauding the morals of Confucianism, he enraged modish Pietists by arguing for the redundancy of Christian revelation in the cultivation of good social order. Some saw Wolff's disgrace as a result of his exaggeratedly stiff sense of self and, as Clark points out, the expulsion was entangled with issues of fatalism and automatism. The young natural philosopher Leonhard Euler met Wolff at Marburg University soon after these events in 1723. According to Euler, a member of the Berlin court had told the king "that, according to [Wolff's] doctrine, all soldiers were nothing but machines, and that, should one desert, it would be a necessary consequence of its structure, so that it would be unjust to punish them, as if one might punish a machine for producing such and such a motion." Marburg was already notable for its links with engineering and state planning: the Hesse-Kassel regime encouraged links between new machinery, military discipline, and economic reform. Meanwhile Wolff's admirers, led by Ernst von Manteuffel, founded a Berlin club in his honor, struck a medal bearing Wolff's image, adding the telling phrase "sapere aude" in 1736. This was just the phrase Kant picked up in his 1784 essay.[62]

On the accession of Friedrich II, the Marburg exile was attracted back to his chair at Halle. The themes of determinism and political order then dominated the often violent debates that raged at Berlin and elsewhere on the implications of Wolffian philosophy and its alternatives in the newly powerful

61. Kant, "What is Enlightenment?" 54–55, 56, 59–60: emphasis in the original. Compare Foucault, "What is Enlightenment?" 36.

62. Euler, *Lettres,* vol. 2, letter 84 (December 1760). For Wolff's predicament in the 1720s, see Wundt, *Der deutsche Schulphilosophie,* 199–230; Saine, "Who's Afraid of Christian Wolff?" 102–33. For Hesse-Kassel and technology, see Philippi, *Landgraf Karl von Hessen-Kassel,* 609–15. For "sapere aude," see Venturi, *Utopia and Reform,* 6–9.

Prussian military state. It was for this reason that Euler, a staunchly Calvinist protagonist of these academic fights, chose in the early 1760s in a set of published letters to the king's young cousin Charlotte to revive memories of Wolff's troubles and, in particular, of the ambiguities of automatism. Chief among these ambiguities was, again, the puzzle of discipline: "[I]f the bodies of men are machines," Euler alleged, "like a watch, all their actions are a necessary consequence." As Mary Terrall and William Clark both point out elsewhere, Euler was a preeminent expositor of the reality of final causes in well-founded sciences. Materialism reeked of antiteleological dogma.[63] The king himself rejected the view that his subjects might be "a sort of machine, only marionettes moved by the hand of a blind power," for then all political discipline would be supererogatory. In 1784, Kant asserted a strong distinction between loyal subjects, who must behave like automata, and rational citizens, who must be allowed public freedom of conscience. Conservatives under the new Prussian regime insisted that there could be no such distinction. "A tolerant governor," wrote the arch-reactionary Johann Wöllner in 1785, "looks upon all his subjects from a single point of view, namely that they should be citizens of the state." Thus, if Wolff had once been accused of limiting state power by exaggerating subjects' internal (and mechanical) determination, Kant proposed to limit state power by redefining that form of determination.[64]

So Kant's account of true enlightenment and its relation with governmentality hinged on a careful specification of the mechanization of the subject. His final remark against La Mettrie was itself rather pointed, since his monarch had penned the eulogy of the refugee physician. La Mettrie's book, the king wrote, "was bound to displease men who by their position are declared enemies of the progress of human reason." "All those who are not imposed upon by the pious insults of the theologians mourn in La Mettrie a good man and a wise physician." Kant's insistence on the falsehood of La Mettrie's central thesis thus appeared reflexively as an example of his claim to free use of public reason, in despite of piety and politics, for which his 1784 essay argued.[65] The appeal to automata, so often a theme in biographical reminiscences of Kant's daily habits—William Clark links Kant's obsessive timetables with his attack on metaphysics—was also a fundamental theme in his political critique in this decade. In his second *Critique* (1788), he asserted transcendental freedom, the

---

63. Euler, *Lettres*, vol. 2, letter 84. For Wolff's polemics, see Polonoff, *Force, Cosmos, Monads*, 68–92, and Calinger, "Newtonian-Wolffian Controversy," 329. For anti-Wolffianism in Berlin see, especially, Terrall, "Culture of Science," 354, 359.

64. Friedrich II against Holbach in Cassirer, *Philosophy of the Enlightenment*, 71; Wöllner in Henri Brunschwig, *Enlightenment and Romanticism*, 167; compare Epstein, *Genesis of German Conservatism*, 140–54.

65. Friedrich II, "Julien Offray de la Mettrie," 8–9.

precondition of moral law, in contrast to determinists' notions of automatism. Humans subject to mere physical or psychological causality would enjoy "nothing better than the freedom of a turnspit." He attacked notions of time and space as "attributes belonging to the existence of things in themselves" because then "man would be a marionette or an automaton, prepared and wound up by the Supreme Artist. Self-consciousness would indeed make him a thinking automaton, but the consciousness of his own spontaneity would be mere delusion if this were mistaken for freedom." When his critique turned to the polity, in the *Groundwork of the Metaphysics of Morals* (1785), Kant insisted that a monarchy "corresponds to a living body when ruled by the inherent laws of the people, and to a mere machine when ruled by a single absolute will." Tyranny made its subjects machines and was itself one.[66]

A Naked Automaton Ends the Enlightenment

The very moment at which Kant's essay appeared coincided with a remarkable show of the power and meaning of an automaton, the Turkish chess player. In that month of September 1784, this astonishing machine appeared first at the great Michaelmas Fair in Leipzig, then elsewhere in the German lands. Spectators were shown into a darkened chamber in which stood a large cabinet running on castors, behind it an impressive full-scale model of a seated Turk smoking a pipe. On top of the cabinet was screwed a chessboard, the object of the Turk's fixed attention. The showman would open the front and back of the cabinet, revealing a complex array of gearwheels, barrels, and pulleys. The custom was to shine a candle into the cabinet to show that nothing could possibly be hidden, and the Turk's torso and legs would also be stripped bare. "You see at one and the same time, the naked Automaton, with his garments tucked up, the drawer and all the doors of the cupboard open." After the automaton had been wound up, giving it enough power to run for about a dozen moves, the games would begin, the Turk gracefully moving pieces with his left hand, nodding his head when giving check, tapping the table and replacing any piece if a false move was made by his opponent, and bowing to the spectators when the game ended, almost always with the Turk's triumphant victory. One intriguing and suggestive accomplishment of the automatic chess player well illustrated how it helped link public show and enlightened reason. The academies' mathematicians had devoted some attention to the puzzle of the Knight's Move, the path of a knight touching every square on a chessboard just once. Baptized

---

66. Kant's use of the automaton in these texts is described in Mayr, *Authority, Liberty, and Automatic Machinery*, 133, 136. For a contemporary Berlin account of Vaucanson's automata, see Martius, "Vaucansons Beschreibung eines mechanischen Flötenspielers" (1779), in Völker, *Künstliche Menschen*, 103–12. For the political background, see Hinske, *Kant als Herausforderung an die Gegenwart*, 67–78; Beiser, *Enlightenment, Revolution, and Romanticism*, 32–6.

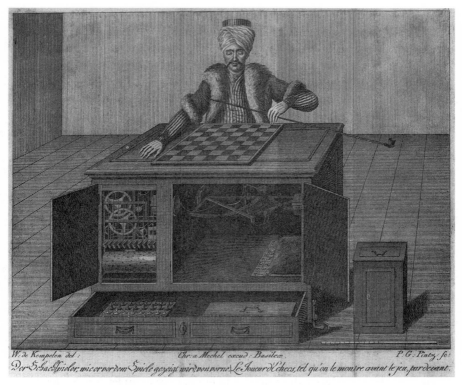

Fig. 5.4. Wolfgang von Kempelen's Turkish chess player, showing its internal gearing and chessboard. From Karl Gottlieb von Windisch, *Inanimate Reason* (London, 1784).

"Euler's problem" after the great mathematician's analysis published by the Berlin Academy in 1759, it became a cynosure of rational skill. Chess puzzles mirrored the evolutions of Potsdam parade-ground drill. The Turk's manager would challenge the audience to choose an initial square; the Turk would then show the right route round the board. Matching Euler's analysis demonstrated the machine's intelligence and played on the materialist possibilities of mid-century mathematics. "Never before did any mere mechanical figure unite the power of moving itself in different directions as circumstances unforeseen and depending on the will of any person present might require." Nowhere else in Europe did the relation between intelligence, mechanism, and concealment become such a matter of public interest.[67]

67. [Windisch], *Inanimate Reason*, 25–26, vi. For the show at Leipzig, see Carroll, *Chess Automaton*, 22–23. For the history of the Knight's Move, see Frolow, *Le problème d'Euler*. For the enormous contemporary literature on the Turkish chess player, see Dotzler, Gendolla, and Schäfer, *Maschinen-Menschen*, 14–22.

156   Simon Schaffer

This automatic Turk was built for display at the Habsburg court in 1769–70, in the wake of Knaus's earlier writing automata, which merely represented the mechanical rote of bureaucracy—the Turk instead mechanized higher faculties. Its designer was a Slovak engineer and courtier, Wolfgang von Kempelen, a prominent civil administrator and devotee of a range of schemes ranging from mining technology and steam pumps to the mechanization of the human voice. His chess player was first put on show in competition with magnetic tricks at Schönbrunn and then at Kempelen's house in Bratislava in the early 1770s. The automaton went on its travels round Europe in early 1783, partly as a result of the commands of the Emperor Joseph II, then rather involved in a major anti-Turkish diplomatic campaign, partly through the commercial ambitions of its promoters.[68] A game was staged at the Paris Academy of Sciences in May 1783 against the preeminent chess player Philidor (the Turk, for once, lost). By the following winter, the Turk had reached London and its notoriety had been assured by the appearance of a catchpenny booklet penned by Kempelen's aide, Karl von Windisch. This work, whose English translation bore the significant title *Inanimate Reason,* cheerfully acknowledged that some "deception" must be involved in the automaton's design, reported that there were those who suspected Kempelen of dealing with the devil, and chorused that "mathematicians of all countries have examined it with the most scrupulous attention without being able to discover the least trace of its mode of operation." Some Londoners straightforwardly guessed that within the Turk's cunning cabinet lay hidden a diminutive and human player. In Paris in September 1783, Friedrich Melchior von Grimm's *Correspondance littéraire,* chief organ of a decisive if restricted public of enlightened taste, agreed that the Turk "could not perform such a large number of different moves, whose determination could not be known in advance, without being subjected to the continual influence of an intelligent agent." Despite, if not because of, this view of the Turk's true nature, the Parisian journalists drew the moral that "physics, chemistry and mechanics have produced in our times more miracles than those believed by fanaticism and superstition in the ages of ignorance and barbarism."[69] As the Turk moved between absolutist courts, public fairs, and salon journalism, Kempelen's design and Windisch's publicity for it thus linked the fate of the Enlightenment with that of their intelligent machine.

News of the automaton spread through the German lands during 1784.

68. Faber, *Der Schachautomat,* and Hankins and Silverman, *Instruments and the Imagination,* 190–97, for Kempelen's career.
69. Windisch, *Inanimate Reason,* 12–13, 35; [Grimm], *Mémoires Historiques,* 3:70 (September 1783). For the role of Grimm's *Correspondance,* see Bryson, *Word and Image,* 154; Habermas, *Public Sphere,* 41; Chartier, *Cultural Origins,* 36.

One report, sent by Karl Friedrich Hindenburg to the eminent head of the Berlin observatory, Johann Bernoulli, reckoned that since the Turk must need both a directing and motive force to play, it depended on a combination of mechanical gearing and some array of magnets. An even more enthusiastic analysis, under the name of Johann Jakob Ebert, mathematics professor at Wittenberg, was published in Berlin in early 1785. Both texts seemed convinced rather that the automaton was devoid of human agency than worried by just how the human intelligence on which it relied was concealed. Another observer, Johann Lorenz Böckmann, who saw the automaton during its visit to Karslruhe, accepted Hindenburg's idea that some form of magnetism was involved, but urged instead that magnetic counters might be used to communicate to the hidden chess player inside. He published this "hypothetical clarification of the famous chess player" in his local *Magazin für Auklärung* in early 1785. It was even credulously reported that Friedrich II himself had been beaten by, then had tried to buy, the Turkish machine.[70]

The persistent link between the Turk and magnetism, ranging from Kempelen's original attempt to quash magnetic tricks at Vienna in 1769 to the theme that some hidden magnets must drive the Turk in the 1780s, suggested an obvious comparison with the contrivances of another Viennese wizard, Franz Anton Mesmer, whose animal magnetic fluids were supposed to govern the human economy. Like Kempelen, Mesmer wowed French and German audiences between the later 1770s and 1784, and, as with the contemporary Turk, mesmerized subjects were seen as puppets under the guidance of a hidden and cunning master. In summer 1784 Benjamin Franklin, rather a fan of his acquaintance Kempelen's chess player, joined Lavoisier and other government commissioners in damning Mesmer's victims: "They are entirely under the government of the person who distributes the magnetic virtue." These commissioners concluded that Mesmer's victims were subject to the power of their own imagination, not to the effects of any genuine fluid. The methods Lavoisier used to calibrate mental and mechanical labor were here used to scotch irrational vaporings. One of Mesmer's fiercer critics was the reformist Paris physician Jean-Jacques Paulet, who published his attack at the same moment as the commissioners. According to Paulet, the tendency of mesmerism and a host of similar projects, whether masonic or physiocratic, was to teach "that destiny itself is determined by particular genies who guide us without our knowing and without our seeing the strings which hold us; at last that in this lower world we are all like real puppets, ignorant and utterly blind slaves."

70. Carroll, *Chess Automaton*, 23–25, 36; for the doubtless spurious purchase by Friedrich II, see [Walker], "Chess Automaton," 725.

Paulet carefully noted the mesmerists' insistence that each victim should "enlighten himself, that man must enjoy his rights, or at least make out the hand
which guides him." [71]

In short, their critics alleged, mesmerists convinced their victims that they
were slaves by turning them into real automata under the power of imagination, then proclaimed a spurious path to liberty in which boulevard showmanship was substituted for enlightened freedom. The year 1784 was marked
in Paris by the establishment of the secret societies that Paulet attacked,
through which, as Dena Goodman tells us, women were excluded and marginalized as subversive of truly rational philosophy—their alleged vulnerability to
Mesmer's blandishments only reinforcing this offensive. Mary Terrall explains
elsewhere in this book just how academic analysis worked to exclude putatively
unsound female mentality from the security of rational judgment, and specifies
how the commissioners investigating mesmerism deployed their own status as
state-sponsored judges of illusion. Robert Darnton has seen mesmerism and
cognate marvels as a mark of the French Enlightenment's final descent into an
occultist culture of the spectacle, most appealing to disaffected radicals and
hacks. Henri Brunschwig similarly analyzes the collapse of the Prussian Enlightenment from the 1780s in terms of young (and unemployed) intellectuals'
belief in the miraculous quality of everyday life. He found mesmerism, and
other enterprises that seemed to turn bodies into marionettes under hidden
power, among this new grouping. Indeed, Brunschwig connects the conservative reaction against the Enlightenment, of which Kant was an eminent victim,
and the experiences of the new king with a notorious somnambulist and clair-
voyante at the decade's end. [72] In these eschatological settings, therefore, the
distinction between the self-proclaimed community of enlightened philosophers and those of their critics and enemies was decisive. In 1784 Lavoisier and
his allies asserted a strong boundary between plebs, women, the aged, and the
weak, who were the unconscious subjects of spurious imagination, and rational savants, who alone could make out the realities of nature and morality.
One mesmerist then complained that if one believed these commissioners
then imagination "would almost be our normal state. If that is so, they should
have told us how they protected themselves when they judged magnetism."
They did so with the logic of the automata: subjects were to be seen as gov-

71. Franklin, *Report*, 27; Paulet, *L'Antimagnétisme*, 3–4. For mesmerism's fate in Paris in 1784,
see Gillispie, *Science and Polity*, 279–83. For Franklin's contact with Kempelen, see Hankins and
Silverman, *Instruments and the Imagination*, 197.

72. Goodman, *Republic of Letters*, 240–41; Brunschwig, *Enlightenment and Romanticism*,
165–66, 181–222; Darnton, *Mesmerism*, 162–65.

erned machines, and this was uniquely obvious to the citizens of the late Enlightenment.[73]

On Kempelen's death in 1804, the Turk was soon acquired by a clever Viennese musical engineer, Johann Maelzel, court mechanician for the Habsburgs and a close ally of one of their favored composers, Beethoven. Maelzel swiftly saw the patronage he could win by trading on Kempelen's automaton, and the Turk became a temporary habitué of the new Napoleonic courts in Germany. Automata like this one seemed to symbolize what enlightenment might achieve. This may help explain the appearance of automata and androids in immediate post-Enlightenment fictions such as those of Jean Paul (in *Titan*), Georg Büchner *(Leonce and Lena)* and E. T. A. Hoffmann.[74] Hoffmann, a fellow musician, found the figure of a mechanical Turk a suitably exotic subject for his pen and in 1814 sent a Leipzig musical magazine a story entitled *The Automata,* in which he "took the opportunity to express [himself] on everything that is called an automaton," teasingly hinting that the Turk might work by setting up a musical harmony with the mind of its audience. Hoffmann used his story to debate the most up-to-date views of occultist German philosophies of nature, much devoted to the inner rhythms of human mental life. The following year, he completed *The Sandman,* his most remarkable exploration of the "fraudulent imposition of an automaton upon human society" by a brilliant and demonic impresario-philosopher. Like Kleist's *Marionette Theatre,* which appeared in the Berlin newspapers in 1810, one salient theme of Hoffmann's story was that from the viewpoint of post-Enlightenment critique, automata might become all too perfect. To be truly human in an age when mechanics had engrossed so much was to be either inexact and faltering or a visionary genius akin to God.[75]

Maelzel threw himself into a lucrative mechanical project for regulating genius with a machine he named the "metronome." The metronome was, of course, a rather more potent means of mechanizing errant human capacities than any mere chess player. After furious patent suits with rival inventors, and complex negotiations with Beethoven, Maelzel established himself as the monopoly distributor of these newfangled musical timekeepers and repurchased

73. Deslon in *Supplément aux deux Rapports de MM. les Commissaires,* 9. For savants' insistence on imagination at the expense of mesmerists, see Azouvi, "Sens et fonction épistémologiques."

74. Sauer, *Marionetten,* 65–119; Gendolla, *Lebenden Maschinen,* 76–103.

75. Carroll, *Chess Automaton,* 43–46; Hoffmann to Rochlitz, 16 January 1814, in Hoffmann, *Selected Letters,* 217. The citation of the *The Sandman* is from Hoffmann, *Tales,* 212. Compare Hankins and Silverman, *Instruments and the Imagination,* 225–26; Völker, *Künstliche Menschen,* 476–79.

the Turk from the Bavarian court, where the occultist philosopher Franz von Baader and his colleagues were intrigued by automatism and mesmerism. Then, in 1818, Maelzel set off on a marketing and publicity tour of Paris and London. He helped realize his metronomic dream of an automatic orchestra, the Panharmonicon, for which Beethoven had composed his ghastly *Battle Symphony*. One London paper praised such an orchestra, which "displayed none of the *airs* of inflated genius, but readily submitted to being wound up." [76] Maelzel faithfully followed Kempelen's recipe and his metropolitan audience reproduced their earlier enthusiasm for the show. Ever considerate to this public, Maelzel even announced that the automaton would purposely make bad moves so as deliberately to lose if the company seemed bored with over-lengthy games, while the Turk's opponents were ordered to move as fast as possible to alleviate the tedium. A pamphlet authored by a pseudonymous Oxford graduate alleged the whole trick relied on a hidden piece of wire or catgut: "[I]t seems to be a thing absolutely impossible," the Oxonian alleged, "that any piece of mechanism should be invented which possessing perfect mechanical motion should appear to exert the intelligence of a reasoning agent." [77]

By the end of 1820, the Turk's nemesis had appeared in the person of Robert Willis, an ingenious young Londoner, heir to a distinguished medical family—his father famously attended George III during the monarch's madness. In later life Willis himself became a distinguished Cambridge mathematician and professor of applied mechanics, introducing an innovative teaching strategy using entire museums of model machines. [78] The young mechanic bought copies of the games the Turk had played and made sure to visit Maelzel's show while it occupied a cramped space, "more favourable to examination as I was enabled at different times to press close up to the figure while it was playing." He smuggled an umbrella into the room so as to measure "with great accuracy" all the dimensions of the Turk's celebrated cabinet. Willis's accurate umbrella and his command of wheelwork made all the difference. The chest, he demonstrated, was much larger than it seemed, giving enough room for a fully grown (and skillful) human chess player to fit inside. "Instead of referring to little dwarfs, semi-transparent chess boards, magnetism, or supposing the possibility of the exhibitor's guiding the automaton by means of a wire or piece of catgut so small as not to be perceived by the spectators," Willis's *Attempt to Analyze the Automaton Chess Player*, finished in December 1820, proffered the simplest possible scheme of the Turk's hidden intelligence. The noisy gearwheels were there so that their sound would conceal any noise made by the

76. Harding, *The Metronome*, 22–28; Altick, *Shows of London*, 357–59.
77. Carroll, *Chess Automaton*, 47–51; *Observations on the Automaton*, 7.
78. Hilken, *Engineering at Cambridge*, 52–54.

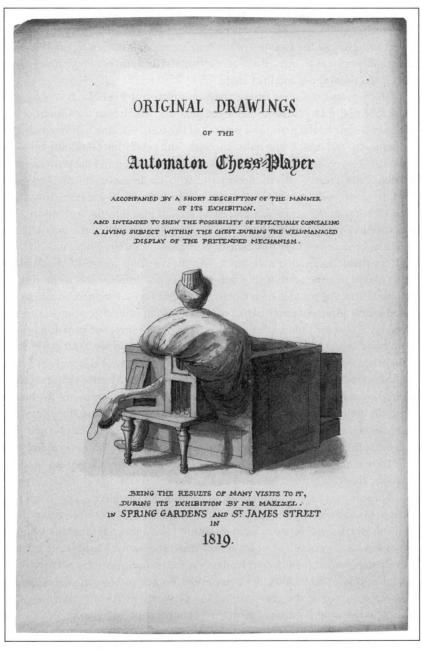

ORIGINAL DRAWINGS

OF THE

Automaton Chess-Player

ACCOMPANIED BY A SHORT DESCRIPTION OF THE MANNER
OF ITS EXHIBITION,

AND INTENDED TO SHEW THE POSSIBILITY OF EFFECTUALLY CONCEALING
A LIVING SUBJECT WITHIN THE CHEST DURING THE WELL-MANAGED
DISPLAY OF THE PRETENDED MECHANISM.

BEING THE RESULTS OF MANY VISITS TO IT,
DURING ITS EXHIBITION BY MR MAELZEL.
IN SPRING GARDENS AND St JAMES STREET
IN
1819.

Fig. 5.5.   Robert Willis's initial attack on the problem of the Turkish automaton. From Robert Willis, *Original Drawings of the Automaton Chess Player* (London, 1819).

concealed player. Willis explained what mechanism could not do: "[T]he movements which spring from it are necessarily limited and uniform, it cannot usurp and exercise the faculties of the human mind, it cannot be made to vary its operations so as to meet the ever-varying circumstances of a game of chess. This is the province of intellect alone."[79]

Many found the story of the chess automaton irresistible. It was a commonplace that such machines belonged at court, whether in the Orient of the Arabian Nights or the grandiose palaces of the tsars. "To the half-bred savages of the north," a London journalist sneered, "the exhibition could not fail to be striking." Novels and reviews told how the Turk had conned the powerful and humbled the great: "[E]ven Bonaparte, who made automata of Kings and Princes at his will, was foiled in an encounter with the automaton chess player." Later in the century, a successful French play put the Turk on stage in a victorious contest with Catherine the Great. The genteel and the superstitious could all be deceived by a mechanism unmasked in public prints directed at a new confident rational readership.[80] The most telling lesson of the Turkish chess player was the relationship between machine intelligence, technological progress, and the puzzles of concealment. This was a moment when, as the automaton's admirers never hesitated to remark, "political economists amuse themselves and the public with the nicely-balanced powers of man as a propagating and eating *animal* and philosophers and divines often assure us that he is, in other and higher respects, but a *machine* of a superior description." One economic journalist, describing the rapid growth and progress of automation in the Lancashire cotton industry, told the story of the invention of the power loom, in the famous year of 1784, by Edmund Cartwright. Just months before Vaucanson's swivel looms reached the Lancashire factories, Cartwright had heard about the mechanical Turk, trusted its purely mechanical origin, and thus been convinced that a weaving machine could scarcely be harder to make than one that could play chess so well.[81]

These innovative conjunctions of automatic machinery and confident analysis were even apparent in the Turk's capacity to solve Euler's Problem. Willis devoted some pages of his demolition job to showing how the Knight's Move worked; a couple of years earlier his erstwhile colleague Babbage had already expounded for London readers the ingenious operational method involved in this trick. In 1803, the preeminent French military engineer, Lazare Carnot, indicated at the very start of his *Geometry of Position* that a mathematical science he baptized "geometry of transposition" would include the

79. Willis diary, 78 (6 June 1820), 88 (16 October 1820), 106 (21 December 1820); Willis, "Original Drawings"; Willis, "An Essay," 2; Willis, *An attempt*, 11.

80. Walker, "Anatomy of the Chess Automaton," 723; Chapuis and Droz, *Automata*, 365–67.

81. "Scientific Amusements," 441; Baines, *Cotton Manufacture in Great Britain*, 229.

Knight's Move within its scope. Carnot explained that such puzzles would form one aspect of a general technique that would allow "the passage from geometry to mechanics" and thus help forge a science of machines, such as knitting frames, where "the aim is to establish such-and-such relations between the directions or the speeds of the different points of a system." Carnot, Babbage, and Willis were among those scientific protagonists who now showed just how military, industrial, and mathematical concerns converged in puzzles like this. In using the Knight's Move to display its rational power, the impresarios of the Turkish chess player selected a key element in new physicalist sciences of machine behavior, important accompaniments of Enlightenment analysis, as Alder shows elsewhere in this book.[82] These new sciences were lauded by the Edinburgh natural philosopher David Brewster in his best-selling work *Natural Magic* (1832), which was devoted to teaching his fellow citizens the inner secrets on which all apparently miraculous and surprising mechanical devices depended. Part of the point was characteristic of a certain kind of enlightened demystification: gaudy tricks dangerously conned the ignorant into idolatry. Part of the argument, however, was economic. In his chapter on automata, Brewster summarized the unmasking of Kempelen's chess player and juxtaposed it with James Watt's engineering. "Those mechanical wonders which in one century enriched only the conjurer who used them, contributed in another to augment the wealth of the nation. Those automatic toys," he concluded, "which once amused the vulgar, are now employed in extending the power and promoting the civilization of our species."[83]

Aftermath: We Have Never Been Enlightened

Automata thus provided an apt topic for the Enlightenment's concerns with power and civilization and its attitudes to the vulgar and the worker. The Encyclopedists portrayed talented artisans as visible automata. "Most of those who engage in the mechanical arts have embraced them only by necessity and work only by instinct . . . [I]n a workshop it is the moment that speaks, and not the artisan."[84] Smith and Ferguson also defined for themselves a place whence they could analyze the efforts of labor, thus discriminating between philosophers' freedom and the mechanics of manufacture. La Mettrie, Desaguliers, and Vaucanson set out the mechanisms governing the animal economy and the social economy, specifying as they did so the crucial role they should discharge. Coulomb and Lavoisier then designed techniques for the precise

82. Babbage, "Euler's Method," 72–77 and Carnot, *Géométrie de Position*, xxxvi. For machines and physicalism, see Gillispie, *Lazare Carnot*, 132; Daston, "Physicalist Tradition," 282–83; Séris, *Machine et communication*, 343–76; and, especially, Guillerme, "Network," 154.
83. Brewster, *Natural Magic*, 336.
84. D'Alembert, *Preliminary Discourse*, 125–26.

evaluation of labor on the presumption that the animal economy—and thence the moral economy— could be understood as a self-regulating system. These British and French views provide examples of ways in which philosophers specified their own predicament. During the 1780s, when the very character of the Enlightenment was in question, further cases of this self-analysis were provided by Bentham's simultaneous production of new personae for the wardens and the inmates of productive institutions, by Kant's discrimination between the free citizen-philosopher and the automated soldier-subject, and by French savants' distinction between the victims of imagination and the rationality of the enlightened. The automatic chess player provided a dramatic case of these relations between the automaton, the enlightened, and the self because for some it was a machine that displayed remarkably human attributes, while for others, including its designer, it was a human who performed in a strikingly machinelike manner. The intelligent automaton has remained a salient theme in analyses of the Enlightenment's aftermath.

Inspired by Willis and Brewster's stories, Poe set out the mechanical principles of poetic composition, and Babbage used games-playing automata to theorize the mechanization of intelligence, thence establishing the principles of a political economy of intellectual labor. For Marx the culmination of manufacture was just a system "set in motion by an automaton, a moving power that moves itself; this automaton consisting of numerous mechanical and intellectual organs, so that the workers themselves are cast merely as its conscious linkages." Thus science "compels the inanimate limbs of the machinery to act as an automaton" and ultimately "the machine is itself the virtuoso."[85] The automaton stays in place as a symbol of modernity because it helps us see the effects of this expropriation of virtuosity, science, and reason. Enlightened science imposed a division between subjects that could be automated and those reserved for reason. Such a contrast between instinctual mechanical labor and its rational analysis accompanied processes of subordination and rule. It seemed as if most subjects had never been, could perhaps never be, enlightened. The concern here has been to see how an enlightened public produced this grim view of society's mechanics.

Primary Sources

Babbage, Charles. "An Account of Euler's Method of solving a problem relative to the Move of the Knight at a Game of Chess." *Journal of Science and the Arts* 3 (1817): 72–77.
Baines, Edward. *History of the Cotton Manufacture in Great Britain*. London: Fisher, 1835.
Bentham, Jeremy. *Panopticon; or, the Inspection House*. London: Payne, 1791.
Brewster, David. *Letters on Natural Magic*. 1832. Reprint, London: Chatto and Windus, 1883.
Carnot, Lazare. *Géométrie de position*. Paris: Crapelet, 1803.

85. Wimsatt, "Poe and the Chess Automaton"; Babbage, *Passages*, 465–71; Marx, *Grundrisse*, 692–93 (February 1858).

Coulomb, Charles. "Mémoire sur la force des hommes." In *Théorie des machines simples.* Paris: Bachelier, 1821.

D'Alembert, Jean. *Preliminary Discourse to the Encyclopaedia of Diderot.* Edited and translated by Richard Schwab. Indianapolis: Bobbs Merrill, 1963. First French edition published 1751.

Desaguliers, John Theophilus. *A Course of Experimental Philosophy.* 2 vols. London, 1734–44.

Deslon, Charles. *Supplément aux deux Rapports de MM. les Commissaires.* Amsterdam, 1784.

Diderot, Denis, and Jean le Rond d'Alembert, eds. *Encyclopédie, ou Dictionnaire raisonné des Sciences, des Arts et des Métiers.* Paris, 1751–72.

Edmonds, Charles, ed. *Poetry of the Anti-Jacobin.* London: Sampson Low, Marston, Searle and Rivington, 1890.

Euler, Leonhard. *Lettres à une Princesse d'Allemagne.* 3 vols. Paris, 1787–89. First edition published 1768–74.

Ferguson, Adam. *An Essay on the History of Civil Society.* Edited by Duncan Forbes. Edinburgh: Edinburgh University Press, 1966. First edition published 1767.

Franklin, Benjamin, et al. *Report of Dr. Benjamin Franklin and other Commissioners charged by the King of France with the Examination of the Animal Magnetism.* London, 1785.

[Grimm, F. M. von]. *Mémoires historiques littéraires et anecdotiques.* London, 1813.

Hoffmann, E. T. A. *Tales of Hoffmann.* Edited by E. F. Bleiler. New York: Dover, 1967.

———. *Selected Letters.* Edited by Joanna C. Sahlin. Chicago: University of Chicago Press, 1977.

Kant, Immanuel. "An Answer to the Question: What is Enlightenment?" (1784). In *Kant: Political Writings,* 2d ed., edited by Hans Reiss, 54–60. Cambridge: Cambridge University Press, 1991.

Kenrick, William. *Account of the Automaton constructed by Orffyreus.* London, 1770.

———. *Lecture on the Perpetual Motion.* London, 1771.

La Mettrie, Julien Offray de. *Man a Machine.* Edited and translated by Gertrude Carman Bussey. La Salle: Open Court, 1912. First French edition published 1747.

Lavoisier, Antoine. *Oeuvres.* 6 vols. Edited by J. B. Dumas and E. Grimaux. Paris: Imprimerie nationale, 1864–96.

———. *De la Richesse Territoriale du Royaume de France.* Edited by Jean-Claude Perrot. Paris: C.T.H.S., 1988. First edition published 1791.

Observations on the Automaton Chess Player. London, 1819.

Ozanam, Jacques. *Recreations in mathematics and natural philosophy.* 4 vols. Edited by Jean Montucla, translated by Charles Hutton. 1778. Reprint, London: Kearsley, 1803.

Paulet, Jean-Jacques. *L'Antimagnétisme.* London, 1784.

"Scientific Amusements—of Automata." *New Monthly Magazine* 1 (1821): 441–48, 524–32.

Smith, Adam. *An Inquiry into the Nature and Causes of the Wealth of Nations.* 2 vols. Edited by R. H. Campbell, A. S. Skinner, and W. B. Todd. Oxford: Clarendon Press, 1976. First edition published 1776.

———. *Lectures on Jurisprudence.* Edited by R. L. Meek, D. D. Raphael, and P. G. Stein. Oxford: Clarendon Press, 1978.

———. *Essays on Philosophical Subjects.* Edited by W. P. D. Wightman and J. C. Bryce. Oxford: Clarendon Press, 1980. First edition published 1795.

Vaucanson, Jacques. *An Account of the Mechanism of an Automaton.* London, 1742.

[Walker, George]. "Anatomy of the Chess Automaton." *Fraser's Magazine* 19 (1839): 717-31.

Willis, Robert. Diary, 1819-1821. Cambridge University Library MSS ADD 7574.

———. "Original drawings of the Automaton Chess Player." 1821. Cambridge University Library Adv.c.98.13.3.

———. "An essay on the Automaton Chess Player." London, 1821. Cambridge University Library Adv.c.98.13.4.

———. *An attempt to analyse the Automaton Chess Player.* London, 1821.

[Windisch, Karl Gottlieb von]. *Inanimate Reason, or a Circumstantial Account of that Astonishing Piece of Mechanism, M. de Kempelen's Chess-Player.* London, 1784.

3

Humans and Natures

Frontispiece from *A New Universal History of Arts and Sciences* (London, 1759), vol. 1.

The Enlightenment saw the rise of novel conceptions of nature and of humanity; in this section, we consider some of the connections between these domains. We explore the new understanding of human nature that was characteristic of the eighteenth century; and we examine how this informed, and was informed by, new approaches to the study of the natural world. Two dimensions of this nexus emerge with particular prominence: First, because humans were viewed as part of nature, changes in ideas of human nature were bound up with changes in the conception of nature itself. Second, inquiry into the natural world was seen to be framed by human nature and its limitations, including those thought to affect one sex especially or only certain cultural or social groups. "Man" was both object and subject of knowledge, so what was taken to be universal and what was taken to be specific about human nature shaped scientific inquiry. Alongside the universal attributes of humans, certain specific features also bore upon what could be known. These included a particular sensibility or "taste," thought of as being diffused among the polite sections of society, and also the weaknesses of mind thought of as characteristically feminine.

Michael Hagner opens consideration of these issues, picking up themes from Simon Schaffer's chapter in the previous section. Schaffer considered how enlightened thinkers dealt with automata, as regards what could distinguish humans from machines. For Hagner, the boundary creatures of the Enlightenment were its "monsters." Like automata, the "oddities" of the period indicated fundamental notions of humanity and nature, because of their status as challenging anomalies and deviants, creating troubles for systems of categorization. The supposedly universal reach of these systems in the eighteenth century wrenched monsters from their baroque state as curiosities and objects of wonder. Now their classification, anatomy, and generation had to be scrutinized intensively to "domesticate" them within the purview of enlightened rationality. Thus, Hagner considers how monsters were conceptually manipulated, culminating in late-Enlightenment Germany, when cabinets of curiosities were opened up as repositories of dead organisms that could serve as resources for inquiries into the processes of life and generation (including human generation).

Hagner shows that the interpretation and representation of monsters in the eighteenth century was also a political deployment of enlightened rationality, as in the "monster politics" of Peter the Great and the Russian court. Subsequently, monsters are shown to have played a critical part in the debate between "preformation" and "epigenesis" accounts of generation, where the issue concerned the preexistence of a formed embryo in the parental germ cells or its autonomous development from unorganized material. The German physiologist Caspar Friedrich Wolff was one of those who argued that monsters displayed deviations from the spontaneous processes of epigenesis, rather than being the result of divine providence or external interference in the course of generation. The connection between monsters and theories of generation is carried forward by Hagner to the rise of the human sciences in the late eighteenth century, represented by the work of the German anatomist Samuel Thomas Soemmerring.

Hagner illuminates aspects of the understanding of human nature in the German *Aufklärung* that are touched on later in this volume by William Clark. Immanuel Kant's "anthropology," or the knowledge of human nature, was taken to establish the conditions for human knowledge of the natural world. Anthropology came to be identified also with aesthetics, the science of "taste," which became the central category in a new understanding of art or culture as an expression of human life. Hagner invokes aesthetics when he notes that Wolff's work on the epigenesis of monsters suggested an aesthetic quality of each step in the development process. This was an instance of the enlightened interest in how human observers perceived aesthetic qualities in the natural world, in other words, what might be called the "sensibility of nature." The notion that particular kinds of "taste" underpinned varieties of natural knowledge was one with wide currency elsewhere in the eighteenth century.

As Clark notes, Kant used the dismissive term "women's sciences" for those based upon notions of taste, thereby linking aesthetics and anthropology with the Enlightenment interest in the differentiating characteristics of the sexes and the distinguishing features of male and female mental and physical attributes. A number of scholars, including Ludmilla Jordanova, Londa Schiebinger, and Thomas Laqueur, have recently scrutinized the late-Enlightenment articulation of a discourse of sexual difference, in which a wide range of mental and physical characteristics were ascribed specifically to one or the other sex. This development unfolded against the background of what G. J. Barker-Benfield has called "the culture of sensibility"—a realm of polite action and discourse that partook of a general "feminization" of manners. Certain kinds of scientific discourse had assumed a place in this culture by the mid-eighteenth century, through the work of such writers as Francesco Algarotti, Bernard de Fontenelle, and Benjamin Martin. Their literature made the case for science as part of enlightened culture by associating it with a "politeness" that was gendered feminine and proposed as an alternative to male rudeness or pedantry. Thus, women and female attributes were crucial to the public profile of enlightened science, even though women's active participation in science was strictly circumscribed by men's ascription to them of mental attributes thought to disqualify them from original scientific investigation.

Marina Frasca-Spada's chapter investigates the connections between polite learning and aesthetics with respect to the common situation of philosophical writings and the novel in eighteenth-century Britain. She sketches a world of philosophy (both natural and moral) that was infused with the values of polite sensibility, with its governing notion of enlightenment as conversation. Both philosophy and novels presented versions of a knowledge of human nature, which, even when putatively "scientific," was narrative in form and emotionally evocative for readers. While it might be possible to distinguish the two genres, Frasca-Spada argues that the boundaries were never entirely clear. Although such writings as David Hume's *Treatise of Human Nature* notoriously failed to reach polite readers, a spectrum of more "conversable" works leads all the way to Samuel Richardson's novels, which were renowned for their emotional impact on readers. Even in works of systematic philosophy, narratives of episodes (factual or fictional) were required to convey knowledge of human nature. (It thus seems telling that John Locke's *Essay Concerning Human Understanding* was presented to the reader as having arisen from a series of conversations, and that Hume described himself as an "ambassador" from the learned to the conversable world, in which women were "sovereigns.") The sensibility of readers was invoked by human action and suffering, as it was by scenes or objects of nature. While the sensational novels of the period have been designated by modern critics as "weeping kits," the same sympathy could be drawn forth (if usually to a lesser degree) by works of moral philosophy. It was in this context that the "science" of human nature was forged, a science that remained historical, narrative, and thoroughly conversable.

As Frasca-Spada notes, knowledge of human nature requires study both of the typical and of the unique, of what is common to all humans and what is specific to sex, social condition, habitation, nation, and so on. If human nature is understood as a condition for knowledge of the external world, then the question arises as to what particular kinds of knowledge might be conditioned by specific human characteristics. The role of women in enlightened science raised these issues to prominence. This is a theme of Mary Terrall's chapter, which makes use of feminist scholars' problematization of the Enlightenment rhetoric of universalism. The category of rationality has been shown to have been unequally applied to men and women, being frequently invoked in a discriminatory fashion to marginalize female attributes and achievements.

Terrall shows how this rhetoric was in play in the validation of mathematical analysis and the exclusion of metaphysical principles of physics (such as Maupertuis's least-action principle) in the Académie des Sciences in the last decade of the old regime. Maupertuis's teleological mechanics was condemned as "metaphysical" and expelled from science along with metaphysics in general. Terrall explores the range of demarcational criteria invoked by those who sought to expunge such metaphysics from rational mechanics. She also delineates the gendered aspect of the implicit anthropology that undergirded new legitimations of scientific knowledge. She shows how members of the academy drew upon gendered language and categories (among others) to demarcate "analysis"—understood both as a specific deployment of algebraic procedure and as a type of methodical thinking in general—from metaphysics and to validate a new kind of "rational" mechanics. This process coincided with a rupture with the polite female audience previously cultivated by members of the academy in the Parisian salons. The investigation of the mesmerism fad in the mid-1780s, by a commission of the academy, made available newly powerful discursive resources for categorizing the defects of female cognition. J. L. Lagrange, among others, echoed the denigration of the imagination directed against the female followers of Mesmer to justify his version of analytical mechanics.

While Terrall describes how the discriminating potential of Enlightenment notions of rationality was used to demarcate academic specializations, Emma Spary returns us to the world of polite, conversable culture in which natural history was practiced. She further enriches the picture of traffic between notions of nature and notions of humanity, displaying how conceptions of human nature reflected and were reflected in visions of the natural world. For natural history in France in the central decades of the Enlightenment, Spary stresses (as do Frasca-Spada, Lissa Roberts, and Jan Golinski elsewhere in this volume) that the polite culture of taste and conversation is the relevant context for much Enlightenment science, not the world of professionalization or institutional formation. The "beau monde" was also the world, as other scholars have noted, that was served by the growing commercial development of leisure and cultural production; although Spary rightly cautions against assuming that this "public sphere" achieved any real autonomy from court and state, to which links of patronage continued to bind it.

Spary notes how the terms "enlightenment" and "nature," both putatively universalizing in their function, were both less than universal in their practical application. She emphasizes how frequently schemes for hierarchical classification and discrimination were shifted between natural and social worlds, generating distinctions that segregated (for example) classes, nations, or genders. The separate works of François-Alexandre Aubert de la Chesnaye on animal classifications and aristocratic genealogy are thus emblematic of connections made even more immediately in Gilles-Augustin Bazin's work on bees and in Ignaz von Born's satirical natural history of monks. Spary argues that these works are more typical of the mode in which the French deployed "nature" in this period than the fringe speculations of atheists and materialists. The aristocratic and satirical taxonomies emerge as more symptomatic of the ways in which the enlightened sensibility was deployed prescriptively.

The taste for natural history in enlightened Paris was evidently not universally distributed. As Spary shows, it was understood to be something that distinguished its possessors from those who lacked such advantages. A taxonomy of the natural world was thus complemented by an implicit taxonomy of the social world, a scale on which prospective students of nature were graded. This is a point made also by the other contributors to this section: The notion that an understanding of nature and humanity required a particular taste or sensibility was applied in a differential manner, empowering for certain groups, and disempowering for others. Finally, Spary's focus on these issues in the geographically delimited world of a small area on the left bank of the Seine raises the issue of the specifically local character that natural knowledge could assume, which is treated in the chapters in the final section, "Provinces and Peripheries."

6

Enlightened Monsters

MICHAEL HAGNER

Monsters provoked contradictions and tensions for the Enlightenment. Remarkably, this challenge started at the end of a century in which they had become omnipresent. In the seventeenth century, monsters were the subject of conversations, discussions, and anecdotes, appearing as spectacles in court culture, at marketplaces, and at fairs. Numerous case studies filled erudite journals, and last but not least, natural oddities were desired objects for collectors' cabinets. It would be an exaggeration to say that monsters were completely unproblematic in the seventeenth century. But that century did demonstrate a relaxed attitude toward monstrosities unknown hitherto. It viewed them mainly as curiosities and sports of nature, as extraordinary products of nature's artfulness. In academic as well as in court circles, admiration and wonder guided the approach to monsters. They had the same status as rare beasts or corals in cabinets. In an atmosphere of baroque curiosity, monsters became interesting in themselves. In its most exaggerated form, this interest led to an aestheticizing of monsters, as, for example, in the paintings and drawings of the famous Gonzalez family of hairy humans, whose various portraits hung in the *Wunderkammer* in Ambras and in the gallery of the Farnese household in Rome.[1]

The seventeenth century had tried to tame monsters by aesthetic presentation and by the use of such categories as curiosity, wonder, rarity, and the playful productivity of nature. The turn of the century, however, saw an increasing demand for systematic classification of monsters, along with an aesthetic contrast between beauty and ugliness, proportion and deformation. Such demands were part of a general plea for a more systematic approach to monsters and marked a shift in view from an active and skillful nature to a more regular and prestabilized one. The image of a playful and tricky nature did not disappear completely, but it was overshadowed by an image of nature

---

1. On the role of the extraordinary and on the notion of monsters in the sixteenth and seventeenth century, see Park and Daston, "Unnatural Conceptions," and *Wonders,* esp. 173–214. See also Daston, "Marvelous Facts." On the painting by Agostino Carracci and the Gonzales family, see Zapperi, "Ein Haarmensch," and on the portraits in Ambras, see Scheicher, *Die Kunstkammer,* 149–50.

based on laws, "preformation," and deterministic processes that nevertheless incorporated deviations and the extraordinary.

In the seventeenth-century view, still powerful in the eighteenth, the monster was admired as a rare spectacle and curious object. After visiting the court's *Kunstkammer* in Cassel in November 1709, the distinguished scholar Zacharias Conrad von Uffenbach noted in his diary: "Furthermore there is a very curious biheaded monster, a six-month-old fetus in the womb which looks pretty good. Everything is well proportioned except the two heads. . . . Next to it is a *monstrum vitulinum* also with two heads, of perfect size and shape."[2] Uffenbach's description is an example of erudite pleasure in and fascination with monsters, regarded as highlights in a collection. They were estimated as interesting per se, and they did not shock or disturb the informed visitor. Fear and superstition were unknown to the connoisseur, more likely to be found in uneducated and uncultivated people. Caspar Friedrich Neickelius told the story of an erudite doctor who had spent a lot of money to buy "an extra rare skeleton of a seven-month-old [human] fetus . . . with the natural head of a pig and a muzzle." On his way home, the doctor forgot the box containing the skeleton in the house of a peasant. When the poor fellow opened the box, he was terribly shocked and ordered his farmhand to destroy "this diabolical and magical image of death, which in itself was a rare curiosity for the connoisseur."[3] Here we see the intimate linkage of connoisseurship and the overcoming of fear, anatomical conservation and trade, "disenchantment" and incorporation into a collection—concerns quite independent of the question of monstrous generation or details of classification.

If for Uffenbach, Neickelius's doctor, and many other erudite collectors and visitors of cabinets, monsters presented a pleasant diversion, such opinions did not go unquestioned in the years around 1700. In his handbook on museums, Michael Bernhard Valentin declined to publish the image of a monstrous calf with two heads, because "that is an unpleasant sight." Challenging Uffenbach's admiration, this was an implicit criticism of the thoughtless representation of monsters in the *Wunderkammer* and other "natural" cabinets. Valentin did not suggest removing curiosities from the cabinets, but he attributed a precise function to them. "A beautiful and well-proportioned image will be even more beautiful if something ugly and deformed is posed beside [it], because two contrary things standing together elucidate one another"[4] The extraordinary and the rare are no longer admirable objects; they are, rather, the background against which the ordinary stands out. Valentin's classification

2. Uffenbach, *Merkwürdige Reisen,* 1:13.
3. Neickelius, *Museographia,* 7.
4. Valentin, *Museum Museorum,* 158.

affirms an unbridgeable gap between proportioned and deformed beings. The latter have no use in themselves. Only in the construction of a polarity can the monster become useful and thus be legitimized.

Bernard de Fontenelle, secretary of the Parisian Académie des Sciences, himself frequently published reports about monstrous births in the *Mémoirs de L'Académie des Sciences;* nonetheless, he expressed his dissatisfaction with those innumerable accounts. The enormous knowledge about monstrosities remained "little instructive," Fontenelle argued, if it was not possible to find general laws by which monsters could be classified. Regularity, classification, and order—these were the categories through which Fontenelle wanted to implement studies on monsters. His uneasiness with curious case histories must be seen in the light of a programmatic sketch of the utility of natural philosophy that he had formulated a few years earlier as a guideline for the new century. In a short foreword to the *Mémoirs* of 1699, he defined anatomy as a science with theological ramifications, since it studied the structure of animals in order to reveal God's "infinite reason." [5] Greater understanding of the divine plan was the project of "physico-theology." The usefulness of the "normal" parts of the body obviously served as examples of God's almightiness, but where did monsters fit in this scenario? An answer to that question was not impossible. Leibniz had solved the basic problem of theodicy by proclaiming that monsters "follow rules and are in conformity with the general will [of God] although we are unable to perceive that conformity." [6] He compared the existence of deformities in nature to irregularities in mathematics, which ultimately fit in an ordered system. Contemporaries might have found this analogy comforting, but it was in fact not very helpful for their attempts to find rules that accommodated nature's irregularities.

The quotations cited above might indicate that monsters—as problematic cases for any social and natural order—posed a challenge to Enlightenment culture from the beginning. There is the contradiction in Fontenelle, who published numerous case studies in the *Mémoirs* while bemoaning their uselessness in the annual history of the academy. There is also certainly a tension between Fontenelle's and Leibniz's statements on the one hand and Valentin's on the other, the former demanding an integration of monsters into the regular course of nature, the latter excluding them from this regularity, at least in an aesthetic sense. These tensions should not obscure, however, the themes common to all three positions. Fontenelle, Leibniz, and Valentin all acknowledged the problems monsters raised in their understanding of nature, and they

5. Fontenelle, *Histoire,* 39, and "Préface sur l'utilité," 56.
6. Leibniz, *Théodicée,* 261.

all conceded that monsters could be regarded as a disturbance and threat to patterns of order. Finally, they all tried to fix the representational space where monsters should be located and integrated.

The phenomena of classification and order, of regularity, beauty, and usefulness, of representation and dissection, and of the generation of monsters can only superficially be described with the word "naturalization," commonly used in modern studies of this matter.[7] The problem was not so much how to define monsters as natural phenomena, but rather, as I argue in this chapter, how to integrate, incorporate, and domesticate them in the material and discursive arsenals of enlightened rationality. The various difficulties of this attempt were connected to a radical change in the epistemic foundations of understanding life as a process and human beings as results of that process. Moreover, these difficulties led to attitudes toward deviation and otherness that constituted the dark side of the Enlightenment and outlasted the Age of Reason, casting a huge shadow on our modernity.

The underside of enlightenment was nowhere put into brighter light than in Adorno and Horkheimer's *Dialektik der Aufklärung*. In a crucial passage, Adorno and Horkheimer defined the central aim of enlightenment as "delivering humans from fear and installing them as masters." For this it was necessary to follow the "route of disenchantment," and the goal was reached "if there is nothing unknown left." Adorno and Horkheimer established the link between knowledge production, overcoming fear, and possession by a spatial metaphor: "Nothing is to be left out there, because the mere idea of an out there is the real source of fear."[8] This suggests that enlightenment means possession, integration, domestication, and the establishment of order as modes of overcoming the unknown. From this viewpoint, building a baroque natural cabinet as a space of representation was as well a way of rationalizing the world through the classification of natural beings or the theory of generation. The problem with such a generalization lies in the fact that it fails to notice the historical specificities and changes in the process of enlightening monsters. Therefore, it is useful to look at the various spaces of representation and integrative practices by which monsters were pushed to "get in."

Although the eighteenth century invented fundamentally new categories for monsters, there are undoubtedly strong elements of continuity with earlier attitudes. Living monsters continued to be displayed in public shows and

7. An epitome of this view can be found in Canguilhem, "La monstruosité." For a critique of this view, see Daston and Park, *Wonders*, 176, 205.
8. Horkheimer and Adorno, *Dialektik der Aufklärung*, 13, 27. All citations in this chapter are from the 1947 edition of this work.

marketplaces.[9] They were exhibited in natural cabinets and provided the subject matter for countless case studies in erudite journals. In fact, the eighteenth century witnessed a massive increase in the representations of monsters. They became subject to natural classifications and taxonomy; they were discussed in physico-theology and in preformationist theories of generation; they were the objects of laws and edicts; they appeared as items in encyclopedias; they were argumentative tools in the hands of materialistic and epigenetic attacks against the view of a prestabilized order of nature. Throughout the eighteenth century, simpler categories for monsters as singular examples of nature's capriciousness or as intriguing curiosities for the connoisseur disappeared from the scene.

I have divided my examination of this process into seven sections. I begin by reconstructing the foundation of and the epistemological and representational changes in the monster collection in St. Petersburg during the course of the eighteenth century (section 1). St. Petersburg exemplifies the places monsters were collected, displayed, and thus practically integrated. The more theoretical aim of finding rules and order in deviations and extraordinary structures had its two principal domains in the classifications of natural history and in the linkage of generation theories and other explanations for the existence of monsters (section 2). Significantly, the various attempts at explaining and classifying monsters would crucially affect the theory and practice of generation theories and natural history in the eighteenth century. Just how becomes clear when we consider the role of monstrosities in the debate between preformationist and epigenetic theories after the 1750s (section 3). Monstrosities proved instrumental in arguments for epigenesis, as the decline of preformationist theory went hand in hand with their reconceptualization. Monsters lost their status in natural cabinets and were displayed in other spaces of representation, including aesthetic notions of monsters (section 4). The work of Samuel Thomas Soemmerring provides a crucial example of such visualizations, which linked monstrosities to anthropological and physiognomical discourse for the first time (section 5). These changes are a starting point for the incorporation of monstrosities into the emerging human sciences around 1800, and in particular, their role in embryology had crucial consequences for a new, highly problematic understanding of human development (section 6). Finally,

9. In this paper I cannot delve further into the popular display of monsters and its relation to scholarly forms of representation. I try to analyze aspects of this relationship in Hagner, "Monstrositäten in gelehrten Räumen." On popular display of monsters from the sixteenth to the twentieth centuries, see Scheugl, *Show Freaks und Monster;* Fiedler, *Freaks;* and Bogdan, *Freak Show.*

I conclude this chapter with a few remarks on the relation between monsters and reason (section 7).

Monsters in Baroque Spaces of Representation

In the seventeenth century, monsters were desired, spectacular, and delightful objects in *Wunderkammer,* where they were put together with other curiosities and objects of nature and art. Because of their rarity, they were not to be found in every cabinet. The catalogs of the famous collections of Ole Worm and of Levinus Vincent, for instance, did not mention monstrosities.[10] Also, the registers of famous collections incorporated by Valentin as appendices to his *Museum Museorum* show that monstrosities were extremely rare. Among others, Valentin included catalogs of collections in Cassel, where Uffenbach had admired the bicipital fetus. The Theatrum Anatomicum in Leiden and the anatomical collection of Frederic Ruysch in Amsterdam both contained monstrous specimens as well. The rarity of monsters and their diverting appearance bore practical consequences. They were dissected, but whenever possible, they were anatomically prepared, preserved in spirits, and presented in public.

The place of monsters in a baroque space of representation may be best illustrated by the anatomical collection of Ruysch. The fame of the collection was due to Ruysch's celebrated "injection technique," which gave the objects preserved in spirits a lively expression. Ruysch used this "liveliness" in spectacular allegorical arrangements, in which a singular organ, limb, or skeleton was part of a whole tableau. Working as a young apothecary in his shop at the Hague, Ruysch had taught himself anatomy before going on to study medicine at Leiden. In Leiden, the moralizing dimension of anatomical study was reinforced by the allegorically arranged building and equipment of the lecture hall itself.[11] After Ruysch had been appointed as teacher in anatomy by the Amsterdam surgeons' guild in 1666, he established the manufacture of anatomical preparations in his home, where his whole family was involved in the business. In his preparations, he adapted the allegorical, moralized representations of (parts of) the body displayed so prominently in the Leiden lecture theater. For example, he put the skull of a prostitute under the leg of an embryo, so that it looked as if the fetus had kicked the skull away. The message of the arrangement was that the prostitute had become infected (and died) because of her reprehensible profession.[12] This and other arrangements were primarily done for the benefit of the interested public regularly viewing Ruysch's

10. See Vincent, *Elenchus Tabularum,* and Worm, *Musaei Wormiani Catalogus.*
11. See Lunsingh Scheurleer, "Un amphithéatre."
12. Luyendijk-Elshout, "Death Enlightened," 125. On the interference of anatomical images and moral allegories, see also Chastel, "Le baroque."

cabinet. As Uffenbach remarked, Ruysch had told him proudly that synthesizing the preparations would lead to two hundred bodies in all. Not surprisingly, Ruysch had access to monstrous bodies and dissected a remarkable number of them.[13] Monsters were part and parcel of his allegorical world theater. A *Sceleton polydactylon monstrosum* was supplied with the symbol "Oh fate, oh terrible fate" *(Ah fata, ah aspera fata!)*[14] Although Ruysch also displayed unusual livers, spleens, and penises, in most cases single parts of a monstrous body were not included in arrangements with other organs or limbs. Rather, the whole creature in its singularity was displayed.

This is nowhere more obvious than in the copperplates of monsters in the *Thesaurus anatomicus* (in eleven volumes), where Ruysch published and described his most valuable and beautiful preparations. The *Thesaurus* contains numerous plates of single parts of the body, the only exceptions, two monstrous embryos drawn entire. The third volume of the *Thesaurus* contains the image of a hermaphrodite sitting on a towel, and the second a hydrocephalic boy, holding part of a placenta in his hand (see fig. 6.1). In the comment to this plate, Ruysch noted that the placenta was "roughly delineated," because he did not want to display the placenta in its real constitution on this particular plate.[15] The explicit reference to the *wrongness* of the placenta—which was only recognizable to the anatomically experienced reader—can be interpreted as an acknowledgment of the unbridgeable gap between the monstrous boy and the object in his hands. If the placenta was usually a symbol for embryonic development, then its incommensurability with the monster is the subject of this plate. The plate shows the ambiguity inherent in Ruysch's representational system. On the one hand, monsters were part of an emblematic transfiguration of the body. They had in common with all other corpses the notion that physical death was the necessary condition for the allegorical overcoming of death. In this sense, Ruysch integrated monsters into his moralized anatomy. On the other hand, Ruysch tended not to dissect monsters, but rather exhibited them *in toto;* and the remarkable reference to the placenta reflects their difference from normal bodies. Monsters were still represented as singularities of nature. In 1716, large parts of Ruysch's collection were transported to St. Petersburg, where they became a cornerstone of Tsar Peter I's natural cabinet.[16] St. Peters-

---

13. Uffenbach, *Merkwürdige Reisen,* 3:640. In a synoptic overview written late in his life, Ruysch listed not less than fifteen types of monstrous deformations he had dissected and incorporated into his collection. See Ruysch, *Adversiorum,* 20–22.

14. Ruysch, *Observationum,* 186.

15. Ruysch, *Thesaurus Anatomicus,* 91.

16. For details of the relationship between Ruysch and Peter and the circumstances of selling and transporting the collection, see Müller-Dietz, "Anatomische Präparate," 761–63. On the Petersburg collections in general, see Neverov, "His Majesty's Cabinet."

Fig. 6.1.   Frederic Ruysch, drawing of a hydrocephalic child,
from *Thesaurus Anatomicus* (1702), vol. 2. Photo courtesy of the
Niedersächsische Staats- und Universitätsbibliothek, Göttingen.

burg was a place of exaggeration. Founded by Peter I in 1703, it was built in a
swampy area near the Baltic Sea. The incomprehensible violence of this gigantic project of city building was emblematized by the deaths of around a hundred thousand workers—Russian peasants and soldiers and many captured
Swedish soldiers—in the first few months of construction.[17] Historical developments in St. Petersburg—despite periods of stagnation between Peter's
death in 1725 and Catherine II's seizure of power in 1762—serve as a magnify-
ing glass, bringing out certain tendencies of the century with great clarity. I do
not want to suggest that St. Petersburg was a center of enlightenment in the

17. Reimers, *St. Petersburg*, 1:27. In 1703, St. Petersburg was a huge fort; only one year later
the first private houses were built.

reign of Peter I. But when we look at the ways Peter brought together absolutist and cruel power with technologies and collections, experts and scholars—mainly from France, Germany, and the Netherlands—it becomes evident that certain developments commonly regarded as enlightened were not separable from specific court interests. As we have seen, in the first half of the eighteenth century, monsters were still regarded as curiosities, still the rare and spectacular highlights of anatomical and natural collections. But at the same time, they were subordinated to the demands of order, discipline, state control, Enlightenment from above and scientific interests. The temporary overlapping of different representations of monsters—court spectacle on the one hand and enlightened subordination on the other—was quite typical of royal capitals such as Berlin, Vienna, or Cassel. St. Petersburg had the largest and most famous collection of monsters, and its different uses in the first half of the eighteenth century are paradigmatic for the historical transformations in this period.

Peter I was an admirer of Ruysch, whom he had twice visited, and he was a passionate collector of monsters. In the 1690s, he had established a collection of monstrosities, anatomical and zoological preparations in the principal pharmacy of Moscow. In 1704, he ordered that midwives were strictly forbidden to kill newborn children with deformations or even to hide them. In 1718, he repeated and expanded the *ukasz,* according to which all monstrous births had to be delivered to the communal clergyman, who would then send them to the cabinet in St. Petersburg. He stated that monsters were collected for cabinets of curiosity in all other countries and criticized the fear and superstition of his own compatriots: "The unlearned hide [monsters] because they believe that such births originate from the devil by witchcraft and sodomy. But this is impossible, because the creator of all living beings is God and not the devil, who has no power over any individual. Inner lesions, fear and swear words shocking the mother can evoke such signs in children. There are many examples for this. The same may happen, if she suffers from mechanical irritations or sickness." [18]

The ambivalence of Peter I's monster politics is grounded in a passion for collecting and spectacle legitimized by his absolute power, as well as in his fight against fear and superstition. The argumentation in the edict was rich with nuance. It did not deny the existence of the devil, but it did make the devil absolutely powerless. Furthermore, the tsar was careful enough not to mention the particular theory, according to which monsters were preformed in germs, because this might have confused his compatriots. He merely stated other common explanations for the birth of monsters, such as external and accidental influences and the mother's imagination. The edict of a tsar was not the place

---

18. *Polnoe Sobranie,* vol. 5, no. 3159, cited in Müller-Dietz, "Anatomische Präparate," 764.

to discuss subtle problems of physico-theology. Hence, there was no need to mention the argument that monsters could serve for scientific purposes. Of course, the educational aspect of the collection was implied, but more in the sense that the preservation of monstrosities in bottles could reduce viewers' fear and superstition than that such exhibits would extend knowledge. More important, the delight and wonder of looking upon the monsters were part of court life. Peter I even used to receive subjects, foreign guests, and diplomats in his natural cabinet.[19]

That the tsar did not trust his subjects to comply with his edict is demonstrated by the fact that he not only ordered severe punishments for hiding monsters but offered rewards for delivering them: three to ten rubles for animal or human cadavers, and up to a hundred rubles for living human monstrosities. Until the 1740s, when the practice seems to have stopped, at least ten children and adults were transported to St. Petersburg, among them a giant, a dwarf, a hydrocephalus, and hermaphrodites.[20] The often mentioned dwarf, Foma Ignatjew, lived in the *Kunstkammer* from 1720. He was not obliged to be around all the time, but of course he had to be present as Peter I wished. According to the tsar's orders, both Foma and the giant, Bourgeois, were stuffed after their deaths and incorporated into the collection. The tragic and perverted aspect of such imprisonments did not go unnoticed. The German diplomat Friedrich Wilhelm von Bergholz reported the sad fate of one "monster" who lacked external sexual organs, male or female, having instead a kind of tumor with an aperture in the middle through which urine constantly dripped.[21] This person had been abducted from Siberia, and his desire to return home was turned down even when he offered to pay a hundred rubles for his freedom.

It is not sufficient to explain such practices simply as late expressions of attitudes known from the Middle Ages and Renaissance, when dwarfs, giants, fools, and "exotic" persons from Africa, Asia, and America were an everyday part of court culture.[22] The fascination with exotica was certainly a major motivation behind Peter's collection, especially as it accorded with a taste for bizarre wedding celebrations, fancy dress balls, and other events of that kind. But Peter's monster politics were at the same time an expression of Enlightenment rationality closely connected to the exercise of power. First of all, the edict of 1704 was part of an attempt to get state control over birth and mid-

19. Stählin, *Originalanekdoten,* 54–57.
20. See Russow, "Die lebenden Missgeburten."
21. Bergholz, "Tagebuch," 115. Apart from the missing external sexual organs, this person was obviously male, so Bergholz did not regard him as a hermaphrodite.
22. See Floegel, *Geschichte der Hofnarren;* Mezger, *Hofnarren im Mittelalter;* Zapperi, "Ein Haarmensch." On "wonders at court," see also Park and Daston, *Wonders,* 100–108.

wifery.[23] It is therefore revealing that the procedures it laid out for the reporting and delivering of monsters to the tsar continued after Peter I's death in 1725, when the cabinet came under control of the Academy of Sciences, founded the same year. The academicians record a telling example of how monstrosity and wildness were treated: On his scientific journey to Siberia in 1742, Johann Georg Gmelin had discovered two pairs of young siblings who were illiterate hermaphrodites. One year later, they arrived in St. Petersburg, together with their fathers. Although the fathers demanded two of the children be set free because they were needed for work at home, four of the St. Petersburg academicians wrote expert opinions in which they argued for detaining them as valuable curiosities. This was, however, more than mere curiosity on the part of the experts; it represented an attempt to impose order on something outside the normal order of life. The children were regularly examined by the physicians and anatomists of the academy; moreover, it was decided to teach them reading and writing and send them to a higher school later on. In the end, one of the children escaped for good, and the others made so little progress at school that they were sent home to Siberia in 1746. But another reason for the children's dismissal may have been that the academicians had decided after continuous observation that they were not real hermaphrodites. Rather, they were males with defective outer sexual organs, and there were no anatomical characteristics referring to the female sex. In short, the children had lost their status as monstrous beings, and therefore they were no longer interesting. A final reason for sending them home was the financial difficulties of the academy.[24]

Despite its unsatisfying outcome, the children's accommodation in the closed space of the cabinet, including medical research and education, meant their integration into a system of order, by which all that was unknown and mysterious in them was to be put under control. In other words, monsters were objects of practices that attempted to transform the deviant and the wild into domesticated human beings, who might or might not serve for further court spectacles and scientific examinations. It is obvious that these attempts failed in 1740s St. Petersburg, but other eighteenth-century attempts went in the same direction. Hence the fascination for "wild children," unkempt and illiterate, perhaps nurtured by animals, who became subject to educational and anthropological experiments throughout Europe. The same procedures were

23. In a way, Russia was the forerunner with such a monster edict. A comparable order was released in Prussia in 1724. See Artelt, "Die anatomisch-pathologischen Sammlungen," 98. I am aware that the formulation of such orders does not necessarily determine their practice. The point is that the state gets an interest in centralizing and thus overseeing monstrous births.
24. Boerhave, "Historia anatomica ovis," 329; Russow, "Die lebenden Missgeburten," 147–49.

undertaken with "wild humans," transported as objects of study, as with exotic animals domesticated in menageries. All these experiments indicate a strong belief in the integrational and disciplinary powers of enlightened practices and thinking despite the fact that the results were all too often a complete failure.[25] The discipline and investigation of living human beings took place in the same cultural milieu as the collection of monsters in bottles. Educational experiments were performed by the same scientist-collectors and likewise represented the translation of a courtly pattern of entertainment and culture of curiosity into the categories of rationality and order.

Monsters and the Preformationist Order of Living Beings

The St. Petersburg example shows that the "disenchantment" of monsters was a practical process, and it was strictly based on visual evidence. It is worth noting that this process took place without any plausible or coherent theory of the generation of monsters. As we saw above, their display in a cabinet implied that such beings could not possibly be the result of demonic or occult forces. Monstrosities in bottles were thus decontextualized from their mysterious aura, as well as from controversies turning around their generation, as we shall soon see. Objects in a collection were permanently at hand and thus lost their alarming potential. In other words, they represented their own subordination. The practice of displaying monsters was not restricted to court culture. In a less spectacular and less violent way, it also characterized bourgeois collections. In his dialogues on the beauty of nature, the Berlin philosopher Johann Georg Sulzer defined the owner of a natural cabinet as a "master" of nature, with "power" over its spectacles.[26] On a practical level, Sulzer had in mind that working in a cabinet meant independence from natural phenomena—seasonal changes, rain, and so on. Moreover, he argued that the collected object could not escape, and the ongoing and concentrated observation of it should ideally lead to its comprehension, including management of the emotions it aroused. In short, preservation in a bottle transformed the rumor of monstrous births and their connection to more or less mysterious circumstances into the material arsenals of reason.

This process certainly started in the late Renaissance and continued in the seventeenth century, but it was not before the eighteenth century that it gained

25. The most well known cases of "wild children," Jean Itard's Victor of Aveyron and Caspar Hauser, lived in the early nineteenth century. Nevertheless, in the eighteenth century such cases were known and widely discussed. Rousseau, for example, gave five examples of isolated children in his *Discours sur l'origine de l'inégalité* (1754). La Mettrie, Voltaire, and other philosophers discussed the case of the "wild girl of Chalon" in the context of the question of the origin of morals. See Malson, *Les enfants sauvages.*

26. Sulzer, *Schönheit der Natur*, 31.

a broad significance. One major reason for this was the enormous expansion of collections: they were no longer restricted to courts, nobility, and a certain number of wealthy scholars, but became also material representations of bourgeois self-understanding. Pharmacists, physicians, clergymen, and businessmen invested time and money in collections, which included monstrosities. The fascination was so great that fakes were common. For example, a merchant in Hamburg paid a lot of money for a seven-headed hydra. When Linnaeus saw the object in 1735, he quickly discerned that it was spurious. This information was so unwelcome that Linnaeus decided to depart.[27] Of course no collection, private or public—even the anatomical theater in Berlin or the cabinet in Cassel, both of which contained a remarkable number of monstrosities—could compete with the collection in St. Petersburg. But all cabinets preserved their monstrosities well in the bright light of their collections, even though these centerpieces remained problematic in theoretical terms.

A debate on the generation of monstrosities was underway in another center of the sciences, namely, in Paris. As already mentioned, numerous case studies on monsters had been published in the journal of the Académie des Sciences. To the consternation not only of Fontenelle, the simple recording of these cases did not resolve the issue of God's responsibility for such creatures. The favored theory of generation in the late seventeenth century was the theory of preformation, according to which all living beings were divinely "formed" in the original creational act, growing from a homunculus into a living being. Consequently, God must be directly responsible for the creation of monsters. One possible way out of this dilemma was the argument that, despite preformation, hazard might operate to ill effect upon the growing homunculus. For Malebranche, the laws of nature established by God left space for such accidental developments. Thus God was not entirely responsible, and nature became a rather imperfect and unreliable agent.[28] Much more challenging than the devaluation of nature was the implication that if God was not responsible, and if he was all-good, then he could not be almighty at the same time. This was the problem Leibniz had tried to solve with the argument of the undiscovered order of monsters. The rules of nature did not deviate from God's will, and neither did anomalous developments. If this was so, then all attributes of normal beings also had to be ascribed to monsters: they were perfect, regular, harmonious, purposeful, and beautiful. The only problem now was to unveil these attributes.

27. See Jahn and Senglaub, *Carl von Linné*, 32–33. Holländer reports about an eighteenth-century mermaid from a collection in Berlin, which was only identified as fake by x-ray much later. See Holländer, *Wunder, Wundergeburt, und Wundergestalt*, 187–88.

28. Malebranche, "Eclairecissement," 325; see Roger, *Les sciences de la vie*, 398–401.

The theoretical concerns of natural philosophers and theologians thus inaugurated a series of anatomical debates on the questions of perfectibility and
the preformation of deviation. The anatomists Benignus Winslow and Louis
Lémery were the main opponents in the dispute: whether monsters were preformed as defective germs, which in their deficiency nevertheless displayed the
perfection of divine creation (Winslow), or whether they were caused by accidental components such as the occasional mixture of two eggs or mechanical
lesions in the womb (Lémery).[29] Winslow and Lémery's debate in the *Mémoirs*
of the academy continued for decades, since they only published when they
had new case histories, and only then if they had had the opportunity to dissect
a new body. Neither of them concealed practical problems. Lémery admitted
that he had to beg a woman again and again before she handed over her monstrous birth preserved in a bottle. Winslow had to dissect the corpse of a
twelve-year-old girl in great haste, because the father had demanded it for
burial.[30] Winslow searched for order in the corpses, because, in accordance
with Leibniz, Arnauld, Fontenelle, and others, he believed that the disposition
of any deformed body was in itself as harmonious and perfect as a normal
body. Lémery, on the other hand, searched for evidence of disorder as proof of
the accidental modulation of normal development. Each case brought new
considerations to the debate, but in the end Winslow had to concede that both
theories might be correct. The state of the art did not allow a definite answer,
he argued, and thus he played the theological card with the rhetorical question,
Which theory is more suitable for the praise of God?[31] This pattern of argumentation recalls that of the museologist Valentin, who had found a place for
monsters as foils for ordinary beings. In both cases, it was not so much a question of which theory was more plausible, or whether the phenomenon was
interesting in itself. The point was, rather, whether the external referent constructed by Winslow and his mentors was God or the order of nature.

Despite these overlaps between anatomical and theological arguments one
should not overlook that both alternatives remained within the framework of
preformation. Accidentalists such as Lémery merely stated that God's plan
could be destroyed by certain external influences. This was scandalous for protagonists of natural theology such as Winslow, who conceded an imperfection
of human intelligence in understanding divine attributes, rather than a weak-
ening of God's omnipotence. The remaining and embarrassing problem for
scientists of the eighteenth century was that there was no unifying concept for
the generation of monstrosities, because every single and contingent case oc-

29. For details see Roger, *Les sciences de la vie*, 408–18; Tort, *L'ordre et les monstres*.
30. Lémery, "Sur un foetus monstrueux," 46; Winslow, "Remarques sur des monstres," 369.
31. Winslow, "Remarques sur deux dissertations touchant les monstres," 118–19.

casioned its own theory. Despite numerous repetitions of Leibniz's hypothesis of undiscovered order in monstrosities and despite Winslow's attempts to find order in their bodily structures, monsters remained a scandal for the perfectibility of the natural order. First, it was hard to find any purpose in monsters except for exhibiting them in cabinets. Second, it was impossible to ascribe aesthetic qualities to them, except perhaps to integrate them into an allegorical world-theater such as Ruysch's collection in Amsterdam and St. Petersburg. But what functioned in the particular space of a cabinet was not transformable into generally accepted judgments. In 1739, Zedler's *Universal-Lexikon* vehemently rejected the opinion that monsters were beautiful.[32] A more moderated view was proposed by Sulzer in his dialogue on the beauty of nature. Standing on the side of preformationism and natural order, he found it nonsense to regard monsters as a source of disturbance. A single deviation from beauty and rule among one hundred thousand perfect counterexamples could not possibly be taken as a proof of disorder. Furthermore, the lesser degree of perfection in monsters did not contradict the intentions of nature simply because we were unable to comprehend them.[33] With this compromise, Sulzer settled the problem of monsters. His solution represented the tendency around the middle of the century to play down the irreconcilability of monsters and natural order. Indeed, in 1743, the new secretary of the Parisian academy, Dortous de Mairan, claimed that he and most of the academicians followed Winslow's opinion.[34]

The tensions and contradictions in preformation theory became most clear in Albrecht von Haller's extensive work on monstrosities. Haller had collaborated with Winslow when he stayed in Paris for a couple of months in 1727. As a follower of Winslow, Haller started to publish his results, vigorously arguing against Lémery's theory of accidence. But Haller encountered the same difficulties with his theory as Winslow had had with his: after the dissection of an anencephalic fetus, he could not but admit that external effects caused that particular lesion. Trouble increased when Haller regarded conjoined twins not as monsters but as an example of a new species. Their capability of living, he argued, was a proof of God's power to create completely new forms of human species.[35] The most obvious example of the dilemma of preformation was Haller's article for Fortunatus de Felice's *Encyclopédie*. Haller attempted to create order by distinguishing between different classes of monstrosities,

32. Zedler, "Missgeburt," 486.
33. Sulzer, *Schönheit der Natur,* 88–92.
34. Mairan, "Sur les monstres," 64–65.
35. Haller, "De feto humano septimestri sine cerebro edito" (1745), in *Opscula,* 290; "Anatome fetus bicipitis ad pectora connati" (1735), in ibid., 150.

describing the structure of their bodily organs, and finally dealing with their causes. Again he defined himself as a follower of Winslow and argued for a "primitive structure, different from the ordinary structure." But although he could enumerate a much larger group of anatomists in support of this opinion, he was unable to use it to explain most of his own cases. Color changes of skin were environmental, hence accidental; left-right exchanges were not. Excessive growth of one part of an organ, while another part remained small, was accidental; a sixth finger or a surplus muscle was due to the original structure. Inhibitions of growth were caused by mechanical pressure; double monsters conjoined at certain points were not accidental. Haller was so uncertain of his ground that he did not repeat his idea that conjoined twins were an example of a new species in the article for the *Encyclopédie*.[36]

In conclusion, attempts to classify monsters had to be put in relation to two competing theories of generation. This situation caused a problem of ordering and classifying. The general classification, which was far from new and was accepted by Georges Buffon, Haller, Charles Bonnet, and many others, had three categories: monstrosities with defect, monstrosities with excess, and monstrosities with a faulty and wrong position of the organs.[37] If the first group was accidental, the second one either accidental or caused by a defective germ, and the third one indicative of either a defective germ or a new species, then the whole system of classification was doubtful because it dealt with different entities.[38] The preformationist theory could explain generation, but in the case of regeneration and of deviation it had its limitations. It would be an exaggeration to claim that difficulties with a plausible explanation of monstrosities caused the decline of preformationist theory and in a broader sense the decline of eighteenth-century natural history. But the fact that monsters eluded classification, as well as confounding preformation, gave them the status of an "epistemic instrument" in the controversies on the theory of generation. The same status can be given to the polyp *Hydra*. After Abraham Trembley had discovered the power of regeneration in the little polyp around 1740, critics of preformation regularly referred to this phenomenon in order to illustrate the insufficiency or even absurdity of the theory. Likewise, the paradox of monsters in preformationist theory was highlighted. Maupertuis, for example, repeatedly hinted at this point.[39] Nowhere else, however, were monsters as much

36. Haller, "Jeu de la nature," 207–9.

37. See, e.g., Buffon, *Histoire naturelle,* 578; Haller, "Jeu de la nature," 197–200.

38. On the confusion in defining monstrosity in the seventeenth and eighteenth centuries, see Moscoso, "Vollkommene Monstren."

39. Maupertuis, *Vénus physique* (1746), in *Oeuvres,* 81–82, and *Système de la nature* (1751), in ibid., 159–60.

used as an epistemic "instrument of science," as in the epigenetic approach of Caspar Friedrich Wolff.[40]

Monsters as an Instrument in the Hand of Epigenesis

In 1768, Haller published a paper in which he explicitly stated that the question of whether monstrosity was caused by a faulty germ or by hazard had crucial implications for the theory of generation. At that time, he was already involved in a dispute with the St. Petersburg anatomist and physiologist Caspar Friedrich Wolff.[41] Wolff had propagated "epigenesis" as the generative principle in the vegetative and animal world in a famous doctoral dissertation, *Theoria generationis,* written at the University of Halle in 1759. Wolff was not the inventor of epigenesis, but on the basis of wide experimentation, he argued that generation was based on principles that were equivalent in all species, including human beings. Epigenesis rejected a singular creative act. Instead, the various parts of the body were said to have developed from unorganized matter, so-called vesicles. The organs developed in succession and partially resulted from one another. Hence, for Wolff, generation was due to a temporal succession, whereby all structural and formal alterations and developments of bodily parts were driven by a common force, the *vis essentialis.*[42]

In the very last paragraph of his dissertation, Wolff applied his theory to monstrosities. The example he gave was, ironically, the conjoined twins dissected by Haller more than ten years before. Although Wolff generously acknowledged Haller's work, he attributed irregularly doubled parts of the body neither to an original conjunction nor to external influences. Nor did he regard them as a new species. Instead he argued for a superactivity of the *vis essentialis* in a crucial moment of organ formation, which led to the erroneous double growth.[43] This account assigned monstrosities a place within the process of generation: a deviation because of a fault of generation itself, not predetermined and not caused by mechanical or other external influences. For the time being, Wolff did not delve further into this topic. On the contrary, when he published an extended revision of his dissertation in 1764, he did not even mention monstrosities. A probable reason for this carefulness is that Wolff—as he wrote to Haller in 1761—wanted to face Haller's critical remarks with new

40. The term "instrument of science" is taken from Canguilhem ("La monstruosité," 179), who has argued that monsters were used as an instrument in the debates between preformationists and epigeneticists.
41. Haller, "De Monstris," 131. See also Haller's "De monstrorum origine mechanica," in *Opscula,* 299–314. On the debate with Wolff, see Roe, *Matter, Life, and Generation,* 45–89.
42. Wolff, *Theoria generationis,* 72–73.
43. Ibid., 134–35.

observations and experiments.[44] In the case of monstrosities, however, he was completely inexperienced. His note from 1759 was based on secondhand information, and thus its promise had yet to be confirmed by his own evidence. Part of the problem was that, in the middle of the eighteenth century, it was a rare privilege to dissect monstrous bodies. The difficulties of the Parisian anatomists have already been mentioned. Haller had dissected the conjoined twins in his early days in Berne. In Göttingen, Haller popularized anatomical dissections, and because of his extraordinary position in the town he mobilized the administration, so that he had more than enough corpses at his disposal. Between 1736 and 1753, when he left Göttingen, Haller had dissected 350 human bodies, a very remarkable number. Conditions in Göttingen were ideal compared to other cities. The number of monstrosities (at least six, animal and human) Haller had at his disposal was also quite unusual for that time.[45] Haller certainly profited from the fact that he was an unquestioned authority in Göttingen, where there was no court. The university was the all-powerful institution in town, and therefore there was no competition in acquiring and using monstrosities. In Berlin, however, where scientific life was more decentralized and competition was an issue, the situation was quite different.

Wolff moved to Berlin in 1763, where he applied for permission to deliver public lectures at the Collegium Medico-Chirurgicum. He had hoped to start an academic career, but the professors of the collegium voted against him. Wolff finally received permission, thanks to the intervention of an influential military physician, but several renowned anatomists, such as Johann Friedrich Meckel the elder and his disciple Johann Gottlieb Walther, remained his enemies. Not only was Meckel, as a disciple of Haller and translator of Winslow's works, a follower of preformation theory, but he and Walther also controlled the anatomical theater collection, which contained a respectable number of preparations in bottles.[46] According to the unwritten law whereby one could represent monstrosities but not dissect them, these objects were taboo. Furthermore, Meckel, Walther, and Nathanael Lieberkühn themselves had built up private anatomical-pathological collections, so that any new monstrosity was likely to be reserved for one of them.[47] For this reason, and because of Wolff's problematic status in the collegium, Berlin was not the place to per-

44. See Wolff to Haller, 29 December 1761, in Schuster, "Der Streit um die Erkenntnis," 204.

45. This situation made Göttingen an attractive place for medical students, because Haller offered the students the chance to perform dissections. For details see Gierl, "Vom Wühlen der Aufklärung," 7–8; Thode, Die Göttinger Anatomie; Voss, Das pathologisch-anatomische Werk. For Haller's own enumeration of the monstrous bodies he dissected, see his "De monstrorum," 300.

46. See Schaarschmidt, Verzeichniss der Merkwürdigkeiten, 11–14.

47. Artelt, "Die anatomisch-pathologischen Sammlungen," 97–99.

form dissections on monstrous bodies in order to answer questions concerning their generation.

Despite that, Wolff was not so isolated and unappreciated as is often claimed in the historical literature. His most distinguished supporter was Leonhard Euler, who had already nominated him for a position in St. Petersburg in 1760. Euler's renewal of a reference in 1766 led to Wolff's appointment to the Academy of Sciences in St. Petersburg as a professor of anatomy and physiology.[48] Wolff became the curator of the anatomical cabinet and theater, and—as Euler wrote to Jacob von Stählin, the secretary of the academy—he was expected to build the scientific reputation of the academy. The material of the monster collection was ideal for a continuation of his epigenetic studies and Wolff did not hesitate to use it. In 1776, he wrote to Haller: "The very rich storehouse of monsters that has been collected and preserved over a long series of years in the Imperial museum has now been handed over to me, so that I can compose a description of them and perform anatomies where I decide to. Therefore it will be necessary to deal once more with the origin of monsters as well as with generation in general."[49] This letter shows that Wolff saw a close connection between monsters and generation, and it suggests as well that the the Academy of St. Petersburg was no longer interested in collecting and displaying monsters. The process of marginalizing monsters had begun in the 1740s. After 1746, no more living humans were incorporated into the cabinet. Around the same time, activities in expanding the collection came to a halt. This was largely due to a quarrel in the academy that led to the resignation of its anatomist, Jean George Duvernoy. The last major activity in the cabinet was the publication of a catalog of the anatomical collection in 1742. This catalog was mainly the work of Duvernoy's colleague, the anatomist Josias Weitbrecht, and it listed the large number of monstrous human specimens (forty-one) and a smaller number of monstrous pigs and sheep.[50] Moreover, drawings of some monstrosities were planned as part of a huge illustrated catalog that was never finished. After Weitbrecht's death in 1747, his position remained unoccupied. The collection no longer expanded, and even the "monster edict" fell into oblivion.[51] This decline was hastened by a huge fire in 1747 that destroyed parts

48. Euler to G. F. Müller, 4 December 1760, in Juskevic and Winter, *Die Berliner,* 1:164–65; Euler to Jacob von Stählin, 4 May 1765, 3:234–35. Euler's admiration for Wolff was connected to the generation theory, because Euler strongly argued for epigenesis, as his ongoing dispute with Charles Bonnet, the most renowned defender of preformation besides Haller, demonstrates. In his letters to Bonnet, Euler repeatedly referred to Wolff. See the abstract of these letters in Euler, *Briefwechsel,* 60–62, and the selected edition of the letters in Euler, *Letters to Scientists,* 29–53.

49. Wolff to Haller, 27 September/8 October 1776. To my knowledge this letter has been, up to now, only published in English by Roe, *Matter, Life, and Generation,* 170.

50. [Weitbrecht], *Musei Imperialis Petropolitani,* 293–304, 349–351.

51. For details see Baer, "Über den jetzigen Zustand," 134–38.

of the building where the collections were housed. The anatomical collection was not affected directly, but the restoration of the building led to removal of the preparation hall, and no other room was built for it. In consequence, corpses for forensic dissection were no longer sent to the academy, but rather to the surgical school. Dissection came more or less to an end, and anatomical topics were seldom to be found in the annual *Commentarii* of the academy.[52]

For decades monsters had been preserved as objects of representation. In the hands of Wolff they received a new material existence and a new function. They were dissected and thus became an instrument with which the new theory of generation was tested. Wolff was not shy in his practice. He removed specimens from the collection and initiated a republication of the "monster edict" so that he could be provided with new material. Of Siamese twins who had died shortly after their birth and were sent to St. Petersburg, Wolff stated: "The death of this monster is much happier for anatomy and physiology than its survival. [In the latter case] perhaps there would have never been a chance to dissect it."[53] Moreover, Wolff ordered drawings of all specimens before he dissected and disturbed them. Although Wolff investigated monstrosities instead of displaying them, his policy did not find the unequivocal applause of his successors. As Karl Ernst von Baer complained one century later, the collection lost a remarkable number of extremely rare and interesting objects.[54]

Wolff planned a unifying *Theoria monstrorum* as the culmination of his scientific career and worked on this topic for many years. He published several contributions, but did not finish his opus.[55] Again the failure does not seem to be coincidental. Wolff was prepared to attack preformation. For instance, cyclopism could not possibly be God's will because it would have been implausible to create a perfect eye without an optical nerve. Instead, he argued, "nature tried everything to create a perfect eye, but she could not overcome the obstacles."[56] Wolff understood these obstacles as integral parts of the generation process. This meant that generation became something uncertain and unforeseeable. Although Wolff did not give up the idea of a telos, he did not

52. Between 1726 and 1747, forty-eight anatomical contributions were published in the *Commentarii;* the *Novi Commentarii* between 1749 and 1765 contained only eighteen anatomical papers. See Brandt, "Kurze Übersicht der Fortschritte," esp. 85–89.

53. Wolff, "Notice touchant un monstre," 43.

54. Baer, "Über den jetzigen Zustand," 114–15; see also Burdach, "Über die anatomische Sammlung," 435.

55. The preserved documents and manuscripts of Wolff in St. Petersburg contain about a thousand hand-written pages related to that project. One manuscript was published in a Latin-Russian edition in 1973. See also Baer, "Litterärischer Nachlass"; Raikov, "Caspar Friedrich Wolff," 593–604; Lukina, "Caspar Friedrich Wolff," 415–21; Roe, *Matter, Life, and Generation,* 124–26, 133–37.

56. Wolff, "Descriptio vituli bicipitis," 562.

regard the obstacle as being separate from the course of nature. In radical deviation from preformation theory, life was constructed as a process. It was provided with inner dynamics and showed a temporal succession of development.[57] Unshaped and fluid matter was transformed into shaped bodily structures. The details of those processes were to be found out. But this was Wolff's problem. In the *Obiecta meditationum,* he explained monstrosity differently than he had in his dissertation, introducing the term "mutationes," and distinguishing between structural deviations and alterations of the property of an organ, for example, a different color. In the first case, the deviation was due to a quantitative process. That is, the *vis essentialis* produced too much or too little nutrition, while in the latter case nutrition was qualitatively disturbed.[58] The assumption of nutritional factors in the generation of monstrosities was not entirely new. For example, Alexis Littré and Gottlieb Friderici had proposed comparable opinions earlier in the century without any remarkable response.[59] The cornerstone of Wolff's approach, however, was a direct link between nutrition and growth and the idea of a proportional relation between physiological processes and the development of the bodily shape. In terms of a physiological understanding of anomalies, this represented a turning away from exclusively visual evidence of the object under investigation. Haller had distinguished between monsters and varieties as a quantitative difference between minor and major deviation. Wolff did not reject this premise, but the visibly deformed anatomical structure was no longer sufficient for an understanding of bodily growth and its deviations, for the invisible dynamics of the generation process had also to be taken into consideration. The visible body was regarded from the perspective of bodily change and development. For Wolff, this process epitomized the creative activity of nature. In contrast to older assumptions concerning nature's contingent playfulness, Wolff aimed at finding out its regularity.

This approach was inseparable from a completely new practical handling of monstrosities. They were no longer necessarily preserved in a bottle *in toto.* Rather, they were "consumed," and the bodily presence was replaced by a pictorial representation. Not coincidentally, this new handling of monsters no longer took place in a public space in the academy or any other public dissection hall. Wolff dissected in his own house in a remote part of St. Petersburg. Although this choice of location may have been due to lack of space, it shows

57. The assumption of such a rupture is a cornerstone of Foucault's "archéologie des sciences humaines." Cf. Foucault, *Les mots et les choses,* 170–76, 238–45.

58. Wolff, *Obiecta meditationum,* 229–30.

59. Littré, "Observation sur un foetus monstrueux," 286–87; Friderici, *Monstrum humanum rarissimum,* 27–28.

that, for Wolff, scientific work on monsters was not a public affair. This epis-temological and social transformation of monstrosities was in perfect harmony with the politics of the St. Petersburg collectors. In a quasi-official description of the collection, which also served as a guide for visitors, Jean Bacmeister wrote that anatomical preparations were extremely important for natural scien-tists but they did not belong to those objects that served "for fun of all kinds of spectators." This was not the case for some of Ruysch's preparations, which Bacmeister described in detail, but he ignored the monstrosities because "the celebrated academician Mr. Wolff has undertaken the task of an anatomical description; certainly this will throw bright light on the theory of genera-tion." [60] Only forty years earlier, monsters had been subordinated and "disen-chanted" via incorporation into a collection displayed at the court or in the private cabinet of an anatomist. This practice had been replaced by the handing over of monstrosities to distinguished scientists investigating life processes. Witnesses were no longer necessary, at least not the delicate spectators for whose eyes monsters were more painful than delightful. In other words, mon-sters did not now pass the aesthetic criteria of the natural cabinet.

What Bacmeister regarded as ugliness, as caprices of nature, and hence as a disturbing of the order in a cabinet found a very different reception in epi-genetic discourse. Besides Wolff, Johann Friedrich Blumenbach quickly real-ized the epistemic potential of monstrosities. Blumenbach argued that they were linked to certain developmental laws that led to their "admirable equifor-mation." [61] He conceded that in the case of monsters the formative drive was led into a wrong direction, but this was not an offense against natural order. Although Blumenbach and Wolff were far from giving sufficient answers on monstrous generation, they had translated monstrosities to a new field of sci-entific investigation. This was at the same time a theoretical and a practical shift. As a result, bodily deviations became a constitutive part of the life sciences.

The Aesthetic Turn of Monstrosities

The question of regularity, order, and beauty in monstrosities was, even in the 1770s, not separate from moral and theological connotations. Haller's and Bonnet's resistance to accepting epigenesis sprang from the problem that the assumption of generation out of nothing or out of shapeless unorganized mat-ter was incompatible with belief in the divinely bestowed moral nature of hu-mans. Haller used this argument in a letter to Wolff, whose response shows a certain embarrassment. Wolff accepted that epigenesis might become a dan-

---

60. Bacmeister, *Essai*, 207.
61. Blumenbach, *Über den Bildungstrieb*, 57–58.

gerous instrument in the hands of atheists, but he assured himself that recognition of the powers of nature and natural causes were no argument against the existence of God.[62] This was precisely the position of Maupertuis, who also had used monsters as an argument against preformation, but who had then argued that God had equipped nature with properties that allowed it to develop independently after the act of creation.[63] The activity of nature was thus posed between God and the visible organism. But even as Wolff linked the *vis essentialis* to God, he formulated other arguments in favor of epigenesis, adding what may be called an aesthetic turn to the idea of generation.

In his *Theorie von der Generation* Wolff criticized preformationists, who had postulated the existence of embryonic structures invisible under the microscope. According to this postulate, the only change of organic beings was the step from invisibility to visibility and the stages of determined development, and then the wear and tear of different periods of life. Such a view of nature, Wolff argued, completely ignored nature's beauty. Nature was not seen as active and creative, but as "lifeless matter, which loses one piece after the other until all that junk *(Kram)* comes to an end. I cannot stand such an awful nature."[64] The subtlety of this argument, which perhaps pushed Haller to play his theological card, was that Wolff linked the idea of creation with a mechanistic concept of fading away. In Wolff's view, the scandal was not the new building of parts of the body; the scandal was rather the lifeless persistence of the homunculus until it came into existence, since existence was nothing more than a decline. Against this deterministic view of life, Wolff proclaimed a telos in nature, which was visible in the beauty of organic structures. Even if a monstrosity did not fully reach its aim, it was not excluded from telos, and hence beauty could also be ascribed to monsters. "In several monsters I have seen the viscera in highest beauty and elegance, so that I have no doubt that the first purpose of nature was to make beauty in these creatures."[65] This did not mean that monsters were perfectly structured. Wolff explicitly rejected Haller's view of double monsters as being a new species as well as his arguments on monstrous preformation: "Monsters are produced by the powers of nature, and they do not evolve from created germs."[66]

The difference between this view and that of older anatomists such as

62. Wolff to Haller, 17 April 1767, in Schuster, "Der Streit um die Erkenntnis," 214. On Bonnet's defense of preformation, see Roger, *Les sciences de la vie,* 712–25.

63. Maupertuis, *Système de la nature,* 145, 154–55; see Roger, *Les sciences de la vie,* 479–87. I shall not delve further into the French position here, but I have analyzed Diderot's version of epigenesis and his theory of monsters in relation to Wolff and Blumenbach in Hagner, "Vom Naturalienkabinett," 90–93. See also Hill, "Materialism and Monsters," and "The Role of 'Le Monstre.'"

64. Wolff, *Theorie von der Generation,* 73.

65. Wolff, *De inconstantia corporis humani,* 234.

66. Wolff, *Descriptio vituli bicipitis,* 570–71.

Ruysch, Santorini, or Albinus was that the latter three did not link the represented body to an invisible dynamics of nature. For them, beauty was the expression of God's perfection. Physico-theologists and preformationists searched for order and regularities in monsters, but it was impossible for them to link this search to an idea of beauty. For them, as for many in the eighteenth century, monsters remained the Other, a constant source of confusion. At best the problems inherent in their existence could be played down, as in the case of Sulzer, but they could never be resolved completely. In the epigenetic view, however, beauty was visible in each step of development. In other words, it was embedded in the process of development. With this understanding, Wolff established epigenesis as an aesthetic view of nature.

Blumenbach went in the same direction and then a step further. After having collected arguments for the existence of the "formative drive" *(nisus formativus)*, he wrote: "I boldly refer to the inner feelings of everyone who has observed reproduction in animals and plants of such a simple texture, and who at the same time has appreciated the arguments in the foregoing section against the preexistence of the embryo of the chick in the yolk of the egg." [67] The naturalist should thus decide which theory he preferred on the basis of both observation and certain feelings. What was the origin of the "inner feeling" that enabled the scientist to understand natural processes? An answer to this question is linked to the return of nature as an active agent after a period of being regarded as passive and determined by divine laws. This switch in how nature was perceived was not restricted to a handful of naturalists. It was part of a general movement, a development for which the arguments of Johann Gottfried Herder provided a focus in the 1780s. Herder certainly was an attentive reader of Wolff. Probably even before he was familiar with him, Herder had used dynamic metaphors in his descriptions of the generation process. [68] In his *Ideen zu einer Philosophie der Geschichte der Menschheit* he argued that searching for "the tools and weapons of the organic forces" was a business for "philosophical dissectors." Herder's noble elevation of anatomy did not point to preformationists such as Haller or Bonnet, but to those who, like Pieter Camper, Wolff, and Samuel Thomas Soemmerring, were the "researching dissectors on this intellectual physiological way of comparing in various species the forces of tools in their organic life." [69]

Eulogies like that supported a self-consciousness in anatomists about their business. Anatomists actively participated in the intellectual movements of the late Enlightenment; in bringing together the philosopher and the anatomist,

67. Blumenbach, *Über den Bildungstrieb,* 90.
68. See Müller-Sievers, *Epigenesis,* 116–17.
69. Herder, *Ideen,* 92, 84.

Herder was describing a sensibility that he hoped would be the key to a further understanding of bodily processes. It is probable that Herder's characterization of the philosopher-anatomist encouraged Blumenbach to develop his notion of the "inner feeling" of the scientist. Blumenbach was a careful reader of the *Ideen* and he "deeply admired" Herder for his wide knowledge and the ability "to put things into bright light." [70] The passage in which Blumenbach referred to the "inner feeling" cannot be found in the first edition of the treatise *Über den Bildungstrieb* from 1781, but only in the second edition of 1789. The publication of Herder's *Ideen* fell between these two dates.

The redefinition of the sensibility of the anatomist complemented the aesthetic turn of epigenesis. The anatomists' concentration on the body came close to the creative activity of the artist. In this context, the beauty of the body in general, and of monstrosities in particular, was no longer detectable by the connoisseur in a natural cabinet; rather, it would be the knowing sensibility of the anatomist and physiologist that would decipher beauty and order. This shift is not reducible to an intellectual redefinition of anatomy; it was only possible with the creation of new forms of representation. The most important change was that the natural cabinet was no longer the primary space of representation for monstrosities. In those cabinets, monsters had represented the variety of natural productivity and the wisdom of the Almighty, the ugly and deformed creature in contrast to the harmonic form, or merely the Other as a singularity of nature. All this vanished in the epigenetic discourse. I do not want to argue that monstrosities were removed from collections. On the contrary, in the nineteenth century they were put in anatomical and pathological collections with increasing energy. However, in representing monsters as momentary snapshots of the life process, other spaces of representation were necessary.

Monsters and the New Visual Representation of Bodies

After their "contamination" by epigenesis in late eighteenth century, monstrosities came to represent the invisible and dynamic processes of life. The epigenetic approach that saw the temporal heterogeneity of organ growth as a necessary condition of bodily development incorporated monstrous deformation into that development. Formation and its deviations were subsumed under the same laws. As mentioned above, proposing laws for monstrous births as such was not entirely new. But the physico-theological attempt to apply regularities to monsters was part of the larger project to perpetuate a divine order of nature. This enterprise failed because every single monster suggested

---

70. Blumenbach to Soemmerring, 3 May 1785, in Soemmerring, *Briefwechsel, 1784–1786*, 178.

its own theory and because physico-theologists hesitated to draw the logical conclusions of their incorporation of monsters into natural order, that is, to ascribe purpose and beauty to monsters. One solution was to put monsters on the dark side of ordered nature: their main purpose was to clarify and strengthen the beauty of regular structures.

The linking of normal and pathological bodies within the framework of epigenesis was radically different from any former attempts to explain monsters. As Wolff and Blumenbach had argued from different angles, monstrosities were part and parcel of formation. The goal now must be to establish relationships between different parts of one organism in their temporally succeeding order and between different organisms. Monstrosities thus became momentary steps in the course of development. Before this new epistemic role for monstrosities was fully developed in Romantic embryology, however, the new order of monsters was tried out in a new kind of visual representation.

The anatomist Samuel Thomas Soemmerring was a close friend of Blumenbach and another fascinated reader of Herder's *Ideen.*[71] In 1791, Soemmerring published his *Abbildungen und Beschreibungen einiger Missgeburten.* The title suggests that visualization was primary, that here indeed was a new way of representating monsters. In general, Soemmerring regarded images as the most important tools for anatomical investigation. More specifically, his conviction was driven by the epigenetic redefinition of monstrosities. The approaches of Wolff and Blumenbach had marginalized the numerous anatomical case studies of monstrosities, but Soemmerring argued that such studies were still crucial. However, he was also suspicious of any existing generation theory: he merely wanted to describe monstrosities without drawing conclusions for physiology.[72] The question was how far anatomical representation could contribute to ongoing investigations. After epigenesis had located relationships between monstrosities and regular beings in the invisible black box of the dynamics of life, it was not sufficient for an anatomist like Soemmerring to describe a singular specimen in lesser or greater detail. The challenge was to translate these relationships into visual evidence. This meant detailed comparison of monstrous bodies with one another and with normal bodies. For doing that Soemmerring could no longer rely upon the method he had chosen in order to display a female skeleton—the "schöne Mainzerin."[73] Following the Leiden anatomist Bernhard Siegfried Albinus, Soemmerring had used parts

71. See Soemmerring to Herder, 16 January 1785, in Soemmerring, *Briefwechsel, 1784–1786,* 97–98.

72. Soemmerring, *Abbildungen und Beschreibungen,* 4.

73. See Choulant, *Geschichte der anatomischen Abbildung,* 134–35; Schiebinger, *Mind Has No Sex?* 198–203.

of several corpses and created an idealized type, or *femina perfecta*. In displaying monstrosities, Soemmerring's goal and procedure were quite the opposite. Here the point was to represent individual types, which showed characteristic properties, lumping together various kinds of deformation and thus leading to a more unified understanding of monstrosities.[74]

To establish a wide comparison of organisms, Soemmerring relied upon the methods at hand—physiognomy and physical anthropology—to depict connections and to highlight small differences. This was not surprising, as he himself had heavily contributed to both fields. For instance, he had applied the physiognomical method to brain research and anatomy, when he claimed to have found a bodily, moral, and intellectual difference between Africans and Europeans, on the evidence of a macroscopic investigation of the brains of three Africans.[75] He believed that the physiognomical inspection of the skull and the brain would deliver the parameters for the establishment of a new natural and social order dividing humans and animals, "blacks" and "whites," men and women, intelligent and dull, healthy and sick. Although Johann Caspar Lavater's idea of a direct correspondence between the outer characteristics of the face and the inner moral and intellectual constitution of the individual was attacked by most European intellectuals, the more subtle version of physiognomy—investigating the facial angle, the skull, or the brain by ethnic group or species—was an important issue of anthropological, physiological, and anatomical debate. The role of physiognomics in the establishment of modern racism and sexism has overshadowed another aspect of the physiognomical method. It not only delivered parameters for distinctions, but also gave ways for finding similarities and continuities in living beings. What seems at first glance to be a contradiction is rather a mutual relationship in the emergence of modern human sciences: the more investigations of the physical order (*l'homme physique*) blurred the border to the moral order (*l'homme moral*), the more pressure was felt to establish new borders. The more facets of humans, from genius to monstrosity, were brought upon the stage, the more the hierarchies of the intellectual and the emotional, of good and bad, were built up. Although anatomists such as Soemmerring insisted on the moral neutrality of their research, their writings were full of ambiguities, which served to build up new discriminations in the anthropological hierarchies of human beings.

The Dutch anatomist and drawing teacher Pieter Camper was widely known to anthropologists and anatomists in the last decades of the eighteenth

74. Here I follow the differentiation between ideal and characteristic images employed by Daston and Galison "Image of Objectivity," 88–94.

75. Soemmerring, *Körperliche Verschiedenheit*, 57–67. For details, see Hagner, *Homo cerebralis*, 63–71.

century. After a visit to Camper's lessons, Johann Heinrich Merck wrote to Amalie, duchess of Sachsen-Weimar: "It is unbelievable what this man is able to do. With two strokes he alters the physiognomy of a Roman into a Hotten-tott or into an ape, of a horse into an ox, of an adult into an old man. Recently he has drawn all this publicly with chalk on a blackboard. The effect was that the people were moved by him and they began to cry when he drew with two strokes a dying person."[76] It was the suggestive power of Camper's ability as an artist that disturbed the spectators. But it was also not accidental that Camper used his lessons for demonstrating the close relationships between human and animal, regularity and irregularity, life and death. The problem of close rela-tionship and of continuity was tried out in the visual field of physiognomy before it was implemented into a broader embryological concept in the early nineteenth century. The reliance on visual evidence went hand in hand with one of Camper's important innovations, namely, the architectural technique of drawing anatomical subjects.

The common technique up to then, favored and masterfully performed by Albinus and his artist Jan Wandelaer, was a drawing in perspective, based on a "fixed viewpoint . . . and the mapping of [a] visual field onto [the] plane of representation" so that the artist was disciplined to follow an exact square-to-square correspondence.[77] Camper, in contrast, argued that an anatomical object should be drawn in an architectural way, that is, with a flexible view-point, so that each part of the object was viewed from the same distance. This technique did not allow drawing a part of the object in smaller or larger pro-portion than it really was in order to settle singularities in proper relation to the whole. The implications of this shift were not overlooked. Goethe asked his friend Merck for details of Camper's technique. Merck responded: "[Camper has] so to speak an ambulant viewpoint. He looks at the object, as if it would be situated on the floor in front of him, so that he has a complete overview."[78] It was within this architectural reconstruction of the body that Camper devel-oped his most influential and notorious contribution to anatomy and anthro-pology, namely, the so-called facial angle, conceived as a parameter for the hierarchy of the chain of being from the frog up to Apollo Belvedere. Accord-ing to Camper's admirer Herder, nature had invented this angle to differentiate not only "between animals, but also between different nations" and to elevate living beings "up to the most beautiful human being."[79] In his study on the

76. Merck to Herzogin Anna Amalie von Sachsen-Weimar on 3 June 1784, in Merck, *Briefe,* 452.
77. Daston and Galison, "Image of Objectivity," 90; see also Choulant, *Geschichte der ana-tomischen Abbildung,* 113–15.
78. Merck to Goethe, 29 April 1784, in Merck, *Briefe,* 440; see also Choulant, *Geschichte der anatomischen Abbildung,* 119.
79. Herder, *Ideen,* 134.

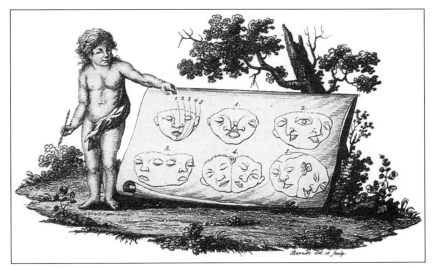

Fig. 6.2.   Samuel Thomas Soemmerring, title page vignette from *Abbildungen und Beschreibungen einiger Missgeburten* (1791). Photo courtesy of the Institut für Medizin- und Wissenschaftsgeschichte, Medizinische Universität zu Lübeck.

brain, Soemmerring referred to Camper as well as to Herder in order to legitimize the conjunction of anatomy, anthropology, physiognomics, and aesthetics that also guided his investigation of monstrosities.[80]

From his early career onward, Soemmerring was regarded as a disciple and close ally of Camper. Around 1790, Soemmerring was working both on his own book on monstrosities and on a German translation of Camper's seminal treatise on the facial angle.[81] The common physiognomical context of these two works was unveiled in the title vignette of Soemmerring's book, which depicts a boy displaying a board on which six monstrous heads have been drawn (fig. 6.2). The first head is missing parts of the front skull and brain; the second displays rudiments of double-headedness, as the nose is divided and the eyes are positioned far from another. The third to fifth heads show more and more distinct characteristics of division; the sixth one is almost completely divided. In the genealogy of motifs, Soemmerring's vignette is a reference to Lavater: The title vignette of the first volume of the *Physiognomische Fragmente* displays a tableau of several faces shown from several perspectives (fig. 6.3), the order

80. Soemmerring, *Körperliche Verschiedenheit*, 16–19.

81. Merck, "Über einige Merkwürdigkeiten von Cassel," 218; Soemmerring visited Camper in 1778. Afterward they had an intense correspondence. The book Soemmering was translating was Camper's *Unterschied der Gesichtszüge*.

Fig. 6.3.   Johann Caspar Lavater, title page vignette from *Physiognomische Fragmente, zur Beförderung der Menschenkenntniss und Menschenliebe* (1775), vol. 1. Photo by author.

of which highlighted the similarities and differences between the subjects.[82] More important, however, is Soemmerring's notation of the "natural order or elevation" of the monstrosities, a direct reference to Camper's representations of skulls and faces from the ape to Apollo Belvedere (fig. 6.4). The geometrical precision enabled the anatomist to compare the various skulls and faces. The possibility of reconstructing the elevation in the chain of being with the aid of the facial angle was thus an ideal strategy for visualizing similarities and differences, whether between humans and animals or between humans. The same procedure should also lead to an understanding of the monstrous nature.

Most of the monstrous specimens presented and described in his book were examined by Soemmerring when he was professor of anatomy in the town of Cassel, capital of the principality of Hesse-Cassel, from 1779 to 1784. Since the opening of the Collegium Carolinum in Cassel in 1709, medicine had been part of the curriculum. Its importance was rather marginal until the 1760s, when Landgrave Friedrich II tried to establish the reputation of the col-

82. On physiognomics in general see, e.g., Magli, "Face and Soul"; Schmölders, *Das Vorurteil im Leibe;* Schneider and Campe, *Geschichten der Physiognomik;* on Lavater in particular, see Louis "Der beredte Leib"; Brittnacher "Der böse Blick des Physiognomen"; Gray, "Sign and Sein," and "Aufklärung und Anti-Aufklärung."

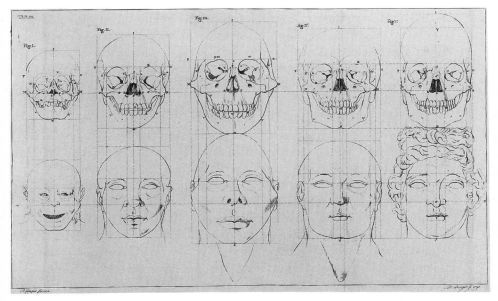

Fig. 6.4.   Pieter Camper, drawings of skulls and faces from the ape to Apollo Belvedere, from *Über den natürlichen Unterschied der Gesichtszüge in Menschen verschiedener Gegenden und verschiedenen Alters* (1792). Photo courtesy of the Niedersächsische Staats- und Universitäts-bibliothek, Göttingen.

legium, which had had to serve as a military academy, as part of a general rebuilding and modernization of the town after its decline in the Seven Years' War. Distinguished scholars such as Georg Forster and the historian Johannes von Müller were hired. When Soemmerring received a call to the collegium in 1779, he moved into a completely new and luxurious anatomical building, offering high standard facilities for dissection and equipped with enough space for an anatomical collection.[83] The anatomical collection had been part of the court *Kunstkammer* since the early eighteenth century. Soemmerring had to take care of the old objects, and he was also obliged by his contract to assemble new preparations. All anatomical objects were part of the collection of the landgrave. If Soemmerring wanted to prepare an object for himself, he even had to pay for the necessary conservation fluids from the pharmacy. The strictly illustrative and courtly character of the anatomical collection explains

83. Heinemann, "Aus der Blütezeit," 88–90. See also Enke, "Soemmerrings erste Professur," 97–107. The building, constructed by the architect Simon Louis du Ry, was dismantled after the decline of the Collegium Carolinum and then rebuilt in Marburg, to where some of the now unemployed professors were transferred.

why Soemmerring was restricted in his manipulations of the specimens. A few years before the publication of his drawings, he mentioned that Cassel's was "the most complete and the most beautiful collection of children with defective brains."[84] But he could dissect them only so far and no further, lest they be ruined as representative parts of the collection. He was strictly forbidden to skeletonize them.

These restrictions, together with the premises of the physiognomical approach, guided Soemmerring's strategy of investigation and visualization. Instead of Camper's skulls and his own later *femina perfecta*, objects that were idealized and thus conceptualized as typical, Soemmerring was confronted with individual objects. First, he had but one specimen to investigate for any given category, and second, it was possible that the age of the specimens had led to morphological and chemical alterations. How could one represent double-headedness and its typical properties when there was no second specimen at hand, and thus no clear notion of what was typical? The strategy to overcome such problems was to construct images. Soemmerring referred to this point in a long paragraph and stated that not every artist was able to fulfill such a task: "Most of them take a great deal of meticulous care with minor matters, such as wrinkles, unnaturally caused by alcohol, shriveling up, or a wrong position, caused by compression in the glass. This damages the main characteristics. You cannot train certain artists at all merely to see what really should be expressed and to omit minor points or coincidences which do not matter."[85] This was not intended to be an antiaesthetic statement; it was, rather, directed against graphic artists who produced beauty rather than portraying the anatomical significance of specimens. The beauty of a monstrous object, indeed any object, could not truly be grasped by the artist, but only by the anatomist, who knew by experience "what exactly shall be expressed." The junction between anatomy—or, better, the physiognomical method, which linked the outer characteristics with inner properties—and the architectural technique was the lesson Soemmerring took from Camper. To be sure, Soemmerring did not criticize Albinus, whom he deeply admired and whose carefulness and perfection he mentioned repeatedly. But in his visualization strategy, he preferred the architectural technique. By chance, Soemmerring's graphic artist had worked as a surveyor and was mathematically trained, so Soemmerring was

84. Cited in Monro, *Bemerkungen*, 20 (Soemmerring was the German editor of Monro's *Observations on the structure and functions of the nervous system*, and he added a number of his own remarks and observations). When Soemmerring worked in Cassel, the collection contained ten monstrous bodies (Soemmerring, "Präparate," 22–23). On the costs and responsibilities of maintaining and preparing specimens, see Dienstvertrag, 20 August 1779, Staatsarchiv Marburg 1144, fol. 13, cited in Aumüller, "Geschichte der anatomischen Institute," 72–74.

85. Soemmerring, *Abbildungen und Beschreibungen*, 5.

able to ask him to measure the objects, to construct the main parts with the aid of a circle, and to draw them in one-to-one correspondence. The idea was not to create *monstra perfecta,* but to represent an object, which could be compared with the original or with any other embryo with a similar defect. The idea of comparison between a two-dimensional drawing and a three-dimensional specimen delivers the key to the question of why Soemmerring preferred Camper's architectural method. In principle, the drawing or engraving would serve as a draft for a wax model. In the case of the monstrosities, this did not happen; but Soemmerring's images of the eye and the ear and of normal embryos were rebuilt in wax.[86] This modeling explains the lively appearance of the objects in his drawings. In figure 6.5, a monstrous child sits on a step in a lifelike pose. The skin seems fresh and vivid; shadows and light are carefully drawn. All possible artificial deformations—the aging from conservation in spirits and physical distortions from storage in the bottle—have been completely removed. The combination of objectivity and artificiality of this procedure becomes obvious when one compares it with a drawing of another monstrous deformation, a specimen Soemmerring needed for a crucial comparison that was not from his Cassel collection. Figure 6.6 is Soemmerring's copy of an early-eighteenth-century drawing originally possessed by the anatomist Hieronymus David Gaubius in Leiden. Although it employs the architectural method, the difference to the other drawings is remarkable. There is no setting in which the object is integrated; the limbs hang down lifelessly; the structure of the skin remains invisible; and it is undecidable whether specific wrinkles and bumps are hereditary or caused by conservation. Soemmerring remarked that the copperplate was brilliantly done, but he regretted that the child was not drawn in its original magnitude. Furthermore, he complained, the arms seemed to be too long and the head too small in relation to the rest of the body. "All this would have been probably avoided, if one would have preserved the natural magnitude."[87]

The reason Soemmerring incorporated such an old and insufficient drawing into his work leads back to his goal of presenting a hierarchy of monstrosities. The serial order was more important than the perfection of a specific drawing. "If you compare these cases with others in other collections or writings, [you will realize] that nature principally follows a certain order, a certain route and uniformity even in monsters."[88] In conclusion, Soemmerring had

86. Choulant, *Geschichte der anatomischen Abbildung,* 133. The reference to Camper is also obvious in Soemmering's work on the beauty of ancient children's skulls. See Oehler-Klein, "Anatomie und Kunstgeschichte," 209–212.
87. Soemmerring, *Abbildungen und Beschreibungen,* 18.
88. Ibid., 38.

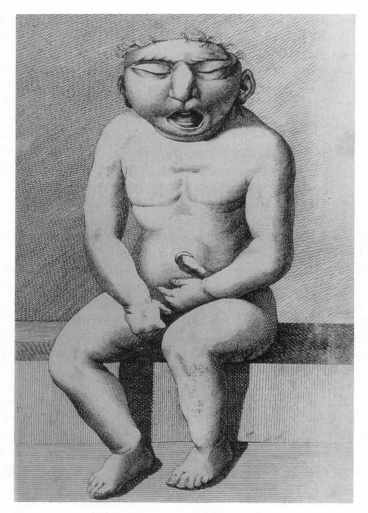

Fig. 6.5.   Samuel Thomas Soemmerring, drawing of an anencephalic child, from *Abbildungen und Beschreibungen einiger Missgeburten* (1791). Photo courtesy of the Institut für Medizin- und Wissenschaftsgeschichte, Medizinische Universität zu Lübeck.

two closely linked intentions. First, the drawings should be used by other scientists as a substitute for the specimen itself. If a drawing could be compared with a real body, then the anatomical image was established as a tool in the hands of anatomists. Second, Soemmerring intended a serial order of deformation from anencephaly to double-headedness. This order suggested a com-

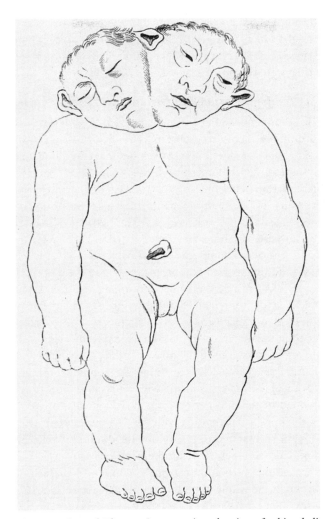

Fig. 6.6.   Samuel Thomas Soemmerring, drawing of a bicephalic child, from *Abbildungen und Beschreibungen einiger Missgeburten* (1791). Photo courtesy of the Institut für Medizin- und Wissenschaftsgeschichte, Medizinische Universität zu Lübeck.

mon explanation without referring to generation theories, and it transformed the image into a crucial representational device in the investigation of monstrosities. The use of images was adopted in physiognomics and physical anthropology at the same time and in phrenology a few years later. The study of skulls or faces *in vivo* and the study of their drawings *in vitro* were now seen as

equivalent techniques. That Soemmerring had redefined images and the physiognomical method in anatomy with the aid of monstrosities is highly significant. In the late eighteenth century, monstrosities had been integrated into and
at the same time were shaping the new discourse of life.

The Temporalization and Gendering of Monstrosities

As I argued in section 2 of this chapter, the physico-theological view of nature
demanded a unifying theory of monsters. The incorporation of monsters in
the epigenetic course of development was far from fulfilling that unifying
theory. Early-nineteenth-century embryologists could not provide more than
a rough differentiation between a surplus and lack of generative forces or formative drive. And it was furthermore assumed that specific monstrosities were
caused by mechanical reasons, others by deviant mixtures of maternal and pa-
ternal fluids. Gottfried Reinhold Treviranus stated in his *Biologie* that "several
completely different causes could lead to malformations."[89] The heterogeneity
of monsters was no longer seen as a problem because the epigenetic assumption of the early existence of the embryo as a half-liquid, unorganized, and
unshaped matter made it impossible and also unnecessary to construct a relationship between an original form and its deviation. Instead, such relationships
were embedded in the dynamics of development, and this led to a network of
classifications and connections in the understanding of living beings.

For Johann Friedrich Meckel, Soemmerring's physiognomical chain of
monsters was a real hierarchy: "starting with a first imperfect rudiment of an
embryo, inhibited in the first period of its development; the perfect normal
formation in the middle; and ending with the perfect double-monster."[90] The
question "What is regular?" was replaced by the question "What is regular at
a certain point of development?" Around the same time, Meckel and Friedrich
Tiedemann argued that certain monstrosities could be explained on the basis
of an "inhibition of the whole organism or of single organs on earlier normal
steps of development."[91] In Tiedemann's words, monstrosity was an inhibition
on "a lower level of organization."[92] Meckel went into further detail and explained the so-called inhibited formations *(Hemmungsbildungen)*. "[There is]
a state in the development of the fetus, in which the organs, which are absent
or abnormal in monstrosities, behave normally in this kind. Monstrosities
are due to the fact that these organs do not further develop in the usual way,

89. Treviranus, *Biologie*, 442; see also Prochaska, *Lehrsätze*, 255.
90. Meckel, *Handbuch*, 70; on Meckel's teratology, see Clark, "Contributions."
91. Meckel, *Beyträge*, 34. A similar point had already been made by Johann Ferdinand Autenrieth a few years earlier, but for Autenrieth it was not more than a mere consideration. See Autenrieth, *Supplementa*, 39.
92. Tiedemann, *Zoologie*, 177.

while others do more or less run through all periods [of embryological development]."[93] The connection between insufficient productivity of generative force and inhibition at a prior stage of development was the first step of argumentation. It was still more important that "any deviation of the embryo from the human form is a fall-back in animal formation, and thus any monstrosity is like an animal, not always in its outer shape, but more or less in its inner one."[94]

From an exclusively physiognomical point of view, any embryo was a monster, or at least a beast, at a certain point of its inner history. With the introduction of time patterns in embryology, the contiguity of the chain of being was rewritten as a continuity of life. That is, the processes of development were themselves responsible for the differentiation between humans and beasts. In the worst cases, this difference had been neutralized. A monstrosity was thus the right phenomenon at the wrong moment, because in an earlier state of the embryo such a form could have been normal. Early-nineteenth-century embryologists thus embedded Wolff's notion of a succession in the development of the single parts and organs of the body in a comprehensive temporal process. In this process, the individual development, from the act of procreation to birth, and the development from primitive living beings to humans constituted one discursive field. Monstrosities became a powerful tool in this field. They represented an intermediate state on the way from beasts to humans, but not in the old-fashioned sense of missing links. Rather, they were used to calibrate a new order of life and a new order of human beings. It is widely known that late-eighteenth-century physical anthropology, anatomy, and physiology worked hard to define bodily characteristics for differences among humans, mainly between different races and between women and men. The new epistemological status of monstrosities simultaneously led to a connection with a gendered notion of embryology. Tiedemann and Meckel in particular supported a scientifically legitimized order of sex in their embryological and teratological studies.

During the eighteenth century, various anatomists had noted, without explanation, that the majority of acephalic embryos were female.[95] Tiedemann picked up this point and argued that all embryos had exclusively female sexual organs in their earliest states. The male organs only developed later on. Accordingly, hermaphroditism was "an inhibition in the development of sexual organs on various earlier levels, normal for the embryo at that time." Likewise, the normal female sexual organs were an "inhibition of the embryo on a lower

93. Meckel, *Beyträge*, 159.
94. Tiedemann, *Zoologie*, 178.
95. See, e.g., Soemmerring, *Abbildungen und Beschreibungen*, 9.

level of development." From this, it was only a short step to the claim that "obviously woman is much more similar to the fetus than man, and thus the woman stands on a lower level of development than the man." [96] In a similar way, Meckel interpreted an inner and elevated position of the testicles in a monstrous male embryo as an inhibited formation and at the same time as an "approach of male to female formation." [97] Similar to the way physiognomy took the brain and skull as parameters for sex differences, embryology inscribed entities such as higher and lower development, normality and anomaly, formation and malformation, male and female organs. And as nature was liberated from the chains of predetermined order and God lost power as a reference point, nature acquired incalculable and dangerous properties. Nature itself became the problem, and monstrosities its most elucidating witnesses.

Conclusions: Monsters and Reason

In his magisterial study *Die Philosophie der Aufklärung,* Ernst Cassirer pointed out a transformation of the conceptualization of reason in the eighteenth century. In the seventeenth century, reason was "the region of eternal truths, i.e. those truths which are common to human and to divine mind. . . . Any act of reason reassures us of its participation in the divine nature." For the eighteenth century, however, reason had become "a fundamental power leading to the discovery of truth and its definition and establishment . . . an energy, a power which can only be grasped fully in its practice and its consequences." [98] In other words, there had been a transition from a static and material understanding of reason as possession to a more functional and dynamic one of reason as process and activity. One could argue that Horkheimer and Adorno implicitly rejected this optimistic notion in their critical analysis of reason. According to them, the Kantian notion of reason was ambiguous: on the one hand, "reason as the transcendental super-individual self contained the idea of the free living together of human beings." On the other hand, reason was the "instance of calculating thinking which arranges the world for the purpose of self-preservation." In their view, it is precisely this unreflective use of reason for getting into the unknown, this "subordination of all natural *[alles Natürliche]* under the autocratic subject that finally leads to the rule of the blind objective and natural *[des blind Objektiven, Natürlichen]*." [99]

96. Tiedemann, *Anatomie,* 84, 87.
97. Meckel, "Beschreibung," 113. On the history of the notion of women as incompletely developed men, see Laqueur, *Making Sex,* esp. chaps. 2 and 5. German Romantic embryology shows the importance of the commensurability of the two sexes in early-nineteenth-century discourse. The modern politics of difference did not rely solely upon the construction of incommensurabilities, but also on the emphasis of minimal differences.
98. Cassirer, *Die Philosophie der Aufklärung* (1973), 15, 16.
99. Horkheimer and Adorno, *Dialektik der Aufklärung,* 102–3, 10.

What does the confrontation of Cassirer's grand narrative with that of Horkheimer and Adorno reveal for the analysis of monsters and the Enlightenment? Although they could hardly be further from one another, they both offer an important framework for looking at the period—and in doing so reveal their possibilities and limitations. Adorno and Horkheimer do not deal with a specific period, and Cassirer's chronology does not fit very well with the transformations of the late eighteenth century. Indeed, one questions his periodization of the transformation of reason because much of what he points out for the eighteenth century did not come to hold sway before the second half, and the reconceptualization of categories like nature, generation, and beauty came even later. Nevertheless, one indication of a shift in thinking is the way the process of enlightening monsters uncovered radical changes in their epistemic localization. For the first two-thirds of the eighteenth century, "getting into the unknown" meant order, incorporation, and domestication in the closed space of a collection. There monsters served for delight, wonder, and erudition, or they served as a contrast to the beauty of ordered formations. In this role, they were so valued that Louis-Sébastien Mercier reserved the most prominent place in his utopian natural cabinet for them, just under the dome.[100] The "comfortable" and public residence of monsters in cabinets, which was the primary place of their "disenchantment," was contrasted by their resistance to incorporation in the natural order of things. All attempts to classify them and to explain monstrous generation confounded the physico-theological view of a determined, ordered, purposeful, and beautiful nature. In this domain, monsters could not be possessed by the enlighteners—a scandal for the authority of a system that intended to naturalize, subordinate, and classify all living beings. Thus, when enlighteners themselves saw the immensity of their task in overcoming mythology and superstition—and monsters had been a popular subject for this enterprise—they created a whole field of problematic inquiry involving the "domestication" of monsters.

The last third of the eighteenth century revealed a new order of monsters. In the dynamic system of epigenesis, they received an epistemic status that helped to transform the contiguity of living beings into the continuity of life. Certain forms of monstrosity were regarded as survivals of transitory embryonic forms. This was a crucial moment for the temporalization of humans, because human existence was no longer infinitely far away from deformation, disorder, hybrids, and transitory forms. Humans no longer began their

---

100. Mercier, *L'an 2440*, 292. Of course, Mercier here referred to the final passage of Diderot's article "Cabinet d'Histoire Naturelle" in the *Encyclopédie* in which he dreamed of a gigantic museum with several halls. The largest one in the middle should be reserved for the monsters of earth and sea. However, Diderot did not have in mind deformed beings, but crocodiles, elephants, and whales ("Cabinet d'Histoire Naturelle," 492).

existence as perfect creations, but as vulnerable embryos and potential monsters. Predetermination was replaced by chance and risk, and life was provided with its own history. At that moment, monsters lost their status as threats to classification and taxonomy, and as singularities. Although they were still collected and put into anatomical and pathological collections, they were no longer regarded as untouchable and sensational icons in natural cabinets. They were there to be dissected, examined, and finally discarded. Consequently, drawings of monstrosities increased enormously. They replaced the objects and were used as tools in the quickly expanding fields of embryology and pathology.

Did the process of "enlightening monsters" really lead to an enlightened attitude toward monstrosity? Certainly the integration of monsters into the processes of life completed a long period of "disenchantment" and defeated concepts like predetermination, passive nature, and fate. From this viewpoint, the shift analyzed in this chapter can be linked to the intellectual shift in the understanding of reason that Cassirer took as an example for the progression of Enlightenment culture. Horkheimer and Adorno clearly have not much to say on historical transformations within the Enlightenment. But their analysis of parallels between Kant and de Sade certainly opens our eyes to the ambiguities of reason. The instrumentalization of monsters in the human sciences and in gender discourse marked a new chapter in the highly problematic relation between "deformation" and the order of life. The subordination of the "natural" under the rule of autocratic reason changed: the scientific notion of deformations as representatives of older and more primitive biological forms of existence that did not reach the defined standard created new monsters.

Acknowledgments

A preliminary version of this paper was presented to seminars at the Max-Planck-Institut für Wissenschaftgeschichte in Berlin, the Science Museum in London, and the Wellcome Unit in Cambridge. I have benefited enormously from discussion after these lectures. In addition to the authors involved in this volume, I would like to thank Nani Clow, Anke te Heesen, and Lily Kay for their comments and criticisms. While working on this chapter, I was supported by a Heisenberg grant from the Deutsche Forschungsgemeinschaft.

Primary Sources

Autenrieth, Johann Ferdinand. *Supplementa ad Historiam Embryonis Humani.* Tübingen, 1797.
Bacmeister, Jean. *Essai sur la bibliothèque et le cabinet de curiosité d'histoire naturelle de l'Academie des Sciences de Saint Petersbourg.* St. Petersburg, 1776.

Baer, Karl Ernst von. "Über den litterärischen Nachlass von Caspar Friedrich Wolff, ehemaligem Mitgliede der Akademie der Wissenschaften zu St. Petersburg." *Bulletin de la classe Physico-Mathématique de L'Académie Impériale des Sciences de Saint Pétersbourg* 59 (1847): 129–60.

———. "Über den jetzigen Zustand und die Geschichte des Anatomischen Cabinets der Akademie der Wissenschaften zu St. Petersburg" (1850). *Publications du Musée d'Anthropologie et d'Ethnographie de l'Académie Impériale des Sciences de St. Petersbourg* 1 (1900): 111–45.

Bergholz, Friedrich Wilhelm von. "Tagebuch, welches er in Russland von 1721 bis 1725 als hollsteinischer Kammerjunker geführet hat." *Bueschings Magazin für die neue Historie und Geographie* 19 (1785): 3–202.

Blumenbach, Johann Friedrich. *Über den Bildungstrieb.* Göttingen, 1781.

———. *Über den Bildungstrieb.* 3d ed. Göttingen, 1791.

Boerhave, Abraham Kaau. "Historia anatomica ovis pro hermaphrodito habiti" (1747–48). *Novi Commentarii Academiae Scientiarum Imperialis Petropolitanae* 1 (1750): 315–36.

Brandt, Johann Friedrich. "Versuch einer kurze Übersicht der Fortschritte, welche die Kenntniss der thierischen Körper den Schriften der Kaiserlichen Academie der Wissenschaften zu St. Petersburg verdankt." In *Recueil des Actes de la Séance Solennelle de L'Académie Impériale des Sciences de St. Pétersbourg,* 51–117. St. Petersburg, 1832.

Buffon, Georges-Louis Leclerq de. *Histoire naturelle, générale et particulière.* Supplement, vol. 4. Paris, 1778.

Burdach, Karl Friedrich. "Über die anatomische Sammlung in der Kunstkammer zu St. Petersburg." In *Russische Sammlung für Naturwissenschaft und Heilkunde,* 1:423–56. Riga and Leipzig, 1816.

Camper, Peter [Pieter]. *Über den natürlichen Unterschied der Gesichtszüge in Menschen verschiedener Gegenden und verschiedenen Alters . . . Üb. v. S. Th. Soemmerring.* Berlin, 1792.

Diderot, Denis. "Cabinet d'Histoire Naturelle." In *Encyclopédie ou Dictionnaire raisonnée des Sciences, des Arts et des Métiers,* 2:489–92. Paris, 1776.

Euler, Leonhard. *Pis'ma k ucenym* (Letters to scientists). Moscow: Izdatel'stvo Akademii Nauk USSR, 1776.

———. *Briefwechsel, Beschreibung, Zusammenfassung der Briefe und Verzeichnisse.* In *Opera Omnia,* edited by Adolf P. Juskevic, Vladimir I. Smirnov, and Walter Habicht, ser. 4, vol. 1. Basel: Birkhäuser, 1975.

Floegel, Carl Friedrich. *Geschichte der Hofnarren.* Leipzig: Liegnitz, 1789.

Fontenelle, Bernard de. *Histoire de l'Academie Royale des Sciences.* Paris, 1712.

———. "Préface sur l'utilité des mathématiques et de la physique" (1699). In *Oeuvres de Fontenelle,* 1:47–60. Paris, 1825.

Friderici, Gottlieb. *Monstrum humanum rarissimum recens in lucem editum in tabula exhibet simulque observationibus pathologicis.* Leipzig, 1737.

Haller, Albrecht von. *Opuscula sua Anatomica.* Göttingen, 1751.

———. "De Monstris Libri II." In *Operum anatomici argumenti minorum,* vol. 3. Lausanne, 1768.

———. "Jeu de la nature et monstres." In *Encyclopédie ou Dictionnaire universel raisonné des connoissances humaines,* edited by Fortunatus de Felice, 24:197–212. Yverdon, 1773.

Herder, Johann Gottfried. *Ideen zur Philosophie der Geschichte der Menschheit* (1784–91). In *Sämmtliche Werke,* edited by B. Suphan, vol. 13. Berlin, 1887.

Juskevic, A. P., and Eduard Winter, eds. *Die Berliner und die Petersburger Akademie der Wissenschaften im Briefwechsel Leonhard Eulers.* 3 vols. Berlin: Akademie, 1959–76.

Leibniz, Gottfried Wilhelm. "Essais de Théodicée sur la bonté de Dieu, la liberté de l'homme et l'origine du mal" (1710). In *Die philosophischen Schriften,* edited by C. J. Gerhardt, 6:25–461. Berlin, 1885.

Lémery, Louis. "Sur un foetus monstreux" (1724). *Mémoires de l'Academie Royale des Sciences* (Paris), 1726, 44–62.

Littré, Alexis. "Observation sur un foetus monstrueux qui n'avoit qu'un oeil" (1717). *Mémoirs de l'Academie Royale des Sciences* (Paris), 1719, 285–87.

Mairan, Dortous de. "Sur les monstres" (1743). *Histoire de l'Academie Royale des Sciences* (Paris), 1746, 53–68.

Malebranche, Nicolas. "Eclairecissment sur le 6e livre de la Recherche de la Vérité." In *La Recherche de la Vérité*, 5th ed., vol. 3. Paris, 1700.

Maupertuis, Pierre-Louis Moreau de. *Oeuvres de Maupertuis.* Nouvelle édition. 4 vols. Lyon, 1768.

Meckel, Johann Friedrich. *Beyträge zur vergleichenden Anatomie.* Vol. 1. Leipzig, 1809.

———. *Handbuch der pathologischen Anatomie.* Vol. 1. Leipzig, 1812.

———. "Beschreibung zweier menschlicher schädelloser Missgeburten." In *Anatomisch-physiologische Beobachtungen und Untersuchungen*, 79–146. Halle, 1822.

Mercier, Louis-Sébastien. *L'an 2440: Rêve s'il en fut jamais.* 1771. Bordeaux: Ducros, 1971.

Merck, Johann Heinrich. "Über einige Merkwürdigkeiten von Cassel." *Der Teutsche Merkur*, 4 Vierteljahr (1780): 216–29.

———. *Briefe.* Edited by Herbert Kraft. Frankfurt am Main: Insel, 1968.

Monro, Alexander. *Bemerkungen über die Struktur und Verrichtungen des Nervensystems.* Leipzig, 1787.

Neickelius, Caspar Friedrich. *Museographia oder Anleitung zum rechten Begriff und nützlicher Anlegung der Museorum oder Raritäten-Kammern.* Leipzig and Breslau, 1727.

Prochaska, Georg. *Lehrsätze aus der Physiologie des Menschen.* 2d ed. Vol. 2. Wien, 1811.

Reimers, Heinrich von. *St. Petersburg am Ende seines ersten Jahrhunderts: Mit Rückblicken auf Entstehung und Wachsthum dieser Residenz unter den verschiedenen Regierungen während dieses Zeitraums.* 2 vols. St. Petersburg, 1805.

Ruysch, Frederic. *Thesaurus Anatomicus secundus.* Amsterdam, 1702.

———. *Adversiorum Anatomico-medico-Chirurgicorum: Decas Prima.* Amsterdam, 1729.

———. *Observationum Anatomico-Chirurgicarum Centuria: Accedit Catalogus Rariorum, quae in Museo Ruyschiano asservantur.* Amsterdam, 1737.

Schaarschmidt, August. *Verzeichniss der Merkwürdigkeiten, welche bei dem Anatomischen Theater zu Berlin befindlich sind.* Berlin, 1737.

Schuster, Julius. "Der Streit um die Erkenntnis des organischen Werdens im Lichte der Briefe C. F. Wolffs an A. von Haller." *Sudhoffs Archiv* 34 (1941): 196–218.

Soemmerring, Samuel Thomas. *Über die körperliche Verschiedenheit des Negers vom Europäer.* Frankfurt and Mainz, 1785.

———. "Präparate, welche Herr Hofrath Sömmerring (jetzt zu Mainz), dem anatomischen Theater zu Cassel 1784 zurück liess." *Baldingers Medicinisches Journal* 16 (1787): 14–23.

———. *Abbildungen und Beschreibung einiger Missgeburten die sich ehemals auf dem anatomischen Theater zu Cassel befunden.* Mainz, 1791.

———. *Briefwechsel, 1784–1786.* In *Werke*, edited by Jost Benedum and Werner Friedrick Kümmel, vol. 19, part 1, edited by Franz Dumont, Stuttgart: Fischer, 1997.

Stählin, Jacob von. *Originalanekdoten von Peter dem Grossen.* 1785. Munich: Winkler, 1968.

Sulzer, Johann Georg. *Unterredungen über die Schönheit der Natur.* Berlin, 1750.

Tiedemann, Friedrich. *Zoologie: Zu seinen Vorlesungen entworfen.* Book 1, *Allgemeine Zoologie: Mensch und Säugethiere.* Landshut, 1808.

———. *Anatomie der kopflosen Missgeburten.* Landshut, 1813.

Treviranus, Gottfried Reinhold. *Biologie, oder Philosophie der lebenden Natur für Naturforscher und Ärzte.* Vol. 3. Göttingen, 1805.

Uffenbach, Zacharias Conrad von. *Merkwürdige Reisen durch Niedersachsen, Holland und Engelland.* 3 vols. Ulm and Memmingen, 1753.

Valentin, Michael Bernhard. *Museum Museorum.* Vol. 2. Frankfurt am Main, 1714.

Vincent, Levinus. *Elenchus Tabularum, Pinacothecarum, atque Nonnulorum Cimeliorum, in Gazophylacio Levini Vincent.* Harlemi Batavorum. 1719.

Wagner, Rudolf. *Samuel Thomas von Soemmerrings Leben und Verkehr mit seinen Zeitgenossen.* Edited by Franz Dumont. Vol. 2 of *Soemmerring-Forschungen*. 1844. Stuttgart and New York: Fischer, 1986.

[Weitbrecht, Josias]. *Musei Imperialis Petropolitani.* Vol. 1, *Pars prima qua continetur res naturales ex regno animali.* St. Petersburg, 1742.

Winslow, Benignus. "Remarques sur des monstres . . . Première partie" (1733). *Mémoires de l'Academie Royale des Sciences* (Paris), 1736, 366–89.

———. "Remarques sur deux dissertations touchant les monstres" (1742). *Mémoires de l'Academie Royale des Sciences,* (Paris), 1745, 91–120.

Wolff, Caspar Friedrich. *Theoria generationis.* Halle, 1759.

———. *Theorie von der Generation in zwo Abhandlungen erklärt und bewiesen.* Berlin, 1764.

———. "Descriptio vituli bicipitis cui accedit commentatio de ortu monstrorum" (1772). *Novi Commentarii Academiae Scientiarum Imperialis Petropolitanae* (St. Petersburg) 17 (1773): 540–75.

———. "Notice touchant un monstre biforme, dont les deux corps sont réunis par derriere" (1778). *Acta Academiae Scientiae Imperialis Petropolitanae* (St. Petersburg), part 1 (1780), 41–44.

———. "De inconstantia fabricae corporis humani, de eligendisque ad eam repraesentandam exemplaribus" (1778). *Acta Academiae Scientiae Imperialis Petropolitanae* (St. Petersburg), part 2 (1780), 217–35.

——— *Obiecta meditationum pro theoria monstrorum.* Edited by Tatjana A. Lukina. Leningrad: Academiae Scientiarum URSS, 1973.

Worm, Ole. *Musaei Wormiani Catalogus.* 1642.

Zedler, Johann Heinrich. "Missgeburt." In *Grosses vollständiges Universal-Lexikon,* vol. 21. Leipzig and Halle, 1739.

7

The Science and Conversation
of Human Nature

Marina Frasca-Spada

Speculation had not yet attempted to analyze the mind, to trace the passions to their sources, to unfold the seminal principles of vice and virtue, or sound the depths of the heart for the motives of action. All those inquiries, which from that time that human nature became the fashionable study, have been made sometimes with nice discernment, but often with idle subtility, were yet unattempted. . . . Mankind was not then to be studied in the closet.

Samuel Johnson, preface to the *Plays of Shakespeare* (1765)

The "science of human nature" to which this chapter is devoted was the attempt to make sense of human nature on the part of moral philosophers, studying the human mind and passions from their closets. It was this intellectual enterprise that inspired in Johnson, in 1765, such nostalgia for the old times when human nature was a subject for poets, rather than for acute but often too subtle inquirers. Its heroic phase spanned three or four generations of British men of letters: from the prehistory in Francis Bacon, to the English Locke, Shaftesbury, Mandeville, and Joseph Butler, and the Scottish Francis Hutcheson, Henry Home, Monboddo, David Hume, and Adam Smith. This "science of human nature" no longer exists, and study of its eighteenth-century exponents is now parceled out to historians of different disciplines: to historians of psychology, of anthropology, of natural history, and of philosophy. Of course, what a historian is a historian of makes a difference in the emphasis put on this or that side of the same enterprise: so a historian of anthropology is likely to concentrate on writings on the origin of languages and on the differences between peoples, while a historian of psychology will presumably be interested in the descriptions of mental phenomena, and a historian of natural history will consider the studies of human nature as a possible link between natural history and moral philosophy.[1]

1. For history of Enlightenment anthropology, see Wokler, "Apes and Races," and "From *l'Homme Physique* to *l'Homme Morale*." For history of psychology, see Richards, "Absence of

Being a historian of philosophy, I consider the "science of human nature"
as it appears in eighteenth-century writings on metaphysics and moral philosophy.[2] For my purposes, Johnson's lines quoted above define the scope of
this "science of human nature" with admirable accuracy; and the terms he
chose to express his disapproval of this side of the Lockean tradition remind
one of his condemnation of the most abstract and "abstruse" among the
"scientists" of human nature: David Hume, the hated Scottish skeptic who was
more subtle than deep, the man for whose vanity "truth will not afford sufficient food."[3]

I discuss the treatment of the workings of passion and perception in books
of metaphysics and moral philosophy, and explore the relations between these
books and representatives of another literary genre altogether, namely, the
novel. In particular, I consider on the one hand the works of David Hume and
Adam Smith, published between 1739 (books 1–2 of Hume's *Treatise of Human
Nature*) and 1762 (completion of Hume's *History of England*), and on the other
hand, some of the literary productions of Samuel Richardson, Sarah Fielding,
and Laurence Sterne—works published between 1744 (volumes 1–2 of Sarah
Fielding's *Adventures of David Simple*) and 1768 (Sterne's *Sentimental Journey*).
This will, I hope to show, cast some light on features of the science of human
nature that have so far remained relatively unexplored.[4]

For a start, I discuss the relations of the moral philosophers' "science of
human nature" with history. It has been convincingly argued by historians
of natural history that the science of human nature typically compiled and
ordered histories of the human mind, passions, and ideas just as natural history compiled and ordered histories of the kingdoms of nature.[5] But I explore

---

Psychology," and *Mental Machinery*. And for history of natural history, see the following works by
Wood: "Natural History of Man" (with reference to the pioneering Bryson, *Man and Society*); "Science and the Aberdeen Enlightenment"; "Hume, Reid, and the Science of the Mind"; and "Science
of Man." For a balanced discussion of the notion of "human nature" in the eighteenth century and
a review of its uses in Enlightenment historiography, see Smith, "Language of Human Nature."

2. A clear history-of-philosophy slant also characterizes, for example, most of the studies
contained in Jones, *Science of Man*.

3. "Truth is a cow which will yield such people no milk, and so they are gone to milk the
bull," is what Johnson said to Boswell about Hume and his arguments (Boswell, *Life of Samuel
Johnson*, Thursday, 21 July 1763). Hume himself was always ready to acknowledge vanity as his
capital defect. See, for example, his "Character of . . . , written by himself," in Mossner, *Life of
David Hume*, 569 ("Fancies he is disinterested because he substitutes vanity in place of all other
passions"); and "My Own Life," in ibid., 611–15. His letter of 11 May 1758 to Gilbert Elliot of Minto
(*Letters of David Hume*, 1:278) contains a delightfully ironic and playful passage on the vanity of
authors. See also below, note 47.

4. On the fruitfulness of a philosophical standpoint for the interpretation of novels, see the
stimulating Jones, *Philosophy and the Novel*.

5. See, for example, Wood's works cited above, note 1. Smith's *Theory of Moral Sentiments*
and book 2 of Hume's *Treatise*, with their classifications of the passions, are indeed in certain
respects reminiscent of contemporary natural historical treatises.

two further respects in which the science of human nature was a form of historical knowledge: first, the extent to which it appealed to history in the traditional rhetorical and sententious manner;[6] second, the extent to which human nature, that is, human selves, ideas, and passions, were themselves regarded as genealogically constituted entities. I then consider the novelists' "knowledge of the human heart," emphasizing the similarities between the two from the point of view of present-day readers. I suggest that some novels were representations of their own readers' conversation and social life, in which they were in turn used with a view to amusement, instruction, and moral improvement. At this point, I complicate the picture and suggest that sentimental readers are to be regarded as a composite group embodying a variety of reading and conversing habits—from Richardson's extended family with their reading sessions and exchanges, to Boswell's solitary reading and polite and literary conversation. In the background I point to the presence of young David Hume, with his projection of hard, deep, solitary reflection preparatory to delightful human interaction.

The extreme representatives of the "knowledge of the human heart," that is, Richardson's novels, were acknowledged by an immense readership. The most "abstruse" side of the "science of human nature," represented by Hume's *Treatise of Human Nature,* on the contrary, failed to find readers prepared to acknowledge it through their conversation. To polite and sentimental readers, it appeared so peculiar and so unsentimental that they thought it unconversable as well.[7] My suggestion is that between *Clarissa* and Hume's *Treatise* lies a whole variety of writings, of readers, and of conversation forms—different combinations of abstract reasoning, sentiment, and sociability that, taken together, constitute the very core of the Enlightenment.

The Science of Human Nature, Common Life, and History

In its interest in perception and passion, the science of human nature was a highly abstract study, and as such, as any eighteenth-century inquirer into human nature would have explained, it needed examples and illustrations, for these would make even the most abstruse argument comprehensible, by allowing it to affect the imagination.[8] History and a second inexhaustible resource, namely, experience—"long life and a variety of business and company," Hume

6. Hampton, *Writing from History,* is the standard work on history as repository of moral exemplars in the early modern period; for the continued prevalence of sententious history during the eighteenth century, see Koselleck, "Historia Magistra Vitae," in *Futures Past,* 21–38.

7. For the lack of success of the *Treatise,* see Mossner, "Continental Reception of Hume's *Treatise,*" and *Life of David Hume,* chap. 10, 116–32.

8. See, for example, Price, "Reading of Philosophical Literature" (quoting Hugh Blair, John Gregory, etc.).

calls it[9]—provided the illustrations and confirmations necessary to the abstractions of this science of human nature, thus constituting its imaginative flesh.

In Adam Smith's *Theory of Moral Sentiments,* both history and social life are searched for the illustrations of the workings of human passions for which the book was famous.[10] For example, in the chapter "Of the origin of ambition, and of the distinction of ranks," we are referred to the human tendency to sympathize more strongly, both in happiness and in distress, with the lucky, rich and powerful—a tendency so strong that "[a] stranger to human nature, who saw the indifference of men about the misery of their inferiors, and the regret and indignation which they feel for the misfortunes and sufferings of those above them, would be apt to imagine, that pain must be more agonizing, and the convulsions of death more terrible, to persons of higher rank than those of meaner stations."[11] There follow, as illustrations of this general observation, a whole collection of cases and stories: episodes of Roman and more recent English and Continental history, as well as vivid vignettes of contemporary behavior. Louis XIV was, Smith reminds us on the basis of Voltaire's description, famous for his good looks, his pleasant voice, his dignified deportment, and in general his truly regal appearance. As Smith puts it: "These frivolous accomplishments, supported by his rank, and, no doubt too, by a degree of other talents and virtues, which seems, however, not to have been much above mediocrity, established this prince in the esteem of his own age, and have drawn, even from posterity, a good deal of respect for his memory." And he comments:

> Compared with these, in his own times, and in his own presence, no other virtues, it seems, appeared to have any merit. Knowledge, industry, valour, and beneficence, trembled, were abashed, and lost all dignity before them. But it is not by accomplishments of this kind, that the man of inferior rank must hope to distinguish himself. Politeness is so much the virtue of the great, that it will do little honour to any body but themselves. The coxcomb, who imitates their manner, and affects to be eminent by the superior propriety of his ordinary behaviour, is rewarded with a double share of contempt for his folly and presumption. Why should the man, whom nobody thinks it worth while to look at, be very anxious about the manner in which he holds up his head, or disposes of his arms

9. Hume, *Enquiries,* 65.

10. See, for example, Edmund Burke's comments in his letter to Smith of 10 September 1759: "I own I am particularly pleased with those easy and happy illustrations from common life and manners in which your work abounds more than any other that I know by far" (cited in Smith, *Theory of Moral Sentiments,* 28).

11. Smith, *Theory of Moral Sentiments,* part 1, sec. 3, chap. 2, 52.

while he walks through a room? He is occupied surely with a very superfluous attention, and with an attention too that marks a sense of his own
importance, which no other man can go along with. The most perfect
modesty and plainness, joined to as much negligence as is consistent with
the respect due to the company, ought to be the chief characteristic of the
behaviour of a private man. If ever he hopes to distinguish himself, it
must be by more important virtues).[12]

Note the combination of history and common life, represented respectively by
the king and by the coxcomb. Smith is at his most sententious for the occasion,
as is evident already in his calling the king's virtues "frivolous accomplishments"; and the passage culminates with the inverted snobbery of the condemnation of the snob: the virtues of a king are plain stupid in a "man of
inferior rank." Royalty is to be admired and sympathized with, but at a distance; if "we" are not to be coxcombs, "we" are meant to be frank and generous, solid and prudent, patient and resolute, industrious and knowledgeable—
in short, no-nonsense bourgeois.[13]

So history is a gold mine for the science of human nature. But this is not
all—their relation runs deep in another sense as well, for as we find, for example, in a famous page of *Tristram Shandy,* in the Lockean tradition the human mind is the object of a radically historical form of knowledge:

Pray, Sir, in all the reading which you have ever read, did you ever read
such a book as *Locke*'s Essay upon the Human Understanding—Don't
answer me rashly—because many, I know, quote the book, who have not
read it,—and many have read it who understand it not:—If either of
these is your case, as I write to instruct, I will tell you in three words what
the book is.—It is a history.—A history! Of who? what? where? when?
Don't hurry yourself.—It is a history book, Sir, (which may possibly recommend it to the world) of what passes in a man's own mind; and if you
will say so much of the book, and no more, believe me, you will cut no
contemptible figure in a metaphysic circle.[14]

In the case of Hume, the "abstruse" philosopher who also writes history,
there is a further twist. In his *Treatise of Human Nature* our own self is based
on memory and passions: "As memory alone acquaints us with the continuance and extent of this succession of perceptions, 'tis to be considered, upon

12. Ibid., 54–55.
13. See Pesante, "An Impartial Actor," 174–75.
14. Sterne, *Tristram Shandy,* 2:2. See also Voltaire, *Letters Concerning the English Nation,*
letter 13, 56, who commends Locke's empiricism by, among other things, contrasting his history of
the soul with Descartes's and Malebranche's romances: "Such a Multitude of Reasoners having
written the Romance of a Soul, a Sage at last arose, who gave, with an Air of the greatest Modesty,
the History of it."

that account chiefly, as the source of personal identity. Had we no memory, we never shou'd have any notion of causation, nor consequently of that chain of causes and effects, which constitute our self or person." Memory is, in this sense, what connects together "that succession of related ideas and impressions, of which we have an intimate memory and consciousness," what ties up "the bundle or collection of different perceptions" constituting our self. The science of human nature is therefore a genealogical science—it reconstructs the very process which constitutes personal identity. The study of the human mind is historical because, from this point of view, it reflects the state and the very constitution of its object.[15]

On the other hand, the human nature studied by this "science" is considered as in principle universal: "The same motives always produce the same actions," as Hume puts it in a famous passage. Indeed, even when reading history, we do rely to a great extent on this sort of human nature, and on our acquired wisdom about it: "What would become of history had we not a dependence on the veracity of the historian according to the experience that we have had of mankind?"[16] Of course, this is a universality of predispositions and tendencies. The ways people are do vary, because people are shaped by their circumstances, and the whole point of studying human nature is precisely to have as wide a variety as possible of the operations of the same invariable principles, that is, of the innumerably varied patterns of human behavior stimulated by innumerably varied circumstances.[17] But between these two sides of human nature, historicity and universality, there is an obvious tension, which the inquirers into human nature share with their colleagues writing about natural history proper: the tension between the tendency to classification and the interest in the individual—indeed, in exceptional individuals, rare, sublime or monstrous, sometimes seen as the best key to understanding the ordinary.[18]

As a result of this tension, in the writings of eighteenth-century moral philosophers human nature appears under several types: sometimes there is

15. I quote here from pages 261, 277, 252 of the *Treatise*. On Hume's conception of time and the temporality of the self see Livingston, *Hume's Philosophy of Common Life*, for example, 126–27.
16. Hume, *Enquiries*, 65 and 99. See also Smith, *Wealth of Nations*, 1:19–20.
17. See Hume, *Enquiries*, 66: "[F]rom observing the variety of conduct in different men, we are enabled to form a greater variety of maxims, which still suppose a degree of uniformity and regularity. Are the manners of men different in different ages and countries? We learn thence the great force of custom and education, which mould the human mind from its infancy and form it into a fixed and established character." On Hume, human nature, and history, see Pompa, *Human Nature and Historical Knowledge*; Phillipson, *Hume*; Preti, *Alle origini dell'etica contemporanea*; and the classic Lévy-Bruhl, "L'orientation de la pensée philosophique de David Hume."
18. For discussions of this tension in natural history see also Hagner, chap. 6 above, and Spary and Jardine, chaps. 10 and 15 below.

the type of humanity at its most familiar and most easily sympathized with—
we may call it our ordinary acquaintance—and sometimes there are exceptional individuals. I have already cited Smith on a coxcomb and a king; we
shall see both other ordinary people and other rare, sublime, or monstrous
characters of kings, queens, and intellectuals. Finally, I shall show that in the
study of perception there is also, as representative of humanity, either a solitary
"I," or Adam. Let me start, now, with some ordinary acquaintances.

One's acquaintances are familiar and taken for granted in all their complexities without need for reflection. They constitute the ordinary landscape of
one's life: they are the people populating drawing rooms and cafés, and are the
normal objects of both one's sympathy and envy. The "acquaintance" typically
appears in Montaigne-style essays, as a part of the familiarity between author
and reader typical of the genre—we, the author and the reader, are learning
something together from the behavior or fortune of a common acquaintance.
In Hume's "Of the study of history," for example, there is a lady who refuses
to read the Plutarch she has borrowed when she realizes it is not a romance,
but a history book:

> I remember I was once desired by a young beauty, for whom I had some
> passion, to send her some novels and romances for her amusement in the
> country; but was not so ungenerous as to take the advantage, which such
> a course of reading might have given me, being resolved not to make use
> of poisoned arms against her. I therefore sent her PLUTARCH's lives, assur-
> ing her, at the same time, that there was not a word of truth in them from
> beginning to end. She perused them very attentively, 'till she came to the
> lives of ALEXANDER and CAESAR, whose names she had heard by accident;
> and then returned me the book, with many reproaches for deceiving her.[19]

Here a relatively uneventful occurrence—the reaction of a (real or fictional)
"young beauty" to a book and to a small deception—is the occasion for moralizing reflections on women's taste for fiction, and on the instruction and entertainment to be drawn from the reading of history. A common specimen of
humanity, Hume's young beauty is also made, in this passage, an emblem of
female imagination and its reactions to reading, and a foil to solemn and dignified female personifications of history.[20]

The same sort of mankind is also represented in the common-life "we"

19. Hume, *Essays,* 564. The essay was first published in 1741, and suppressed after the 1760
edition of the *Essays.*
20. On the differences between reading history and reading fiction, and on reading and the
imagination, see also Hume, *Treatise,* e.g., 97–98, 122–23, 630–32. See Meyer Spacks, *Desire and
Truth,* 26 and 46–47, for other contemporary opinions on the effects on the imagination of history
and fiction. For reading and women's imagination, see Barker-Benfield, *Culture of Sensibility.*

stories of Smith's *Moral Sentiments.* Our sympathy with somebody else's joy, Smith writes, is less lively a feeling than our sympathy with somebody else's sorrow; and yet, joy by sympathy and original joy are closer in intensity than sorrow by sympathy and original sorrow. It is worth reading in full the illustration of this maxim, which shows "us" during a visit of condolences:

> Our sorrow at a funeral generally amounts to no more than an affected gravity: but our mirth at a christening or a marriage is always from the heart, and without any such affectation. Upon these, and all such joyous occasions, our satisfaction, though not so durable, is often as lively as that of the persons principally concerned. Whenever we cordially congratulate our friends, which, however, to the disgrace of human nature, we do but seldom, their joy literally becomes our joy: we are, for the moment, as happy as they are: our heart swells and overflows with real pleasure: joy and complacency sparkle from our eyes, and animate every feature of our countenance, and every gesture of our body. But, on the contrary, when we condole with our friends in their afflictions, how little do we feel in comparison of what they feel? We sit down by them, we look at them, and while they relate to us the circumstances of their misfortune, we listen to them with gravity and attention. But while their narration is every moment interrupted by those natural bursts of passion which often seem almost to choak them in the midst of it; how far are the languid emotions of our hearts from keeping time to the transports of theirs? We may be sensible at the same time, that their passion is natural, and no greater than what we ourselves might feel upon the like occasion. We may even inwardly reproach ourselves with our want of sensibility, and perhaps, on that account, work ourselves up into an artificial sympathy, which, however, when it is raised, is always the slightest and most transitory imaginable; and generally, as soon as we have left the room, vanishes, and is gone forever. Nature, it seems, when she loaded us with our own sorrows, thought that they were enough, and therefore did not command us to take any further share in those of others, than what was necessary to prompt us to relieve them.[21]

This is the other side of the moralistic allegory—the "science of human nature" as a variant of middle-class civil exchange. Hume's texts, and even more so Smith's, project a whole world of polite social interaction. Let us now

21. Smith, *Theory of Moral Sentiments,* 90–92. But according to Smith, the asymmetry between sympathetic joy and sorrow cuts across social classes: "How hearty are the acclamations of the mob, who never bear any envy to their superiors, at a triumph or a public entry? And how sedate and moderate is commonly their grief at an executions?" he observes in the same page. As we shall see, Smith's "common acquaintance" figures (or "we") are remarkably similar to many of the characters David Simple encounters in book 1 of his *Adventures,* while David himself has a very different reaction to other people's distress.

consider another side of enlightened sociability—the side of conversation aimed at moral and cognitive self-improvement—and its links with reading and books as conversational objects.

Knowledge of the Human Heart

The figure I have called "our ordinary acquaintance" has immediate counterparts in the universe of novels. In Sarah Fielding's *Adventures of David Simple* experience is, as in Hume and Smith, a collection of cases—in fact, it is acquaintance with a sample, as wide as possible, of human types. *David Simple* has no true plot. In volumes 1–2, published in 1744, David starts off a good youth, but poor and miserable, traveling "in the search for a real friend," and ends up wiser, well-off and happy, surrounded by little children in the company of his dear wife and two real friends, his brother-in-law and his wife. Volume 3, the collection *Familiar Letters between the principal Characters in David Simple,* followed in 1747. Finally, at the end of *Volume the Last,* published in 1753, David dies poor again and sad (after the death of his wife, and of most of his friends and children), but serene and a much better soul than he started.[22]

The protagonist's pilgrimage through society may be seen as a variation on Bunyan, the story of a journey and of moral improvement through experience. David's quest for a friend, together with his curiosity for human nature, bring him to explore different social environments, using his own social identity as a means and a guarantee of his freedom of movement from one class to another as he collects his pieces of experience. The persons he briefly associates with at the beginning of his journey, only to be bitterly disappointed at their shortcomings of mind or heart, have typically transparent names such as Mr Orgueil, Mr Spatter, and Mr Varnish, and for various reasons remind one of the protagonists of Adam Smith's vignettes.[23] One after the other, they chaperone David in his explorations of the high life among players of whist, of the "conversable society" in a salon, of the learned society in a tavern, of the vain men of the coffeehouses. Finally (in the last quarter of volume 1) David meets some real friends: first the unlucky Cynthia, and then Camilla and her brother

22. Fielding, *David Simple.* On Sarah Fielding's life, see Battestin and Battestin, *Henry Fielding,* and Battestin and Probyn, *Correspondence of Henry and Sarah Fielding;* for further reading see Kensall's "Bibliographical Essay" in Fielding, *David Simple.* On *David Simple* see Barker, "*David Simple:* The Novel of Sensibility in Embryo," and Probyn, *English Fiction,* 20–21.

23. David is, as remarked in Todd, *Sensibility,* 97, "an ardent anti-capitalist, despising those who hoard and increase money and dispensing his own wealth liberally and with speed. The vicious Mr Orgueil in *David Simple* is a kind of capitalist in both emotional and economic ways, lending only when sure of a gratifying return for his money and kindness. David simply expends." See also Skinner, "Economic Sense and Sensibility."

Valentino. In both cases friendship stems from a *récit*—the girl, invited by David, tells him her story; and its essence crucially includes sharing a deep interest in stories and human nature.

Let us consider a piece of experience Cynthia presents in the form of a moral story. It is about two sisters she once knew—one wise and prudent, the other, called Corinna, flirtatious and vain—and contains, among other things, a set of seven portraits of the vain sister's admirers. The first six are arranged in couples of opposite characters, and each is also associated with the misguided judgment of the world. The first couple consists of a man of sense with a reputation for artfulness—"because he did not talk *like a fool,* he must act like *a Villain*"—and of a low and cunning but foolish man famed for his naivety. Next come a cheerful, indifferent man capable of hyperbolic expressions of sympathy and accompanied by a predictable fame for his good nature and, as his counterpart, a truly good-natured man who, however, "had so much Tenderness in him, that he was continually hurt, and consequently out of humour." The third pair are a superficial fellow called Le Vive, always transported by his passions and pride, and his mediocre opposite, a man incapable of decision and therefore famous for his prudence and wisdom. Finally, the seventh suitor is introduced, a vain Frenchman of "very high station," who admires Corinna for her great crowd of admirers.[24] This is the result of combining these ingredients in a drawing room:

> I once visited *Corinna*—Cynthia says—when all her Lovers happened to be there together. . . . The grave Man of Sense appeared diffident of himself, and seemed afraid to speak to her. The artful Man sat silent, and seemed to be laying some very *deep Plot.* The Man who was so apt to be hurt by the Behaviour of others, could hardly forbear breaking out in Reproaches. The gay, good-humoured Spark, *caper'd* and *sung,* and was never better pleased in his Life. The Balancer attempted to speak several times, but broke off with half a Sentence, as not having considered enough whether he was going to speak *wisely* or no. *Le Vive* had no patience, and could hardly be civil to her; but perfectly stormed at her, and left the Room in a violent Passion. But the vain Man was all *Joy* and *Rapture:* for, in some particular Civilities she shewed him, he concluded he was the *happy Man.*[25]

The flirtatious Corinna deals with them all with admirable dexterity: respectful and serious to the first man, she counterplots against the second, thus impressing him with her own cunning; always cheerful with the indifferent and

24. See Fielding, *David Simple,* book 4, chaps. 3–4, 255–71.
25. Ibid., 270–71.

cheerful man, she keeps the sensitive one constantly on edge by behaving inconsistently; showing great reserve and delicacy to the impetuous Le Vive, she asks the indecisive one for advice on endless trifles. She will end up, unsurprisingly, marrying the vain man (and having a perfectly good time ever after).

The way this story is told displays both the author's moralistic intention and her interest in the "Labyrinths of the Mind." Contemporary readers must have easily recognized this combination, and this is how an exceptionally authoritative one among them, namely, Richardson, writes about it: "What a knowledge of the human heart! Well might a critical judge of writing say, as he did to me, that your late brother's knowledge of it was not (fine writer as he was) comparable to yours. His was but as the knowledge of the outside of a clockwork machine, while your's was that of all the finer springs and movements of the inside." [26] It is interesting that Richardson found *David Simple* so remarkable: this tells us something about what he meant by "knowledge of the human heart." Clearly, moralistic intention also is part of his picture of what that knowledge constitutes, so that in his view the author of *Tom Jones* turns out to be less perceptive than his sister precisely because of his mechanistic, materialistic attitude to morals. Let us look into this more closely.

Bright and well-read, Cynthia is the intellectual of the group.[27] She tells the story of the two sisters to the other three in order to entertain them in the evening, and to pursue their common program of moral self-improvement. This telling of stories, as well as discussing their own experiences and drawing morals out of them, is a habit of the little group. In this case, we have several levels of experience and communication: There are Corinna and her suitors, and one level up there are their acquaintances, busy misjudging everybody's character and situation. One further level up, there are Cynthia, telling the story, and the little group around her meditating and discussing it. Then, there is the level of Sarah Fielding and her contemporary readers. Finally there is us, now, trying to make sense of how all these levels—particularly the last two—relate to each other: How is a book such as *David Simple* to be read? What do its readers do with it? How do they use it? How much of David, Cynthia, Camilla, and Valentino is actually to be found in the real people reading this and other similar books?

Sympathetic Tears and a Bit of Gossip

For a start, it may be worth remembering that moralizing sentimental books such as *David Simple* are also weeping kits—they work on the imagination and

---

26. Battestin and Probyn, *Correspondence of Henry and Sarah Fielding*, 132.
27. Indeed, a relevant part of her story has to do with her oppressed intellectual ambitions, and is clearly to be read as a criticism of the typical female education: see, for example, Battestin and Probyn's introduction in ibid., xxxii.

stimulate our sympathy, and tears are the most typical expression of sentimental sympathy. Tears are shed profusely by the main characters in *David Simple*. For example, David's first meeting with Camilla and Valentino is a terrible scene of poverty and distress, and the girl is in tears for their predicament: her brother is ill, they are penniless. She begs their heartless landlady to keep them on credit: "During the time she was speaking, *David*'s Tears flowed as fast as hers; his Words could find no Utterance, and he stood motionless as a Statue." In fact, David is so much the hero of sentimental sympathy that his very self has an evident tendency to disaggregation; to put it in Humean-Smithean terms, he is a bundle of perceptions always on the verge of being untied by the motions of sympathy. Shortly after their first meeting, Camilla, while telling David the sad story of her life—how she had lost her dear father's affection thanks to the action of her wicked stepmother—is overcome by grief and passes out: "David catched her in his Arms, but knew not what to do, to bring her to life again; for he was almost in the same condition himself." In both cases, David's sympathetic heart and behavior are proposed as a model to the reader, who is supposed to join in and weep for the virtue in distress represented by Camilla and Valentino.[28]

So not only are writings of this kind studies of human nature, but they are also meant to develop and improve their readers' nature while entertaining them. Moreover, these sentimental novels may be used to test the quality of their readers' hearts. "God forbid that the Man who reads this with dry eyes should be alone with my daughter when she hath no Assistance within Call," Henry Fielding wrote to Richardson to comment on volume 5 of *Clarissa*.[29] Or, as Clara Reeve put it even more clearly, "I should want no other criterion for a *good* or a *bad* heart, than the manner in which a young person was affected, by reading *Pamela*." According to her, if you have "a good heart" you are meant to weep with *Pamela*, and also, presumably, with *Clarissa* and *David Simple*.[30]

Samuel Richardson, the printer-novelist, and the reading practices centered on his *Pamela* and *Clarissa* provide a key to answer my former questions. It is interesting that now *Clarissa* is only commonly available in the form of a wobbly dictionary-sized paperback. A recent admirer of Richardson, Terry

---

28. *David Simple,* 1:126, 154. On tears see Vincent-Buffault, *History of Tears* (on French material). For the physiological side of sentiment and sympathy, see Figlio, "Theories of Perception and the Physiology of Mind," and Mullan, *Sentiment and Sociability,* esp. chap. 5, 201–40. See also Rodgers, "Sensibility, Sympathy, Benevolence," esp. 98, and Lawrence, "Nervous System and Society."

29. Fielding to Richardson, 15 October 1748, cited in Battestin and Battestin, *Henry Fielding,* 443.

30. Reeve, *Progress of Romance,* 1:135.

Eagleton, has suggested that postmodern theories of textuality, a feminist and psychoanalytic approach, and historical materialism can make Richardson readable again and inform a new reception of *Clarissa*. This provocation is presented with tremendous flair, and is attractive in all sorts of ways, but here I would like to pause only on the difference Eagleton's book highlights between now and then, and concentrate once more on some well-known facts about what sort of cultural objects Richardson's novels were in their own time. Eagleton notes that these texts are "kits, greatly unwieldly containers crammed with spare parts and agreeable extras, for which the manufacturer never ceases to churn out new streamlined improvements, ingenious additions and revised instruction sheets." They are, as it were, eighteenth-century versions of interactive texts, "plural, diffuse kits of fiction," which "belong with his collaborative mode of literary production" and make it possible for him to "accommodate others' texts alongside his own, expand to encompass revisionary material, adapt to account for the latest critical feedback."[31] Eagleton has in mind the several editions of the novel and its published accessories—prefaces and introductions, tables of contents, endlessly added footnotes, appendices and postscripts, additions published both with the main text and on their own, remarkable sentiments and religious meditations, all showing the signs of Richardson's cooperation with his friends and correspondents.[32] But this is not all—there is plenty of printed as well as manuscript material by contemporaries testifying to the interactive construction of the text. The materials examined by Tom Keymer in his recent study of the readership of *Clarissa* look very much like minutes of a society of *Clarissa* readers: Richardson's readers were linked to each other by a complex network of communications. They had their own opinions on the characters' qualities and behavior. They were also quite militant and active. When they felt unhappy with the progress of the heroine or were not convinced by Richardson's choices, they would, if not directly question Richardson's authority, at least ask him to explain himself, or indeed they would even produce alternative narratives and moral lessons of their own. And one has the impression that Richardson must have to some extent negotiated with them.[33]

The readers with whom Richardson was in direct or epistolary contact, and with whom he shared ideals and interests, constituted a kind of extension of his family life—a bit like David Simple and his friends, but on a larger scale.[34] But of course Richardson's readership was much larger than David's

---

31. Eagleton, *Rape of Clarissa*, 20–22.
32. See Richardson, *Clarissa: Preface, Hints of Prefaces, and Postscript*, for some of these texts.
33. Keymer, *Richardson's Clarissa*, esp. chap. 4.
34. On some of Richardson's friends-readers, especially women, see also Mild, "Mr Richardson and Mr Highmore," chap. 8 in *Joseph Highmore of Holborn Row*, 254–324 (esp. 302ff.), and

little group. When they were published, *Pamela* (1740–41) and *Clarissa* (1747–48) occasioned real crazes: there appeared theater pieces, illustrated tea sets and fans, cycles of pictures and prints—as Eagleton puts it, in the case of *Clarissa* as in that of *Superman* "the literary text . . . is merely the occasion of organising principle of a multimedia affair." [35] Other comparisons of the same kind are also possible: Richardson's characters seemed to spill out, as it were, of the domain of fictional writing into that of real life—so that we may now regard his novels as the center of a whole form of life now completely lost, something similar to the clubs of friends of Sherlock Holmes, or even of Elvis Presley (whose real existence may of course be regarded as irrelevant in this respect), and requiring for its recreation not a critical edition, but at least a Walt Disney adaptation with irresistible musical scores, all the associated promotional paraphernalia, and a trail of further installments on video.

Eighteenth-century England was a nation of readers (as well as authors), and there must have been innumerable small groups like David Simple's, composed of literate and comfortably-off people who liked spending their free time conversing and reading aloud together at once for edification, for instruction, and for entertainment. [36] The picture emerging from the research of historians of the book shows that the range of readings of such groups included works of different genres and various quality. For example, Thomas Turner, draper in Sussex, used to read either on his own (when business was bad) or with his wife and his occasional evening guests passages from a wide variety of books—from *Clarissa* to *Peregrine Pickle*, from the *Spectator* and the *Freeholder* to Young's *Night Thoughts*, from newspapers and magazines to *Paradise Lost*, from Tillotson's sermons to collections of plays, from the *Peerage of England* to historical works, from *The Whole Duty of Man* to *Tristram Shandy*. [37]

In August 1756, Thomas Turner bought, for 4s. 6d., a copy of Locke's *Essay;* and in the 16 December 1756 entry of his diary he noted: "Read part of Locke's

Myers, *Bluestocking Circle*. For reading as a "sociable activity" in the Richardson household, see Tadmor, "Women, Reading, and Household Life in the Eighteenth Century," esp. 170–74. It may be interesting to remember, by the way, that Sarah Fielding was herself part of Richardson's group, even though a bit marginally: see Battestin and Probyn's introduction to the *Correspondence of Henry and Sarah Fielding,* xxxi–xxxii.

35. Eagleton, *Rape of Clarissa,* 5. Mild suggests, for *Pamela* tea-sets, etc., looking at the *Daily Advertiser* of 23 April, 30 July, 19 November, and 21 December 1745. For the pictures and prints, I have in mind in particular Joseph Highmore's twelve paintings for *Pamela* (and their reproductions), and his famous *Harlowe Family:* see Mild, *Joseph Highmore of Holborn Row;* Duncan Eaves, "The *Harlowe Family* by Joseph Highmore," and "Graphic Illustrations of the Novels of Samuel Richardson"; and Gowing, "Hogarth, Hayman, and Vauxhall Decorations."

36. See Watt, *Rise of the Novel,* especially chap. 2, for a classic treatment of reading, its use as a pastime and a means of improvement, and its part in conversation.

37. See Turner, *Diary of a Georgian Shopkeeper,* appendix D, 347–53. On the Turner household's reading practices, see Tadmor, "Women, Reading, and Household Life in the Eighteenth Century."

Essay on Human Understanding, which I find to be a very abstruse book. I find trade very dull and money prodigious scarce and everything very dear."[38] So Turner's attempt at reading the *Essay* did not give him much consolation. Yet the Turner household's type of literate sociability in the early and mid–eighteenth century may be seen, I suggest, as directly connected with the Lockean legacy. Consider the famous beginning of Locke's *Essay,* with its open declaration of amateurism: the book, we are told, originated from Mr Locke's conversations with a small group of friends, and the search after truth is thus presented as a leisurely activity pursued by conversing gentlemen.[39] But what Locke writes in the *Essay* about knowledge and ideas is doubly social and "conversable," for it is both the result of conversations between Locke and his associates about knowledge and the origin and association of ideas, and a suggestion of how the polite society should handle knowledge and ideas, certainty and evidence, when talking about, say, the weather.[40] Turner's readings, successful or otherwise, suggest that this deliberate "conversability" of the *Essay* is curiously and typically mirrored in the reception of the "way of ideas": Locke's works could be read in the original or, in the case of the *Essay,* in an abstract (such as, for example, the very popular one prepared by the Oxford don John Wynne and approved by Locke himself, which was first published in 1696).[41] Lockeism was also accessible through other, less demanding sources—for example, through dictionary entries and periodical literature, or through the writings of such Lockean authors as Addison and Steele, Voltaire, Sterne, and Johnson. These writings, available to drapers like Turner and their friends and families, that is, to many members of the reading public at large, made Lockeism a model of conversation and a way of talking, a behavior and a fashion, as well as a "philosophy."[42]

All these habits of reading and discussion come into play when we turn to history books, and to their readers, their authors, and the human characters and experiences constituting their subjects.

38. Turner, *Diary of a Georgian Shopkeeper,* entries for 16 August 1756 and 16 December 1756. Turner did not try only the *Essay:* the *Thoughts Concerning Education* is in the list of books mentioned by him in parts of his diary not included in the 1979 Oxford University Press edition (see appendix D, 351).

39. See Richetti, *Philosophical Writing,* 48ff.

40. See Golinski, chap. 3 above, for the weather as a most apt matter for (enlightened and polite) conversation.

41. Wynne, *Abridgment of Mr Locke's Essay* (1696). This book was reprinted several times during the eighteenth century (1700, 1731, 1737, 1752, 1767, 1770 . . .), as well as being translated into Italian and French.

42. See Yolton, *Locke and the Way of Ideas,* for the eighteenth-century reception of Locke, and MacLean, *Locke and English Literature,* for Lockeism in literature. Box, *Suasive Art of David Hume,* 94, calls the *Essay* "the exemplar of popularly successful philosophy, and therefore . . . a likely model for [Hume's] *Treatise.*"

The History of a Philosopher, and Its Readers

Readers closer to the upper end of the intellectual scale tended to read books on their own, even for entertainment. James Boswell was a polite gentleman, friend of men of letters and intellectuals, and himself an author—a man now seen as depressed, morbid, and gossipy, while he was presumably in fact forming himself as fashionably melancholic, sentimental, and conversable.[43] He was also a sinner, even though a constantly repentant one. In January 1763, he was forced to stay put by a venereal distemper—and in his diary we find how he occupied himself:

> *Saturday, 22 January.* . . . Then, as Smith used to observe, a time of indisposition is not altogether a time of misery. There is a softness of disposition and an absence of care which attend upon its dolent confinement. Then, I have often lamented my ignorance of English history. Now I may make up that want. I may read all Hume's six volumes. I may also be amused with novels and books of a slighter nature.

> *Saturday, 29 January.* . . . I have now one great satisfaction, which is reading Hume's *History.* It entertains and instructs me. It elevates my mind and excited noble feelings of every kind.

> *Sunday, 20 February.* . . . I employed the day reading Hume's *History,* which enlarged my views, filled me with great ideas, and rendered me happy. It is surprising how I have formerly neglected the study of history, which of all studies is surely the most amusing and the most instructive.

The gentleman who was to become such a close friend of Dr. Johnson (they were to meet for the first time a few months later, in May) was reading Hume for his own distraction and edification.[44]

43. See, for example, his letter of self-introduction to Rousseau of 5 December 1764: "I was born with a melancholy temperament. It is the temperament of our family. Several of my relations have suffered from it. Yet I do not regret that I am melancholy. It is the temperament of tender hearts, of noble souls" (Boswell, *Journals,* 1). On Boswell see the articles in Clingham, *New Light on Boswell* (especially those by R. B. Schwarz and S. Manning); and the sensitive pages in Mullan, *Sentiment and Sociability,* 211–12.

44. See *Boswell's London Journal.* It is curious that this connection between Boswell's sexual misfortunes and edification from Hume appears at least once more in Boswell's papers, in a letter written to him by William Temple, a former schoolmate and a clergyman, on 7 August 1767. Boswell was again suffering from his usual complaint, and Temple wrote to him: "But you see the consequences of such connections, and how dare I call the punishment unjust? It is according to the order of nature and Providence, and you will ever find, my orthodox friend, that *faith* without *works* is nothing, that virtue is happiness and vice misery; henceforth, never have the audacity to refuse drinking David Hume in my company, and learn to reverence his name till you can imitate his example." Another close friend and biographer of Johnson, Mrs Thrale, was also, in later years, to become a reader of Hume's *History:* see Piozzi, *Correspondence,* 2:498, 505, 507.

If we have a look at Hume's *History,* we find that for once there is no inconsistency in Boswell's taste and behavior. This is most apparent when we consider some of the famous character sketches in the *History:* the most beautiful are well-balanced combinations of human insight and sententious mood, and are still well worth reading on this account.[45] Hume's style rises very high for these exceptional specimens of mankind—men and women who are always rare, sometimes sublime, sometimes monstrous, and occasionally both at once. Of Queen Mary, for example, Hume writes that "[a]n enumeration of her qualities might carry the appearance of a panegyric; an account of her conduct must, in some part, wear the aspect of severe satire and invective." And here is the final assessment of the personality of Queen Elizabeth:

> [W]hen we contemplate her as a woman, we are apt to be struck by the highest admiration of her great qualities and extensive capacity; but we are also apt to require some more softness of disposition, some greater lenity of temper, some of those amiable weaknesses by which her sex is distinguished. But the true method of estimating her merit, is to lay aside all these considerations, and consider her merely as a rational being, placed in authority, and entrusted with the government of mankind. We may find it difficult to reconcile our fancy of her as a wife or a mistress; but her qualities as a sovereign, though with some considerable exceptions, are the object of undisputed applause and admiration.[46]

The most interesting are, perhaps, the paragraphs devoted to the great "philosophers," Newton and Boyle. Boyle "trod with cautious, and therefore the more secure steps, the only road, which leads to true philosophy." Among other things (mainly having to do with his work on the air pump), we find the following wonderful brush stroke combining a focus on the individual person and his achievement with the generalizations of the moralist: "Boyle was a great partizan of the mechanical philosophy; a theory, which, by discovering

45. Some, especially in the Stuart volumes, are near-aphorisms. This is how he sketches the character of the first duke of Buckingham: "Headlong in his passions, and incapable equally of prudence and of dissimulation: Sincere from violence rather than candour; expensive from profusion more than generosity: A warm friend, a furious enemy; but without any choice or discernment in either" (5:103); King James "was able to preserve fully the esteem and regard of none. His capacity was considerable; but fitter to discourse on general maxims than to conduct of any intricate business"(5:121); King Charles "deserves the epithet of a good, rather than of a great man" (5:542). On the tradition of the character sketch upon which Hume is drawing here, see Braudy, *Narrative Form,* 39–44.

46. On Mary, *History,* 5:252; Elizabeth, 4:353–54. See Braudy, *Narrative Form,* 70ff., on the interactions between the public and the private sides of personality in the Tudor volumes, and in particular 72–75 for the cases of Queen Elizabeth and of Mary, queen of Scots. It is interesting to remember that in 1788 the character sketch of Queen Elizabeth was included by Mary Wollstonecraft in her miscellaneous collection *The Female Reader,* 389: see Barker-Benfield, *Culture of Sensibility,* 137.

some of the secrets of nature, and allowing us to imagine the rest, is so agreeable to the natural vanity and curiosity of men." And Newton, "the greatest and rarest genius that ever arose for the ornament and instruction of the species," and whose reputation "broke out with a lustre, which scarcely any writer, during his own lifetime, ever before attained," was "from modesty, ignorant of his superiority above the rest of mankind; and thence, less careful to accommodate his reasonings to common apprehensions: More anxious to merit than to acquire fame." The personal tone coloring this panegyric is a clear projection on the part of a writer who shared his enemies' opinion about his own vanity, and who openly acknowledged both his "love of literary fame"—which in fact he called, together with a "passion for literature," his "ruling passion"— and its frustrations.[47]

The principle underlying this kind of writing is a variation on the theme of the universality of human nature: "Would you know the sentiments, inclinations, and course of life of the Greeks and Romans? Study well the temper and actions of the French and English: You cannot be much mistaken in transferring to the former *most* of the observations which you have made with regard to the latter." [48] If studying the English today may help to understand the Romans of ancient history, it is clear that what happens today in the heart of David Hume may help to see (in this case, by contrast) what was happening yesterday in the heart of Isaac Newton: history teaches about human nature, and knowledge of human nature is of help in reenacting history.

Let us read a few more lines of Hume's portraits—in that of of Anne Boleyn, for example, we find her quiet acceptance of her own death sentence:

> She probably reflected that the obstinacy of Queen Catherine, and her opposition to the King's will, had much alienated him from the Lady Mary. Her own maternal concern, therefore, for Elisabeth, prevailed, in these last remarks, over that indignation which the unjust sentence, which she had suffered, naturally excited in her."

And of James II, Hume writes:

> He burst into tears, when the first intelligence of [his daughter Anne's desertion] was conveyed to him. . . . The nearer and more intimate concern of a parent laid hold of his heart; when he found himself abandoned in his uttermost distress by a child, and a virtuous child, whom he had

---

47. See "My Own Life," in Mossner, *Life of David Hume*, 611–15. Quotes on Boyle and Newton, *History*, 6:71, 541. For Johnson's comments on Hume's vanity see above, and note 3. See also, for a comment by Goldsmith in a similar mood, Boswell's diary, Sunday 26 July 1763: "He said David Hume was one of those, who, seeing the first place occupied on the right side, rather than take a second, wants to have a first in what is wrong."

48. Hume, *Enquiry*, 65.

ever regarded with the most tender affection. . . . It is indeed singular, that a prince, whose chief blame consisted in imprudences, and misguided principles, should be exposed, from religious antipathy, to such treatment as even Nero, Domitian, or the most enormous tyrants, that have disgraced the records of history, never met with from their friends and family.

These are at once beautiful examples of sentimental history, and close relatives of typical moralistic and didactic allegories.[49] The sentimentality of the portraits of Anne Boleyn and of James II is, in the middle of their terrible and sublime life stories, a human touch depending on, and presumably meant to arouse, sympathy. A king's tears are still tears. In this sense Hume's portraits crucially rely on universal human feelings, in the specific case, the affection for one's children, and therefore on projection as a plausible way of talking about fellow humans, even when they are kings and queens.

But with this, it is also evident why the *History of England* has something in common with other kinds of writing—"novels and books of a slighter nature," as Boswell so aptly put it—such as, one may guess, *Clarissa* and *David Simple:* a part at least of their readerships' looking for varied combinations of amusement and instruction. In this period the line dividing genres, in this case history from fiction, must have been more mobile and negotiable than we would imagine today, and the key to make sense of genre in this case must be the readers.[50]

Human Nature and the Philosopher's Searching Eye

In Boswell's diary, we find another reference to Hume's writings, this time not the historical ones:

Saturday, 28 July 1781. I borrowed today at the Advocates' Library, Hume's Treatise of Human Nature, but found it so abstruse, so contrary to sound

49. On Anne and James, *History,* 3:238, 6:513. For a fascinating examination of Hume's sentimental historiography, see Siebert, "In Search of the Hero of Feeling," chap. 1 in *Moral Animus of David Hume,* 25–61. See also Hilson, "Hume: The Historian as Man of Feeling." On the sentimental readers of Hume's *History,* in particular on Madame de Boufflers, see also Christensen, "Example of the Female," chap. 4 in *Practicing Enlightenment,* 94–119.

50. The relations between the history, the romance, and the novel in the seventeenth and eighteenth centuries are an intricate matter: see Watt, *Rise of the Novel,* and McKeon, *Origin of the English Novel.* Among the primary sources see, for example, Samuel Johnson's *Rambler,* no. 60, and *Idler,* no. 84. Sarah Fielding herself was to publish, in 1757, *The Lives of Cleopatra and Octavia,* a work in a mixed genre, half historical biography and half sentimental novel; and it is a well known general tendency of the historiographical and biographical writing of this period to explore, as Christopher Johnson has very appropriately put it, the "characters' inner lives" (see his introduction to *The Lives,* 15 and n. 7; Johnson also discusses whether *The Lives* was read as history or as fiction, 22–24). Interesting remarks can be found in Walling Howard, *Influence of Plutarch.*

sense and reason, and so dreary in its effects on the mind, if it had any, that I resolved to return it without reading it.[51]

Apparently this sentimental reader did not enjoy reading "abstruse metaphysics." It is not impossible that Boswell was actually fibbing here, and that he had, in fact, read the book: after all, his explanation for not wanting to read it—his reference to the "dreary effect on the mind"—is surprisingly close to the very literary topos with which Hume ends book 1 of the *Treatise* (his "metaphysical agony," as it has been called).[52] Also, the summary of Hume's philosophy with which Boswell came up was not only far from superficial, but very much to the point: "David Hume, who has thought as much as any man, who has been tortured on the metaphysical rack, who has walked the wilds of the speculation, wisely and calmly concludes that the business of ordinary life is the proper employment of man."[53] Still, it is a fact that he was at least not prepared to admit having read the *Treatise* and, rather, expressed his polite disapproval for this offensively unsentimental book. We also know that this judgment about Hume was not personal, and that it was a fully articulate one: for instance we know, again from Boswell, that Lord Kames talked about Hume as a man without any taste or sensibility.[54]

I suggest that it was the philosopher's way of looking at human nature that must have been upsetting to polite and sentimental readers. In the works of Hume, there is a whole set of abstract types representing a humanity estranged from the world. Consider the solitary young philosopher as he appears at the beginning of the *Treatise of Human Nature,* introspecting in his closet, and observing himself experiencing. The very first move, the distinction between impressions and ideas, is introduced as something obvious, with recurrent appeals to the reader's own experience:

All the perceptions of the human mind resolve themselves into two distinct kinds, which I shall call IMPRESSIONS and IDEAS. The difference betwixt these consists in the degrees of force and liveliness, with which they strike upon the mind, and make their way into our thought or consciousness. . . . I believe it will not be very necessary to employ many

51. From Boswell papers collected in *Boswell Laird of Auchinleck,* ed. Reed and Pottle.

52. For interesting discussions of the conclusion of book 1 of the *Treatise,* see Baier, *Progress of Sentiments,* 15–27; Richetti, *Philosophical Writing,* 227ff.; Sitter, *Literary Loneliness,* 32–33; and Box, *Suasive Art of David Hume,* 104–8.

53. Cited in Mossner, *Forgotten Hume,* 173.

54. From Boswell papers collected in *Boswell: The Applause of the Jury,* ed. Lustig and Pottle, 31: "My Lord said, 'David's atheism was owing to his want of sensibility. He did not perceive the benevolence of the Deity in His works. He had no taste, and therefore did not relish Shakespeare. His criticism was in the lump, all general. He could not see the beauties. He could see the want of order in Shakespeare's dramas because that is obvious. But could not see the fine poetry in his works.'"

words in explaining this distinction. Every one of himself will readily perceive the difference betwixt feeling and thinking.

In fact, this distinction is very unobvious—so much so that its origin, exact
meaning, adequacy, use, scope, and so on, have been discussed and examined
in all sorts of ways in the literature on Hume.[55] This first page of the *Treatise* is
a good illustration of a radical contrast that has been observed between Hume's
reassuring rhetoric and the complexity of the conceptual contents of his writing.[56] The experience Hume continually refers to as if it were available to everybody does in fact require a very practiced abstracting and analytic mind;
otherwise, objects are as familiar and obvious to us as our acquaintances, and
so they remain as long as we look at them in our ordinary way. They only
become philosophically interesting when we put on the philosopher's spectacles and learn how to stop taking them for granted. But then, the most uneventful experience turns out to be filled with wonder:

> The first circumstance, that strikes my eye, is the great resemblance be-
> twixt our impressions and ideas in every other particular, except their
> degree of force and vivacity. The one seem to be in a manner the reflexion
> of the other; so that all the perceptions of the mind are double, and ap-
> pear both as impressions and as ideas. When I shut my eye and think of
> my chamber, the ideas I form are exact representations of the impressions
> I felt; nor is there any other circumstance of the one, which is not to be
> found in the other. In running over my other perceptions, I find still the
> same resemblance and representation. Ideas and impressions appear al-
> ways to correspond to each other. This circumstance seems to me re-
> markable, and engages my attention for a moment.[57]

Hume is taking himself as the specimen of humanity under investigation in
order to reconstruct the genesis of ideas in everyday experience. Following the
history of the mind through the development of a child leads to a highlighting
of the chronology, as it were, of ideas; this philosopher's introspection, performed in real time for the reader, on the contrary shows their genesis by
bracketing the mind's past and concentrating on the present operations of consciousness.[58] Certainly Boswell would not have liked to look upon his own

55. *Treatise*, 1–2. For a variety of positions, see the classic Kemp Smith, *Philosophy of David Hume*, chap. 10, 205–26; Bennett, *Locke, Berkeley, Hume*, chap. 9, 222–34; Stroud, *Hume*, chap. 2, 17–35; Livingston, *Hume's Philosophy of Common Life*, chap. 3, 60–90; Garrett, *Cognition and Commitment in Hume's Philosophy*, chap. 2, 41–57.
56. See Richetti, *Philosophical Writing*, 185ff.
57. *Treatise*, 3.
58. See Salmon, "Central Problem of David Hume's Philosophy," esp. 310ff., on Hume's use of introspection and his history of the mind. It is well known that the mature Hume presented a similar, but less intense philosophical message in his *Enquiry*, which was more reader-friendly and "conversable," and therefore more successful.

experience like this; and it was easy for an unfriendly contemporary reviewer to criticize the author's "egotism" and to make a parody of the entire enterprise by turning his anonymous experimenting "I" into highly identifying third-person pronouns:

> The first Circumstance that strikes his Eye, is the great Resemblance betwixt our Impressions and Ideas in every Particular, except their Degree of Force and Vivacity. When he shuts his Eyes and thinks of his Chamber, the Ideas he forms are exact Representations (he tells us) of the Impressions he felt. I fancy most other People have made the same Observation. However, this Circumstance seems to our Author remarkable, and engages his Attention for a Moment.[59]

And yet there is a connection between the human nature that the philosopher discovers by introspection and the human nature described in sentimental history, or observed with sympathetic if gossipy interest (that is, sentimentally) in our own acquaintances. For when it is the philosopher who is looking at it, experience is always in some measure reflexive—in order to find out about human nature, he looks at himself experiencing:

> I am here seated in my chamber with my face to the fire; and all the objects, that strike my senses, are contain'd in a few yards around me. My memory, indeed, informs me of the existence of many objects; but then this information extends not beyond their past experience, nor do either my senses or memory give any testimony to the continuance of their being. When therefore I am thus seated, and revolve over these thoughts, I hear on a sudden noise as of a door turning upon its hinges; and a little after see a porter, who advances towards me. This gives occasion to many new reflexions and reasonings.[60]

In this respect, the "experience" of the philosopher is, it turns out, similar to that which was to be described years later by the most philosophical of the novel writers, Laurence Sterne. During his *Sentimental Journey,* Yorick is in one sense close to the scientist of human nature: he observes human nature. For example, he goes to see Maria, the unhappy girl driven insane by love, and meditates on her sorrows. But the most important point of his experience is not what he observes outside. This casts some light on the mechanism behind the workings of a crucial principle in the "science of human nature," that is, sympathy. Let us consider sympathy as it makes its appearance in the heart of

---

59. Review of Hume's *Treatise* in *History of the Works of the Learned,* 357. On this review, see Mossner, *Life of David Hume,* chap. 10, 116–32; and Frasca-Spada, *Space and the Self in Hume's "Treatise,"* 108–116, for a discussion of Hume's "egotism" and of the rhetorical strategy of the reviewer.

60. *Treatise,* 196.

Yorick, during his *Sentimental Journey,* in the chapter entitled "The captive. Paris." It all begins with a bird in a cage:

> The bird in his cage pursued me into my room; I sat down close by my table, and leaning my head upon my hand, I began to figure to myself the miseries of confinement. I was in a right frame for it, and so I gave full scope to my imagination. I was going to begin with the millions of my fellow-creatures born to no inheritance but slavery; but finding, however affecting the picture was, that I could not bring it near me, and that the multitude of sad groups in it did but distract me—I took a single captive, and having first shut him up in his dungeon, I then looked through the twilight of his grated door to take his picture. I beheld his body half wasted away with long expectation and confinement, and felt what kind of sickness of the heart it was which arises from hope deferred. Upon looking nearer I saw him pale and feverish: in thirty years the western breeze had not once fanned his blood—he had seen no sun, no moon in all that time—nor had the voice of a friend or kinsman breathed through his lattice—his children—But here my heart began to bleed—and I was forced to go on with another part of the portrait. He was sitting upon the ground upon a little straw, in the furthest corner of his dungeon, which was alternately his chair and bed: a little calendar of small sticks were laid at the head, notched all over with the dismal days and nights he had passed there—he had one of these little sticks in his hand, and with a rusty nail he was etching another day of misery to add to the heap. As I darkened the little light he had, he lifted up a hopeless eye towards the door, then cast it down—shook his head, and went on with his work of affliction. I heard his chains upon his legs, as he turned his body to lay his little stick upon the bundle—He gave a deep sigh—I saw the iron enter into his soul—I burst into tears—I could not sustain the picture of confinement which my fancy had drawn . . . [61]

The sentimental Yorick induces his own state of distress by the multiplication of minute details in the life of the individual prisoner—details deeply affecting the very same imagination that created them. The most important part of Yorick's experience is the sentiment that it produces in him: his journey is "sentimental" in precisely this sense. This is also why, by the way, one is meant to enjoy weeping with *Clarissa* and *Pamela:* tears are the sign of a good heart, and of the operating miracle of self-knowledge.[62]

61. Sterne, *A Sentimental Journey,* 201–3.
62. See Van Sant, *Eighteenth-Century Sensibility,* 100–102; and Todd, *Sensibility,* 90: "Yorick is a connoisseur of feeling rather than a man of feeling." The sinister side of this is obvious (see Brissenden, *Virtue in Distress*). For a psychoanalytic definition of sentimentality in eighteenth-century fiction and theater, see Ellis, *Sentimental Comedy,* chap. 1, esp. 3–8. For Boswell's persistence in attending executions, which upset him and made him feel terribly depressed for days on end, see Turnbull, "Boswell and Sympathy."

In Place of a Conclusion

Human nature is the subject of all the writings examined here, from *Clarissa* and *Pamela* to the *History of England,* from *David Simple* and the *Sentimental Journey,* to the *Theory of Moral Sentiments* and the *Treatise of Human Nature.* Having pointed out the similarities between treatises of moral philosophy and metaphysics, history books and novels, it may now not be out of place to conclude by focusing on some of the differences between them. These differences must have been very evident at the time, in spite of the similarities that are evident to us now—otherwise it would be difficult to see how Dr. Johnson, who is so opposed to the "science of human nature," could have written such works as *Rasselas* or *The Lives of the Poets,* where the "knowledge of the human heart" plays such a crucial part.[63] Let us once more focus on Richardson and Hume.

Richardson started his writing about "the human heart" following a commission from two booksellers for a collection of sample letters to instruct and help letter writing. His career as a writer of novels thus opened with a book that fit safely in an already well established practice—a book planned with a careful eye to the market.[64] Richardson ended up leaving this book aside for a while and writing an epistolary novel instead, *Pamela,* which, apparently to the author's surprise, became the beginning and the center of a whole network of activities.

In Hume's *Treatise,* the subject matter is still human nature and the vagaries of human behavior. In this sense it is the same as in Richardson's letters and novels; but there are two crucial differences. The first is that while Hume sits at his table reflecting and analyzing the depths to be found in the most ordinary facts of common experience, Richardson's characters, writing "in a new Manner—to the Moment,"[65] are always so busy catching up in their letter writing with their instantaneous emotions, that they seem to have no time to reflect at all; all their experience is minutely transposed on paper without ever acquiring any depth. The second difference is that Hume's "science of human nature" is "knowledge of the human heart" forced into the form of a systematic "anatomy of the mind." And a systematic, analytical treatise of anatomy,

63. For example see, in *Rasselas,* the story of Imlac, especially in chap. 10. For an interesting reading of Johnson focusing specifically on his explorations of human nature, see McKenzie, "Systematic Scrutiny of Passion in Johnson's *Rambler.*" On *Lives of the Poets,* see Rogers, *Johnson.* Johnson's "great knowledge of the human mind" is praised by Oliver Goldsmith in the preface of his abridgment of Plutarch's *Lives* (cited in Wendorf, *Elements of Life,* 214).

64. Richardson, *Letters Written to and for Particular Friends.* A situation similar to Pamela's is to be found in letters 138–39, "A Father to a Daughter in Service, on hearing of her Master's attempting her Virtue," and "The Daughter's Answer."

65. The expression quoted is from a letter to Lady Bradshaigh, 9 October 1756. It also appears in a letter of Lovelace to his friend (letter 224) in *Clarissa,* 721.

no matter whether of the body or of the mind, is linked with ways of reading different from that of novels. But where the master printer Richardson had a practical purpose—selling the book and, later on, keeping his printing shop in a healthy state—Hume was a projector, and had only a message.[66]

Hume's behavior can be regarded as opposite to Richardson's: he was famous even during his lifetime for his consistent refusal ever to interact with his readers, that is, for example, to discuss his work in letters. His usual answer to private solicitation about his thought was to point out that his written work was there, that he had said all he had to say in the printed pages.[67] Of course, the *Treatise* interacted with a lot of other literary products, but not with anybody's correspondence. From this point of view, Hume was eccentric enough as to make it difficult, now, to think of him as an "author" in the sense of Johnson, Richardson, or Sarah Fielding. Even in the most abstract parts of the *Treatise*, his writing was, like theirs, a reflection of a real-life practice; but it was meant to make readers think deep and hard, and to talk to their reason, not to their heart. It was too ambitious and too demanding, and it ended up failing to feed anything back into that practice.

I have suggested that between the "abstruse" metaphysics of Hume's *Treatise* and Richardson's sentimental epistolaries there is a variety of intermediate writings and corresponding reading practices. The readers (vainly) addressed by Hume in the *Treatise* are potential treatise writers, and polite essay and letter writers—people like Boswell, who did, after all, borrow Hume's *Treatise* from a library, and who, ten years later, was to publish his *Life of Johnson*, a true tour-de-force as a study of human nature in conversation. Readers of this kind in fact chose to learn from and be entertained by the "knowledge of the human heart," and to be delighted by paintings of the human soul, rather than turning to the hard study of the anatomy of the mind. They ended up reading, and being instructed by, less peculiar, less extreme works than Hume's *Treatise*: by novels and collections of poems, by historical works, including Hume's own *History of England*, by Locke's writings or their epitomes, by Dr. Johnson's externations, and by the less alarming moral and metaphysical essays and trea-

66. For a discussion of Hume's *Treatise* from the point of view of contemporary genres (esp. "anatomy" and "essay"), see Box, *Suasive Art of David Hume*, 90–96. See Watt, *Rise of the Novel*, 52–59, for the role of booksellers in the establishment of literary standards, and their influence on genre and readership; in particular on Richardson see 57. It may be interesting to add that John Noon, the bookseller who published 1,000 copies of the two octavo volumes of Hume's *Treatise*, books 1 and 2, is unlikely to have suffered from the lack of success of the book: see Mossner, *Life of David Hume*, 113–14.

67. There are plenty of examples of this attitude. See, for example, the letter of 2 December 1737 in response to Henry Home's curiosity about the general plan of Hume's work, and the letter of February 1754 announcing to James Stewart his unwillingness to engage in responding to his criticisms of the *Treatise* and of *Essays*. See *Letters of David Hume*, 1:23–25 and 185–88.

tises written by such contemporaries as Kames and Adam Smith (and even by the mature Hume himself). The books thus chosen are perhaps less deep, and certainly they often contain little abstract science of human nature; but they are more readable and more "conversable" works. These writings and the reading practices around them are therefore central to our definition of Enlightenment, representing enlightened conversation and sociability both more adequately than young Hume's solitary, abstract "anatomy of the mind" and more appropriately than Richardson's highly conversable, but also suspiciously pious, "knowledge of the human heart." [68]

Hume's *Treatise* embodied, and in fact I am convinced it still embodies, a notion (or a projection) of how to combine deep abstract thought and lively conversation, enlightened solitude and enlightened sociability different from and at odds with the ideals, needs, and capacities of the polite and sentimental readers and conversationalists of his own time and place. Hume's "abstruse" writing and thought needed a different context to be read and properly appreciated and conversed about. But this is another story; here it will suffice to say that a few years later, in a university where the philosophical study of human mind was turning into a discipline, a solitary, gentle, methodical middle-aged German professor of (among other things) natural history found them quite inspiring.[69]

Acknowledgments

I wish to thank Peter Lipton for delightful conversations on the more "abstruse" matters touched on in this chapter; the members of the Kolloquium des Instituts für Wissenschaftsgeschichte, Georg-August-Universität, Göttingen, and of the Cambridge Historiography Group, for asking me friendly, but nettling questions on early versions; all those involved in this volume, and Cristina Chimisso, Serafina Cuomo, Lucia Dacome, Peter Jones, Sachiko Kusukawa, Susan Manning, Olivia Santovetti, Anne Secord, Jim Secord, Paul White, and John Yolton, for variously reading, encouraging, commenting, and suggesting at different stages.

Primary Sources

Boswell, James. *The Life of Samuel Johnson, LL.D. Comprehending an Account of His Studies, and Numerous Works, in Chronological Order; A Series of His Epistolary Correspondence and*

68. See Bechler, "Samuel Richardson and Christian Dialectic," for a fascinating interpretation of the exceedingly suspicious religious and political views of Richardson and his circle of (male) friends—George Cheyne, William Law, John Heylin, John Freke, etc.

69. On the German reception of Hume, see Gawlick and Kreimendahl, *Hume in der deutschen Aufklärung* (esp. chap. 9 for Kant's reading of Hume).

244 MARINA FRASCA-SPADA

Conversations with Many Eminent Persons; and Various Pieces of His Composition, Never Before Published. London: Printed by Henry Baldwin, for Charles Dilly, in the Poultry, 1791.

———. *Boswell's London Journal, 1762–1763*. Edited by Frederick A. Pottle. London: William Heinemann, 1950.

———. *Boswell Laird of Auchinleck, 1778–1782*. Edited by J. W. Reed and F. A. Pottle. New York: McGraw-Hill, 1977.

———. *Boswell: The Applause of the Jury, 1782–1785*. Edited by I. S. Lustig and F. A. Pottle. New York: McGraw-Hill, 1981.

———. *The Journals of James Boswell, 1762–1795*. Selected and introduced by John Wain. London: Mandarin, 1990.

Fielding, Henry, and Sarah Fielding. *The Correspondence of Henry and Sarah Fielding*. Edited with an introduction by M. C. Battestin and C. T. Probyn. Oxford: Clarendon Press, 1993.

Fielding, Sarah. *The Adventures of David Simple, Containing an Account of His Travels Through the Cities of London and Westminster in the Search of a Real Friend*. Edited with an introduction by Malcolm Kensall. Oxford: Oxford University Press, 1987. First edition published 1744–53.

———. *The Lives of Cleopatra and Octavia*. Edited by Christopher D. Johnson. Lewisburg: Bucknell University Press; London: Associated University Press, 1994. First edition published 1757.

Hume, David. *A Treatise of Human Nature*. Edited by L. A. Selby-Bigge. 2d edition revised by P. H. Nidditch. Oxford: Clarendon Press, 1972. First edition published 1739–40.

———. *Enquiries Concerning Human Understanding and Concerning the Principles of Morals*. Edited, with an analytical index, by L. A. Selby-Bigge. 2d edition revised by P. H. Nidditch. Oxford: Clarendon Press, 1975. First edition published 1749.

———. *The History of England*. 6 vols. 1754–62. Reprint, Indianapolis: Liberty Classics, 1983.

———. *Essays Moral, Political, and Literary*. Edited by E. F. Miller. Indianapolis: Liberty Classics, 1985.

———. *The Letters of David Hume*. Edited by J. Y. T. Greig. 2 vols. Oxford: Clarendon Press, 1932.

Johnson, Samuel. *The Yale Edition of the Works of Samuel Johnson*. New Haven: Yale University Press, 1958–.

Locke, John. *An Essay Concerning Human Understanding*. Edited by P. H. Nidditch. Oxford: Clarendon Press, 1975.

Piozzi, Hester Lynch. *The Piozzi Letters: Correspondence of Hesther Lynch Piozzi, 1784–1821 (formerly Mrs Thrale)*. Edited by E. A. Bloom and L. D. Bloom. Newark: University of Delaware Press, 1991.

Reeve, Clara. *The Progress of Romance through Times, Countries, and Manners*. 2 vols. Colchester: Printed for the author by W. Keymer, 1785.

Review of Hume's *Treatise of Human Nature*. *The History of the Works of the Learned*, 1739, 353–90, 391–404.

Richardson, Samuel. *Pamela*. Edited by Peter Sabor, with an introduction by Margaret A. Doody. London: Penguin Books, 1985. First edition published 1740–41.

———. *Letters Written to and for Particular Friends, on the Most Important Occasions*. London: Printed for C. Rivington, in St Paul's Churchyard; J. Osborn, in Pater-noster Row; and J. Leake, at Bath, 1741.

———. *Clarissa; or, The History of a Young Lady*. 1747–48. Reprint, London: Penguin Books, 1985.

———. *Clarissa: Preface, Hints of Prefaces, and Postscript*. With an introduction by R. F. Brissenden. William Andrews Clark Memorial Library, Augustan Reprint Society Publication no. 103. Los Angeles: University of California, 1964.

Smith, Adam. *The Theory of Moral Sentiments* (1759). In *The Glasgow Edition of the Works of Adam Smith*, vol. 1, edited by D. D. Raphael and A. L. Macfie. 1759. Oxford: Clarendon Press, 1976.

———. *An Inquiry into the Nature and Causes of the Wealth of Nations* (1776). In *The Glasgow*

Edition of the Works of Adam Smith, vol. 2. General Editors R. H. Campbell and A. S. Skinner. Textual editor W. B. Todd. Oxford: Clarendon Press, 1976.

Sterne, Laurence. *The Life and Opinions of Tristram Shandy, Gentleman.* 3 vols. Edited by M. New, J. New, R. A. Davies, and W. G. Day. Gainesville: University Presses of Florida, 1984. First edition published 1759–67.

————. *A Sentimental Journey through France and Italy.* Edited by G. D. Stout, Jr. Berkeley: University of California Press, 1967. First edition published 1768.

Turner, Thomas. *The Diary of a Georgian Shopkeeper.* A selection by R. W. Blencowe and M. A. Lower, with a preface by F. M. Turner. 2d edition with a new introduction by G. H. Jennings. Oxford: Oxford University Press, 1979.

Voltaire. *Letters Concerning the English Nation.* Edited with an introduction by Nicholas Cronk. Oxford: Oxford University Press, 1994. First edition published 1733.

Wollstonecraft, Mary. *The Female Reader; or, Miscellaneous Pieces in Prose and Verse; Selected from the Best Writers . . .* 1788. Reprint, New York: Delmar, Scholar Facsimile and Reprints, 1984.

Wynne, John. *An Abridgment of Mr Locke's Essay Concerning Human Understanding.* London: Printed for A. and J. Churchill . . . , and Edw. Castle, 1696.

8

Metaphysics, Mathematics, and the Gendering of Science in Eighteenth-Century France

Mary Terrall

When Condorcet was inducted into the Académie française in 1782, following a bitterly contested election, he took the opportunity to proclaim the triumph of reason over "ignorance and error": "The method of discovering truths has been reduced to an art, one could almost say to a set of formulae. Reason has finally recognized the route that it must follow and seized the thread that will prevent it from going astray. . . . Every scientific discovery is a benefit for humanity."[1] This optimistic assessment of the power of reason to save humanity from darkness rested on the conviction that nature endows every individual with the capability to recognize truth, and hence to become enlightened. In Condorcet's rhetoric, reason took on a life of its own, fighting the good fight by force of systematic method, strangely independent of individual philosophers. But in the elite world of the Académie française and the Académie des Sciences, reason and the analytic spirit belonged especially to the members of this meritocracy, which served the absolutist state as well as humanity in general.[2] This double service is one root of the tension between exclusive privilege and universalism that underlies so many of the texts that make up the canon of the Enlightenment in France. In the abstract, all people—men and women, peasants and merchants—have access to reason and truth; in practice, the authorization of discoveries and their meanings was reserved for a small group of men.

The domain of the Académie des Sciences encompassed the tools and methods of science as well as its discoveries. As a privileged body with special ties to the crown, the academy worked to distinguish its expertise from the popular science of the marketplace and the enthusiasms of naive outsiders. This essay examines the role played by algebraic analysis as a language and an

---

1. Condorcet, "Reception Speech," 5–6.
2. On science in the service of humanity, d'Alembert, *Preliminary Discourse*.

instrument of the academy's power to sanction certain forms of knowledge. In mathematical physics, "analysis" referred to the method of reducing problems to equations, which could then be solved using integral and differential calculus. Rational mechanics, which expressed idealized physical situations in systems of equations, was the prime field for the application of analysis in this sense.[3]

By the 1780s, the highly specialized language of analysis had come to represent the rarefied intellectual atmosphere of the academy. This was partly because only a select few could understand the language, but also because of the way it was explicitly tied to the more general meaning of analysis, as the privileged method of reasoning and guarantee of philosophical truth.[4] One feature of the academy's growing commitment to analytic methods in physics over the course of the eighteenth century was the erasure of teleological metaphysics from rational mechanics. This process of erasure contributed to the definition of analysis as the form of mathematics appropriate for understanding nature, and it also came to represent the exclusive right of its practitioners to demarcate truth from falsehood.

The term "metaphysics" appeared in quite a range of epistemological and theological contexts, in discussions of God, the human soul, space, time, force, and mind. I focus here on just one corner of this discourse relevant to natural knowledge, on principles like conservation or economy that served as metaphysical foundations for the equations and experiments of physics. I look at how mathematicians brought questions of foundations to bear on debates about what constitutes good science, or what kind of knowledge is true and useful. Negative statements about "unhealthy" metaphysics abound in the writings of such philosophers as Condillac, d'Alembert, and Voltaire. When used as a term of disparagement, as when d'Alembert described forces as "obscure and metaphysical," it became a generic buzzword for all that lay beyond the reach of human understanding and reliable scientific investigation. However, until the 1760s, such metaphysical principles still entered openly into some areas of rational mechanics, especially in debates about gravity, *vis viva,* and the principle of least action. By the 1780s, explicit claims about metaphysics had been written out of mechanics in the unadorned language of algebraic analysis. In these decades, the academy was also strengthening its ties to the government, and especially to officials intent on reforming traditional political practice. What was it about metaphysical principles that bothered mathematicians? How did

3. For the transition from geometry to analysis in eighteenth-century mathematics, see Hankins, *Science and Enlightenment,* 17–23.

4. The overlap in meanings of analysis became a commonplace, but see especially Condillac, *La logique.* On analysis as scientific method, Baker, *Condorcet,* chap. 2.

they go about excising such principles from the domain of mathematics and physics? I will argue that the denigration of metaphysics, culminating in Lagrange's formalized analytical mathematics and Laplace's program for a reductionist calculus of point forces, was elaborated by drawing parallels between the dangers of metaphysics and those of ignorance, imagination, delusion, and blind religious faith. In rejecting metaphysics, the mathematicians of the academy also set themselves up as the foes of superstition, imagination, and enthusiasm of all kinds, evils they construed as threats to social stability, as well as to their own authority.

Questions of intellectual authority and its links to political authority came into public view at a time when the very definition of publicity was in flux.[5] At the same time, the cultural significance of science, as a royally sanctioned enterprise, was expanding and stabilizing. The academy presented science as something done by men with certain talents and skills to benefit a "public" made up of people without direct access to scientific knowledge. The crown could call on academic expertise to safeguard a vulnerable populace from the effects of ignorance, and the academy trumpeted the results of such consultations as evidence of its usefulness. But the academy's public face, visible in its journal and semiannual public meetings, was always backed up by the exchanges that took place in a more restricted space, behind closed doors. Analysis and rational mechanics occupied the most esoteric region of the academy's preserve.

Although the technical work of analysis was done only by a few "men of genius," the ideal of a powerful and universally applicable analysis was put forward by mathematicians as evidence for the utility and elite status of science. This ideal contributed to the enhancement of the cultural and political standing of science and of the academy. In the process, gender was insinuated into mathematical discourse in rather subtle ways. The meaning of gender differences was a central topic in Enlightenment debates, especially since claims and counterclaims about difference highlighted the tensions in discussions of universal human reason and the rejection of traditional hierarchies of authority.[6] These debates about difference provided ideological resources for men trying to secure a stable and perhaps even a powerful place for science on shifting political ground.[7] Mathematicians brought gendered connotations into the

5. Baker, *Political Culture of the Old Regime;* Gordon, *Citizens without Sovereignty.*

6. Diderot, "Sur les femmes"; Thomas, *Essai;* Rousseau, *Émile.* See also Jordanova, *Sexual Visions;* Outram, *The Enlightenment,* chap. 6.

7. Joan Scott articulates the promise of gender analysis for the practice of history in "Gender: A Useful Category." See also Keller, *Reflections on Gender and Science;* Jordanova, "Gender and the Historiography of Science"; Schiebinger, *Nature's Body;* Terrall, "Gendered Spaces."

definition and legitimation of their practices, as they worked to purify enlightened mathematics from the taint of metaphysics and imprecise thinking.

Metaphysical Principles in Mechanics

Many of the technical disputes that embroiled mathematicians and fascinated readers of the periodical press in the Enlightenment entailed discussions of metaphysics as well as the question of what constituted proper scientific practice. Perhaps the most explicit use of metaphysics was in the principle of least action as propounded by Pierre-Louis Moreau de Maupertuis as part of a reformulation of the goals and methods of mechanics. In a paper on refraction read to the Paris Academy in 1744, he introduced the term "quantity of action" as the core of a new minimum principle governing the motion of light.[8] He subsequently extended the notion of action to analysis of elastic and inelastic collisions, defining action as the product of mass, velocity, and distance traveled in a given time. In its most general form the principle stated: "The quantity of action necessary for any change in nature is the smallest possible."[9] Once action was defined mathematically, the principle generated equations that could be solved by more or less simple differentiation. Maupertuis intended this principle to be both mathematically precise and metaphysically meaningful. Because it applied to optics, dynamics, and statics, he claimed to have brought diverse phenomena under the aegis of a universal principle that reflected the true economy of nature. But he went further, deriving a proof for God's existence from the economy principle.

> I could have started with the laws [of motion] given by the mathematicians and confirmed by experience and looked there for marks of God's wisdom and power. But . . . I thought it more sure and more useful to deduce these laws from the attributes of an all-powerful and all-wise Being. If those that I find in this way are the same as those observed in the universe, is that not the strongest proof that that Being exists and that he is the author of those laws?[10]

Nature had been designed so that every change required a minimal expenditure of "action," and God calculated that quantity continuously. In basing mechanics on an extremum principle, Maupertuis brought a mathematical

8. Maupertuis, "Accord des différentes loix." On the principle of least action, Brunet, *Étude historique sur le principe de la moindre action;* Guéroult, *Dynamique et métaphysique leibniziennes;* Terrall, "Maupertuis and Eighteenth-Century Scientific Culture"; Pulte, *Das Prinzip der kleinsten Wirkung;* Boudri, *Het mechanische van de mechanica.*
9. Maupertuis, "Lois du mouvement," 290.
10. Ibid., 279.

version of final causation into physics, as all motions fulfill God's purpose in minimizing action and maximizing efficiency.

The principle of least action carried immediate implications for problem solving in mechanics. For example, instead of analyzing a motion in terms of the forces acting at each instant, the mathematician applying this principle defines the quantity that is minimal for the whole path, and solves for the path that satisfies this condition. For specified endpoints, the method optimizes an abstract quantity ("action") over the whole range. The final position must somehow be taken into account, by the moving body as well as by the mathematician, so that the problem is conceived globally rather than reductively. Since the actual path instantiates the general economy principle, the solution expresses God's purposes for the physical world. In principle, the same mathematical result should be obtainable by adding up the increments of force at all points along the path; in practice, this may be a more cumbersome calculation. Hence the method had practical as well as metaphysical advantages for the mathematician.

Metaphysical considerations cannot be separated from the physical or the mathematical and hence cannot be ignored by the mathematician. "Action" is quantifiable, but not directly deducible from the concepts of matter and motion; Maupertuis took this as an indication of the critical importance of metaphysics for mechanics: "There will always remain some obscurity in the science of dynamics. While one portion of its phenomena are subject to mathematics, the other undoubtedly depends on a superior science; and it seems we cannot hope to link them by a purely geometrical chain." [11]

Metaphysics surfaced intermittently in other disputes that concerned academicians as well. In the 1740s, Clairaut's work on the three-body problem led him to challenge the accuracy of the inverse square law of gravitation based on apparent discrepancies with observational data concerning the moon's orbit.[12] Among others, Buffon objected to this clouding of Newtonian clarity, citing "metaphysical reasons." On metaphysical grounds of simplicity, a single force must be represented by a single mathematical expression, Buffon asserted: "A physical and general quality in nature is always simple and must in consequence have a simple measure. . . . [A physical law] is in effect only the expression of the simple effect of a simple quality. This law cannot be expressed by two [mathematical] terms, because a quality which is unitary can never be measured by two quantities." [13]

11. Maupertuis, untitled ms., Académie des Sciences, Paris (Fonds Maupertuis).
12. Clairaut, "Du système du monde."
13. Buffon, "Seconde addition," 580. Buffon's original memoir objecting to Clairaut's conclusions about gravity appeared in the same volume: "Réflexions sur la loi," 493–500.

This is not the only argument Buffon made to defend the integrity of the inverse-square law, nor is it a particularly rigorous use of metaphysical reasoning, but it shows that metaphysics was still playing a role in mathematical debates in 1749. In his response, Clairaut articulated a more skeptical line on metaphysics, insisting that calculation and observation must take precedence in deciding disputes: "Metaphysics is without doubt appropriate for enlightening us and for validating the real assistance furnished by Physics and Geometry, but if we let ourselves be guided only by its flame, we can lose our bearings at any moment." [14]

D'Alembert concurred. Although Clairaut finally found an error in his calculation and retreated from his challenge to Newton's law of gravity, d'Alembert could not condone Buffon's retreat into metaphysical arguments. "Attraction must be judged by rigorous analysis and not by metaphysical reasoning which could be used just as easily to destroy a hypothesis as to establish it," d'Alembert noted.[15] Again, the issue is whether metaphysics can be a reliable guide to the truth, or whether its principles must remain inherently ambiguous. In juxtaposing rigorous analysis and metaphysical reasoning, d'Alembert set up the conceptual framework for future attacks on metaphysics by academy mathematicians.

The brief, unsigned article "Métaphysique" in the *Encyclopédie* took metaphysics to mean simply the codified rules underlying practices.

> Everything has its metaphysics and its practice. Practice, without the reason for the practice, and the reason without the exercise, form only an imperfect science. When we limit the object of metaphysics to empty and abstract considerations about time, space, matter and spirit, it is a despicable science; but when we consider it from the right point of view, it's another matter.[16]

Thus metaphysics is relevant to scientific knowledge as a system of reasons and practical rules, rather than abstract principles ambiguously connected to the empirical world. Other forms of abstraction are not so ambiguous; d'Alembert readily applied the abstractions of geometry to real-world problems. Geometry deals with "truths of pure abstraction . . . but these truths are nonetheless useful." [17] And they are useful because they enable calculations about empirical phenomena that would otherwise remain opaque to reason. The simplicity of the abstraction means it can be grasped and manipulated, made subject to the

14. Clairaut, "Réponse aux réflexions de M. de Buffon," 538.
15. D'Alembert, *Recherches sur la précession des équinoxes,* 450. On this episode, see Hankins, *Jean d'Alembert,* 80–81.
16. "Métaphysique," 440.
17. D'Alembert, "Géometrie," 632.

mathematician's technique. A real circle (whether an orbit or the shape of an object) may deviate from the perfection of its geometrical counterpart, but it can approach those properties "to a sufficient degree for our uses." Precision, then, need not be absolute to successfully predict motions of planets, pendula, or vibrating strings. Geometry makes the perturbations and complexities of nature accessible to analysis and measurement. The abstractions of metaphysics cannot serve the same function, in d'Alembert's view, because they cannot translate into unequivocal representations of phenomena.

In the context of evaluating scientific claims, d'Alembert represented metaphysics as an obfuscating cloud that interfered with the potential clarity of calculation and understanding. He developed this point in his eulogy of Johann Bernoulli, portrayed as a heroic figure whose one flaw was an unfortunate propensity for metaphysics. Bernoulli's reliance on Leibnizian principles led him seriously astray when, for example, his commitment to the law of continuity prompted him to deny the existence of hard bodies in nature. In d'Alembert's view, this was tantamount to denying properties without which matter cannot be intelligibly conceived. Describing Bernoulli's solution of the classic brachistochrone problem, d'Alembert argued that metaphysics can never be unequivocal; only hard-nosed calculation results in demonstrably true solutions. (The brachistochrone, being a path of minimum time of descent, lent itself to metaphysical reflections on the economy of nature or the principle of sufficient reason.) "To present vague notions as exact demonstrations is to substitute false glimmerings for light; it is to retard the progress of the mind in the guise of enlightening it." D'Alembert likened the metaphysical formulation of problems in mechanics to confused perception, susceptible to error, as when a blind man interprets shapes by tapping with his stick. Mathematics brings problems into focus, and what's more, can do so perfectly well without the aid of metaphysics. The latter is "sterile," "infertile," or, in Bacon's metaphor, "a virgin consecrated to God who produces nothing." Not only feminine, but defectively so, metaphysics cannot even produce viable offspring. The infertile virgin evokes uselessness, promise unfulfilled, and waste. Good mathematics, for d'Alembert, should be productive and, so to speak, worldly, rather than "consecrated to God." [18]

D'Alembert's search for a simple and general foundation for mechanics led him to abandon the customary language of force because he saw in it unwarranted assumptions about causality.[19] Instead, he reduced mechanics to three principles: inertia, equilibrium, and the composition of motion. Any

18. D'Alembert, "Éloge de Bernoulli," 55, 18, 23–24.
19. Hankins, Jean d'Alembert; Casini, "D'Alembert."

problem in dynamics could, at least in principle, be formulated as a problem of equilibrium where motions, construed as geometrical lines, balanced each other out. This reduction made mechanics "rather a science of effects than of causes."[20] But this did not mean a retreat from geometrical abstraction. "D'Alembert's principle," as it became known, relied on the purely mathematical deconstruction of motions into components not observable directly, in the name of avoiding unknowable forces or causes. By breaking the motion resulting from a collision into one component of initial motion and another of motion destroyed or lost in the collision, d'Alembert sidestepped the question of what caused the change in motion at the instant of impact. The component of motion destroyed is a precise quantity that can meaningfully appear in a descriptive equation, regardless of how it came to be lost or where it went.

In his *Treatise on Dynamics,* d'Alembert substituted his own allegedly unequivocal mathematical principles for the equivocation of metaphysics. He laid out "the route that a philosopher should take" as an alternative to that of the "metaphysician," who gets sidetracked by unsubstantiated claims about God's wisdom and purposes. These latter remain as unknowable as the forces that fueled the *vis viva* controversy. Descartes had been led astray by his conviction that God's infinite wisdom implied the conservation of motion; more recently, others had "run the risk either of being mistaken like [Descartes], or of taking for a general principle that which holds only in particular cases, or finally of regarding as a primitive law of nature that which is nothing more than a purely mathematical consequence of some [more basic] formulae."[21] Here d'Alembert clearly had in mind Maupertuis's claims about the universality of the principle of least action and the relevance of teleology for mechanics. Maupertuis saw metaphysics and mathematics as mutually reinforcing components of natural knowledge; d'Alembert undertook to clear away the metaphysical cobwebs that interfered with the clarity of mathematics. As a result, they drew the boundaries of their vocations as "philosophers" quite differently.

The Disappearance of Metaphysics

Maupertuis's teleological physics never gained the kind of prominence in France that he had initially anticipated.[22] D'Alembert, in contrast, effectively laid the groundwork for the subsequent development of rational mechanics. By the 1780s, metaphysics no longer played an explicit role in mathematics or

20. D'Alembert, *Traité de dynamique,* 405; Fraser, "D'Alembert's Principle."
21. D'Alembert, *Traité de dynamique,* 404.
22. In Berlin, with the support of Leonhard Euler, he fared somewhat better, though not for long. See Terrall, "Culture of Science in Berlin."

physics in the Académie des Sciences. The only references to such principles are found in historical introductions, contrasting outmoded approaches with the more enlightened and general analysis that had superseded them.

The gradual disappearance of metaphysics coincided with the consolidation of the strength and status of the Académie des Sciences. The number of potential members grew decade by decade as more men acquired the technical prerequisites for making a career in science, so that elections became increasingly competitive and the institution ever more exclusive. Although members of the academy accepted their ultimate dependence on the good graces of the crown, their expanding control of potentially useful knowledge allowed them to assert a degree of autonomy not available to them earlier, and the philosopher-secretary Condorcet was able to pronounce publicly in 1782 (with an excess of optimism) that "the greatest benefit princes can bestow on the sciences has been to render them independent of princely power." [23]

As it became more exclusive, the academy generated several schemes for structural reforms. These proposals for marginally greater autonomy and lessened internal hierarchy were only possible because of the institution's solid position with respect to the state. Even so, the first such plan (in 1769) failed to gain royal approval. But after Condorcet's highly contentious election as secretary in 1785, revised regulations were pushed through. The new secretary took the opportunity to reflect on the rising status of his chosen profession: "We have seen . . . a greater number of men devoting themselves more exclusively to the cultivation of the sciences, and this occupation has become, by virtue of its influence on the general welfare, an honorable estate, and in a way a public function." [24] The "greater equality" among members instituted in the new regulations followed from this recognition of their honorable condition as members of a body valued by the state.

In response to its rising fortunes, the academy stepped up the policing of its borders, asserting its jurisdiction over truth by formally disallowing claims made by certain categories of outsiders. In 1775, the academicians announced publicly that they would no longer waste their time considering claims having to do with squaring circles, trisecting angles, duplicating cubes, or designing perpetual motion machines. [25] In relegating these "chimerical" solutions to the no-man's-land of fantasy, the academy claimed to be acting for the welfare of the "public" as well as reducing demands on its experts. "Humanity required

23. Condorcet, introduction to *Histoire*, 8; also Hahn, *Anatomy of a Scientific Institution*, 97–102.

24. Hahn, *Anatomy of a Scientific Institution*, 1–2, also 126–34.

25. Ibid., 145. Decision announced in *Histoire de l'Académie Royales des Sciences* [1775], 1778, 61–66. On perpetual motion, Schaffer, "Show That Never Ends."

the Academy, persuaded of the absolute uselessness *[inutilité]* of examining solutions to the squaring of the circle, to destroy by public proclamation these popular opinions that have been harmful *[funeste]* to many families." Unlike real mathematicians, who drew on the collective experience of "the most celebrated men," undisciplined visionaries deluded themselves into thinking they were operating under special protection from Providence. "It is but a single step from this idea to thinking that all the bizarre combinations of ideas that present themselves are so many inspirations."[26] From the outside, the condemnation of unauthorized mathematicians looked like unjustifiable tyranny. One self-professed circle squarer attacked the academy for its abuse of power when they deemed his solution invalid: "Do you possess the authority to enslave genius to your whims, by making reason bend to your opinions? . . . What right have you to intellectual knowledge over anyone else? Is your genius to be taken as a measure of what is just, of the highest efforts of the human mind?"[27] Although its authority was not immune to challenge, as this case shows, the academy did in fact aspire to setting the standard for what qualified as genius. True mathematical knowledge was the fruit of the self-discipline required by the hard work of building rationally and unequivocally on existing results. When the academy proclaimed certain problems insoluble, and hence out of bounds for serious mathematicians, it was exercising control over the line separating talent from false inspiration, truth from error.[28]

Academicians augmented their authority by exercising what Condorcet called their "public function." This was most visible in the institution's responses to governmental requests for evaluations of a wide range of pressing industrial, economic, and social problems. The results of these inquiries went first to the ministers of the crown, but were also presented to the "public" in the pages of the academy's journal.[29] The academy willingly made its expertise available to the state for the public good, but maintained its right to define the terms of its interactions with its various publics. In his capacity as secretary, Condorcet went to some trouble to stress the treacherous barriers between the expert elite and a "public" in need of its help. When, for example, commissioners were appointed to evaluate the efficacy of vinegar to counteract noxious fumes from cesspools, Condorcet presented their work as a necessary antidote to the "public enthusiasm" for anything "marvelous." Although the chemists claimed to know that the remedy would not have the desired effect,

26. *Histoire de l'Académie* (1775), 65.

27. G. R. de Vausenville, *Essai physico-géometrique* (Paris, 1778), 164, cited in Hahn, *Anatomy of a Scientific Institution,* 146–47.

28. See Alder, chap. 4 above.

29. Gillispie, *Science and Polity;* Brian, *La mesure de l'état.*

they supervised a test to debunk the "purported secret" because it had excited such a clamor. When fumes duly asphyxiated a workman in the course of the test (in spite of the vinegar), "public enthusiasm" was to blame.[30] In recounting this story, Condorcet set the marvels and secrets of charlatans against the systematic techniques and logical thought processes of chemists. The social value of the collective judgment of the academy was thrown into relief by the collective gullibility of the public, ready to believe in any panacea, however unreasonable. The tragic consequences of the experimental trial illustrated the academy's vital role as protector of a vulnerable public, unable to judge claims about nature unassisted. The academy asserted its right to the public trust, based on its record for the successful production of knowledge and the unmasking of false claims.

The interchange between academy and state extended beyond the delivery of expert judgments in exchange for remuneration. As Keith Baker has argued, reforming ministers dealing with a widening circle of crises invoked scientific rationality as an alternative to traditional formulations of authority. Administrators cast their attempts at making policy in terms of a social order grounded in the scientific principles familiar to readers of the academy's publications. Baker notes that these men invoked the authority of science to give their decisions and policies the legitimacy of rationality and specialized knowledge. The scientific elite had worked hard to establish the superiority of the natural knowledge emanating from the academy; now that they had done so, they made it available to reforming ministers looking for new forms of legitimacy. The royal ministers also found themselves in need of a variety of numerical calculations, tabulations and projections. Although these numbers and tables did not rely on the highly articulated methods of rational mechanics or the calculus of variations, they were performed by the same men whose authority stemmed from their experience with esoteric analytic methods. Eric Brian has worked out these connections, situating the increasing hegemony of analysis (as a category of academic mathematical practice) in the 1770s and 1780s, with the consolidation of d'Alembert's legacy in the work of Condorcet and Laplace. The dominance of analysis as a "scientific model" followed from the power struggle within the academy between factions centered on d'Alembert and Buffon, which came to a head with Condorcet's election as secretary.[31]

The most rigorous analytical mathematics that filled the pages of the acad-

30. Condorcet, "Sur le méphitisme," 13. The full account of the commission's work was read to the public session of the academy in April 1782 and published as Fougeroux de Bondaroy, "Mémoire sur un moyen proposé pour détruire le méphitisme des fosses d'aisance," *Mémoires de l'Académie Royale des Sciences*, 1782, 197–204.

31. Baker, "Scientism"; Brian, *La mesure de l'état*, 199.

emy's journal was not necessarily useful in terms of practical applications or public displays. The equations of celestial mechanics or hydrodynamics did not translate easily into engineering design or industrial production.[32] However, they did represent the most highly articulated form of the ideal of abstract and rigorous rationality grounded in knowledge accessible to a small elite. This rationality purported to represent the truth about nature in its purest form, purged of metaphysical connotations, and especially of final causes. The erasure of metaphysics accompanied the emergence of an ideal of analysis that was to be the basis of understanding of natural phenomena as well as the governance of society. This ideal was also put to work in the academy's efforts to control the parameters of acceptable scientific practice, as a distinguishing mark of a new kind of elite service to state and public.

Women, Gender, and Enlightened Rationality

Cultural historians have located another shift in the intellectual landscape at the end of the old regime, an undermining of the standing of elite women as facilitators and participants in the philosophical ferment in the decades before the revolution. Dena Goodman has argued that the *salonnières* and the sociability they promoted in their homes were vital to the articulation of the philosophic project of enlightenment. When the philosophic party got "stronger" in the 1780s, men found they could dispense with the "female governance" of the salons, and new institutions replaced the salons as the locus of enlightened sociability. The story of the "end of salon culture," in Goodman's words, is the story of the marginalization of women with respect to a political culture that reified public opinion as the voice of reason.[33]

As a corporation with an internal hierarchy of merit, the Académie des Sciences was the antithesis of the salons. Still, many academicians moved easily between the distinct but overlapping worlds of salon and academy, between the formal declamatory presentations of academic meetings and the modulated exchange of conversation. The institution maintained the inner sanctum of its own exclusively male space, but it also looked outward to a wider audience that included women. These women served as prestigious admirers and spectators whose attention validated the scientific enterprise, especially in the academy's early years.[34] At just the time when the salons, characterized by polite conver-

---

32. Ken Alder discusses the use of analysis by military engineers, especially in their training. See chap. 4 above.
33. See Goodman, "Governing the Republic of Letters," and *Republic of Letters*, chap. 6. Also Gordon, *Citizens without Sovereignty*, and Landes, *Women and the Public Sphere*.
34. Terrall, "Gendered Spaces."

sation, were being displaced by men promoting less aristocratic forms of association, the academy was consolidating its role as arbiter of rationality and
utility and relying less on its polite, and female, audience.

At the same time, the undifferentiated public that figured as the dupes of
unsanctioned knowledge claims was commonly described in feminine terms.
If the consolidation of the academy's position, with its attendant denial of
the value of metaphysical principles, coincided both with a backlash against
women's participation in the intellectual life of the Republic of Letters and with
a feminizing of ignorant victims of charlatans, we might expect to find a link
between metaphysics and femininity in the rhetoric of mathematics. In this
context, feminizing metaphysics made sense as a way of excising it from scientific practice. The positive values assigned to calculation and precision measurement were subtly gendered as masculine, while metaphysics was castigated
in feminine terms. The definition of analysis and mechanics as products of
carefully defined masculine rationality underscored the status of the academy
as the source of esoteric knowledge of practical and ideological utility to reformers in the government. The gendering of rationality and method collapsed
the distinctions among overlapping but disparate groups outside the magic
circle of the academy: women in general, with their delicate nerves and fibers;
intellectual women more particularly, who knew how to converse but not to
calculate; and the gullible men and women, of laboring or propertied classes,
who believed the promises of charlatans or conjurers.

The promotion of a vital and masculine analytic mathematics drew on a
discourse of gender oppositions that crop up alongside the universalizing language of enlightenment that stressed symmetries between men and women.
According to Condorcet, for example, men and women alike are born with a
natural capacity for enlightenment; in erotic literature, men and women share
equal capacities for sexual pleasure.[35] Oppositions were articulated, however,
in an evolving discourse of sexual difference, often based on purported physiological differences. Some of this discourse dwelt on the limitations of female
minds, and especially women's natural incapacity for forms of knowing such
as generalization and active discovery. So reason was frequently gendered as
masculine in opposition to imagination, sensibility, or feeling. Strength and
activity were attributes of reason; weakness and passivity were often linked
to sentiment. In his dialogues on Newtonian optics, for example, Francesco
Algarotti noted the restrictions on his female readers, who could never reach

35. On universalizing language, see Ellrich, "Modes of Discourse"; Crampe-Casnabet, "Sampling of Philosophy"; Steinbrügge, *Moral Sex;* and Daston, "Naturalized Female Intellect." On
pornography, Norberg, "Libertine Whore," and Jacob, "Materialist World of Pornography."

the most profound truths because they belonged to "that sex which prefers to feel rather than to know."[36] This does not mean that reason was always more highly valued than feeling or imagination, nor that women were devoid of reason. In the parlance of the salons, rooted in the feminist movement of the seventeenth century, feminine virtue went along with sensibility and civilization, through which women regulated the rude force of masculine nature.[37]

The politeness cultivated in salons did not preclude the use of reason; however, *philosophes* and *salonnières* agreed in associating particularity and the concrete with female intelligence. Women's reading and writing habits echoed this characterization. The witty character portrait, a genre developed in the salon, recorded the idiosyncrasies of individuals, with no pretension to general reflections of wider import. It became a cliché to observe that women preferred reading fiction to history, the richer in fabulous and passionate detail the better. References to the dangers of novel-reading were also commonplace, as women were assumed to be especially susceptible to the suggestions and seductions of fictional characters.[38] In Condillac's sensationalist epistemology, the imagination plays a crucial role in reviving past perceptions and combining them in new ways. But this process can go astray, leading to misconceptions that can become delusions or madness. Female brains seem to be more easily taken over by the imagination in this way. Trouble arises when the mind cannot distinguish between the real and the imaginary. "If we do not master this operation, it will lead us astray infallibly; but it will be one of the main resources of our knowledge, if we know how to regulate it."[39] Women can be taught to manage their dangerous imaginations by adopting masculine standards of enlightenment (which many women were quite willing to do). The labile imagination, ready to slip into enthusiasm or delusion, remained the feminine counterpart to masterful reason.

The same connection between soft tissues, fluid senses, and imagination appears in a treatise on women by a member of the Académie française, Antoine Thomas: "Their mobile senses travel over objects and take away the image of them. . . . The real world does not suffice for them; they love to create

36. Algarotti, *Le newtonianisme pour les dames,* xxxix. By making the spectator/interlocutor a woman, Algarotti feminized his whole readership, men and women alike. Cf. Desmahis, "Femme," 472: "La nature a mis d'un côté la force et la majesté, le courage et la raison; de l'autre les graces et la beauté, la finesse et le sentiment."

37. Goodman, *Republic of Letters,* 7–9. See Lougee, *Paradis des femmes,* for the seventeenth century.

38. Condillac, *Essai,* 128–29. On the alignment of male-female oppositions along an abstract-concrete axis, see Daston, "Female Intellect," 218–19.

39. Ibid., 119. The imagination itself is figured by Condillac as "une coquette, qui, uniquement occupée du désir de plaire, consulte plus son caprice que la raison" (135).

for themselves an imaginary world; they inhabit it and embellish it. Specters, enchantments, prodigies, all that departs from the ordinary laws of nature, are their work and their delight. Their soul is exalted and their spirit is even closer to enthusiasm." Thomas links the delusions of enthusiasm with a kind of hypersensitivity to sensations, but also to the instability of female attention, which cannot focus long enough to discern truth. The language of enchantment also belongs in the nexus of terms that associate superstition, magic, and false metaphysics. In the *Encyclopédie,* we read that superstition is a kind of enchantment, "the unhappy daughter *[fille malheureuse]* of the imagination." And Condillac refers to the deceptions of a metaphysics that searches for hidden essences as *enchantements.*[40] The *Encyclopédie* also associated women with equivocality and dissimulation and stressed the mystery and inscrutability of women. The author of the article "Femme" bemoaned the difficulty of defining women at all: "In truth, everything speaks in them, but in an equivocal language."[41] This cannot be the language of science, of course. Confessing himself helpless in the face of feminine tricks and seductions, the philosopher equates equivocation and femininity. The project of enlightenment, and the rational understanding of nature, must dispense with the temptations of feminine discourse and stick to the language of rationality and control.

The Academy and the State

In the first half of the century, science had benefited by association with the feminine domain of pleasant conversation and particularity. Electrical and mechanical demonstrations, lecture courses on chemistry or botany, public sessions of the Académie des Sciences—all looked to the cultured nobility for approval and validation. The boundaries dividing these genres of exposition and venues for display were far from rigid, and people moved from one setting to another easily.[42] Practitioners of science represented their work to the urban elite as useful, amusing, and above all, difficult. Some decades later, however, the tenor of the interactions between academicians and these audiences had altered, as the academy distanced itself from the display of curiosities tailored for polite consumption. This distancing accompanied a strengthening of the institution's ties to the state, as its members were called on to use the authority of scientific rationality to combat various perceived dangers.

The canonical example of this academic function was the commission ap-

40. Thomas, *Essai,* 84; Jaucourt, "Superstition," 669; Condillac, *Essai,* 3.
41. Desmahis, "Femme," 472; Steinbrügge, *Moral Sex,* 30–34.
42. On audiences for science in the early eighteenth century, Sutton, *Science for a Polite Society;* Harth, *Cartesian Women.*

pointed in 1784 to investigate the phenomena of animal magnetism that had taken fashionable Paris by storm.[43] As Jean-Sylvain Bailly noted in the commission's presentation to the academy, "the sciences, which grow by the acquisition of truths, gain even more from the suppression of an error; an error is always bad leavening that ferments and eventually corrupts the mass where it is introduced." In the same session, Condorcet pointed to the palpable dangers of error: "[Error] produces both those passions that trouble the order of the world and the weakness that makes those passions dangerous." The attack on error was, of course, to be mounted by the discerning and dispassionate experts of the academy. The institution's commitment to unmasking error became an increasingly public task as it addressed controversies with social as well as intellectual ramifications.[44]

The commission's official report to the government stressed the contrast between the systematic reasoning of the commissioners and the contagious gullibility of the subjects of magnetic treatment. As the first step in their investigation, the commissioners observed a group session of mesmeric treatment, where patients routinely passed into and out of violent seizures. Faced with the confusion of the scene around the therapeutic tub, the men of science admitted to astonishment: "Nothing is more surprising than the spectacle of these convulsions; if one has not seen it, one cannot have any idea of it. . . . Such a spectacle seemed to transport us to the time and reign of enchantment *[la féerie]*." As detached observers, these men successfully resisted the effect of the magnetic fluid. Their rationality protected them from the infectious enthusiasm of the sick as well as from the seductive power of the healer, represented as an enchanter or a conjurer putting on a spectacle. They quickly realized that this quasi-public scene was not accessible to their methods and could never yield definitive answers about magnetic phenomena. Too many variables cluttered up the picture: "[O]ne sees too many things at once to see clearly any one thing in particular." But they "kept their *sang-froid* in the middle of the enthusiasm and were able to listen to their reason and search for enlightenment."[45] We might read academic sangfroid as the emblem of analytical method, opposed to the susceptible bodies and imaginations of the magnetized subjects. To his credit, the magnetizer, M. Deslon, agreed to cooperate in experiments designed by the commission and conducted in a private residence

43. The commission was formed as a joint venture by the Académie des Sciences and the Académie de Medicine. See Gillispie, *Science and Polity,* 257–89; Sutton, "Electric Medicine"; Darnton, *Mesmerism;* Schaffer, "Self-Evidence."

44. Bailly, "Exposé des expériences," 8 (this is a short version, stressing the experiments, of the *Rapport des commissaires,* published separately); Condorcet, "Discours," 3. For another example of academic expertise defusing public panic, see Schaffer, "Authorized Prophets."

45. Bailly, *Rapport des commissaires,* 9, and "Exposé des expériences, 9.

away from the press of patients seeking cures. By moving to a space they could control, the commissioners were able to replace Deslon's mastery with their own. Experiments relied on the relentless untangling of an embarrassing confusion of factors, each to be isolated for special scrutiny. In due course, the commissioners found they could not isolate or quantify the magnetic fluid, and this led finally to the conclusion that it had no real existence.

One of the striking features of the vibrant scene witnessed initially by the commissioners was the preponderance of women among those in "crisis." As the investigation proceeded, zeroing in on the imagination as the explanation of magnetic phenomena, this gender imbalance appeared less surprising, though no less worrying. So worrying, in fact, that Bailly did not mention the gender of the subjects in the report to the academy. He saved the most explicit discussion of this danger for a secret report to the crown on the moral implications of healing by animal magnetism. This report started with the relevant differences in male and female physiology. Women are especially susceptible to impressions of all kinds because their nerves are "more mobile," their imaginations "more lively, more exalted." The physical, the mental, and the moral are here inextricably bound up together, since the imagination, linked via an unknown mechanism to the nerve fibers and fluids, can be excited physically by touch as well as by nonmaterial cues. As Bailly put it, "In touching them [women] in any part whatever, we could say that we touch them everywhere at once."[46] What could be less linear, less subject to analysis, more equivocal? What's worse, women are "like resonating strings perfectly tuned in unison," so that if one is put in motion, all the others instantly begin to vibrate. Hence female nerve states can literally spread from one individual to another, creating a contagious mass phenomenon similar to religious ecstasies, public panics, or bread riots.

Simon Schaffer points out that animal magnetism was perceived as a threat because it "challenged important boundaries in enlightened culture."[47] One of those boundaries, recognized by the commissioners immediately, was the line dividing real physical agents from the imagination. Even the academicians admitted that the imagination could produce detectable bodily effects. Being unpredictable and beyond the reach of calculation, these effects could only be regarded (by the state and by mathematicians) as dangerous. The link to femininity enhanced the danger. Under the influence of their powerful and excitable imaginations, women could produce effects on their own bodies and the bodies of others, effects that easily slipped outside the bounds of rational

46. Bailly, "Rapport secret," 511. The baseline of comparison is of course the male standard.
47. Schaffer, "Self-Evidence," 353.

control. The moral implications of magnetic therapy followed from the imagination's distressing tendency to cause sexual crisis. "The imagination spreads disorder in the whole machine [the body]," and the affected woman does not even recognize what is going on. The "total disorder of the senses" that accompanies sexual desire is not noticed by the subject, although observers see clearly what is going on.[48]

The secret report insisted that these female subjects needed protection from the effects of their own imaginations. Their experience of the crisis, after all, was strictly pleasurable. Their orgasmic convulsions, followed by a state of languor and happiness, left them with nothing but agreeable memories. "[They] feel better, and have no repugnance for feeling the same thing again."[49] But they do not suspect that anything immoral transpired, and the more respectable they are, the less they tend to suspect the real cause of their sense of well-being. If mesmeric treatment is equivalent to seduction, with an implied downward spiral into moral degradation, the state and its agents (whether police or scientists) have an interest in safeguarding innocent subjects from its power.[50]

From the commissioners' point of view, their ability to see and warn of moral and metaphysical danger was rooted in their facility with a way of knowing where mathematics, reason, and experiment reinforced each other. Their special access to these methods in turn reinforced their status as an expert body with the power to sanction questionable forms of natural knowledge. They set out to bring a threatening situation under control, whether by certifying the magnetic fluid as real and measurable or by exposing a hoax. Either way, the effects of imagination would be isolated and subjected to academic scrutiny. The investigation threw into relief the contrast between reasoned analysis and the work of the imagination, especially the female imagination. Although the methods used in this case were not mathematical, there are parallels with the program of rational mechanics: to reduce the confusion of nature to general laws in the form of equations. As we shall see below, these equations, if sufficiently general, would eliminate the need for metaphysical principles. The animal magnetism episode displays quite graphically a network of terms linked to the feminine: error, imagination, susceptibility to deception, equivocation, mystery, and lack of control. The commission demonized these aspects of femininity in performing its "public function"; in other contexts the same terms were linked to metaphysics.

48. Bailly, "Rapport secret," 513.
49. Ibid.
50. Under questioning from the lieutenant-general of police, Deslon denied any misbehavior, but he did agree that such abuse of power was possible (ibid., 514–15).

The Gendering of Analysis

Returning to the story of the demise of metaphysics in mechanics brings us back into the protected space of the academy and the mathematical physics of such men as Lagrange and Laplace. This science was characterized by confidence in the power of algebraic analysis and the conviction that the universe was *in principle* knowable as the sum of physical forces that could be represented exactly in mathematical form. This vision extended from the cosmological scale of the stars and planets to the submicroscopic attractive and repulsive forces responsible for phenomena on the human scale. It rested on the mathematical reduction of complexity to simplicity, so that dynamics reduced to statics, and the macroscopic to the microscopic.[51]

What happened to the metaphysics of economy and the principle of least action in this reductionist analytic scheme? I look here only at Lagrange, who dealt with this question explicitly, though a similar argument could be extended to Laplace and his protégés. Lagrange had published his first papers on the application of the calculus of variations to extremum problems suggested by the principle of least action. Writing to Maupertuis in 1756, he described his work as "demonstrating with the greatest possible universality how your principle [least action] always supplies, with marvelous facility, the solution of all the most complex problems . . . in dynamics as well as hydrodynamics."[52] Although he pointedly ignored the teleological features of the principle inherent in the work of both Maupertuis and Euler, Lagrange claimed he could "easily resolve all the questions of dynamics" with a new variational method he developed for solving extremum problems. Starting from the principle of least action, he derived familiar equations of motion and showed that well-known theorems concerning linear and angular momentum followed mathematically from the same extremum principle.[53]

By the time he wrote his magisterial *Mécanique analytique* (published in Paris in 1788), he had demoted the principle of least action, along with the rest of dynamics, into a consequence of the statical principle of virtual velocities. Without denying the validity of the action principle in certain contexts, he reversed its foundational status and explicitly purged it of all connotation of final causes. At the same time, he elevated an ideal of deductive rigor that bolstered an implicit metaphysics of determinism. The metaphysical interpreta-

51. Fox, "Rise and Fall."

52. Lagrange to Maupertuis, 4 November, 1756, Archives of Académie des Sciences, Paris (Fonds Maupertuis).

53. Lagrange, "Application de la méthode," 365. See Fraser, "Lagrange's Early Contributions"; Filho and Comte, "Formalisation de la dynamique." Lagrange moved to Berlin in 1766 and stayed at the Berlin Academy of Sciences until moving to Paris in 1788.

tion of the principle of least action he branded "vague and arbitrary," just the opposite of the precise, analytic, and unequivocal truth embodied in the principle of virtual velocities. Though it could be used to solve certain classes of problem, these applications, he now claimed, were too particular to serve to establish the truth of a general principle. "Vague and arbitrary" might seem weak terms of castigation, but for Lagrange they represented the worst negative qualities that his reductionist and formalized version of mechanics avoided. Where rational deduction and calculation produce generality, imagined principles could only give rise to arbitrary results.[54]

Lagrange demanded the suppression of imagination in his readers, even the rather limited act of imagination involved in picturing equations with curves. He regarded his refinements of algebraic notation as sufficiently powerful to warrant dispensing with the diagrams that routinely accompanied rational mechanics in this period. He undertook to extend the generality of equations and to crystallize processes like differentiation or variation into simple algorithms. Diagrams limited generality by inscribing relations among variables into two dimensions. Lagrange claimed that recourse to spatial representation was a crutch that could be discarded with the advent of sufficiently powerful analytical methods. In effect, equations expressing relations among arbitrarily many variables were truer representations of nature than geometrical curves that undercut the promise of generality held out by systems of equations. The formalism of his mathematics canonized the values of generalization, ordered calculation, and rational reduction of the complexities of nature.

In the *Mécanique analytique*, Lagrange set out to make mechanics (incorporating dynamics and statics) into the discipline he called analysis. "Those who love analysis," he boasted in the preface, "will take pleasure in seeing mechanics become a new branch of it, and will thank me for having thus extended its domain." This was the culmination of a lifelong project, and it required the eradication of final causes or any other such metaphysical principles. General principles need no grounding in anything other than the equations that give them expression and make them available to the manipulations of the mathematician. "It is *in these equations* that the nature of the principle [in this case, the conservation of angular momentum] consists."[55] For Lagrange, the very equations from which everything else can be derived served as the true foundations of mechanics, and he did not want to invest them with

---

54. Lagrange, *Mécanique analytique*, 261. Condorcet used similar language in criticizing Fontaine's dynamic principle as "metaphysical and vague" compared with the precision of d'Alembert's principle. See Condorcet, "Éloge de Fontaine," *Histoire de l'Académie Royale des Sciences*, 1771, 109 (cited by Greenberg, *Problem of Earth's Shape*, 632).

55. Lagrange, *Mécanique analytique*, 261.

special significance as either metaphysical or physical principles. He offered up simplicity, facility, and generality as "all that could be desired" for a truly analytical mechanics. (The irony of the simplicity and facility being that only he, and perhaps one or two of his correspondents, could follow what he was doing.) Lagrange often used "ingenious" to indicate the promising but limited value of the work of other mathematicians. Referring to Euler's work on extremum problems, for example: "However ingenious and fertile his method may be, it must be admitted that it does not have all the simplicity that one could desire in a purely analytical subject." [56] Euler's acceptance of final causes meant that his work was not sufficiently analytical, or sufficiently pure, for Lagrange's taste. Ingenuity reeked of the same kind of open-ended speculation that characterized metaphysics and imagination. Just as the female imagination needed to be controlled, the ingenuity of Euler's mathematics needed to be refined, formalized, and purged of diagrams and of final causes before it could deliver the power of real analysis.

Of course, opposition to metaphysics did not just burst unannounced onto the scientific scene in the 1780s when Lagrange came to Paris. D'Alembert had campaigned against the dangers of this way of thinking decades earlier, as we have seen. But the theological and epistemological questions that metaphysical principles answered for Maupertuis in the 1740s evolved into strictly calculational issues for the mathematical physicists of the 1780s. This trend has sometimes been read as the inevitable progress of science from initial gropings to ever greater generality and clarity. But generality itself underwent a shift in meaning: for Maupertuis, metaphysical underpinnings made the science of physics general and also gave it a purpose, in demonstrating the existence of a God with the intelligence to build optimal efficiency into the laws of physics. The mathematics of determining maxima and minima provided a tool for solving problems, but also for deriving a proof for God's existence that Maupertuis presented as a rational alternative to a physico-theology of wonder. Arguments about God, on this view, fell within the purview of the man of science, and especially the mathematician. Lagrange, along with Laplace and others in the academy, saw the pursuit of mathematics as an enterprise whose success could be entirely defined in its own terms, without reference to theology or philosophy. The man of science had no use for claims, however rational, about the nature of God. Generality resided in the equations from which Lagrange could derive all particulars, in themselves trivialities. For Laplace, it was the demon-

---

56. Lagrange, "Essai d'une nouvelle méthode," 336. For another example of this use of "ingenious," see *Mécanique analytique*, 253: "Quelqu'ingénieuse que soit l'idée . . . , cette idée n'est ni si naturelle ni si lumineuse que celle d'équilibre entre les quantités de mouvement acquises et perdues." On Euler's metaphysics, see Terrall, "Culture of Science," and Clark, chap. 13 below.

strable power of mathematics that made God an unnecessary hypothesis. An economy of calculation, presumably achievable by enlightened governments as well as skillful analysts, replaced the economy of final causes.

Conclusion

French mathematics at the time of the Revolution rested on the kind of anti-metaphysical rationalism promoted by Lagrange and Laplace. The definition of rational thinking as a form of calculation, whereby ideas and sensory input are "compounded and decompounded," became commonplace in the Enlightenment.[57] Mathematical equations, especially when they expressed general truths, were the most formal and visible expression of this analytic ideal. The value Lagrange placed on general formulations was rooted in the work of calculation. No prior metaphysical principles were necessary to guarantee truth or generality; in this Lagrange agreed with Condillac. Furthermore, successful calculations produced discoveries that did not refer back to principles of a different order, and remained independent of metaphysics in a way that Maupertuis's principle of least action did not. In order to make analytical calculations, the mathematician manipulated symbols, but this was not a blindly mechanical process. Condorcet saw the exposition of analytic discovery as a contribution to the program of enlightenment:

> Every analytic operation tends to change the form of a given equation, to bring it into the form being sought. It is necessary to divine [deviner] which operations can most easily accomplish this change. But this kind of divination, which is only possible for genius, has its course and its motives in each particular case. In following the inventors through the process, one can, if not bestow genius, at least encourage its development in those who are born to have it.[58]

This mathematical divination is not the flash of inspiration associated with Romantic genius; it is the product of sustained attention and application by a specially endowed mind. As such, it is not suited to the flightiness often associated with female intelligence in the Enlightenment.[59]

No one suggested that the implicitly feminine attributes of metaphysics marked it as something women could do well. Gender functioned rather to separate metaphysics from the real work of analysis, reinforcing the masculine

57. Condillac, *Essai;* on calculation as a model for intelligence, see Daston, "Enlightenment Calculations."

58. Condorcet, quoted in "Avertissement," *Supplément,* i.

59. Daston, "Naturalized Female Intellect;" Steinbrügge, *Moral Sex.*

value of this arcane and formal mathematics. Calculation was represented as a useful technology that maximized the human potential to understand the world. The article on analysis in the *Supplément* to the *Encyclopédie* compared modern advances to the synthetic mathematics of the ancients:

> To rise above ordinary knowledge, the ancients had to painstakingly pile reason upon reason, as the giants piled up mountains to climb to the heavens. The moderns, like Daedalus, made themselves wings, with which they ascended easily to the most sublime regions accessible to human understanding. Those who have perfected calculations, and continue to perfect them daily with so much effort *[peine]* and so much wisdom, deserve all our admiration and all our gratitude.[60]

For all their relative efficiency, the operations of analysis still require hard work. But like the wings of Daedalus, calculations transport the mind directly to the heights, and the "most sublime regions" belong to mathematics rather than to metaphysics. Lorraine Daston has shown how the status and meaning of calculation changed between the Enlightenment and the nineteenth century, as calculation ceased to be associated with intelligence and insight and became instead rote mechanical work. She sees Prony's ambitious project to produce logarithmic and trigonometric tables adapted to the metric system as a turning point in this evolution of meanings. This "monument to calculation" employed an integrated hierarchy of mathematical workers, with a few analysts producing general formulae at the top and dozens of people performing arithmetical calculations at the bottom.[61] The analysts inherited the prestige attached to analysis and rational mechanics in the old-regime academy, while the routine work of adding and subtracting foreshadowed the teams of anonymous female calculators working in nineteenth-century astronomical observatories. As we might expect, metaphysics had gone the way of superstition and royal authority in this application of analysis to the goals of the Revolutionary republic.

The dismissal of metaphysics from the realm of analysis elevated rigor, deductive mathematics, and formal precision over equivocation, arbitrariness, and imagination, all the latter being commonly linked to the feminine. The undisciplined and susceptible female imagination provided a model for representing pitfalls associated with metaphysics and its irrational and useless siblings, superstition and enthusiasm. These value-laden terms were called into play in the ongoing attempt to draw secure boundaries around academic expertise and mathematical practice. Another related gendered opposition

60. Castillon, "Analyse," 385.
61. Daston, "Enlightenment Calculations."

worked alongside the imagination-reason fault line. When Lagrange reduced dynamics to statics and transformed mechanics into an algebraic analysis free of visual aids, he advanced an ideal of generality that referred to the gendered contrast between general and particular, abstract and concrete. Here the negatively valued terms were still routinely marked as feminine, though not directly associated with metaphysics. Metaphysics, after all, aspired to both generality and abstraction. Lagrange's attack on metaphysics was not, of course, a plea for particularity, but rather for a forceful and unequivocal generality that would abandon the vague and unproductive generalities of metaphysics. The association of metaphysics with femininity thus worked in delicately nuanced ways. While the general-particular opposition maps onto a male-female axis, metaphysics, although not concerned with particulars, is still feminized because its generality is feminine: its vague and equivocal generality cannot determine the solutions to problems in the observable world. Metaphysics and particularity alike become devalued, since concrete data are useless until they are subsumed under general equations.

The demise of metaphysics in late-Enlightenment France reflected the codification of an approach to analysis that aspired to describe nature transparently in systems of equations. Only accomplished mathematicians could operate on these equations, and access to this form of accomplishment continued to be a mark of elite status well after the demise of the old regime itself. The development of antimetaphysical analysis was intertwined with the elevation of rigorous and systematic rationality. By the time of the French Revolution, this analytic rationality was also clearly associated with masculinity.[62] When they argued for the utility of their particular expertise for the effective working of the state and the well-being of the public, men of science used language that reflected and reinforced contemporary definitions of gender difference. One effect of this rhetoric was to bar the feminine, and hence women, from the world of the mathematician and the academy. But the gendered language of difference was available to academicians intent on marking the special value of their work partly because women were already securely outside the scientific enterprise. The qualities of female intelligence and imagination effectively confirmed the status of women as outsiders to the academy, and this confirmation also boosted the value of other attributes—generality, precision, focused attention, deductive reason, hard work—gendered masculine by contrast. Gender thus played a subtle and somewhat circular part in the complex process by which metaphysics was banished from analysis, and mathematicians asserted the value of their reductive program for natural knowledge.

62. Outram, *Body and the French Revolution.*

Acknowledgments

In addition to the editors and other contributors to this volume, I would like
to thank Tom Broman, Ted Porter, Ayval Ramati, and Norton Wise for helpful
comments and conversation.

Primary Sources

Algarotti, Francesco. *Le newtonianisme pour les dames.* Amsterdam, 1738.
Bailly, Jean-Sylvain. *Rapport des commissaires chargés par le roi de l'examen du magnétisme animal.* Paris, 1784. Reprinted in Bailly, *Discours et mémoires* (Paris, 1790), 2:1–91.
———. "Exposé des expériences qui ont été faites pour l'examen du magnétisme animal" (1784). *Histoire de l'Académie Royale des Sciences* (Paris), 1787, 6–15.
———. "Rapport secret sur le mesmerisme or magnétisme animal." In *Du magnétisme animal en France, et des jugements qu'en ont portés les sociétés savantes,* edited by Alexandre Bertrand, 511–16. Paris, 1826.
Buffon, Georges-Louis Leclerc de. "Réflexions sur la loi de l'attraction" (1745). *Mémoires de l'Académie Royale des Sciences* (Paris), 1749, 493–500.
———. "Seconde addition au mémoire qui a pour titre: Réflexions sur la loi de l'attraction" (1745). *Mémoires de l'Académie Royale des Sciences* (Paris), 1749, 580–83.
Castillon, Jean. "Analyse." In *Supplément à l'Encyclopédie,* vol. 1. Amsterdam: M. Rey, 1776.
Clairaut, Alexis. "Du système du monde dans les principes de la gravitation universelle" (1745). *Mémoires de l'Académie Royale des Sciences* (Paris), 1749, 329–65.
———. "Réponse aux réflexions de M. de Buffon, sur la loi de l'attraction et sur le mouvement des apsides" (1745). *Mémoires de l'Académie Royales des Sciences* (Paris), 1749, 529–48.
Condillac, Etienne Bonnot de. *Essai sur l'origine des connaissances humaines* (1746). In *Oeuvres complètes,* vol. 1. Paris: Charles Houel, 1798.
———. *La logique.* In *Oeuvres complètes,* vol. 2. Paris: Charles Houel, 1798.
Condorcet. Introduction to *Histoire de l'Académie Royale des Sciences,* 1–12. 1782. Paris, 1785.
———. "Sur le méphitisme des fosses d'aisance" (1782). *Histoire de l'Académie Royale des Sciences* (Paris), 1785, 13–15.
———. "Discours prononcé à l'Académie, devant Son Altesse Royale le prince Henri de Prusse" (1784). *Histoire de l'Académie Royale des Sciences* (Paris), 1787, 1–6.
———. "Reception Speech at the French Academy" (1782). In *Selected Writings,* edited by Keith Baker, 3–32. Indianapolis: Bobbs-Merrill, 1976.
D'Alembert, Jean Le Rond. "Géometrie." In *Encyclopédie ou Dictionnaire raisonnée des Sciences, des Arts et des Métiers,* edited by Denis Diderot and Jean d'Alembert, 7:629–38. Paris: Briasson, 1757.
———."Éloge de M. Jean Bernoulli" (1748). In *Mélanges de littérature, d'histoire, et de philosophie,* vol. 2. Amsterdam: Chatelain, 1773.
———. *Recherches sur la précession des équinoxes* (1749). In *Oeuvres philosophiques, historiques et littéraires de d'Alembert,* vol. 1. Paris: A. Belin, 1821.
———. *Traité de dynamique* (2d ed., 1758). In *Oeuvres philosophiques, historiques et littéraires de d'Alembert,* vol. 1. Paris: A. Belin, 1821.
———. *Preliminary Discourse to the Encyclopedia of Diderot.* Translated by Richard Schwab. Indianapolis: Bobbs-Merrill, 1963. First French edition published 1751.
Desmahis, Joseph. 1756. "Femme (morale)." In *Encyclopédie ou Dictionnaire raisonnée des Sciences, des Arts et des Métiers,* edited by Denis Diderot and Jean d'Alembert, 6:472–75. Paris: Briasson, 1756.
Diderot, Denis. "Sur les femmes" (1772). In *Oeuvres,* edited by A. Billy. Paris: Gallimard, 1951.
Jaucourt, Louis. "Superstition." In *Encyclopédie ou Dictionnaire raisonnée des Sciences, des*

Arts et des Métiers, edited by Denis Diderot and Jean d'Alembert, 14:669–70. Neufchâtel: Faulche, 1765.

Lagrange, Joseph-Louis. "Essai d'une nouvelle méthode pour déterminer les Maxima et Minima des formules intégrales indéfinies." In *Miscellanea taurinensia.* 1760. Reprinted in *Oeuvres,* edited by J.-A. Serret (Paris: Gauthier-Villars, 1867), 1:335–62.

———. "Application de la méthode . . . à la solution de differents problèmes de dynamique." In *Miscellanea taurinensia.* 1760. Reprinted in *Oeuvres,* edited by J.-A. Serret (Paris: Gauthier-Villars, 1867), 1:363–468.

———. *Mécanique analytique* (1788). In *Oeuvres,* edited by J.-A. Serret, vol. 11. Paris: Gauthier-Villars, 1888.

Maupertuis, Pierre-Louis Moreau de. "Accord de différentes loix de la nature qui avoient jusqu'ici paru incompatibles." *Mémoires de l'Académie Royale des Sciences* (Paris), 1744, 417–26.

———. "Les loix du mouvement et du repos, déduites d'un principe de métaphysique." *Histoire de l'Académie Royale des Sciences de Berlin,* 1746, 267–94.

———. Untitled ms. Archives de l'Académie des Sciences (Fonds Maupertuis). Paris. N.d.

"Métaphysique." In *Encyclopédie ou Dictionnaire raisonnée des Sciences, des Arts et des Métiers,* edited by Denis Diderot and Jean d'Alembert, vol. 10. Neufchâtel: Faulche, 1769.

Thomas, A. L. *Essai sur le caractère, les moeurs et l'esprit des femmes dans les différens siècles.* Paris: Moutard, 1772.

❧ *9* ❧

The "Nature" of Enlightenment

E. C. SPARY

A generation ago, studies of nature and the natural in eighteenth-century France concentrated upon the scope of the repertoire for interpreting nature, from Platonism to pantheism, from the classical to the calculating, from the agricultural to the Arcadian.[1] By contrast, histories of the Enlightenment in the same period and country were long equated with a particular philosophical tradition, epitomized by the *Encyclopédie* of Diderot and d'Alembert. The relations between nature and enlightenment were used to illustrate the radicalism of individuals from the Enlightenment canon such as d'Holbach, Helvétius, or Diderot, by showing their innovative applications of notions of nature and reason to human society and the absolutist state.[2] Here, appeals to nature were associated with anticlericalism, even atheism, and with sweeping social reform. More recently, the "history of biology" has yielded a body of writings suggesting that scientific pursuits performed by a circumscribed, usually institution-based, group of "scientists" encompassed certain views and practices taken to epitomize the "real" purpose of enlightenment.[3]

These traditions, however, restricted the commentary on nature-culture-enlightenment to a tiny and closely scrutinized group of individuals. Moreover, they assumed the closure of debates about the meaning of the scientific, the natural, and the enlightened that were far from being settled at the time. What happens to the generally received understanding of the terms "enlightenment" and "nature" if, instead, one takes as the subject of a historical study individuals who themselves traded in notions of nature? Here, my aim is not to explore the reasons why eighteenth-century writers frequently invoked nature in the same breath as enlightenment, but rather the function that such invocations

---

1. Ehrard, *L'idée de la nature;* Charlton, *New Images of the Natural;* Pilkington, "'Nature' as Ethical Norm"; Delaporte, *Nature's Second Kingdom.*

2. Hawthorn, *Enlightenment and Despair,* 13; Gay, *The Enlightenment;* Kors, *D'Holbach's Coterie.*

3. Callot, *La philosophie de la vie;* Glass et al., *Forerunners of Darwin;* Salomon-Bayet, *L'institution de la science;* Sloan, 'Buffon's Idea of Biological Species'; Dawson, *Nature's Enigma.* My comments here extend only to the literature on French natural history. Buffon looms large in the combined tradition of the history of ideas and the search for the origins of the 'life sciences', as exemplified in Beaune, *Buffon 88;* Roger, *Buffon,* 89–90; Roger, *Les sciences de la vie.*

performed. The account that follows is largely limited to natural history texts in French, published between 1740 and 1790. As such, it only represents a preliminary sketch of an alternative to the existing secondary literature on this question, a cultural history of French natural history.[4] Nonetheless, it will become clear that such a history can help to demonstrate that "nature" was not merely an "idea" that could be utilized at will by all participants, but a symbol whose contested usage connoted different social worlds and (naturalized) power hierarchies.

Natural history's scope extended well beyond the boundaries of mathematization, experimentalization, and even rationalization that have often been taken to characterize the enlightenment project. Even the empiricism upon which natural historical practitioners prided themselves did not trivialize, in their view, the study of the effects of nature upon the subjective interplay of the passions. Rather than seeing the eighteenth century in the light of the nineteenth—that is to say, as a process of development toward an ideal of experimental, historical inquiry into the natural world—perhaps one might treat natural history in broader terms, as it was understood and practiced by a large and enthusiastic section of polite society from 1740 until 1790. Rather than assuming automatic hierarchies of institutionalized over noninstitutionalized practices of natural history, or of professionals over amateurs, one can explore how contemporaries interpreted the boundaries of expertise in natural history.

Histoire naturelle in eighteenth-century France existed to a large extent outside any institutional context. It occupied a social place that is now hard to categorize unless contemporary society is understood in terms of the tangled and slippery networks of patronage that simultaneously enabled the practice of *histoire naturelle* and placed limitations on the kinds of inquiry naturalists could pursue. If enlightenment is translated into an account of the development of disciplines and institutions, then the meaning of *histoire naturelle* within eighteenth-century French society is lost. For most of its practitioners were not institutionally employed; and the lucky few who, by the end of the century, had a coveted post that paid rather than required money for the pursuit of natural history, sought to construct a new natural history stripped of all aesthetic and commercial aspects. These aspects were, however, what distinguished "natural history" in the *siècle des Lumières* from earlier and later projects by the same name. Standard accounts of growing institutionalization and professionalization do not sit easily with these midcentury practices. In 1773, for example, when the Académie Royale des Sciences discussed the possibility

---

4. I borrow here particularly from Chartier, *Cultural History,* introduction; but see also Jardine and Spary, "Natures of Cultural History."

of a reshuffle that would have created a class of natural history within the institution, the young botanist and physician Antoine-Laurent de Jussieu defended the existing categories of anatomist, botanist, and mineralogist. The name *histoire naturelle*, he insisted, would prohibit specialization in particular topics; the generality of the subject would harm the scientificity of academic knowledge.[5]

But it was those who already possessed institutional posts who wished to redefine the scientific as that which was pursued in institutions. Such attacks should not be taken to reflect the status quo in France at this time, but rather as an indication of the hopes of a small group who would successfully negotiate positions of power during the French Revolution.[6] To view natural history from this perspective forces us to reinterpret the significance of the Foucaldian "gaze" at the end of the century. The transformation in ways of seeing and ordering natural objects in the period 1775–1825, which Foucault identified as a shift from "classical" to "modern" epistemes, reflected a profound change in the makeup of the group shaping the content of natural history, a change that entailed a reevaluation of the criteria of scientificity, order, taste, and display in natural history. The Revolution would have its most devastating effects upon precisely the categories of people who had practiced natural history during the Enlightenment, stripping them of wealth and rank, and thus overturning the hierarchy of emulation, possession, and taste that had previously given value and meaning to natural history specimens and practices. Thus the shift observed by Foucault as an epistemic phenomenon can only be explained if one moves outside the institutional context to consider how the practice of natural history reflected the diversity of the educated levels of eighteenth-century French society.[7]

Within this culture, both nature and enlightenment were functional terms, rather than descriptive ones. Both were normalizing and universalizing: providing standards for behavior that extended far outside the realm of polite society, unlike manners or erudition. In twentieth-century accounts, enlightenment is still sometimes portrayed as a process of liberation, learning, and utility that was democratically accessible to all. But, whether viewed as process

5. "Observations sur le projet de changer le nom de la classe de botanique en celui d'histoire naturelle" (1774), MS 1197, pièce 17, Bibliothèque Centrale du Muséum National d'Histoire Naturelle.

6. Spary, "Making the Natural Order."

7. Foucault, *Les Mots et les choses*. On the question of this shift, see Lepenies, *Das Ende der Naturgeschichte*, who, however, like Larson, *Interpreting Nature*, takes the victory of the nineteenth-century scientific method for granted. The project of natural history that I describe, despite its diversity, was restricted to the polite world or beau monde of midcentury France; collecting was apparently not practiced by laborers or even in artisanal families prior to the Revolution.

or event, it has just as readily been demonstrated to be a phenomenon of cultural imperialism, riven by its own inconsistencies. This chapter explores some of the Manichaean processes that were bound up in all discourses about nature and enlightenment, as expressed within some rather less well studied manifestations of that relationship.[8]

Monarchical Myology

In 1746, Jacques Gautier d'Agoty (d. 1785) published his *La Myologie complete, ou description de tous les muscles du corps humain* as a vehicle for a new printing technique of which he was the owner and which involved layered printing of the same sheet of paper with a succession of up to three different colors. The *Myologie*, a folio work, with each plate produced using the new technique, was d'Agoty's showpiece; using it, he hoped to attract attention from the elite world for his invention and for himself, and thereby move further up the patronage scale that dominated the lives of educated Parisians. This work, and a subsequent *Anatomie complete de la tête*, both with texts by the academician anatomist Duverney, did indeed attain a modest success, the *Anatomie* receiving a royal gratification of 600 livres. Six years later, d'Agoty began to publish a short-lived journal, the *Observations sur la Physique, l'Histoire Naturelle et la Peinture*.[9] D'Agoty included a number of plates in the journal, several of which had been printed with the color technique. Among the natural historical topics covered were the problem of human generation, the cultivation of silkworms, and the anatomy of a variety of small animals, including the snail, mole, and frog.

The journal appeared at a time of intense interest in the problem of generation in the savant world, some ten years after the discovery by a Swiss tutor, Abraham Trembley, that a small freshwater polyp possessed the faculty to replicate from severed parts—in other words, that a new individual could be created by a mechanical operation, without divine intervention. Vartanian has discussed the ways in which various "representatives" of the Enlightenment such as Buffon, Diderot, and La Mettrie responded to the polyp problem, in many cases by abandoning their adherence to versions of ovist or spermaticist preformation in favor of epigenetic (sometimes materialist) versions of

8. See Bender, "A New History of the Enlightenment?" The drawing of ethical conclusions from studies of "the Enlightenment" seems still to be a powerful imperative for historians, encouraging the continued reification of the Enlightenment as historical object. For recent perspectives upon the debate, see Outram, *Enlightenment;* Porter, *Enlightenment;* Porter and Teich, introduction to *Enlightenment in National Context;* Lively, "Europe of the Enlightenment"; Crocker, "Interpreting the Enlightenment"; Jacob, "Enlightenment Redefined"; Foucault, "What is Enlightenment?"; Hulme and Jordanova, *Enlightenment and Its Shadows;* La Vopa, "Conceiving a Public."

9. Continued by d'Agoty and Toussaint, 1756–57.

generation.[10] D'Agoty is far from being part of this Enlightenment canon. In fact, his attacks on Newton, particularly upon the theory of colors (manifested in his color printing techniques) caused him to be marginalized both within his lifetime and in subsequent biographies.[11] It is thus all the more interesting to find him criticizing ovist preformationist explanations of generation by appealing to enlightenment. In an enlightened century, he argued, no rational person could possibly believe that the female could play the principal role in generation. The simple evidence of one's own eyes revealed that the fetal structures were all present and perfect within the male ejaculate, as Hartsoeker had demonstrated. As proof, d'Agoty supplied a color picture of semen (presumably his own) in a glass of water, shown in figure 9.1.

For d'Agoty, it was thus an error of perception to suggest, as "oviparists" including his old collaborator Duverney (the only living individual named in his list) were doing, that the mother contributed more than mere matter to the fetus. Although he opposed the claims of both "oviparists" and "vermiculists," he directed his main attack against the former. The danger of oviparism lay in its social consequences.

> At all times, and all over the earth, fathers have been regarded as the true pro-creators of their children; all the honors of generation have been given to them. Even the language of sacred Scripture supports this doctrine. There it is always said that such [a man] begat another, and it is never said of women that they begat. They would however beget in effect, if they were to furnish their part in the substance of the fetus. I might go so far as to say that it is upon that universal belief that the preeminence of our sex over the other is founded; in particular here in France the Salic law concerning the succession of the Crown.[12]

In part 11 of the *Observations* (1754), d'Agoty described carrying out experiments on male and female frogs at Cressy, before an audience of the king and members of the Paris medical faculty, to show the living fetus in the seminal

10. Vartanian, "Trembley's Polyp"; Gasking, *Investigations into Generation*, outlines the differences between ovist preformation (fetal parts preformed at generation, with the male providing only the initial spark of life), spermaticist preformation (fetal parts preformed in the individual sperm, or—as here—in the semen as a whole, with the mother providing only inert matter), and epigenesis (fetus formed ex nihilo from matter, sometimes with the regulated play of forces). For a fuller discussion, I refer the reader to this work.

11. D'Agoty's alternative was separately published in a range of works such as his now rare *Chroagénesie, ou génération de couleurs, contre le système de Newton*. He argued for the existence of four primary colors, black, blue, yellow, and red. The *Nouvelle Biographie Universelle* article on d'Agoty distributed equal amounts of ridicule on his printing technology and his scientific views.

12. D'Agoty, "Découvertes sur la génération, contre les Oviparistes et les Vermiculistes," in *Observations sur l'Histoire Naturelle* (1752), 1:16. Quotations from this work hereafter cited by page in the text.

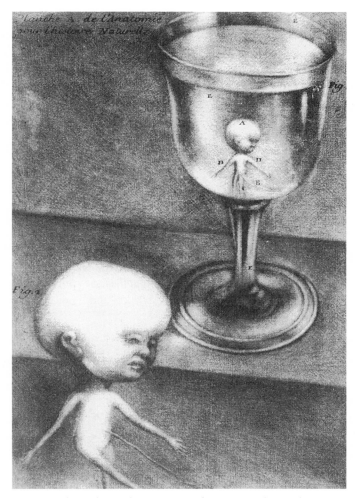

Fig. 9.1. D'Agoty's ejaculate in a wine glass. A to D denote the various fetal body parts, including an umbilical cord ready to plug into the maternal womb. From Jacques Gautier d'Agoty, *Observations sur la Physique, l'Histoire Naturelle et la Peinture* (1752), vol. 1, opposite 16. By permission of the Syndics of Cambridge University Library.

vesicles of the male. Unfortunately, the experiments were inconclusive, for similar fetuses appeared in the female. Since frogs lacked a penis, it was not, d'Agoty admitted, easy to understand how the living fetuses could have entered the female body from the male. Although his experiments were proving difficult to replicate, he insisted that it was not the result of errors in his vision:

"[T]he quality of Painter and Draughtsman that I possess preserves me from illusion where the observation of shapes is concerned" (50). Only further experiments on frogs and snakes could explain the mystery of generation for these animals. At the end of this issue, in a discussion of extracts from other journals, d'Agoty outlined his cosmology: the earth was the passive principle, the "vagina" into which trees and plants deposited their seeds; likewise, female animals were the earth into which males deposited their seed. This power of generation and succession of individuals proved the reality of creation (151).

For d'Agoty, there was the most intimate link between paternity, monarchism, and natural history. Others likewise appealed to natural historical evidence to support the monarchical, paternalist model of society.[13] The d'Agoty family continued to represent this view of the social order, as far as it is possible to judge. Because of the marginality of the family and its printing techniques, only fragmentary materials concerning the family or published by its members survive. We know very little about d'Agoty's background or income, and nothing about two of his sons, nor (significantly) anything about his wife and daughters. He himself continued until the end of his life to publish prints using his own techniques on a range of natural historical and medical subjects, and on painting and color theory.[14] His eldest son, Arnaud-Eloï, revived his father's journal as *Observations périodiques sur l'Histoire Naturelle, la Physique et les Arts* for one year, 1771. After his death in the same year, it became the model for the far more successful journal edited by the agronomist Rozier, generally known under the title *Journal de Physique*.[15] Two younger sons produced some incomplete and now very rare series of portraits of French monarchs: Jean-Baptiste (d. 1786) published a *Monarchie françoise, ou recueil Chronologique des portraits en pied de tous les rois & les chefs des Principales familles* (Paris, 1770). Jean-Fabien (b. 1730) published both anatomical and natural history prints, and portraits of Louis XV, Cardinal Fleury, and others. The d'Agoty family's technology was thus the material link between natural history and paternalist monarchy.[16] Indeed, the continuing affirmation of monarchism through printing and natural history was the very basis for the continuing success of the family in polite society.

13. See, e.g., Buffon, "Les animaux carnassiers," in *Histoire naturelle,* vol. 7 (1758): "An Empire, a Monarch, a family, a father, here are the two extremes of society: these extremes are also the limits of Nature." Unlike d'Agoty, Buffon stressed the equal role of male and female in generation at this time.

14. But see his candid acknowledgment of his own marginality in "Lettre de M. Dagoty père," *Mercure* 109 (August 1775): 175–79.

15. Its real title was, however, *Observations sur la Physique, sur l'Histoire Naturelle et sur les Arts.*

16. See the entries on various members of the d'Agoty family in the *Nouvelle Biographie Universelle.*

Systematizing Society

It might be thought that I have played devil's advocate here by choosing as my model for enlightenment a family of marginal, paternalist, anti-Newtonian, royalist color printers. But d'Agoty is not the only natural historical writer who directly appealed to enlightenment, reason, and modernity in order to promote a view of French society that differs sharply from our expectations of what is entailed by those terms. François-Alexandre Aubert de La Chesnaye des Bois (1699–1784) is another such individual, about whom even less is known, except the fact that he apparently abandoned a Capuchin order to scrape out a living writing and translating in Paris. De La Chesnaye's *Système naturel du Regne animal, par classes, familles ou ordres, genres et especes* (Paris, 1754), which received the royal privilege for publication, arranged quadrupeds, birds, and amphibians according to the method of Jakob Klein. It was a standard example of the kind of project undertaken by eighteenth-century writers: testing the water with a short work, planned merely as a preliminary to a longer study on the same topic, in this case the *Dictionnaire raisonné et universel des animaux, ou le regne animal* (Paris, 1759). In the preface, the former monk aligned himself firmly with the moderns: only their writings could help to correct the errors of the ancients. He even endorsed Buffon and Daubenton's attack on the cognitive principles underlying methods of classification in their *Histoire naturelle* of 1749, although like many others, he had resorted to a patchwork of methods borrowed from other naturalists to organize the book: Klein for quadrupeds, birds, and amphibians; Artedi for fish; Linnaeus for the insects and worms in volume 2.[17] Being a modern, however, he promised to correct any errors in his sources whenever they might arise.

In the same year, de La Chesnaye also translated the *Doutes ou observations de Mr. Klein,* concerning several of Linnaeus's classes of animals, from the Latin. His preface succinctly summarized the Linnaeus-Klein controversy: Klein's publication of a critique of Linnaeus's *Systema naturae* in Danzig in 1743 and his fame in the German lands for his alternative systems of quadrupeds and birds. Ultimately, however, such debates served the progress of natural history: it was by means of the proliferation of systems "that all the Sciences are modeled and remodeled by new pens, which never change anything fundamental about them, but may perfect them." For the translator, the trajectory of modernity was the product of compilation and revision.

> As long as there are philosophers uniquely occupied in admiring the Creator's marvels, and in observing them in the three Kingdoms as if with

17. Sloan, "Buffon-Linnaeus Controversy."

the eyes of Argos . . . , this science [which is] so natural for man, and so worthy of his curiosity, will make perpetual progress. Every day for more than a century, there have been new Observations and new Discoveries, and consequently new Riches in favor of those who feel that they have enough taste and patience to cultivate [natural history].[18]

De La Chesnaye's renown as a writer derived principally from his classifications of French society, however. The notion of an underlying "right" toward which human disagreements strove underpinned these also. In the second edition of the *Dictionnaire de la Noblesse* (1770), a work judged so aristocratic by the Jacobins that it was condemned to be destroyed, de La Chesnaye promised— as his title spells out—the "genealogies, the History and the Chronology of the Noble Families of France, the explanation of their arms, and the condition of the great Estates of the Realm today possessed as Principalities, Duchies, Marquisates, Counties, Viscounties, Baronies, etc. whether by creation, inheritance, alliances, donations, substitutions, mutations, purchase or otherwise." In natural law, men were equal, de La Chesnaye claimed. If monarchy was rejected in the name of liberty by the ancient Greeks, "Nobility however was not unknown among them" in the form of descendants of the first Heroes. The Romans and Gauls, likewise, recognized a nobility. The conquering Franks suppressed this Gallic nobility, and high-ranking Gauls turned to the ecclesiastical estate instead. Although the first French were rude barbarians, greedy and restive for all their valiance and love of liberty, there were lords among them to whom the people devoted itself by virtue of their birth or outstanding actions.[19] Thus de La Chesnaye reversed the distinction between the right to nobility by birth versus that attained through achievement: since even the most ancient civilizations possessed the latter, it was actually *more* ancient than the former, which only originated with the French monarchy. As soon as the dignity of nobles was no longer attached to achievement, it ceased to excite others to glory.

De La Chesnaye's brief history of French nobility was addressed principally to the nobility (or would-be nobility) itself. Particularly in the second half-century, there was considerable interchange between some sections of the French nobility and wealthy commoners in the beau monde, through marriage and the possession of common centers of cultural exchange.[20] The well-to-do, particularly landowners, were almost certainly the principal consumers of natural history, the only group which could afford the hand-colored folio

18. "Avis," in *Doutes ou observations de Mr. Klein,* 3–4.
19. "Avant-Propos," in *Dictionnaire de la Noblesse,* ii. On noble histories, see Baker, *Inventing the French Revolution,* chap. 2; Ellis, "Genealogy, History, and Aristocratic Reaction."
20. See Chaussinand-Nogaret, *French Nobility.*

volumes of Pierre-Joseph Buc'hoz and Pierre-Philippe Alyon, for example, or the china dinner services depicting Buffon and Bexon's natural history of birds, which the future comte de Buffon called "my Sèvres edition." [21] The wealthy alone possessed the power to purchase titles or pay dowries for assimilation into the nobility. Many natural history writers framed their projects for this audience in terms of an appeal for a revival of distinction and merit among the nobility, responding to contemporary concerns about the erosion of the noble state by intermarriage. These were certainly not proposals for egalitarian social reform: thus de La Chesnaye extolled the art of genealogy as enabling the classification of families and the generation of a history of heroism, in effect a system for historical classification. "Without the thread of the Genealogical and Heraldic Art, one loses oneself in the labyrinth of names; one confuses everything, and one fails to distinguish the chronological sequence." In fact, genealogy enabled one to distinguish the original "Nobles by race" (families always recognized as being noble) from the "Nobles by birth" (mere creations by the monarch)—thus allowing the all-important distinction to be made between illustrious ancient families, their origins lost in the mists of time, and recent creations. "What science, indeed, is more necessary for social harmony than that of Genealogies?" [22] The function of classification, in all the respects in which de La Chesnaye used it, was always that of generating hierarchies of value: naturalists' chronological progressions toward modernity, animals' bodily complexity, nobles' historical rights. Thus, whether classification was spatial, morphological, or temporal, it was always comparative. [23]

Natural historical classifications adopted a particular naturalizing sleight-of-hand, which slipped between the natural world and the social, simultaneously generating distinctions of class, nation, gender, age, and geography. But before such classifications could be operated universally, naturalists needed to recruit—read "enlighten"—a particular wealthy and powerful audience, and to establish not only the expertise of naturalists over the natural, but also the

21. E.g., Buc'hoz, *Nouveau Traité physique et économique* (1787); Alyon, *Cours de Botanique* (1787). Buffon and the abbé Bexon's *Histoire naturelle des Oiseaux* (1770–83) fetched thirty livres for the folio edition, twenty-four for the small folio, and eighty-eight for the edition with ninety-eight colored plates (*Journal de Paris*, April 10, 1779, 400). Roche, *Les républicains*, chaps. 2 and 3, describes the increases in publication and purchasing of works on the sciences and arts over the course of the eighteenth century.

22. "Avant-Propos," *Dictionnaire de la Noblesse*, xiv. See also Buffon, "Le Lion," in *Histoire naturelle*, vol. 9 (1761): "I understand by species [which are] noble in Nature, those which are constant, invariable, and which one cannot suspect of having degraded."

23. Here my argument differs somewhat from the interesting studies of the history of systematics. For France, in particular, see Stafleu, *Linnaeus*, chap. 8; Duris, *Linné et la France*; nothing approaching Allen, *Naturalist in Britain*, exists for France, but c.f. Gillispie, *Science and Polity*, on the Société d'Histoire Naturelle.

dominance of the natural over the social, and consequently over both polite society and the "people." Thus, by transforming both those who practiced the sciences of nature and those who were classified by them, the reductive power of nature could serve as an instrument of enlightenment whether empowering or disempowering. This transformative role was couched, in many natural history texts, within a narrative about the development of the self in the light of reason.

Civil Conversations

Self-development was a theme elaborately explored in Gilles-Augustin Bazin's two-volume popularization of René-Antoine Ferchault de Réaumur's *Histoire naturelle des insectes*. Bazin was one of an elliptical patronage group of naturalists with two "foci": the first, in Paris, was centered on Réaumur (b. 1683), a leading figure in the Académie Royale des Sciences of Paris from his entry in 1708 as *élève géometre* until his death in 1757. Réaumur was elected *directeur* of the academy eleven times between 1714 and 1753, and *sous-directeur* eight times between 1713 and 1752. The second "focus" was Charles Bonnet (1720– 93), naturalist and philosopher, in Geneva, a relation of Abraham Trembley, the discoverer of the polyp. Bonnet's *Contemplation de la Nature* (Amsterdam, 1764) was one of the most widely read works in France. Prior to Réaumur's death, he had been one of Réaumur's most devoted protégés and correspondents.[24]

Réaumur's early association with the Académie des Sciences in Paris as a geometer was reflected in his subsequent approach to natural history. His most famous work was his *Mémoires pour servir à l'Histoire des Insectes,* published between 1734 and 1742 under the academy's privilege. But most of his publications concerned the animals then known as insects (including what we now call crustaceans) and, to a lesser extent, the birds. Insects, especially, served as the most evident demonstrations of his natural theological geometries. Thus, his natural theological version of natural history, to which Buffon was bitterly opposed, often appeared in the works of contemporaries as inextricably associated with the study of insects. The example par excellence of design in the natural world was the beehive, and Réaumur's accounts of the construction of the honeycomb in volume 5 of the *Mémoires* (1740) became a showpiece of natural theological argument. Réaumur had used geometry to show what the divine *mécanicien* was about: hexagonal cells such as those that bees constructed

24. Grasse, *La vie et l'oeuvre de Réaumur*, 1–11, 17–24; Dawson, introduction to *Nature's Enigma;* Torlais, *Un ésprit encyclopédique*, 157–68. De La Chesnaye was also one of Réaumur's protégés, according to Hérissant, *Bibliothèque physique de la France*. The popularity of Bonnet's work is described in the *Journal de Paris*, 31 August 1779.

from wax were of the shape that exploited the space available most effectively, while using the least wax.

Such was Réaumur's dominance over the realm of insects that subsequent works of natural history drew their descriptions of insects largely from his. Not until the very end of the century, at the hands of Latreille and Olivier, did a different science of insects come to form a part of natural history; Buffon and his protégés neglected the area almost entirely. In his *Pensées sur l'Interprétation de la Nature* (1754), Diderot, a staunch supporter of Buffon's natural history, echoed that naturalist's criticism that insects were too small and too troublesome to be worthy of serious study:

> And you, who take the title of philosophers or of wits and who do not blush to resemble those importunate insects which spend the instants of their ephemeral existence in annoying man . . . What is your aim? . . . Despite your efforts, the names of Duclos, d'Alembert and Rousseau; of Voltaire, Maupertuis and Montesquieu; of Buffon and Daubenton will be honored among us and our nephews; and if anyone remembers yours: "They were, he will say, the persecutors of the foremost men of their times; and if we possess the preface of the Encyclopédie, the Histoire du siècle de Louis XV, the Ésprit des Lois, and the History of nature, it is because luckily it was not within the power of those people to deprive us of them."[25]

This division of the natural world reflected more than just a difference in theoretical priorities. The sharp separation between the patronage groups of Buffon and Réaumur divided the Académie des Sciences for several generations. Former Réaumur supporters fielded the young marquis de Condorcet in opposition to Buffon's candidate for the permanent secretaryship of the academy in 1771, and Condorcet, although not a Réaumur protégé, was responsible for the famously scathing *éloge* of the "great charlatan" Buffon in 1788.[26] However, the construction of the *Encyclopédie* as the principal representative of "enlightenment" has meant that Réaumur's natural history has generally been portrayed as reactionary, because opposed to the reforming ideals of the Encyclopedists. These judgments represent the results of a struggle for audiences and patrons and for control of institutions such as the Académie des Sciences and the Jardin du Roi. Outside the limited circles of those immediately

---

25. Diderot, "Des obstacles," pensée 4 in *Pensées sur l'Interprétation de la Nature*. J.-A. Lelarge de Lignac, a protégé of Réaumur, criticized the natural history cabinet at the Parisian Jardin du Roi (run by Buffon and his protégés) precisely for its lack of adequate representation of birds and insects. By order of the king, however, Réaumur's collections were assimilated into the Jardin's cabinet at the naturalist's death in 1757 (Orcel, "Daubenton").

26. Baker, "Les débuts de Condorcet"; Roger, *Buffon*.

involved in establishing and maintaining these boundaries, however, the different approaches to the natural world advanced by Réaumur and Buffon were commonly amalgamated in the representations of natural history that reached polite society as a whole. For example, Jacques-Christophe Valmont de Bomare's *Dictionnaire raisonné universelle d'histoire naturelle,* which appeared in several editions, was a pastiche of extracts from, among others, Buffon for the quadrupeds and birds, Réaumur for the insects, several botanists on plants, and his own works for the minerals.[27] Concerns with the demands of patronage and audience, the utility of natural history, and the spectacle of nature pervaded natural history publications.

Réaumur's correspondent Bazin (d. 1754), a Strasbourg physician, devoted many of his published writings to adapting Réaumur's works on insects, complete with natural theological message, to a different readership. In particular, he published the two substantial volumes of the *Histoire Naturelle des Abeilles* (Paris, 1744). From the start, his purpose was clearly set out: to popularize the art of beekeeping among French landowners and thus to improve the wax trade. Peasants lacked both the money and time to perfect the art, so that it was the responsibility of the wealthy and leisured landowners to do so. Bazin pinpointed the principal mismatching of savant writing such as Réaumur's and the fears and habits of the "intelligent people" in the countryside who might help to perfect apiculture. The *Mémoires,* Bazin claimed, was a work "which by its price, length and the learning contained within it, only seems to be accessible to Savants and the Curious." Thus those who knew bees best and spoke of them most knowledgeably "have only addressed themselves to those who are not able to profit from their discoveries and teachings."[28]

Bazin's solution to the problem of enlightening the polite readership, like that of so many other natural history writers midcentury, was to fictionalize the encounter between a leisured woman, wife to a country landowner, and her friend, who was savant to follow the reasoning of Réaumur (and who, Bazin claimed, represented himself). Clarice and Eugène's exchanges were supplied as lively dialogues, a style of presentation common in natural history works where, in a friendly and informal setting, one or more instructed characters conversed with and improved an unenlightened individual, usually a child, but in this case a woman.[29] Such conversations typically involved mem-

27. Valmont also offered public lectures on natural history from 1756, and his course supplied the raw material for the *Dictionnaire.* See Lacombe's review of one edition in *Mercure* (October 1775, no. 1): 151–56 (vol. 109 of the facsimile edition of *Mercure* published in 1971).

28. Bazin, "Avertissement," in *Histoire Naturelle des Abeilles,* 1:10.

29. See Cope, *Compendious Conversations;* Amies, "Amusing and Instructive Conversations"; Mortier, "Variations on the Dialogue in the French Enlightenment"; Gordon, "Public Opinion"; Goodman, "Governing the Republic of Letters."

bers of the first two orders of French society, the clergy and the nobility. This format was quite standard in the educational writings of the mid–eighteenth century, and perhaps reflected an increasing number of attacks on the irrelevance of the secluded monastic education that had still been the norm in the seventeenth century. Writers such as Voltaire railed against the failure of such instruction to prepare one for life in the beau monde; proper education should reflect the setting in which ordinary life took place.[30]

The work of both Réaumur and Bazin belongs within a corpus of writings on the reform of *éducation* through the polite exchange of learning.[31] In the *Histoire Naturelle des Abeilles,* the acquisition of natural historical knowledge has a theological purpose, teaching the reader about the similarities and differences between God's plans for the animal and human worlds. But the book serves a far broader function, for it makes inextricable the strands of self-improvement, patriotism, and wealthy amusement. Here commerce in the sense of trade and commerce in the sense of social exchange are intricately interwoven, and both are the necessary underpinnings for national prosperity and wealth. Bees, routinely portrayed as exemplifying industriousness, serve as Bazin's models for the ideal *polis,* following God's laws to achieve the good of the whole. The product of natural history is, thus, to be the exercise and understanding of civic duty and the cultivation of the mind. Recent writings on the *siècle des Lumières* have stressed the importance of this development of the model of civic responsibility as an ideal of noble self-development that separated the identity of polite society from the monarchical hierarchy upon which it was dependent.[32]

The transformation suggested by Bazin for the French provincial landowner is above all manifested in Eugène's appropriation of a pack of playing cards—symbols of the shallow hedonism of the leisured world—to model the cells in a honeycomb, shown in figure 9.2. It is also exemplified by the transformation of the frivolous, satirical, mocking Clarice into a serious, public-spirited, responsible philanthropist.[33] Bazin's initial description of her was as a

30. Voltaire especially stressed the importance of matching the place of education to the child's ultimate place in society; see Spencer, "Women and Education," in *French Women and the Age of Enlightenment,* 90–91.
31. Chisick, *Limits of Reform,* 42, indicates the extent to which the expulsion of the Jesuits fueled the debate on education. For recent uses of politeness as a criterion for participation in the social elite, see Lougee, *Paradis des Femmes;* France, *Politeness and Its Discontents,* chap. 4; Goodman, *Republic of Letters;* Goldgar, *Impolite Learning,* chap. 1; Sutton, *Science for a Polite Society,* 103–41.
32. Chaussinand-Nogaret, *French Nobility,* especially chap. 2; Gordon, "Public Opinion"; Goodman, *Republic of Letters;* all drawing upon Elias, *Civilising Process.*
33. My reading of the gender role allocated to Clarice and to the queen bee differs sharply from that of Merrick, "Royal Bees," who sees naturalists such as Bazin and Réaumur as limiting

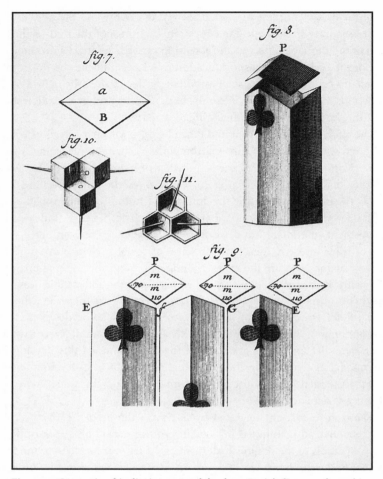

Fig. 9.2.   Converting frivolity into natural theology: Bazin's diagrams for making honeycomb cells out of playing cards. From Gilles-Augustin Bazin, *Histoire Naturelle des Abeilles* (1744), vol. 2, plate 8. By permission of the Syndics of Cambridge University Library.

"mother of a family, living on her estate, and whose mind has no other cultivation than that acquired from a good education, worldly commerce, and the

---

the importance of bees as moral models in order to enforce gender categories confining women within domestic and maternal roles. Clearly Bazin was treating Clarice's maternal responsibilities in a civic sense, generating rather than denying a public existence for her. On card playing as the symbol among enlightened scientific practitioners of noble corruption and of the deception of the public by charlatans, see Stafford, *Artful Science*, 90.

reading of books which are not completely frivolous."[34] Although not incapable of following the abstract sciences, Clarice had not been raised for them,
and was frightened by anything "that sounds and looks like them." She feared
that such knowledge might even be socially inappropriate for her sex, like provincial ladies at court mixing a courtly air with their "region's Idiom" (1:2).
Her efforts to read bee habits in the style of a "Novel" (1:7) gradually crumbled
under Eugène's sententious revisions of what she saw and how she interpreted
it. By the end of volume 2, she had been completely integrated into the language of agricultural improvement, prominent in natural history writings until
well after the Revolution, even suggesting a project for the propagation of apiculture among her peasant tenants: "I think I have found an easy means of
procuring a considerable benefit for my country in this part of its commerce,
both by prodigiously multiplying the species which is its object, and by developing an opportunity for succor for the peoples, and for economy in our
households" (2:340–41). This "useful and salutary project" (2:352) had both
biblical and political implications. Clarice wished "that my land might become, according to the terms of Scripture, *a land flowing with milk and
honey*"(2:348) but recognized the need for the "attention of the ministry for
the good of commerce" (2:352). What made natural history at midcentury into
a project of enlightenment in the eyes of contemporaries was its potential
utility; natural historical practice was profoundly political in serving to under-
pin proposals for "improvement" of the self, of the land, of the nation. In
France, as in other European countries, rational improvement and enlightenment were nearly synonymous for many. Appeals to utility were a pervasive
characteristic of the many versions of natural history being generated in mid–
eighteenth century France, and were especially prominent in writings on the
uses of nature for the self-development of the young noble, of which Pluche's
Spectacle de la nature was one example, and Buffon and Daubenton's *Histoire
naturelle,* another.[35]

Natural History: Consumption or Cultivation?

For individuals such as Bazin, de La Chesnaye and d'Agoty, the writing of
natural history works was itself a form of commerce, even a domestic industry.
D'Agoty sold his colored plates from his home on the rue de La Harpe. The
physician-naturalist Pierre-Joseph Buc'hoz, at no. 119 of the same road, was
one of the most prolific of natural history writers, selling his collections of
plates and treatises on natural historical and medical subjects at around 10 to

---

34. Bazin, "Avertissement," in *Histoire Naturelle des Abeilles*, 1:xiii–xiv.
35. Pluche, *Le Spectacle de la nature.*

14 livres for each booklet, often from home. Aubert de La Chesnaye des Bois, too, sold his *Dictionnaire de la Noblesse* from home, on the rue Saint-André-des-Arts, between the hôtel d'Hollande and the rue des Grands-Augustins. It was very rare for an individual who taught or curated in natural history for a living to be entirely self-sufficient; most, even those with medical or patronage connections, such as J.-P. Buisson at the Collège de Pharmacie, Jean Descemet and his mother at the Jardin des Apothicaires, or Pierre-Philippe Alyon, tutor to the children of the duc d'Orléans, published as well, or sold specimens.[36]

There are no simple lines of distinction between writers, collectors, curators, readers, merchants, and lecturers, nor between the institutional and domestic practice of natural history, at whatever point along the social scale one looks, except perhaps at the very top. For most members of polite society, efforts to obtain patronage, positions, royal or municipal *titres* and *brevets* often involved publication, since the renown of a protégé reflected upon the patron. The most famous of eighteenth-century French naturalists, Georges-Louis Leclerc de Buffon, traded on the success of his *Histoire naturelle, générale et particulière* to obtain his county; Jean-Baptiste-Pierre-Antoine de Monet, chevalier de Lamarck, a former army officer and vendor of shells, gained his post at the Jardin du Roi under Buffon on the strength of his successful *Flore françoise* of 1779, a new and simple analytical key to all known French plant species. Participation in the literary world was increasingly linked to participation in polite society, so that patronage success and public esteem were coming to be ever more closely articulated.[37] However, natural history was never just literary; it was tied to a whole economy of material objects, which included books and prints, seeds, bulbs, birds, and bones. While some immortalized their patrons or protégés by naming species after them, others, such as Buc'hoz, put them into print. The second edition of his *Nouveau Traité physique et économique* of 1787 was a collection of dissertations, each illustrated with a hand-colored plate of high quality, and each costing two to six livres. Pierre-Joseph-Victor, baron de Besenval, not only headed the list of subscribers

36. Buisson, *Classes & Noms des Plantes* (1779). The Descemet family had its own nursery business: *Catalogue des . . . plantes . . . dans les jardins et pepinières du Sieur Descemet* (1782); and see Alyon, *Cours de Botanique.* Not all such publications actually earned money for the author, but they could also earn recognition and hence patronage. The practice of printing a series of booklets that would be purchased individually and bound by the owner reflected the high cost and duration of the publishing process; if the book failed, the author or publisher would not have gambled away a fortune.

37. Darnton, *Literary Underground.* Darnton, *Business of Enlightenment,* contrasts the condemnation of the first *Encyclopédie* with the royal permission accorded its successor, the *Encyclopédie méthodique.* More direct evidence is provided by the regular appearance of writings by natural historical authors such as Guéneau de Montbeillard, Buffon's collaborator and coeditor of a periodical on the natural sciences from 1754 to 1787, in the *Almanach des Muses.*

at the end of the volume but also appeared, accompanied by the full list of his titles, at the foot of the plate illustrating *Centaurea Calcitrappa.*

Much of the contact of wealthy Parisians with nature was in the urban gardens of nobles, which were comparatively small, even on large properties. Few undertook botanical herborizations into the countryside until the late eighteenth century.[38] The late-1780s *Vues des Principaux batimens a Paris* contained a series of plates that reveal that many gardens thought worth visiting were very public spaces indeed, perfectly visible from the road and from adjoining gardens. The house of the marquise de Brunoy, for example, was shown with many visitors accompanied by their small dogs, parasols, or books, perambulating amid the orange trees in their portable planters and through the hothouses. Gardens were thus exemplary sites for the public display of botanical virtue by nobles, especially as the cultivation of exotic plants increased sharply during the 1770s.[39] They were, above all, locations in which botanical, agricultural, and sensible display were intimately linked; as nobles filled their gardens with plants, they were representing simultaneously the symbolic repossession of the territory of France and its revivification or valorization; their own sensibility toward the beauties of nature and, in consequence, their redemption from the state of urban corruption; and their recognition of the role of natural history and its associated sciences and arts in mediating these two forms of improvement, both of themselves and of the nation.

Natural history could only have come to play so large a part in the repertoire of noble improvement by virtue of the fact that those for whom natural history was a source of income or status were successful in shaping natural history to the methods and manners of distinction in mid–eighteenth century French society. In Buffon's *Histoire naturelle,* much use was made of the literary genre of the travel account, which was widely read in the period, and of innovative stylistic forms that became the model for writing about nature for much of a generation, and opened the doors of the prestigious Académie françoise to the Burgundian naturalist.[40] Likewise, merchants of nature in midcentury well knew that they had to cater for the polite perception of natural history collecting as demonstrating taste, *goût.* Robert-Xavier Mallet's *Beauté de la Nature, ou*

38. Adams, *French Garden,* chap. 5, 125. See also Wiebenson, *Picturesque Garden,* chaps. 1 to 4.

39. *Vues des principaux batimens a Paris.* The *Guide des Amateurs et des etrangers voyageurs* (1788) contained a detailed account of the various scientific establishments, their collections of natural history and anatomy, teaching courses and history, in addition to its descriptions of noble gardens. Also Adams, *French Garden,* 124–25, and Spary, "Making the Natural Order," chap. 2.

40. On Buffon's style, see Lepenies, "Von der Naturgeschichte zur Geschichte der Natur," in *Das Ende der Naturgeschichte.* The preferred reading of nobles was belles lettres and history (Roche, *Les républicains,* chap. 3; Chartier, "Urban Reading Practices, 1660–1780," in *Cultural Uses of Print,* 183–239).

Fleurimanie raisonnée, Concernant l'Art de cultiver les Oeillets (Paris, 1775) was published as propaganda for his flourishing nursery business at the barrière de Reuilly. Mallet, a specialist in pinks, claimed in his preamble to have been cultivating plants for twenty years. He obtained considerable attention in widely read periodicals such as the *Journal de Paris,* which also regularly indicated the flowering of new exotic plants at the Paris natural history center, the Jardin du Roi, and at the Jardin des Apothicaires.[41] This was not merely advertising: it gives an indication of the complexity of the social location of natural history in the period. These vendors of nature took account of their readers' self-perceptions. What was on offer by writers, engravers, or nurserymen in the mid–eighteenth century was, it should be noted, *reasoned* florimania, the reflexive rather than the unmediated encounter with nature. No simple dialectic between nature and enlightenment was being evoked here, but rather one that enrolled fashion in the service of reason, patriotism in the service of natural history, and self-development in the service of commerce. The material objects of natural history, from skeletons and tulips to plates and microscopes, embodied these relations to varying degrees.[42]

The showroom and auction room used over a couple of decades by a successful team of merchants of natural history and the fine arts, Helle and Rémy, was in the rue Poupée, just off the rue Haute-Feuille, in the same small area surrounding the place St. André des Arts, which also housed Buc'hoz, d'Agoty, and de La Chesnaye.[43] In their *Catalogue raisonné,* whose frontispiece is shown in figure 9.3, Helle and Rémy's lengthy "Avertissement" discussed the relations between connoisseurship, value, and knowledge. Their act of putting this "considerable and varied series" up for auction would teach "Beginners the rarity, and even the value of Shells."

[O]ne acquires more knowledge about the rarity and cost of Shells in a visit [to an auction], than from the sight of Cabinets. From this [latter] study one can learn the name which is given to Shells, inform oneself of its [*sic*] Class or Family, but one will never know well that one species is rarer and more expensive than another, each Owner being in the habit of puffing off what belongs to him, and of judging the rarity of a morsel,

41. *Journal de Paris,* 15 June 1778, submission of his own invention, a form of cold frame, for royal approval. Mallet followed this up with his *Dissertation sur la maniere de cultiver des plantes choisies dans le Chassis physique du sieur Mallet* (Paris, May 1778). On its "advertising" of exotic plants in gardens, see, e.g., *Journal de Paris,* May 1778, 483; 14 September 1779, 1047.

42. Compare, also, the discussion in Benedict, "Curious Attitude," which reveals how many different levels of meaning and skill were involved in the appreciation of objects in eighteenth-century British collections.

43. On Helle and Rémy, see Laissus, "Les Cabinets d'histoire naturelle," in Taton, *Enseignement et diffusion des sciences,* 342–84.

either by the price which he paid for it, or by the opinion which he has of it.[44]

Helle and Rémy promised more than erudition; they offered selectivity, based upon their experience trading with Dutch merchants. Their knowledge and expertise concerning the rarity of specimens was proffered for the instruction of "Amateurs," who could then spread specimens rare even in Holland into French cabinets. They backed up their appeal to would-be connoisseurs by promising to include references to the plates in Dezallier d'Argenville's *Conchyliologie*, the principal text on the natural history of shells until the early nineteenth century. The copy of the *Catalogue* used here was one into which the owner had inscribed all the prices of the different lots, so that it is possible to see what purchasers valued most. Among the picturesquely named specimens of the "Ducal Robe," "Butterfly Wing," "Cloth of Gold," "Tiger," and "Chinese Button," what fetched the most money, at over 50 livres the lot, were spiny oysters, followed by "Amirals," "Pourpres," "Damiers," "Bécasses Epineux," "Tuilées," "Cames," and a scattering of others.

The single highest-priced lot, at 1,611 livres, was a *Scalata*, depicted in the *Conchyliologie*, listed by Rumphius. Helle and Rémy devoted a page and a half of description to this one specimen, giving its probable classification, its probable origin, and a reference to a misclassification in another work.[45] From all the descriptions of the rarest shells, one characteristic comes across clearly: most money was paid for shells with a particularly regular shape. The *Scalata*, of course, resembled a staircase, more valuable when it had not been pierced. Color was also very important, and, just as was the case for ornamental flowers such as the tulip or pink, its regular distribution was of prime importance in determining the value of particular species and of individuals within a species. In another auction two years later, Pierre Rémy described the difference between the true and false "Arléquine": the former seemed to be composed of "a number of small pieces of different shapes, some oval, others triangular, all intersected by clear and distinct lines. . . . The false *Arléquine* is worked in approximately the same way, but the pieces and lines are muddled and mixed together, there is no clarity of work to be found here." Here, the geometry so prized by Réaumur and his protégés in the working of insects was clearly equally important to the collector with taste. Although the need to classify and order was not ignored or denied, these suppliers to the midcentury connoisseur subordinated that need to the hierarchy of tasteful discrimination that only they could satisfy, a sure guarantee that the purchaser possessed more

44. Helle and Rémy, *Catalogue raisonné*, vii.
45. Ibid., 13.

Fig. 9.3. Frontispiece, Helle and Pierre Rémy, *Catalogue raisonné, d'une collection considerable de coquilles rares et choisies* (1757). This depiction of the natural history merchants' showroom, used in all their catalogs, could have been that of the interior of countless other fashionable Parisian shops during the mid—eighteenth century, but for the shells. Compare the illustrations in Brewer and Porter, *Consumption and the World of Goods*. By permission of the British Library, shelfmark 1267.b.34(2).

than mere money.[46] The dual demands of taste and order, however, were linked, for nature was both the source of order (even if that order was inscrutable to human understanding) and of beauty (even if that beauty was also a creation of custom, climate, and fashion). Thus the abbé Favart d'Herbigny's shell guide of 1775 indicated that only a combination of enlightenment and emotional involvement made classification possible at all: "Nothing escapes the enlightened amateur, [who] rightly finds infinite differences among testaceous beings; while those persons who only see them with indifference, imagine that all univalve shells are snails, and all bivalves are oysters and mussels." [47]

The appeal to order was itself made on the same basis of discrimination; many critiques were launched at those who "heaped up" their collections with no notion of what was in them, by those whose existence was enriched by their expertise in ordering, collecting, writing about, or selling natural history. D'Herbigny contrasted the "vague glance" of those who wished to "astound with sophisms, rather than instruct with knowledge" with the expert gaze of cabinet-forming amateurs and conchologists, which could "perfectly distinguish, for example, the *lépas* of the Magellanic coasts, from those of the seas of Europe." [48] It is thus highly important for our understanding of the relation between enlightenment, sciences, and self-perception in France that both naturalists in institutions and natural historical travelers portrayed themselves, not as "professionals" attracted by a "vocation," but in terms such as those of Jean-Christophe-Fusée Aublet: "From my tenderest youth, I have been virtually mastered by a most lively taste for the various parts of Natural History." [49] Aublet ran away to sea, embarking on a ship bound for Spain "without having communicated my plans to my family, for fear that they would prevent me from executing them." Stories of young men defying parental plans for legal, ecclesiastical, or medical training to follow their natural historical "tastes" cannot be construed as tales of developing professionalism of the sciences. Instead, they show that the narratives of self-development that underpinned much

---

46. Rémy, *Catalogue de curiosités* (1759). Cf. Olmi, "From the Marvellous to the Commonplace"; Pomian, *Collectionneurs*. On the importance of taste from the end of the seventeenth century as engendering a realm in which expertise was possessed by members of polite society, see Saisselin, *Taste in Eighteenth Century France,* chaps. 1 and 4. Even in the Cabinet du Spectacle de la Nature et de l'Art, which for twenty-four sous one could visit every day in the Carrefour de l'École opposite the rue des Prêtres, and which offered such delights as "a Medusa Head adorned with Serpent tails, the Sea Mole, the Salamander which lives in fire, Siamese Butterflies, the sea Horse, a Skeleton of the Hummingbird, the Skin of a Snake, the very fine Bull-Fly," and others, offered a free catalog "for the use of the Public" to increase the instructive effects of viewing. (*Catalogue des diverses curiositez et raretez*).

47. Favart d'Herbigny, *Dictionnaire d'Histoire Naturelle,* lvi.

48. Ibid., xxi, xl.

49. Aublet, preface to *Histoire des plantes de la Guiane Françoise,* 1:i.

natural history writing, and particularly the equation, often made in eulogies, between being in touch with one's own nature and having a taste for nature, were taken as literal programs for action, self-presentation, and self-narration by young men. Both academicians and connoisseurs continually stressed the possession of taste as a valuable quality, the mark of distinction in those who appealed for patronage, or as a way of distinguishing a particular "select," "very tastefully composed" collection. Such judgments were made, of course, on the basis of repeated social visits to other collectors' houses, visits made possible by the continual investment in polite associations, which almost inevitably involved whole families. In the 1740s and 1750s, Aublet was the student and protégé of Bernard de Jussieu at the Jardin du Roi, and of several noble collectors, notably the baron d'Holbach, who "opened his purse and his library to me." [50]

Natural history works were not just guides to ordering, except in a handful of cases; most were also guides for the proper behavior of a naturalist, which was actually the same process as the proper self-development of members of polite society. Alyon, for example, pinpointed botany as the only suitable scientific pursuit for magistrates, military men and financiers as well as the (post-Rousseau) *solitaires* who were led to the sciences by "an irresistible impulse," grounded in their nature, and who were the few destined to make useful discoveries. [51] Many guides to the study of natural history explained the popularity of a science that taught both taste and public utility, civicism, and self-control, by appealing to the innate attraction of nature for humans. All of these aspects were inextricably linked in, for example, the work of Buc'hoz. Referring to his troubled existence following the death of his patron, King Stanislaus of Poland, Buc'hoz borrowed from Rousseau to promise his readers that the study of nature offered "a charm which is only experienced in the full calm of the passions, and which alone suffices to render life happy and sweet; but as soon as a motive of interest and vanity becomes involved, whether in filling posts or in making books, as soon as the goal of learning is only to teach, . . . that of herborizing only to become an Author or Professor, all that sweet charm vanishes: plants just become the instruments of our passions." [52] But the writings of Buc'hoz, Alyon, and others described a particular social world that many readers were eager to enter. Natural history works were often marketed for those who wished to display the attributes of taste and politeness

50. Ibid., iv. See also, for example, Dezallier d'Argenville, *La Conchyliologie.*
51. Alyon, "Des avantages qui résultent de l'étude de la Botanique," in *Cours de Botanique,* 1–4.
52. Buc'hoz, "Dissertation servant de Préface ou de Prospectus," in *Nouveau Traité,* 1–2. This is a verbatim quote from Jean-Jacques Rousseau, "Septième promenade," in *Les Rêveries du Promeneur solitaire,* 131.

but lacked the financial resources or leisure needed for expertise. Thus Favart d'Herbigny, Valmont de Bomare, and Duchesne and Macquer all presented their natural history works as substitutes for the expense of purchasing and tedium of reading countless other works on the subject.

> I thought that I should collect and analyse, in a small number of portable volumes, several methodical systems from the most famous Conchologists of our times, for the benefit of those who might wish to get to know them before purchasing authors whose volumes are expensive, lengthy, [and] often quite rare, which mostly contain all parts of Natural History, and to which one is sometimes forced to resort, even if only to make use of just one part.[53]

Natural historical knowledge was considered a valuable means of self-improvement because its very acquisition repeated the steps of self-development judged necessary for the enlightened individual. One made the transition from natural (the brute) to social (member of polite society) by recapitulating the Adamic process of generating order from an initial perceptual chaos. Here, the trajectory of the individual confronted with nature mirrored that conceptual shift. Buffon's *Histoire naturelle* was heavily recruited by subsequent writers on natural history (including Rousseau) to express the passage from natural to civilized, from chaotic to orderly, from unhappy to happy: "[W]hen one . . . casts one's eyes for the first time on this storehouse filled with diverse, new and strange things, the first feeling which results is an astonishment mingled with admiration." A sense of confusion and inadequacy succeeded, but

> through familiarizing oneself with these same objects, through seeing them frequently, and, so to speak, without design, little by little they form durable impressions, which soon link themselves in our spirit by fixed and invariable relations; and from there we lift ourselves to more general views, by means of which we can embrace several different objects at once; and it is then that one is fit to study in an orderly fashion, to reflect fruitfully, and to clear paths for oneself so as to attain useful discoveries.[54]

One of the critiques of the *Histoire naturelle* scathingly referred to hapless troupes of polite observers wandering around collections, waiting for some "relation" to form out of the overwhelming mass of confusing variety in nature.[55] But such adverse comments had little effect prior to the Revolution. The

53. Favart d'Herbigny, *Dictionnaire d'Histoire Naturelle*, xxxii.
54. Buffon, "Premier discours" from *Histoire naturelle* (1749), vol. 1, in Piveteau, *Oeuvres philosophiques de Buffon*, 8.
55. Lelarge de Lignac, *Lettres à un Amériquain;* and cf. Lepenies, *Das Ende der Naturgeschichte,* whose view of natural historical style is profoundly informed by nineteenth-century interpretations of Buffon.

Histoire naturelle, written for an upwardly mobile readership, was the third most common work in private libraries cataloged by Mornet.[56] Moreover, Buffon, his opus, or his institution were acknowledged in virtually every work on natural history published after 1749 that I have consulted. The *Histoire naturelle* was instrumental in promoting the view that natural history offered a privileged access into the natural foundations of human association, and a route to "natural" development, both physical and moral. The appeal to a "nature" existing beyond society, but forming the bedrock of social interactions and moral and aesthetic standards, was what many readers particularly perceived in Buffon, even if there was little consensus on his political or religious views. However, this separation served a highly social function. Buffon used animals as illustrations of normalized social relations throughout the volumes on the quadrupeds, in descriptions of *moeurs* or habits, which gave prescriptions for "normal" behavior in categories defined according to hierarchies of perfection, gender, climate, or race. Female animals were good or bad mothers; male animals were free or enslaved (domesticated) and therefore perfect or degenerate accordingly. It was this slippage between the natural and the social that defined natural history's power to work on the social world.[57]

Policing the Natural

The natural was the pivot for a range of social maneuvers, both serious and satirical. The conjunction of the universalizing claims of enlightenment and the classificatory authority of natural history writings had, by the 1780s, generated a space in which the natural could be used to make objects unnatural or supernatural. Many who study the French Enlightenment are familiar with the social and political reform proposed on the basis of the laws of nature in such works as d'Holbach's *Système de la nature* (1770) or Helvétius's *De l'Esprit* (1757). Less well known is the French version of Ignaz von Born's satire on monks, *Joannes Physiophili Specimen Monachologiae* of 1783, which was revised and translated by the permanent secretary of the Paris Société Royale d'Agriculture, Pierre-Marie-Augustin Broussonet (1761–1807). In the *Essai sur l'Histoire naturelle de quelques especes de moines, décrits à la maniere de Linné,* Broussonet, alias "Jean d'Antimoine," turned natural history's technologies of natural order to the monastic orders. The satire drew its force in particular through Broussonet's claim that he was writing from the "Cabinet du Grand

56. Mornet, "Les Enseignements des bibliothèques privées." Marion, *Recherches sur les bibliothèques privées,* indicates that within the first decade after publication, Pluche's *Spectacle de la nature* was the only work of natural history to feature in the top one hundred most frequently read works in Paris; Buffon's work did not appear. However, Darnton, *Literary Underground,* chap. 6, questions the value of such figures as evidence of what was actually read.

57. See Thomas, *Colonialism's Culture,* chap. 3; Schiebinger, introduction to *Nature's Body.*

Lama": from a non-European perspective, monks were evidently not human.

But the success of the parody turned on Broussonet's (and Born's) perfect mimicry of the style and content of so many natural history books published over the course of the century. The book thus became at one and the same time a critique of monasticism and an effective mockery of the very foundations of natural history. Broussonet's much expanded preface pointed to the benefits for natural history of the modern flowering of the arts and sciences; to the growth of collections through voyaging; and to the difficulty of obtaining and preserving rare specimens: "Among the rarest species, and which would be most difficult to procure for oneself, I include a Grand Lama, an Abyssinian, a Mufti, an Armenian Patriarch and some others. But a part, such as powdered dung from the Grand Lama, a slipper, &c., would suffice, or one could try and get the stuffed individual, or better still preserved in spirit of wine." "Jean" appealed to the patriotism of naturalists and rulers to promote the study and collection of specimens. Broussonet even added a narrative of emotional commitment to natural history for greater verisimilitude:

> I have loved Natural History since my earliest childhood, and I gave myself over to it with ardor, but I do not possess the means to travel; only being able to make a very mediocre name for myself, I turned my attention toward man, I made an in-depth study of our species. I devoted myself to determining the different varieties of the human race.
>
> At first I studied with care the diverse species of animals which approach man most closely in their form, and I established their differences. Having passed under review the Monkeys, the *Sapajous,* the *Guenons,* the *Satyrs,* the *Fauns,* the *Mermen,* &c., I discovered by accident some species of a very vast genus. The *Monks,* to which I refer, offered me the link which serves to unite man with the Monkeys; the species of this genus have a human face; apart from that, they differ essentially from man.[58]

There followed, in the main body of the text, a series of descriptions in the Linnaean style of different monastic orders, such as the Benedictines, Franciscans, Capuchins, each accompanied by accounts of their "Animal Œconomy" and plates depicting the different "parts" of monks (hoods, beards, sandals, belts, habits, etc.) and by a list of earlier writings on the same subject. The effectiveness of this parody rested, of course, upon the "naturalizing" effect of the techniques of natural history; through systematizing, describing, the study of habits and parts, a social phenomenon was made to seem funny and degraded—in other words, *unnatural.* Only the extreme faithfulness of the

58. Broussonet, *Essai sur l'Histoire naturelle de quelques especes de moines,* ix, iii–iv.

Broussonet/Born satire to genuine natural history treatises enabled the joke to work.

In the year that Broussonet's translation of Born appeared, the same strategies were inverted against the royal family, and later against monarchists in general. In a caricature entitled "The Harpy, a living Amphibious Monster," Marie-Antoinette was under attack from the courtly faction of the comte de Provence, Louis-Stanislas-Xavier, brother to Louis XVI, otherwise known as Monsieur.[59] The accompanying "Historical Description of a symbolic monster taken alive on the shores of Lake Fagua, near Santa Fe . . . " was a fake discovery of a wild beast in Chile that exploited a particular tradition of engraving, commemorating the discovery of real and fake "beasts" and "monsters," existing throughout the century. Naturalists were called in to legislate about the authenticity of such discoveries and of public displays of spectacular specimens. Revolutionary caricaturists exploited nature to the full by composing monstrous representations of the queen, the royal family, and aristocrats out of two or more natural objects, whose juxtaposition rendered their subject unnatural. Such images crossed the dividing lines between symbolic and literal representations of the natural to engender a set of unnatural or supernatural representations that shattered social power. Marie-Antoinette's condemnation to death ultimately rested upon the many accusations of unnatural sexual liaisons with various courtiers and even with her son.

Conclusion

The fact that such different views of, and uses for, the terms "enlightenment," "nature," and "modernity" could be generated within a few years of each other and within a few hundred square yards of the same city should enforce upon us in the strongest sense Daniel Roche's insistence that we should no longer "read the Lumières by the light of the Revolution alone."[60] It is curious that although natural history was the discipline to which almost everyone who wished to comment on nature appealed, and which captured a range of individuals, professional writers, institutional botanists, curators of collections, and the wealthy curious, it has barely formed a part of writings on the relationship between nature and enlightenment. Even in the histories of nature it has been largely absent, being replaced by a mythical science of biology.[61] It is particularly revealing that natural history should be thus edited, cropped, divided, or annihilated, for it of all disciplines formed the common ground for relating claims about enlightenment to claims about nature. Such revisions indicate

59. Described in de Baecque's insightful La caricature révolutionnaire, 186.
60. Roche, Les républicains, 16.
61. See, in particular, Roger, Les sciences de la vie.

the great difficulty modern readers have had in coming to terms with the meanings of nature and enlightenment for eighteenth-century practitioners. In eighteenth-century French natural history, "science" and "sentiment" were not distinct; the division manifest in Daniel Mornet's two classics, *Le Sentiment de la nature en France de Jean-Jacques Rousseau à Bernardin de Saint-Pierre* and *Les Sciences de la nature en France, au XVIIIe siècle*, is of nineteenth-century origin. Only by creating such distinctions have historians been consistently able to affirm the marginality of studies of the social world for an understanding of intellectual practices. Even in social histories of the eighteenth century, writers have felt obliged to treat the sciences as somehow demanding special historiographical tools, separate chapters, a particular "specialist" language: in other words, the conceptual isolation of the sciences is reinforced by a methodological, linguistic, and physical isolation.[62] Since the end of the eighteenth century, the links that existed in natural history between taste and reason, connoisseurship and utility, sensibility and scientificity were progressively cut to fit subsequent categories; and conforming to these nineteenth-century concerns, historians have paid little attention to natural history as a cultural whole. Most of the sources I have used in this paper are by individuals now regarded as marginal or irrelevant to the "real" history of eighteenth-century natural history; going through a substantial list of mid-eighteenth-century French writers on *histoire naturelle,* I came to the conclusion that this judgment included virtually every writer on that subject who published between 1740 and 1790, with the exception of Buffon. Historians had evidently been more interested in what was being written about "specialist" subjects—botany and zoology in particular—than in the works that helped to shape the meaning of *histoire naturelle* for French polite society. It cannot be irrelevant, however, that only the widespread recognition of the existence of such a discipline as *histoire naturelle* by legislators from all parts of France—many of whom explicitly cited Buffon as their hero—opened institutional opportunities, lacking for other sciences over several years, for natural history to be practiced and taught, to be closely associated with national pride, public utility, and the preservation of civility throughout the early nineteenth century.[63]

62. Even in Charlton's otherwise rich social history of the natural in eighteenth-century France, the "history of biology" model was so compelling that in his chap. 4, "Sciences of Nature," the eighteenth century appeared as "the first age—the so-called age of empiricism—when scientists really looked at nature closely, in all its complex and many-sided realities and established the crucial role within science of experimentation." Charlton saw the period as the time of the "rise" of "new sciences" such as "geology, botany, biology and zoology, even if in some cases the name of the new science would not be coined until the following century" (70). Thus, the result of the history of biology is effectively to deny the existence of natural history.

63. See Spary, "Making the Natural Order," chap. 3.

Throughout this account, I have been relying upon materials—published texts, natural history specimens, and a collecting public—that have been taken in different ways to capture the link between the eighteenth century and modernity. The work of a growing body of historians of eighteenth-century consumerism stresses the success of changing patterns of commerce and credit in opening up new marketing strategies and financial opportunities well before the "Industrial Revolution." This approach, although very valuable in many ways, is problematic for the case of natural history. As I have shown above, naturalists, whether institutionalized or not, universally linked commerce with distinction, wealth with civicism. Natural historical practices united taste with utility. For the case of France, at least, it is too simple to suggest, with Brewer, that *financial* trustworthiness was the principal underpinning for the use of polite forms of address; natural history enrolled too many kinds of social and intellectual distinction to be reducible to such a scale.[64]

But can one recognize through a study of eighteenth-century natural history the development of an independent public sphere of the form envisaged by Habermas? The answer is, again, no. For the *practice* of natural history simply inverts or collapses many of the categories Habermas sought to use as a way of explaining the process of generation of modernity. It was precisely the domestic property of certain technologies of art and nature that the d'Agoty family exploited simultaneously as a commodity, as a means of regulating relationships in its "internal space," and as a way of gaining entry into the polite world. The polite world was not just the "market of culture products" but also simultaneously the "courtly-noble society" and (by virtue of the fact that some hierarchies of patronage coincided with the hierarchies of government) the "state."[65] There was no inevitability about the separation of these social spaces inherent within the practice of natural history prior to the revolution. What a mixing of Habermasian categories natural history reveals! The notion of "enlightenment" in France reflects a similar situation. It seems that both enlightenment and nature must be understood on several levels at once: as commodity, technology, sensibility, mode of self-presentation, and social hierarchy. "Enlightenment" could be a descriptive category, an evaluation of the state of being of the self and of the other, but overridingly it was a prescriptive category, in which normative forms of comportment were being established. The appeal to nature facilitated and legitimated these classificatory acts. But that appeal also enabled the sidestepping of competing sources of power, allowing earlier models of nature to be dismissed, old regime forms of

64. Brewer subordinates social to commercial credit-worthiness. See "Commercialization and Politics," in McKendrick, Brewer, and Plumb, *Birth of a Consumer Society*, 214.
65. For all these distinctions, see Habermas, *Structural Transformation of the Public Sphere*, 30.

association to be devalued, and even, ultimately, the disorders of Revolution to be naturalized.

Primary Sources

Alyon, Pierre-Philippe. *Cours de Botanique pour servir à l'Education des Enfans de S.A. Sérénissime Monseigneur Le Duc d'Orleans où l'on à rassemblé les Plantes Indigènes et Exotiques employées dans les Arts et dans la Médecine.* [Paris, 1787].

[Ane]. *De l'homme, et de la reproduction des differens individus: Ouvrage qui peut servir d'introduction & de défense à l'histoire naturelle des animaux par M. de Buffon.* Paris, 1761.

Aublet, Jean-Christophe-Fusée. *Histoire des plantes de la Guiane Françoise, rangées suivant la méthode sexuelle, avec plusieurs mémoires sur différens objets intéressans, rélatifs à la culture & au commerce de la Guiane Françoise, & une Notice des Plantes de l'Isle-de-France.* 4 vols. London and Paris: Pierre-François Didot jeune, 1775.

Bazin, Gilles-Augustin. *Histoire Naturelle des Abeilles.* 2 vols. Paris: Guérin frères, 1744.

———. *Abregé de l'Histoire des Insectes, pour servir de Suite à l'Histoire naturelle des abeilles.* 2 vols. Paris, 1747–51.

Berryat, Jean, et al., ed. [and subsequently Philibert Guéneau de Montbeillard et al.]. *Recueil de memoires, ou collection de pièces académiques, concernant la médecine, l'anatomie & la chirurgie, la chymie, la physique expérimentale, la botanique & l'histoire naturelle.* 15 vols. Paris, 1754–87.

Bonnet, Charles. *Contemplation de la Nature.* 2 vols. Amsterdam, 1764.

[Born, Ignaz von]. *Joannes Physiophili Specimen Monachologiae methodo Linnaeana tabulis tribus aeneis illustratem.* Augustae Vindelicorum. 1783.

[Broussonet, Pierre-Marie-Augustin]. *Essai sur l'Histoire naturelle de quelques especes de moines, décrits à la maniere de Linné.* Monachopolis [Paris], 1784.

Bruhier d'Ablaincourt, Jean-Jacques. *Caprices d'imagination, ou Lettre sur differens Sujets d'Histoire, de Morale, de Critique, d'Histoire Naturelle, etc.* Paris: Briasson, 1740.

Buc'hoz, Pierre-Joseph. *La Nature considérée sous ses Différens Aspects; ou, lettres sur les animaux, les végétaux & les minéraux. Ouvrage périodique.* 8 vols. Paris, 1771–72.

———. *Nouveau Traité physique et économique, par forme de dissertations, de toutes les plantes qui croissent sur la surface du globe.* 2d ed. 2 vols. Paris: Buc'hoz, 1787.

Buffon, Georges-Louis Leclerc de. *Supplément à l'Histoire naturelle, générale et particulière, avec la description du cabinet du roi.* 7 vols. Paris: Imprimerie Royale, 1774–89.

Buffon, Georges-Louis Leclerc de, and Louis-Jean-Marie Daubenton. *Histoire naturelle, générale et particulière, avec la description du cabinet du roi.* 15 vols. Paris: Imprimerie Royale, 1749–67.

Buffon, Georges-Louis Leclerc de, Philibert Guéneau de Montbeillard, and Abbé Gabriel-Léopold-Charles-Amé Bexon. *Histoire naturelle des Oiseaux.* 9 vols. Paris: Imprimerie Royale, 1770–83.

Buisson, J.-P. *Classes & Noms des Plantes pour suppléer aux Etiquettes pendant le Cours de Botanique que fera au College de Pharmacie le sieur Buisson, Démonstrateur d'Histoire-Naturelle-Phytologique.* Paris, 1779.

Catalogue des diverses curiositez et raretez, qui sont contenuës dans le Cabinet du Spectacle de la Nature et de l'Art. . . . 1740.

D'Agoty, Arnaud-Eloï Gautier. *Observations périodiques sur l'Histoire Naturelle, la Physique et les Arts.* Paris, 1771.

D'Agoty, Jacques Gautier. *Myologie complette, en couleur et grandeur naturelle, composée de l'Essai & de la Suite de l'essai d'anatomie, en tableaux imprimés.* Paris: d'Agoty, 1746.

———. *Anatomie complete de la tête, en tableaux imprimés, qui représentent au naturel le cerveau sous différentes coupes, la distribution des vaisseaux dans toutes les parties de la tête, les organes des sens, & une partie de la névrologie.* Paris: d'Agoty, 1748.

———. *Observations sur la Physique, l'Histoire Naturelle et la Peinture. Avec des planches*

*imprimées en couleur: Cet Ouvrage renferme les Secrets des Arts, les nouvelles découvertes, &
les disputes des Philosophes & des Artistes modernes.* 6 vols. Paris: Delaguette, 1752–55.
———. *Chroagénesie, ou génération de couleurs, contre le système de Newton.* 2 vols. Paris:
Antoine Boudet, 1750–51.
———. *Lettre de M. Dagoty père, Pensionnaire du Roi, à l'Auteur du Mercure, 1 June 1775.
Mercure de France* 109 (1971): 175–79.
D'Agoty, Jacques Gautier, and Toussaint. *Observations sur la Physique, l'Histoire Naturelle et
la Peinture.* 3 vols. [Paris], 1756–57.
D'Agoty, Jean-Baptiste Gautier. *Monarchie françoise, ou recueil Chronologique des portraits en
pied de tous les rois & les chefs des Principales familles.* Paris, 1770.
[D'Argenville, Antoine-Joseph Dezallier]. *La Conchyliologie, ou histoire naturelle des coquilles
de mer, d'eau douce, terrestres et fossiles; avec un traité de la zoomorphose.* Edited by Favanne
de Montcervelle, père et fils. 3d ed. Paris: De Bure, 1780.
[D'Argenville, Antoine-Nicolas Dezallier]. *Voyage pittoresque de Paris, ou indication de tout ce
qu'il y a de plus beau dans cette grande Ville, en Peinture, Sculpture, & Architecture.* Paris,
1749.
———. *Voyage pittoresque de Paris, ou indication de tout ce qu'il y a de plus beau dans cette
grande Ville, en Peinture, Sculpture, & Architecture.* 4th ed. Paris: De Bure pere et fils aîné,
1765.
———. *Voyage pittoresque de Paris, ou indication de tout ce qu'il y a de plus beau dans cette
grande Ville, en Peinture, Sculpture, & Architecture.* 6th ed. Paris, 1770.
De La Chesnaye des Bois, François-Alexandre Aubert. *Dictionnaire universel d'agriculture et
de jardinage, de fauconnerie, chasse, pêche, cuisine et ménage.* 2 vols. Paris, 1751.
———. *Système naturel du Regne animal, par classes, familles ou ordres, genres et especes.* Paris:
Cl. J. B. Bauche, 1754.
———. *Dictionnaire raisonné et universel des animaux, ou le regne animal.* 4 vols. Paris, 1759.
———. *Dictionnaire historique des moeurs, usages & coutumes des François.* 3 vols. Paris:
Vincent, 1767.
[———]. *Dictionnaire de la Noblesse, Contenant les généalogies, l'Histoire & la Chronologie
des Familles Nobles de France, l'explication de leurs armes, & l'état des grandes Terres
du Royaume aujourd'hui possédés à titre de Principautés, Duchés, Marquisats, Comtés,
Vicomtés, Baronnies, &c. soit par création, par héritage, alliances, donations, substitutions,
mutations, achats, ou autrement.* 2d ed. 3 vols. Paris: Veuve Duchesne, 1770.
Delille, Jacques. *Les Jardins, ou l'art d'embellir les paysages: Poème.* Paris: Philippe-Denys
Pierre, 1782.
Descemet, Jean. *Catalogue des arbres, arbrisseaux, arbustes, plantes, oignons de fleurs, graines
de fleurs, et potagères, qui se trouvent dans les jardins et pepinières du Sieur Descemet.* Paris,
1782.
Diderot, Denis. *Pensées sur l'Interprétation de la Nature.* London [and Paris], 1754.
Duchesne, Antoine-Nicolas. *Manuel de botanique, contenant les propriétés des plantes qu'on
trouve à la campagne aux environs de Paris.* Paris: Didot le jeune, 1764.
[Duchesne, Henri-Gabriel, and Pierre-Joseph Macquer]. *Manuel du Naturaliste. Ouvrage
Utile aux Voyageurs, & à ceux qui visitent les Cabinets d'Histoire Naturelle & de Curiosités.*
Paris: G. Desprez, 1770.
[———]. *Manuel du Naturaliste. Ouvrage Utile aux Voyageurs, & à ceux qui visitent les Cabi-
nets d'Histoire Naturelle & de Curiosités, en forme de Dictionnaire, pour servir de suite à
l'Histoire Naturelle.* 2 vols. Paris: Imprimerie Royale, 1771.
Dulard, Paul. *La grandeur de Dieu dans les merveilles de la nature. Poeme Par Monsieur Dulard,
de l'Académie des Belles-Lettres de Marseille.* 3d ed. Paris: Desaint et Saillant. 1751.
Favanne de Montcervelle, Jacques de, and Guillaume de Favanne de Montcervelle. *Catalogue
systématique et raisonné ou description du magnifique cabinet appartenant ci-devant à M. le
C. de ***. Par M. de ***.* [Paris], 1784.
Favart d'Herbigny, Abbé Christophe-Elisabeth [or Christ-Elisée]. *Dictionnaire d'Histoire Naturelle, Qui concerne les Testacées ou les Coquillages de Mer, de Terre & d'Eau-douce. Avec la*

nomenclature, la Zoomorphose, & les différens systêmes de plusieurs célébres Naturalistes anciens & modernes. Ouvrage qui renferme la description détaillée des figures des Coquilles, l'explication des termes usités, les propriétés de plusieurs, & les notes en partie des endroits où elles se trouvent. Paris: Bleuet, 1775.

Gersaint, Edné-François. Catalogue raisonné de coquilles et autres curiosités naturelles. Paris: Flahault et Prault, 1736.

Guéneau de Montbeillard, Philibert. Dictionnaire des Insectes. In Encyclopédie méthodique, Histoire naturelle, vol. 4. Paris: Panckoucke, 1782.

Guide des Amateurs et des etrangers voyageurs dans les Maisons Royales, Châteaux, lieux de plaisance, établissemens publics, villages et séjours les plus renommés, aux environs de Paris. Paris: Hardouin et Gattey, 1788.

Guindant, Toussaint. Exposition des Variations de la Nature, dans l'espece Humaine, Ou l'on demande si, posées les Loix Naturelles les plus générales sur lesquelles portent l'ordre & l'harmonie du Corps Humain, la Nature peut quelquefois s'en écarter. Paris: Debure pere, 1771.

Helle, [P.-C.-A.?], and Pierre Rémy. Catalogue raisonné, d'une collection considerable de coquilles rares et choisies, du Cabinet de M. Le ***. Paris: Didot, 1757.

Helvétius, Claude-Adrien. De l'Ésprit. Paris: Durand, 1758.

Hérissant, Louis-Antoine-Prosper. Bibliothèque physique de la France. Paris: Veuve Hérissant, 1771.

[Holbach, Paul-Henri-Thierry de]. Système de la nature; ou, Des loix du monde physique & du monde moral, par M. Mirabaud. London, 1770.

Journal de Paris. Paris, 1777–1827.

Journal encyclopédique. Liège, 1756–75.

Klein, Jakob. Doutes ou observations de Mr. Klein sur la revûe des animaux faite par le premier homme, sur quelques animaux des classes des quadrupèdes et amphibies du système de la Nature de M. Linnaeus. . . . Translated by François-Alexandre Aubert de La Chesnaye des Bois. Paris: Bauche, 1754.

Lamarck, Jean-Baptiste-Pierre-Antoine de Monet de. Flore françoise, ou Descriptions Succinctes de toutes les plantes qui croissent naturellement en France, . . . disposée selon une nouvelle méthode d'analyse. 3 vols. Paris: Imprimerie Royale, 1779.

Lambert, G.-F. Bibliothèque de Physique et d'Histoire naturelle, contenant la Physique générale, la Physique particulière, la Méchanique, la Chimie, l'Anatomie, la Botanique, la Médecine, l'Histoire naturelle des Insectes, des Animaux & des Coquillages. Paris: Veuve David, 1758.

Lebreton, François. Manuel de Botanique, à l'Usage des Amateurs et des Voyageurs; contenant les Principes de Botanique, l'Explication du Systême de Linné, un Catalogue des différens Végétaux étrangers, les moyens de transporter les Arbres & les Semences; la manière de former un Herbier, &c. Paris: Prault, 1787.

Lelarge de Lignac, Abbé Joseph-Adrien. Lettres à un Amériquain sur l'Histoire naturelle, générale et particulière, de M. de Buffon. Hamburg, 1751.

Mallet, Robert-Xavier. Beauté de la Nature, ou Fleurimanie raisonnée, concernant l'Art de cultiver les Oeillets, ainsi que les Fleurs du premier & second ordre, servant d'ornemens pour les Parterres; avec une Dissertation sur les Arbrisseaux choisis: fondé sur une longue expérience. Paris: Didot le jeune, 1775.

———. Dissertation sur la maniere de cultiver des plantes choisies dans le Chassis physique du sieur Mallet. Paris: Mallet, 1778.

Mercure de France, dédié au roi par une société de gens de lettres. Geneva: Slatkine Reprints, 1971.

Piveteau, Jean. Oeuvres philosophiques de Buffon. Paris: Presses Universitaires de France, 1954.

Pluche, Noël-Antoine. Le Spectacle de la nature, ou entretiens sur les particularités de l'histoire naturelle qui ont paru les plus propres à rendre les Jeunes-Gens curieux, & à leur former l'esprit. 2d ed. 8 vols. Paris: Freres Estienne, 1763.

Réaumur, René-Antoine Ferchault de. Mémoires pour servir à l'Histoire des Insectes 6 vols. Paris: Imprimerie Royale, 1734–42.

Regnault, Nicolas-François, and Géneviève Regnault. *La Botanique mise à la portée de tout le monde.* 2 vols. Paris, 1771–84.

Rémy, Pierre. *Catalogue de curiosités en différents genres, dont la Vente commencera le Lundi 5 Mars 1759, à trois heures de relevée & jours suivans. . . .* Paris: Didot, 1759.

Rousseau, Jean-Jacques. *Les Rêveries du promeneur solitaire.* Introduction by Jean Grenier. Paris: Gallimard, 1972.

Rozier, Abbé François. *Journal de Physique [Observations sur la Physique, sur l'Histoire Naturelle et sur les Arts].* 43 vols. Paris, 1773–93.

Valmont de Bomare, Jacques-Christophe. *Dictionnaire raisonné universelle d'histoire naturelle; contenant l'histoire des animaux, des végétaux & des minéraux, Et celle des Corps célestes, des Météores, & des autres principaux Phénomenes de la Nature; avec l'histoire et la description des drogues simples tirées des trois regnes; et le détail de leurs usages dans la Médecine, dans l'Economie domestique & champêtre, & dans les Arts & Métiers.* 2d ed. 6 vols. Paris: Lacombe, 1768.

———. *Minéralogie ou nouvelle exposition du règne minéral.* 2d ed. 2 vols. Paris, 1774.

Vues des principaux batimens a Paris. N.d.

PART

4

Provinces and Peripheries

From Carolus Linnaeus, *Flora Lapponica* (Amsterdam, 1736).

On 15 June 1767, in the Genoan town of Ombrosa, after having had a spat with his father, Baron Arminio Piovasco di Rondò, about eating snails, twelve-year-old Cosimo Piovasco di Rondò, baron apparent, climbed a tree, promised never to come down again, and kept his word. In his aerial abode in Ombrosa, Cosimo nonetheless managed to stay in tune with the mind of the Enlightenment and thus knew of the Lisbon earthquake and the Leiden jar. He studied the works of Linnaeus, Rousseau, Montesquieu, and Voltaire. Signing his epistles, "Cosimo Rondò, Reader of the *Encyclopédie*," he corresponded with Diderot. The baron in the trees even turned the nefarious bandit Gian dei Brughi from a life of crime to one of reading. Indeed, the erstwhile bandit became so addicted to novels, above all Richardson's, reading them in secret weeping, that Gian's two young accomplices, seeking to win him back to their criminal company, were driven to seize a novel from Gian, who cried, "Give it back to me! . . . I want to finish reading *Clarissa!*" Cosimo himself not only read but also composed works, especially on politics. While in Paris, however, his brother, alas, found him described in an encyclopedia as the "wild man of Ombrosa." Cosimo appeared in a chapter on monsters, in between hermaphrodites and sirens, and an illustration depicted him as a being covered with fuzz, with a long beard, a long tail, and eating a crab. Disabused of such fantasies by Cosimo's brother, Voltaire queried whether Cosimo lived in the trees as a philosopher to be nearer the heavens. Toward the end of his life, though remaining largely unappreciated in the provincial world of Genoa, so famous had Cosimo become in enlightened Europe (even beyond his complicated relations with the Freemasons), a great Frenchman made a special effort to meet him and exclaimed, in excellent Italian by the way, "Were I not the Emperor Napoleon, I would most like to be Citizen Cosimo Rondò!"

This aerial perspective on the Enlightenment comes from Italo Calvino's *Baron in the Trees* (*Il barone rampante*, 1957). We set Calvino's Cosimo here as an emblem of a provincial, peripheral, and even exotic perspective on the Enlightenment. *Baron in the Trees* shows as well the reduction of the Enlightenment at its margins to its essence: the Republic of Letters. For confined to his arboreal existence, Cosimo's sole means of contact with enlightened Europe (which meant mostly Britain and France in the novel) was essentially reduced, save Napoleon's pilgrimage to him, to reading and writing. But if a republic of letters or public sphere defined the Enlightenment in its essence, its actual embodiments in European provinces and peripheries indicate the problem of exotic cosmopolitanism: An enlightened scholar in arboreal Ombrosa might look like a monster in Paris.

The preceding two sections, "Bodies and Technologies" and "Humans and Natures," have each in their own way addressed a dialectic of the Enlightenment: the aim of its universal concepts and techniques versus their local embodiments. At the center of "Bodies and Technologies" lay, on the one hand, the new disciplines in the Enlightenment for inscribing persons and objects into rational and uniform spaces of knowledge and governance, and, on the other hand, the resistances and frictions generated by such disciplines in the locales and sites of actual application. The section "Humans and Natures" concerned in great part the dilemma of the universal categories of "nature" and "human" versus the peculiarities of their concrete instantiations, for example, in monsters in cabinets and in women in salons. This section, "Provinces and Peripheries," follows a similar vein. Here we look at the European Enlightenment as reflected in some of its marches and margins — exotic cosmopolitanism, so to say. For orientation, we need a cultural geography of Europe.

Excepting the Reformation, up to the Enlightenment, the center for generating most things studied by European cultural and intellectual historians was the Mediterranean: those parts of Africa, Asia, and Europe, along with their provinces and peripheries, interwoven by this body of water. In the seventeenth century, the economic and military center of gravity shifted from the Mediterranean to transalpine Europe. But one cannot say the same about the intellectual and cultural center: the culture of the Baroque remained anchored in Mediterranean lands. The eighteenth century, under the auspices of its Enlightenment, first witnessed the instauration of a new cultural and intellectual center, and not so much placed elsewhere in Europe, as rather (re)fashioning it. The propagation of the Enlightenment embodied the new cultural and intellectual hegemony of Britain and France, arising about a century after the economic and military center of gravity had shifted to transalpine Europe. Once capitals and centers have been set, provinces and peripheries become clear and distinct ideas.

This section does not aim to portray the sciences in enlightened Europe in national context, minus Britain and France. But recognizing that the European Enlightenment had an Anglo-Franco center, we wish to cast the Enlightenment off-center. The chapters here come from four provincial or peripheral regions of enlightened Europe: disunited provinces, called Italy, from the Mediterranean; united provinces, called the Netherlands, from the North Atlantic; peripheral lands, sharing no name other than the Baltic; a provincial-peripheral land, once called Brandenburg-Prussia, from the Baltic and Central Europe.

For two millennia, Italy had had a central role in Mediterranean culture. Much if not most everything of cultural or intellectual interest in Europe had come from Italy as producer or mediator. As Paula Findlen notes, the Enlightenment was essentially the first cultural and intellectual import into Italy from the "barbarian north." Enlightened Italy found itself at the periphery of Europe. Taking a perspective not unlike Calvino's on Italy's status in the Enlightenment, Findlen sets the focus of her chapter on a till now largely forgotten Newtonian, Cristina Roccati, who lived from 1732 to 1797.

Tracing Roccati's education, Findlen takes us on a tour of some of Italy's most famous universities. But we end up in the provinces, for after the 1750s, Roccati lived the better part of her life, if not in the trees, in relative obscurity nonetheless. From this obscurity we get a fair portrait of provincial Italian enlightenment. For her father and most males, Roccati embodied the enlightened *femme savant* almost as *objet d'art,* an essentially nonutilitarian or "Italian" view of enlightenment. We see too that, through the Republic of Letters, enlightenment was able to travel to humble corners of Europe, as Roccati spent her adult life spreading Newtonianism in the provinces. Findlen further points to Roccati's adherence to physico-theology as a link between science and religion. Running counter to currents at the cosmopolitan centers, no deism and surely no atheism would creep into this enlightened provincial science.

The tie of enlightened science to physico-theology and Newtonianism is taken up as well in Lissa Roberts's chapter, set in the United Provinces, the Netherlands. Roberts's chapter, however, brings to stage center the motif of utility, a central concern of enlightened science. Roberts begins with, in order to criticize, the typical view of the Dutch as mediators between Britain and the Continent. Acceptance of "this prescripted analysis," as she terms it, seems crucial for the projects of Anglo-Franco Enlightenment. The Dutch had played a big hand in shifting the economic and military center of gravity north in the seventeenth century, but if not seeing a decline, the Netherlands of the eighteenth century at most maintained economic and military stasis. For the new intellectual and cultural goods of the eighteenth century, arising in the wake of the shifting balance of power northward, the Dutch became, in the prescripted analysis, but merchants and middlemen of ideas.

Rejecting such a script, Roberts casts four acts, each illuminating a site for the production and dissemination of Dutch science: the anatomy theater, the physics theater, the private lecture, and the private scientific society. None of those were peculiar to Dutch science, which all the more underlines their importance. Indeed, Findlen's Roccati operated in at least three of the four sites. The chapters by Findlen and Roberts nicely illuminate enlightened science from different but complementary perspectives: a single person versus a series of sites. Roberts selects these sites as exemplars for enlightened science, among other reasons, since they provided fora in which various social classes could socialize, "popular" science being a crucial project of the Enlightenment. On our tour, Roberts not only takes us through the Netherlands in space but also in time. Though we see each of the sites across the entire eighteenth century, the temporal center for each shifts from early to mid-Enlightenment, as we move from anatomy to physics theaters to private lectures to societies. And as Roberts opens with the specter of Newtonian Britain looming over the Dutch Enlightenment, she brings our tour to an end with the specter of Napoleonic France.

Lisbert Koerner explains that three main watery regions surrounded Europe: the Mediterranean, the Atlantic, the Baltic. With her chapter on the Baltic we have a study drawn from the last named of these regions, while the chapters by Findlen and Roberts drew from the first and second. Roberts's Netherlands constitutes one of the smaller units in Europe, a microcosm of the Enlightenment, while Koerner's Baltic forms by far one of the largest spaces, a European attic uniting Scandinavia, Russia, Poland, and Prussia. Roberts's tour of the Netherlands kept us essentially indoors, while Koerner's voyage through the Baltic has us mostly outdoors and often in the cold. Nonetheless, the "hero" of both chapters seems to be the land itself. In this light, Koerner indicates similarities between the Dutch and Baltic Enlightenments: rejection of francophone cosmopolitical universalism in favor of almost ethnic localism. If Koerner's chapter has a human hero, it is the person most associated with the land. It is the provincial, the "peasant."

The Italian and Dutch Enlightenments of Findlen and Roberts highlight attempts to enlighten the people as consumers of science. Koerner's Baltic Enlightenment focuses on attempts by Baltic scholars, mostly Germano-Scandinavian, to enlighten themselves about the peasantry, as objects of science. After explicating the importance of physcio-theology as well for the Baltic Enlightenment, this chapter takes us away from the Baltic, through voyages of discoveries, then back through sciences of the land, finally focusing on an anthropology of the peasants and ending with their Romantic apotheosis. Koerner moves from the Enlightenment to the emergence of Romanticism by an astonishing voyage through the sort of science at first blush anathema to it: the utilitarian. From economic and managerial sciences, such as localist mineralogy and ethnography, Koerner shows how one provincial project of the Enlightenment ended with the Romantic motifs of place and ethnicity.

William Clark's chapter, set in Prussia, ends this section on provinces and peripheries. Brandenburg-Prussia was part of the "German nations," which occupied much of Central Europe, a region reaching from the Baltic shore to the Italian Alps. From the High Middle Ages to the Baroque, Central Europe had been one of the richest parts of Europe. The Thirty Years' War (1618–48) left much of it in ruins. The European ascendancy of Britain and France from 1648 to 1848 owed as much to that destruction of Central Europe as to the decline of the Mediterranean. Brandenburg-Prussia was one of the few German nations that emerged stronger after 1648 and, as the "Kingdom of Prussia" after 1700, began to spread like a slow cancer. Like the Italian, Dutch, and Baltic, the Prussian Enlightenment articulated itself in the shadow of Britain and France. Prussian distinction expressed itself in the *Aufklärung* as a provincial variation of the Anglo-Franco Enlightenment.

Complementing Koerner's chapter, Clark's arrives at the same end: the emergence of Romantic anthropology out of the project of enlightenment. While Koerner brings us there via an apotheosis of the "peasant," Clark leads us via an exorcism of the "queen." Metaphysics had been seen as the handmaiden of theology and queen of sciences. Declaring "her" a vestige of superstition, Anglo-Franco Enlightenment called for casting metaphysics out, while Prussian *Aufklärung* sought to redeem her by enlightenment. Though metaphysics declared itself the most important science, Clark's Prussian, like Findlen's Italian Enlightenment, does not pivot around utility. Indeed, proclaimed the apogee of human culture, the Prussian "queen," like an Italian *femme savant,* ended up a sort of *objet d'art.* As Koerner's chapter traces precisely the sort of utilitarian sciences that the Enlightenment favored over metaphysical sciences, the complementarity of these last two chapters is all the more interesting in ending at the same destiny for the Enlightenment: the sciences of "culture." Like Findlen, Roberts, and Koerner, Clark also takes us on a tour of sorts. Here it is through three Prussian academic sites: Halle, Berlin, and Königsberg. In moving from site to site, focus shifts from theology to cosmology to anthropology as metaphysical sciences. As with the three previous chapters, this last one also brings out the importance of physico-theology in enlightened science, or exotic cosmopolitanism, in the provinces and peripheries.

❧ 10 ❧
A Forgotten Newtonian: Women and Science in the Italian Provinces

Paula Findlen

> Women who would like to apply themselves to studying physics should know that whoever seeks knowledge of all natural things is in danger of working in vain, of falling into many errors, and of raising herself against God by her vanity.
>
> Louis de Lesclache, *The Benefits Which Women Can Gain from Philosophy and Especially Ethics* (1667)

Italy has often had a marginal role in accounts of Enlightenment thought and culture. Traditionally viewed as a region in decline after the collapse of various princely regimes that had aided in the flourishing of the Renaissance, and as a region embracing backward-looking intellectual traditions after the Catholic church's condemnation of Galileo in 1633, Italy, like much of the Mediterranean, seemed to represent the antithesis of Enlightenment traditions of intellectual freedom and commitment to modernity. Italy's participation in the principles of *Illuminismo* that were on the lips of many *philosophes* has been assigned primarily to the salons of Milan and Naples, where intellectuals such as Cesare Beccaria and Pietro Giannone flourished, or to the political and economic reforms that came as a result of foreign (Habsburg and Bourbon) rule.[1] In this traditional account, very little that was distinctive about the Enlightenment seemed to be inherent to Italy. It was a region that imitated rather than created, that hoped to be modern but never quite arrived there.

More recent work has begun to highlight a number of distinctive features that characterized the Italian Enlightenment. Foremost among them was the visibility of women in intellectual life. By the middle of the eighteenth century, travelers to Italy had begun to marvel at the number of learned women whom they encountered. "[T]he character of a learned woman is far from being ridiculous in this country," wrote Lady Mary Wortley Montagu to her daughter

---

Lesclache is quoted in Poulain de la Barre, *Equality of the Sexes,* 136–37. I have modified the pronouns in the translation to conform to the sense of the passage.

1. This classic account is embodied in Venturi, *Italy and the Enlightenment.*

in October 1753, "the greatest families being proud of having produced female writers."[2]

Less than two decades later, in 1765–66, the famed French astronomer Jerome Lalande called upon the learned ladies of Italy, visiting them in the major cities between Milan and Naples in between trips to observatories, museums, Roman ruins, Vesuvius, and other places of note. While favoring the talents of the celebrated Neapolitan experimenter Maria Angela Ardinghelli, "known in the Republic of Letters for her Italian translations of the English works of Monsieur [Stephen] Hales," he also praised Italy's famous university graduates such as the Newtonian Laura Bassi, who by then had taught physics and mathematics at the University of Bologna for more than thirty years as its first woman professor. "The women have distinguished themselves in Bologna for their knowledge." So well known was this phenomenon of female intellectual achievement that, at the end of the century, it become the subject of Germaine de Staël's novel, *Corinne; or, Italy* (1807). In the novel, a Scottish traveler describes the learned Roman woman he came to love as "one of the wonders of the singular country he had come to visit."[3]

While learned women were prominent in the French salons, known for their publications in England, and occasionally celebrated in Germany, no other region of Europe so closely bound the ascent of female learning to its scientific and cultural institutions as did Italy. There ambitious fathers and enlightened scholars introduced talented women to the pleasures of knowledge, admitting the most successful to their academies and universities. They celebrated their accomplishments in noisy public spectacles designed to gain the attention of the Republic of Letters and those nobility who collected female prodigies like so many curios for their cabinets. In some cities, women founded new academies and played an active role in shaping debates about the form that modern knowledge should take. In others women joined the ranks of the professoriate, teaching such subjects as mathematics, philosophy, physics, and anatomy.[4]

Science played a particularly important role in the evolution of the Enlightenment tradition of female learning. Just as science became central to education in general, mastering the laws of nature became the most tangible proof of a woman's learning. As the Venetian physician Eusebio Sguario remarked in his *Of Electrification* (1746), the female protagonist of his dialogue was learned

2. Montagu, *Letters and Works*, 2:242.

3. Staël, *Corinne; or, Italy*; Lalande, *Voyage d'un françois en Italie*, 6:369; 2:117.

4. On this subject, see especially Elena, "In lode della filosofessa di Bologna"; Findlen, "Science as a Career"; Berti Logan, "Desire to Contribute"; Ceranski, *Und sie fürchtet sich vor niemanden; Alma mater studiorum*. For a general overview of women in early modern science, the fundamental work is Schiebinger, *Mind Has No Sex?*

because she "had read extensively all sorts of modern philosophical books."[5] Increasingly learning was defined not by antiquity, as the humanist model had presupposed, but by modernity. Salon conversations about Descartes and Newton replaced scholastic debates about Aristotle and humanist discussions of the merits of Petrarch and Ariosto.

Such activities held even stronger meaning in Italy than in northern European countries, since the vast majority of natural philosophical treatises that embodied this new knowledge were increasingly written by foreign scholars, most notably Descartes, Locke, Boyle, Leibniz, and Newton. Many of their books often appeared on the Catholic Church's Index of Forbidden Books, alongside the works of Galileo. During the period following the condemnation of Galileo, the public face of Italian intellectual life had been conditioned by its limited engagement with modern thought, deemed generally incompatible with the tenets of Catholicism. Yet the first half of eighteenth century, as such scholars as Vincenzo Ferrone have described, initiated the beginnings of a widespread movement in Italy to engage with the ideas of those controversial *oltramontani*. In 1703, Ludovico Muratori made scientific thought an essential component of his *First Sketches of an Italian Literary Republic* when he wrote: "Let us place our greatest hopes for our glory in the Philosophy that we call Experimental."[6] By midcentury such plans revolved specifically around foreign ideas, often to the amazement of Italian scholars, who still could not quite imagine that the barbarian countries of the north could have surpassed Italy, the cradle of ancient civilizations and the arbiter of culture. "Who would ever have believed in times past that England, which was reputed a country of dolts," remarked the Venetian noble Francesco Algarotti in 1757, "should so excel and give laws in the sciences?"[7] Yet it was Algarotti himself who made Newton more available to the Italian reading public with the appearance of his popular and controversial *Newtonianism for Ladies* (1737).

The foreign status of the ideas mastered by Italian *filosofesse* added to the novelty of their learning. While it might be miraculous to have one's pupil imbibe centuries of Latin prose and Italian verse and spontaneously debate

5. Eusebio Sguario, *Dell'elettricismo* (Recurati, 1746), in Biagi and Basile, *Scienziati del Settecento*, 862.

6. Ludivico Antonio Muratori, *Primi disegni della Repubblica letteraria d'Italia* (1707), in Maugain, *Étude sur l'evolution intellectuelle*, 92 n. 3. On this subject, see also Manzoni, *I cartesiani italiani*; De Besaucèle, *Les cartésiens d'Italie*; Casini, *Newton*, esp. 173–227; and Ferrone, *Intellectual Roots*. Ugo Baldini offers a general overview in "L'attività scientifica nel primo Settecento."

7. Francesco Algarotti to Father Severio Bettinelli, Padua, 2 February 1757, in Algarotti, *Opere varie*, 2:378. Enlightenment Italy has primarily been the subject of a variety of studies by Franco Venturi and his students. English-speaking readers will find Venturi, *Italy and the Enlightenment*; Ricuperati and Carpanetto, *Italy in the Age of Reason*; and Dooley, *Science, Politics, and Society* to be a good starting point.

Aristotle on the immortality of the soul, training her to discuss the thorny problem of Cartesian materialism and the merits of Leibniz's versus Newton's definition of force placed the debate about the limits of women's education squarely in the midst of an even larger controversy about the role of modern knowledge. The presence of learned women in Italy's academies and universities made them among the most visible emblems of the arrival of modern knowledge in this increasingly provincial corner of the Republic of Letters.[8] Through the publicity surrounding these women's activities, in conjunction with a growing number of Italian treatises that explored the most recent developments in the experimental and mathematical sciences, Italian scholars announced to the world that they had entered the new age of learning.

Not surprisingly, the most active attempts to create communities of women learned in this modern way occurred in cities such as Bologna, seat of the oldest university in Europe and therefore the cultural center with the most to prove to the upstart French and English who had usurped its place in the world of learning. Capital cities such as Milan and Naples also enjoyed a strong presence of learned women as part of a lively, cosmopolitan salon culture that imitated what one found in Paris.[9] Yet the presence of learned women was not simply for the consumption of foreigners but also part of an active and long-standing tradition among Italian elites who perceived knowledge as a form of family patrimony and as an object of urban rivalry. Families and cities competed openly with each other in their efforts to lay claim to the most learned woman in all of Italy. After Padua offered Elena Lucrezia Cornaro Piscopia a degree in 1678, it was inevitable that Bologna would eventually honor one of their local prodigies, as they did Laura Bassi in 1732. By the middle of the eighteenth century, the competition had expanded to include talented women from more provincial towns in such regions as the Veneto, where a lively tradition of female learning had flourished since the late Middle Ages. There was hardly a town with a few thousand souls in northern Italy that did not lay claim to some learned woman, boast a copy or two of various Newtonian treatises, and perhaps an Italian translation of the *Philosophical Transactions of the Royal Society* for good measure. Gradually and often quite erratically, the debates about female learning and the controversies over new philosophies of nature made their way from the major intellectual centers of Italy into the provinces.[10]

Inspired particularly by the tales of the Bolognese Newtonian Bassi, who

8. This argument was first made in Cavazza's *Settecento inquieto,* 203–35.
9. Findlen, "Translating the New Science." For more on French salon culture, see Harth, *Cartesian Women;* Goodman, *Republic of Letters;* and Sutton, *Science for a Polite Society.*
10. On the history of learned women in Italy, see the essays by Margaret King and Paul Oskar Kristeller in LaBalme, *Beyond Their Sex.* For a useful point of reference on provincial science, see Porter, "Provincial Culture."

Fig. 10.1. Cristina Roccati (1732–97) at age eighteen, upon receipt of her degree in philosophy at the University of Bologna, 1751. Ms. 250, Biblioteca dell'Accademia dei Concordi, Rovigo.

was certainly the most famous and successful of Italy's women philosophers, rivaling the reputation of her French contemporary Émilie du Châtelet, ambitious fathers in small towns promoted their daughters as great prodigies of learning in the hope of winning fame and fortune in the Republic of Letters. While most stories of this kind are irrevocably lost, in some instances, the protagonists succeeded long enough for a substantial record to be left behind. Such documentation provides an unusual perspective on the practices of Enlightenment science and the role of women philosophers in publicizing and disseminating new ideas about nature. It offers a microcosm of all the hopes, dreams, and aspirations pinned to the mastery of knowledge in the eighteenth century, in a small corner of a region that was increasingly self-conscious about its marginal status in enlightened Europe.

This essay examines the circumstances surrounding just one case: that of Cristina Roccati (1732–97) of Rovigo (fig. 10.1). Roccati enjoyed the limelight briefly, in the 1750s, and then retreated into virtual obscurity for the rest of her life. Yet it is perhaps this confinement to a small town in the Veneto that allows us to bring her fully back to life, for the academicians of Rovigo lovingly saved almost every scrap of paper that pertained to her existence. The profusion of letters, poems, lecture notes, library catalogs, and academy minutes that recount Roccati's intellectual activities may make her the best-documented female natural philosopher of the eighteenth century. She was certainly one of the most prolific in terms of sheer intellectual production, albeit mostly unpublished. Following Roccati's career allows us to explore the position of women in Enlightenment intellectual culture, at least as it was practiced in Italy, and to pose some basic questions: What did Roccati want out of her career as a professor of physics? What did others expect of her? Bassi taught at the prestigious University of Bologna and the Bolognese Institute for Sciences, Italy's equivalent of the Royal Society and Paris Academy of Sciences. By contrast, Roccati spent at least twenty-six years of her life (1751–77) instructing her fellow citizens in Rovigo in physics as holder of the chair in that subject at the Accademia dei Concordi, the main scientific and literary academy of the town. Following her intellectual development, from Rovigo to the Universities of Bologna and Padua, and her career as the sole female member of a provincial Venetian academy, we can learn something about the possibilities for and limitations placed upon women who wished to engage in scientific activities and their role in provincial science.

Educating a Daughter

In 1751—one year after the Milanese mathematician Maria Gaetana Agnesi had refused the offer of an honorary chair in the mathematics at the University of Bologna—Cristina Roccati became the third woman to receive a university degree in Italy. Following in the wake of her more celebrated contemporary Bassi, who took a degree in philosophy in the year of Roccati's birth, she was the second woman to receive a degree from the University of Bologna and the first "foreigner" (non-Bolognese) to do so. Roccati's appearance signaled a new phase in the evolution of women scholars, who had initially been nurtured within the immediate environment of the most venerable universities. By the 1740s, news that exceptional women might master experimental physics and attain a degree in philosophy had traveled from these urban centers to smaller cities and towns, leading one Rovigo family to send their daughter away from home for the sole purpose of acquiring knowledge.

In all respects, Roccati fulfilled the ideal that a learned woman should be a true prodigy. Piscopia had received her degree in her thirties, and Bassi in her twenties, but Roccati was not yet nineteen at the time her doctorate was awarded. She was not only versed in the standard languages that learned Italians of her day cultivated—Greek, Latin, and French—but also knowledgeable about virtually every aspect of philosophy, ancient as well as modern. Bassi, who accompanied her to the degree ceremony, must have been somewhat awed by this young woman whose education far surpassed her own in terms of its breadth and depth. This distinction, of course, was not simply coincidental; contemporaries actively fostered the comparison, as when Jacopo Biancano called Roccati "another Laura" in 1749.[11] During the decade preceding her degree, Roccati had been subjected to an intensive education designed specifically to make her the most learned woman in Enlightenment Italy, if not of all time. She was to take her place at the head of a select group of learned women who entered the universities and academies during the late seventeenth and early eighteenth centuries, and whose ability to explicate Cartesian and Newtonian natural philosophy to a broader public signaled the arrival of enlightened thought in Italy. At least this was her family's hope when they agreed to invest in her education.

Cristina Roccati was the eldest daughter of a prominent patrician, Giovan Battista Roccati, and his wife, Antonia Campo. The Roccati were not themselves noble but married into nobility: Giovan Battista's mother came from an old family, the Zangirolami, and his wife belonged to one of the most prominent families in the city.[12] By all accounts, Roccati was an ambitious man, eager to make a name for himself in the Republic of Letters. He soon realized that his eldest daughter, Cristina, rather than his son, Alessandro, provided him with the means of realizing this goal. Educating sons was an important but utterly mundane part of ensuring a family's success; like his father before him, Alessandro married a noblewoman. However, educating a daughter was a more unusual thing to do. "[T]hose men are rarest who employ their women in the study of sciences," wrote Paolo Mattia Doria in 1716. At the encouragement of her father, Cristina buried herself in books in the family library, "making learning and erudition her principal indeed only goal."[13]

Numerous moralists in the late seventeenth and early eighteenth centuries counseled fathers to pay attention to the education of daughters, but few recommended that women be taught scientific and philosophical materials. Good reading and writing skills in Italian, a certain knowledge of the Bible and

---

11. Biancano to Roccati, October 1749, in Grotto, *Delle lodi,* 49.

12. For a discussion of the noble families of Rovigo, see Mazzetti and Zerbinati, *Le "iscrizioni" di Rovigo.*

13. Doria, *Ragionamenti,* n.p.; Girolamo Silvestri, "Breve elogio della Sig[no]ra Dottoressa Cristina Roccati" (10 January 1755), in *Memorie appartenenti alla storia topografica e letterarie del Polesine, e della città di Rovigo,* Conc. Ms.19, 200, Biblioteca dell'Accademia dei Concordi, Rovigo (hereafter BACR).

sacred stories, a smattering of arithmetic and inheritance law, and possibly a little Latin and morally uplifting literature was more than enough. "Thus we can dispense with some of the most difficult branches of knowledge," wrote Fénelon in his influential *Treatise on the Education of Girls* (1687), translated into Italian in 1748 and widely read in Italy. "Even the majority of the mechanical arts are not suitable to them." [14] Educating daughters was a means of ensuring that future mothers would be capable of guiding the intellectual and moral development of their children; it was not intended to be an end unto itself.

However, Giovan Battista Roccati had other things in mind. Like a number of ambitious fathers, he encouraged his daughter to study progressively more difficult subjects in order to distinguish her from the common ranks of educated women, whose learning stopped precisely where hers began. Noticing Cristina's precocious interest in reading as a child, he hired a tutor, the cleric and literary dilettante Pietro Bertaglia, to educate his daughter. Left to his own devices, Bertaglia would have given her precisely the sort of education that Fénelon recommended. By her early teens, Roccati had mastered these skills; she wrote poetry in imitation of Tibullus and had begun to have a certain success in the local literary salons. Afraid that his daughter had taken too ordinary a course, Roccati intervened. "But the Father, who aimed for greater things, wanted [Bertaglia] to give her the first rudiments of Greek as well." [15] Very quickly, Cristina exhausted the limits of her tutor's learning.

Studying Greek, Italian, and Latin prepared Cristina for her father's most cherished ambition: to see his daughter take a university degree in Bologna. While Padua might have seemed a more obvious choice, given its proximity to Rovigo, it no longer gave degrees to women. Immediately after awarding Piscopia a degree in philosophy in 1678, the rectors of the university passed a law barring women from studying there. [16] By contrast, Bologna positively encouraged women to come. Under the patronage of Benedict XIV, a handful of women took degrees and received professorships. Sending Cristina to study in Bologna would enhance "her honor and his own reputation, in addition to the dignity of the *Patria*." On 25 September 1747, fifteen-year-old Cristina Roccati,

---

14. Barnard, *Fénelon on Education*, 2.

15. *Elogio dell Sig.a Dottoressa Cristina Roccati di Rovigo* (3 January 1755), Silv. Ms. 369, fol. 2r, BACR. For background information, see also Conc. Ms. 9, n. 24, fols. 2–3, BACR. The secondary literature on Roccati is fairly old and mostly local but is as follows: Grotto, *Delle lodi;* De Vit, "Sulla vita e sulle opere"; Cessi, "Una dottoressa rodigina"; Rossaro, "Cristina Roccati"; and most recently, Savaris, *Cristina Roccati.* I would like to thank Clelia Pighetti for bring the latter to my attention and Paola Savaris for providing me with a copy of it.

16. On 7 February 1678, in response to the request by one of the professors, Charles Patin, that his daughter be allowed to take a degree, the rectors of the University of Padua determined "that they should not admit women of whatever condition to the doctoral degree" (Maschietto, *Elena Cornaro Piscopia*, 134).

accompanied by her aunt and her tutor, made the long journey by carriage to Bologna to enroll in the Faculty of Arts. As the records of the university note, on 9 November 1747: "Roccati. The Most Illustrious Lady Cristina of Rovigo, Student of Logic under the care of Reverend Father Doctor Bonifacio Collina . . . matriculated." [17] What made this event particularly remarkable was its utter normality. No previous female prodigy had ever enrolled in classes, as if she were a male student.

For the next three years, Cristina rarely returned home. She was intensely occupied with her studies, instructed by some of the clerics who occupied chairs at the university. Under the direction of Father Collina (1689–1770), abbot of the monastery of Saints Cosmas and Damian and professor of philosophy since 1722, she studied the traditional philosophy curriculum—logic, metaphysics, and physics—and Cartesian mathematics and philosophy. Indeed, her friend and biographer Girolamo Silvestri (1728–88) remarked that during that period Cartesianism was "still followed by many in Bologna." Such an observation suggests that Roccati did not enjoy many of the fruits of the Newtonian curriculum in sciences then being developed by such professors as Bassi. Yet she broadly mastered the various forms of modern knowledge. Giovan Angelo Brunelli, professor of geometry at the university and an assistant at the Institute for Science's famed observatory, introduced her to astronomy and meteorology, while Pierre Vert of Lyons instructed her in French. [18] In their lodgings, Bertaglia continued to tutor her.

Roccati did not simply take lessons in her private quarters, however. She participated fully, and almost without comment on the part of Bolognese scholars, in the intellectual life of the city. Accompanied for decorum's sake, by her aunt and perhaps Bertaglia, she attended university lectures, private lessons given by faculty, dissections at the famed anatomical theater, and used the facilities at the astronomical observatory. The local nobles who arbitrated the cultural life of Bologna welcomed her to their academies and salons. She even attended classes at the Institute for Sciences "to hear the learned lessons of those professors, and to observe the physics experiments." In participating in the academy culture of science, Roccati already surpassed Bassi, who prior to her degree had studied only at home. It seems the climate had changed by the late 1740s, allowing a properly accompanied woman the freedom to study in

17. Silv. Ms. 369, fol. 2v, BACR; *Università degli Scolari, Artisti, Registri Matricolari* (9 November 1747), Studio. 377b, Archivio di Stato, Bologna (hereafter ASB).

18. Roccati's studies follow the pattern observed by Heilbron, *Elements of Early Modern Physics*, 137. On Collina's life, see Fantuzzi, *Notizie*, 3:195–97; Conc. Ms. 19, 203, BACR. The folder on Brunelli suggests that he was quite a well-respected teacher. Francesco Maria Zanotti made a point of saying that he sent his students to him. See *Assunteria di Studio, Requisiti dei Lettori*, vol.4, n. 26 (Giovanni Angelo Brunelli), ASB.

public. Or perhaps the Bolognese were less concerned about what a Rodigina did in public, since her honor and reputation, as a foreigner, were of less immediate concern. Certainly her fellow students did not seem to worry overmuch about her presence. In April 1749, they unanimously elected her head of the "Venetian nation," with the proviso that she not exercise her voting rights.[19] When Roccati did intervene in the June elections that year, it led to some procedural problems but ultimately did not become a cause célèbre.

The idea of sending a daughter away from home to study might seem unremarkable to the late twentieth century. But in eighteenth-century Italy, it was virtually unthinkable, unless one confined one's daughters to a monastery. Despite the historic tales of women who taught and attended university lectures behind the screen since the late Middle Ages, such images were more myth than reality; few fathers were willing to risk family honor for the sake of learning. Thus, in an important sense, the very publicity of Roccati's education made it unusual, and distinguished her from other learned women of her generation. By the time she took her degree in May 1751, her fame had spread beyond Bologna.

To many contemporaries, Roccati could have stepped directly out of the pages of Giovan Nicolò Bandiera's *Treatise on Studies for Women* (1740). Since her father acquired this book for the family library sometime before 1749, this conceit had some foundation.[20] In fact, the number of well-known works concerning the education of women in the Roccati library—for instance, *Academic Discourses by Various Authors on the Studies of Women* (1724), the product of a famous debate held at the Accademia dei Ricovrati in Padua in 1723, and Pierre LeMoyne's *Gallery of Strong Women* (1647)—suggests that Giovan Battista availed himself of the best resources of his age when he chose to educate his daughter. Advising readers that educating women would not precipitate "the great overthrowal of the Republic," Bandiera encouraged fathers to educate their daughters to the level of ancient women whose learning brought fame to their families.[21]

---

19. Silv. Ms. 369, fol. 3v, also Conc. Ms. 19, 203, BACR. Early modern universities were organized into "nations" that reflected the regional origins of students. See *Università degli Scolari, Artisti, Registri degli Atti dell'Università,* 1744–52 (28 April 1749), Studio. 395, ASB: "Cui urdet et placet D.na Christina Roccati Nob. Rodigina dercriti in Albo DD. consiliariorum omnibus honoraribus . . . , interdictu tamen interventu ad Anniversaris Pellini, et hoc p. sexu disparitate, et cu[m] conditione quod nullam possit, aut debeat jus habere ad activae dict[a]e Universitatis Prioratus offituis, neque ad gnatem SS. Prioru[m] extractione intervenire, det vobu[s] altu[s] affirmativus, cui non nissum negativum." The 28 June 1749 entry describes how "D. Christina Roccati intervenit Univ. pro ellectione Bidelli" and the controversy it caused. See De Vit, "Sulla vita e sulle opere," 56.

20. *Index librorum latinorum Joa. Baptistae Roccatii Rhodigini* (1749), Silv. Ms. 381, 91, BACR.

21. Bandiera, *Trattato,* 1:349; also 1:373–74, 2:34.

In two volumes, Bandiera offered a comprehensive plan of study for the modern woman. She was to be educated in the classics, among them poets such as Tibullus and Virgil, both of whom Roccati studied, imitated, and translated. She was to shun the popular taste for French romances, comedies, and dialogues, reading only chaste works like Fénelon's *Telemachus* (1717), one of the few French books Roccati owned.[22] Similarly, she should be encouraged to read the histories of excellent and noble women in order to have models for such virtues as chastity, prudence, constancy, and piety. The numerous examples of such histories in Roccati's family library signified more than a fashionable interest, however; in 1749, when the Society of Varied Literature in Pistoia invited Cristina to send in a dissertation as part of her candidacy for membership, she chose the history of learned Hebrew, Greek, and Roman women as her subject.[23]

Unlike Fénelon, Bandiera did not exempt women from the study of philosophy. They should proceed through the traditional curriculum as Roccati did—logic and then metaphysics—before turning to physics. There the female student should be offered a "most useful mixture" of "this ancient and new physics." In Bandiera's definition, physics was no longer simply the core of the Aristotelian natural philosophy curriculum, supplemented by oblique references to the ideas of Copernicus and Galilean mechanics; it was the centerpiece of the Cartesian and Newtonian synthesis. By the 1740s, Italians had access to the full range of ideas embodied in Newtonian natural philosophy, even if they had not yet wholeheartedly embraced them. University textbooks were being written on this subject, and Algarotti, though embroiled in controversy with the Catholic Church, was nonetheless read.[24] Thus, Roccati began her studies at a very auspicious moment. Modern editions of Galileo were being published, the writings of Descartes and Newton circulated readily within Italian cities, and Benedict XIV indicated every desire to make Italy a region that not only embraced but added to the modern philosophies of

---

22. Ibid., 2:2; also 2:26, 43. Roccati owned a 1740 edition of *Télémaque* (Silv. Ms. 381, 109–10, BACR). For the importance of this book as probably the most popular, morally edifying novel of its time, see Barnard, *Fénelon on Education,* xxxvii–xlvii; Reiss, *Meaning of Literature,* 186, 232, 235. An unauthorized version of this work appeared in print in 1699, but Fénelon did not publish the complete edition until 1717; hence the later date.

23. Bandiera, *Trattato,* 2:239–41, 253. In addition to LeMoyne's *Galleria di donne forte,* Italian trans. (Modena, 1701), the Roccati library included a *Memorie del Regno di Catterina Imperadrice della Russia* (Venice, 1730) and the *Nobiltà et eccellenza delle Donne: Opera tradotta dal Francese con una Orazione di M. Alessandro Piccolomini in lode delle medesime* (Venice, 1549); Silv. Ms. 381, 77, 79, BACR.

24. Bandiera, *Trattato,* 2:167. See Ferrone, *Intellectual Roots,* esp. 262. For a general overview of the development of physics as an early modern discipline, see Heilbron, *Elements of Early Modern Physics,* 1–88.

nature. No father who educated his daughter well could afford to ignore these developments.

Roccati's correspondence with her friend and confidant Girolamo Silvestri during her studies in Bologna demonstrates the extent to which her philosophical education reflected the mixing of old and new that characterized eighteenth-century Italian intellectual life. Silvestri, a young noble with scholarly inclinations, first met Roccati under the tutelage of Bertaglia in Rovigo. He introduced her to the literary circles of the town and eventually to the Accademia dei Concordi in 1747. By the time Roccati departed for Bologna, they were close companions; Silvestri had become a local patron for Roccati and also undertook the education of her brother, Alessandro. Better educated than either her father or her tutor, he acted as an intellectual mentor, guiding her reading and supplying her with books and references during her stay in Bologna.

During Roccati's three years in Bologna, Silvestri repeatedly inquired about the progress of her studies, corrected her translations of Latin poetry, and acquired the books necessary for her to complete her treatise on ancient learned women. Roccati responded by thanking him for forming her "imbecile and weak mind." [25] Their correspondence provides important clues regarding Roccati's intellectual formation. In February 1748, she wrote that she was reading the French theologian Malebranche and looking for a Greek tutor. Toward the end of that first year, she sent another progress report: "I report to you that I have already finished the entire study of logic, stars and meteors, and I continue diligently that of languages." In 1749, she described the structure of the course in metaphysics. By January 1750, immersed in the preparation of her treatises for the Pistoia academy, she reported happily that she had found examples of "many erudite and learned Women" in her reading of ancient literature. When she did not write as frequently as Silvestri wished, she blamed it on the "fault of most serious studies." [26]

Course by course, Roccati gradually worked her way through the university curriculum in natural philosophy, mastering Aristotelian physics, Euclidean and Cartesian geometry, and discovering the complexities of the new syntheses, such as the work of the Malebranche, that attempted to harmonize Cartesian, Leibnizian, and Newtonian thought with Christianity. This approach appealed particularly to her own intellectual inclinations, which had been shaped by early encounters with such works as Muratori's *Moral Philosophy Espoused and Proposed to Youth* (1735) that praised French moral philoso-

25. Roccati to Silvestri, Bologna, 4 December 1749, in Grotto, *Delle lodi*, 40.
26. Rossaro, "Cristina Roccati," 17; Silv. Ms. 195, n. 2 (Bologna, 4 February 1748); n. 20 (Bologna, 8 January 1750); n. 3 (Bologna, July 1748), BACR.

phy and natural theology as a model for Catholic intellectuals to follow.[27] Simultaneously, Roccati discovered a love for mathematics and mechanistic philosophy that would shape her future work as a professor of physics in Rovigo. At the beginning of her third year of study, she confessed her delight in geometry—a subject that increasingly took her away from more literary pursuits—and her eagerness to begin the study of physics.[28] Gradually, Roccati realized that her vocation lay in science.

As Roccati's intellectual development suggests, the period of her life spent in Bologna was fundamental to her ability to participate in debates about modern knowledge. Bologna enjoyed the reputation of being one of the Italian cities that was most receptive to foreign ideas. While her father sent her to Bologna to enhance the family's reputation and to win fame for Rovigo, Roccati discovered a personal passion for learning that transcended any sense of familial and civic obligation. With great conviction, she reported to Silvestri: "Certainly I confess—nor am I ashamed to say it—my desire that that day, that hour when I must leave Bologna, a city where studies flourish, may never come."[29] Rovigo could have the honor, Roccati implied, but she wished to remain abroad in order to continue her participation in the learned world. The small-town prodigy had indeed become a *filosofessa*.

By summer 1750, Roccati's renown had gained her admission to many leading Italian academies, including the Accademia dei Concordi in 1749, and her period of study was at an end.[30] Her father decided that her public thesis defense should be held in Rovigo to increase her fame in the Veneto. From his perspective, this event was a triumphal conclusion to his lengthy and expensive investment. Invitations were sent far and wide, and professors imported from Bologna, Brescia, Vicenza, and Venice to judge the merits of her theses. Giovan Battista even succeeded in having his daughter's theses printed in the *Literary News,* assuring that the Republic of Letters would take note of the event (fig. 10.2). One wonders what the citizens of Rovigo thought of a woman proclaiming in Latin in their main church: "We condemn and refute the opinion of Descartes doubting everything. For to doubt everything is useless, indeed

27. Roccati owned the 1746 Venetian edition of this work (Silv. Ms. 381, 69). For its impact on natural philosophy, see Ferrone, *Intellectual Roots,* 176–79.

28. Silv. Ms. 195, n. 15 (Idus October 1749); n. 16 (4 November 1749).

29. Ibid., n. 12 (Bologna, 17 June 1749).

30. Roccati was admitted to the Accademia degli Apatisti in Florence (8 June 1749), the Accademia di Varia Letteratura in Pistoia (7 August 1749), the Accademia dei Concordi in Rovigo (30 December 1749), the Accademia dell'Arcadia (1 June 1750, under the name "Aganice Aretusiana"), the Accademia degli Ardenti in Bologna (24 May 1752), the Accademia de' Ricovrati in Padua (25 August 1753), and the Accademia degli Agiati in Rovereto (20 January 1754, under the name of "Artisia"). Notably the Academy of the Institute for Sciences in Bologna did not offer her admission, as they had to Bassi, Faustina Pignatelli, and Châtelet.

Nella Città di Rovigo il dì 4. di questo Mese nella Chiesa della B. Vergine del Soccorso l'animosa ed erudita Donzella Cristina Roccati *Rodigina,* Figlia del Sig. Giambattista, *e della* Nobil Signora Antonia Campo, *con insolito valore e dottrina sostenne pubblica Conclusione di Filosofia, avendo con sommo applauso e virtù, per lo spazio di due ore continue, illustrate e difese magistralmente* le IV. *seguenti Tesi, due prese dalla Logica, e Fisica, ed altre due dalla Metafisica:*

(1) Cartefii de omnibus dubitantis fententiam improbamus & confutamus. Hæc enim de omnibus dubitatio inutilis eft, immo periculofa.

(2) Phyfica, quæ corporum fcientia eft, noftra quidem & recentiorum fententia, ut formam fubftantialem explicet, in corporibus vitæ expertibus non adfert nifi partium harmoniam, ex qua proprietates evidentiffime diverfæ oriantur.

(3) Univerfæ Philofophiæ præftantior pars, quæ Metaphyfica dicitur, Ens neceffarium exiftere, quod *Deum* vocamus, invictis rationibus demonftrat.

(4) Non ex fola fide, fed ratione etiam naturali conftare *Animæ immortalitatem* ita nobis perfuafum eft, ut hominem ratione fua rite utentem immortalitatem hanc latere non poffe, libentiffime definiamus.

Il P. D. *Bonifazio* Collina, Pubblico Professore di Filosofia *nell'* Università di Bologna, *e che nel Mondo letterario vive celebre per molte sue opere stampate, fu l'* Assistente *della nostra concittadina Giovine* Roccati, *la quale sotto un tanto Duce e Maestro otterrà, come si spera, la Laurea Dottorale in Bologna, e con ciò accrescerà alla Patria, al Bel-Sesso, e alle Madame Letterate del nostro Secolo, nuovo fregio, ornamento, e splendore.*

Fig. 10.2.   Roccati's four doctoral theses, defended in the Church of the Blessed Virgin, Rovigo, 4 August 1750. From *Novelle letterarie,* 7 September 1750, 288.

perilous." A rather standard definition of physics as the "science of bodies" followed, with two concluding theses on metaphysics as a demonstrable proof of God's existence and an argument in favor of the immortality of the soul "not only by faith but indeed by natural reason." Such arguments in favor of Christian natural philosophy and rational theology surely must have warmed the hearts of the clerics sitting in church that day, even if they puzzled and probably bored the majority of the listeners, who knew little Latin and had never read Malebranche. Yet this did not prevent the noblewomen of the city from waiting on Cristina throughout the ceremony, a social concession usually reserved only for royalty.[31]

After the successful defense of these four theses on 4 August 1750, Roccati returned to Bologna to prepare for the degree ceremonies. Her letters to Silvestri during this period suggest the pressures that she was under to perform well. "I understand that this carnival the academy won't meet," she wrote on 16 February 1751, "which gives me great pleasure because certainly I have no time to tend to the Muses, my entire soul being busy with philosophical things." The leading science professors in Bologna reexamined Roccati on 27 April 1751 and awarded her a degree on 5 May. As they had done with Bassi twenty years earlier, the Bolognese academicians emphasized the uniqueness of the privilege accorded to a woman. "She entered with the most Illustrious and Excellent Lady Laura Caterina Bassi Veratti, our honorary Collegian," recorded the secretary of the College of Medicine and the Arts, "and no other Women were admitted without urgent necessity."[32] Apparently Bassi had already visited Roccati to congratulate her on continuing the lineage of learned women in Bologna. One wonders what these two women said about the meaning of such events in their lives. Did Bassi offer advice on the problems of being a female natural philosopher and warn Roccati that fame was a double-edged sword? Since what transpired never appeared in print, we will never know.

At the same time, male patrons of female learning rushed to celebrate this latest testament to the intellectual capacities of women. Medoro Rossi had written in the *Literary News* that Cristina's degree would bring "new decoration, ornament and splendor to her *Patria,* her Sex and the Learned Ladies of our Century." Such sentiments found informal expression in the correspondence that accompanied poems written in celebration of her degree. In May 1751, the publisher Giovan Antonio Volpi wrote anxiously to the secretary of the Institute for Sciences, Francesco Maria Zanotti, to inquire whether his "sonnet in praise of Signora Roc[c]ati of Rovigo" had arrived. "As for me, I am pleased that she comported herself well in the Doctorate, since I have not had the fortune to know her except by reputation: thus my effort will not be wasted."[33] Cristina's accomplishments were known and judged by the larger intellectual community that chose to participate in them to the extent that they became meaningful to the public presentation of knowledge.

31. See Medoro Rossi's article in *Novelle letterarie* (7 September 1750), 288; Lazzaro, *Rovigo,* 54.

32. Roccati to Silvestri, Bologna, 16 February 1751, Silv. Ms. 195, BACR; *Collegio di Medicina e d'Arti, Second Libro Segreto di Filosofi,* 1712–1800, c. 80v (27 April 1751), Studio. 228, ASB.

33. Rossi, *Novelle letterarie,* 288; Volpi to Zanotti, Padua, 27 May 1751, Ms. B.160, Biblioteca Comunale dell'Archiginnasio, Bologna.

Her successes elevated the Roccati family in the estimation of their superiors in Venice, giving them an intellectual nobility that yielded tangible benefits for their social status. Nobles who might not normally have sought out their company now paid their compliments. Writing to her kinsman, Count Antonio Lando in May 1750, the Venetian patrician Pietro Gradenigo commented on the glory that Cristina had brought to her city and her family. "I wish the Father and Daughter every happiness, so that those rare events that always render a city more famous may flourish in [our] *Patria,* when Girls, making good use of their own talents, are able to sustain public Conclusions and are admitted to the most celebrated Academies in Europe."[34] Certainly Giovan Battista had succeeded in bringing to fruition the plan that he envisioned in the early 1740s, when he encouraged his daughter's scholarly inclinations. Roccati was now one of the most educated women in Europe, ready to bring fame to Rovigo.

Honor and Shame in the Republic of Letters

What sort of world did Roccati enter upon receiving a university degree? For at least a month or so, she was too busy to consider what prospects the future held. Yet after the visitors who traveled from all over Italy to see her thesis defense had departed from Rovigo, and after she had finally said her farewells to Bologna, she found herself in a very limited and claustrophobic environment. Roccati's fond memories of her time in Bologna certainly did not predispose her to return to a sleepy agricultural town of approximately 5,500 inhabitants, where the sole intellectual life resided in an academy, run by her kinsmen and their close friends, that only begrudgingly acknowledged her accomplishments. From the start, there were tensions between Roccati and the other local academicians, many of whom viewed her fame with suspicion if not outright scorn. Detractors grumbled that her admission had simply been a favor to her friends and family who ran the academy.

Despite Silvestri's personal enthusiasm for her studies—he traveled all the way to Bologna to see her degree ceremonies in 1751—the local intellectual elite were decidedly ambivalent about their most famous citizen. Although the members of the Accademia dei Concordi were aware of Roccati's talent long before she became known elsewhere in Italy, they did not admit her until more prominent academies such as the Accademia dell'Arcadia produced offers of membership. Roccati expressed her exasperation at the academy in private correspondence. She was aware that its members, pleading poverty, had delayed

34. Gradenigo to Lando, Venice, 30 May 1750, in Grotto, *Delle lodi,* 64.

publishing a set of poems for her degree. Their lack of interest in celebrating her accomplishments, she wrote, would tell the Bolognese nobles and professors "that they know nothing or think nothing of me." The official records of the academy state that it was for lack of time, hardly believable since they had almost a year to prepare for the event. Angrily Roccati wrote to Silvestri: "My Signor, you may be certain that I do not say this because I believe myself worthy of such an honor, since God has deprived me of this ambition. . . . They say that they are pleased, which is something I do not doubt. But foreigners don't know their internal sentiments but only certify them by external actions and demonstrations. Enough said!"[35] In the end, they shamed the academicians into producing a few poems honoring her degree, but this did little to alleviate strained relations.

Returning to Rovigo in the early 1750s, Roccati found its intellectual climate particularly unpromising. Despite the growing sense that women had a place in the learned world, this was a conclusion less easily drawn in a small town with few cultural venues. With her father's blessing, Roccati resolved to go to Padua in order to complete her education. "And because she greatly desired to study Newtonian physics, and Mathematics, she resolved to go to the University of Padua to apply herself more seriously there."[36] Continuing her tour of Italy's venerable universities, Roccati planned to conclude her studies of the most difficult and exacting fields of knowledge offered at the time in order to complete her education in ancient and modern physics.

By fall 1751, Roccati was in Padua. Her Bolognese professor Collina wished her "a good journey to Padua, where I know you will find the best pasture for your erudite genius." There she studied conic sections with Giovan Alberto Colombo, professor of mathematics. Roccati was so pleased with the interest Colombo took in her work that she wrote excitedly to Silvestri: "Here is the news: I have begun mathematics and physics and I am under a master who truly invites me to study hard by the good manners and graces with which he teaches." She also enjoyed the company of Padua's most well known physics professor, Giovanni Poleni, and the botanist Giulio Pontadera; she began Hebrew. Roccati was having such a good time that she even forgot to write her father, who frantically sought out Colombo to inquire about his daughter. "You wish your most worthy daughter well," wrote Colombo in February 1752, "and, on my part, rest assured of a caring, indefatigable and well pledged assistance." When Roccati returned to Rovigo in summer 1752, she

35. Silv. Ms. 195, n. 32 (Bologna, 16 February 1751); *Degli Atti Accademici de' Signori Concordi di Rovigo*, 1697–1772 (7 April 1751), Conc. Ms. 27, n. 2, 29, BACR.
36. Silv. Ms. 369, fol. 5v; see also Conc. Ms. 19, 208.

had every expectation that she would spend the next few years studying modern physics and ancient languages at "Il Bò."[37]

Under the instruction of the Paduan professors, Roccati's curiosity about nature and her skill in interpreting it only increased. Her letters home were filled with reports of scientific novelties. When she returned to Rovigo between terms, letters from Colombo followed, encouraging her to complete her reading of Euclid's *Elements* and the Dutch physicist Pieter van Musschenbroek's *Mathematical Elements of Experimental Physics* (1720–21), both of which Roccati apparently mastered with little difficulty. Colombo encouraged her to continue these studies. "I am pleased, from my soul, that nothing defeated you while you were reading through the elements of Algebra," he wrote in August 1752. "I understand that easily from there you will go on to the *Institutions* of Agnesi where they abound, since they seem agreeable to you and easy enough."[38] Based on his pupil's facility for learning, Colombo predicted that she would enjoy "fame throughout the world." In fact, given the rumors that circulated about the idea of persuading Bassi to move to Padua when Poleni died in 1761, it is not unreasonable to speculate that Roccati was well positioned to become Padua's first woman professor.

Fame came most swiftly via publication. Roccati began to entertain the idea of writing a treatise or two on various scientific subjects. After narrowly escaping being struck by lightning, she became interested enough in the phenomenon "to make some important observations on the modern opinion of the Origins of Thunderbolts" and sent her treatise on this subject to Colombo in Padua.[39] The subject was neither a casual nor accidental choice. By the 1740s, electricity had become the most popular subject in the physics curriculum, spilling out beyond the halls of academe to intrigue a much broader audience in the mysteries of nature that experimental physics promised to unlock. Toward the end of that decade, electricity had become popular enough in the Veneto to inspire a local physician, Sguario, to write a dialogue about the pleasures of electrical experimentation that featured a learned female protagonist. That same year, the Veronese antiquary Scipione Maffei published *On the*

37. In Grotto, *Delle lodi,* see Jacopus Blancanus to Roccati, Bologna[?], October 1749, 49; Bonifacio Collina to Roccati, Bologna, 23 November 1751, 69; Roccati to Girolamo Silvestri, Padua, 2 May 1752, 46; and Colombo to Giovan Battista Roccati, Padua, 14 February 1751, 74; also Silv. Ms. 195, n. 36 (Padua, 29 November 1751), BACR. On science in Padua prior to Roccati's arrival, see Ferrone, *Intellectual Roots,* and Dooley, "Science Teaching."

38. Colombo to Roccati, Padua, 6 August 1752, in Grotto, *Delle lodi,* 78; also 75–76. All of the books on Roccati's summer reading list appear in the family library: she owned the Naples 1744 edition of Euclid, Antonio Genovesi's famous 1745 edition of Musschenbroek (also published in Naples), and Agnesi (Silv. Ms. 381, 14–15, 73, BACR).

39. Conc. 19, 209, BACR. In the annals of the Concordi, the loss of Roccati's treatise critiquing Maffei is noted: Conc. 163, 1:283.

Formation of Thunderbolts (1747). In a series of letters addressed to such luminaries as Poleni, Antonio Vallisneri, Apostolo Zeno, and the French naturalist Réaumur, Maffei argued that lightning was a material entity with no supernatural agent behind it. His conclusive proof regarded the trajectory of lightning, which always moved from the earth to the sky and never the reverse.[40] Lightning, in short, was very much on the minds of local *letterati*, a fact that Cristina surely must have known when she put pen to paper.

Unfortunately Roccati's treatise on thunderbolts, no doubt an attempt to correct some of the materialist implications of Maffei's hypothesis, has not resurfaced since she first wrote it, and it never found its way into print. Yet the act of writing this work speaks volumes about this female philosopher's ambitions in the early 1750s. Maffei was one of the most revered intellectual figures of mid–eighteenth century Italy. Through his editorial role in the Venetian *Giornale de' letterati,* which appeared from 1710 to 1740, his steady stream of publications, voluminous correspondence, and political presence as adviser to various rulers, Maffei enjoyed enormous prestige in his society. What are we to make of a twenty-year-old scholar, degree fresh in hand, who dared to take on one of the aging monarchs of the Italian Republic of Letters? Maffei was displeased enough when some criticism of his treatise appeared in a Venetian journal to complain to his friend Giovanni Lami, editor of the *Literary News* in Florence.[41] He was enormously proud of this work, which encompassed some thirty-five years of observations. Admittedly, science was hardly Maffei's forte, so perhaps he was an easy target. More to the point, he was a highly visible adversary, which was precisely what Cristina needed to take the next step in her entry into the world of scholarship.

What Roccati had not counted on was the dishonor of her family. "[B]ut the most baleful misfortunes prevented these plans," commented Silvestri tersely in his biography. In August 1752, her father fled town, one of a number of conservators of the Sacro Monte di Pietà accused of pilfering the coffers. Some historians have suggested that Giovan Battista imprudently loaned himself money to finance the exorbitant cost of his daughter's education, particularly the library that he acquired rapidly between 1749 and 1752—a speculation that seems consistent with his behavior prior to the financial scandal. In exile for two years, Roccati's father eventually returned to Rovigo, shamed and dishonored, only to die a few months later in 1754. With their goods confiscated, the family was left in virtual penury. Thanks to the intervention of Silvestri, who by the time of his own death had a library of more than thirty-

40. Maffei, *Della formazione de' fulmini.* For more on the emergence of electricity as a science, see Heilbron, *Electricity.*

41. Maffei to Lami, Verona, 15 January 1750, in *Epistolario,* 2:1266.

five thousand volumes (the core of the Accademia dei Concordi's holdings today), Roccati, her aunt, her mother, and her siblings were provided with an income derived from the sale of the library. In June 1758, Silvestri paid 4,165 *lire* for most of the books. As if to ensure that no one might ever question the propriety of this sale and the family's role in making this difficult decision, Silvestri had his friend sign at the bottom of each page of the inventory: "I, Cristina Roccati, affirm this."[42]

The family was saved from financial ruin, but what had happened to their status and ambitions? The death of her father, Silvestri remarked, "took away her hope of returning again to the way it was before." Certainly Roccati could no longer study in Padua; it was not even clear whether it would be proper for her and her kinswomen to leave their home, for they were in virtual mourning. Confined to the family palace, Roccati was deprived of the one talisman of her previous existence that might have given her some small comfort: the family library. Containing precious first editions of classical authors, numerous dictionaries, a wide range of scientific and medical texts, as well as locally written works such as the many volumes by Scipione Maffei, not to mention Ludovico Muratori's *Rerum Italicum Scriptores* (which Giovan Battista had acquired in Bologna for his daughter without ever paying the bill), it was the working library of an Enlightenment scholar well versed in ancient and modern authors.[43]

But little of this mattered anymore. In fact, the sale catalog revealed what is perhaps the greatest irony of this story: while Cristina was probably the most educated woman in Italy, her younger sister Marianna was completely illiterate. Just beneath "Dottoressa Cristina Roccati," her former tutor Bertaglia wrote: "I, Doctor Pietro Bertaglia write in the name of Signora Marianna Roccati who, *not knowing how to write*, made the cross here † ." The contrasting levels of education between the youngest daughter and the oldest serve to remind us how little the idea of educating women played a role in Roccati's extraordinary intellectual career. As if to underscore this point, works such as Bandiera's *Treatise on Studies for Women* were included in the sale.[44] Without honor, what did female education signify?

42. Conc. Ms. 19, 210, BACR; Lazzaro, *Rovigo,* 55–58; Savaris, *Cristina Roccati,* 15; Silv. Ms. 369, fol. 7r, BACR. Noticeably few reports—contemporary or historical—mention the actual incident. Instead we are left with the impression that life was difficult because her father was "away" and then died. It is quite likely that Giovan Battista was pardoned because of his daughter's fame as well as his family connections to the nobility.

43. Conc. Ms. 19, 210. Unfortunately I don't yet have the archival citation for the second library catalog (1 June 1758). But I would like to thank Paolo Pezzuoli for kindly providing me with a copy of it.

44. Conc. Ms. 19, fol. 2v (sister's illiteracy; emphasis mine); fol. 3v (sale catalog).

Denied the sort of fame that Laura Bassi enjoyed through her abrupt removal from Italian university culture, Roccati instead made a career disseminating Newtonian natural philosophy in a Venetian backwater, becoming one of the stoic heroines whom she had read about in the family library. Undoubtedly at Silvestri's insistence, she kept those books from the family library that might best aid her in the writing of her annual academy lectures, for instance, her copies of Euclid, Agnesi, and Musschenbroek. For a time, Roccati continued to correspond with natural philosophers in Bologna and Padua, particularly her mentors Collina and Colombo. Both received copies of her critique of Maffei's treatise on lightning. Colombo encouraged Roccati to continue her studies and sent her the certificate of the Accademia dei Ricovrati in Padua, to which she had been admitted in 1753. He cajoled her not to forgot those authors whose work she had read so carefully the previous summer: "What! Well, what of Agnesi, what of Musschenbroek? Now they both sleep soundly!"[45]

Gradually, Roccati's contact with the world beyond Rovigo ceased. In her misfortune, Roccati found a different vocation and discovered a certain strength. "She nonetheless took it all with a strong spirit; and being already able to walk on her own, she planned to continue her studies, especially those which she loved above all others: modern philosophy and mathematics."[46] Rather than representing the city to the world, Roccati now began to contribute what she had learned to the betterment of the local patriciate. In her fame, Roccati had been difficult for the citizens of Rovigo to appreciate. Would they like her better as the disgraced but learned daughter of an untrustworthy public servant? After 1752, Roccati was no longer famous but infamous (at least in Rovigo). All she had left of her former glory was a skill—detailed knowledge of the development and practice of physics as an early modern discipline— that many of her contemporaries increasingly sought to acquire.

Newtonianism in the Provinces

The context in which Roccati displayed her knowledge of physics—the Accademia dei Concordi—had seemed to lead nowhere in the early 1750s when the academicians had refused to celebrate her degree properly. Yet intellectually the academy was the only place in Rovigo capable of accommodating Roccati's brand of learning. Fortunately, its members had aspirations almost as grand as Roccati's own. Participating in that quintessential eighteenth-century activity—the refounding of an academy to modernize its curriculum—they made mathematical physics a centerpiece of their intellectual ambitions. In doing so,

45. Colombo to Roccati, Padua, 8 August 1753, in Grotto, *Delle lodi*, 82; more generally, 75–83.
46. Conc. Ms. 19, 210, BACR.

they contributed to the second wave of institutional reform that transformed Italian academic life in the eighteenth century. In the late 1730s, universities such as Bologna and Padua had introduced chairs in experimental physics. Less ancient and venerable institutions such as Rome, Pisa, and Turin followed in their wake throughout the 1740s, when cabinets of experimental physics became generally popular.[47] Thus, the goals of the Rovigo academicians signaled how far experimental physics had penetrated the consciousness of the Italian elite by midcentury, when they chose to include this subject in the academy curriculum.

In 1751, the Accademia dei Concordi decided to reorganize, modeling itself along the lines of the Bolognese Institute for Sciences. Founded around 1580 by one of Roccati's kinsmen, Gaspare Campo, the academy ran intermittently in the first century and a half of its existence. By the late 1740s, it was a small but flourishing group, largely devoted to literary and antiquarian pursuits but committed to a program of aristocratic cultural edification. Yet the type of knowledge proper to an academy seemed increasingly open to debate. Certainly the Concordi's contact with the lively intellectual world of the Veneto, from which they drew many of their most distinguished members, gave them access to these new ideas. In 1747, the members discussed whether "*belles lettres* and the sciences or chivalrous arts" best suited noblemen. A year later, they wondered "if liberal or mechanical arts are more useful to cities."[48] Apparently Rovigo's intellectuals resolved that both ought to be taught. On 27 July 1751, they appointed fifteen of their members to public lectureships in their newly founded "Institute for Sciences." The topics suggested a desire to cover a range of traditional subjects that an academy might offer in the absence of a local university—two professorships in biblical and church history, three in law, others in logic, metaphysics, mathematics, and theoretical medicine. They also introduced humanistic topics (Italian, eloquence, poetry, "ancient erudition," and moral philosophy) and scientific subjects (geography and physics) that ventured outside the bounds of the traditional curriculum. Though not present at the meeting, Roccati was unanimously elected to the chair in physics.[49]

Roccati's election to this position preceded her father's disgrace. For the

47. Heilbron, *Elements of Early Modern Physics*, 137. The chair in Bologna, established in 1737, was not at the university proper but in the Institute for Sciences.

48. Pietropoli, *Accademia dei Concordi*, 94. Beginning in 1736, the Accademia dei Concordi gave diplomas to many leading intellectual figures in the Veneto, among them, the naturalist Antonio Vallisneri, Apostolo Zeno, and Giovanni Antonio Volpi (ibid., 100).

49. Conc. Ms. 163, BACR; G. B. Roberti, *Annali dell'Accademia dei Concordi*, 1:213–23 (statutes of Istituto delle Scienze); *Giornale dell'Accademia dei Signori Concordi, 1697–1793*, Conc. Ms. 250, 215 (Roccati's election to chair in physics). During the 1750s–80s the academy added such disciplines as anatomy, chemistry, astronomy, geometry, natural history, agriculture, and commerce, bringing the total number of professorships to thirty.

first year of her appointment, she sent her lectures by courier from Padua to Rovigo, where they were read aloud in her absence. Her celebrity only added to the importance of the subject in the academy's public assemblies; posters "affixed in the most convenient places in the city" announced the time and place of her lectures.[50] Family ruin produced no significant change in her academic obligations. Having made the appointment, the Rovigo academicians were determined to stick by it, in good times as well as bad. Perhaps they recognized that mathematical physicists were a rare commodity; as Giovan Battista Beccaria, one of the leading experimental physicists of his generation, told Roger Boscovich in 1762, "For twenty pure mathematicians, one is hardly able to find one mathematical physicist."[51] From this comment we can conclude that Roccati's training was unusual not only because of her sex.

In the wake of the family scandal, Roccati's responsibilities were enhanced, as if to reward her for being the virtuous and stalwart daughter of a fallen father, undone by his passion for female learning. On 7 August 1754, four months after Giovan Battista received his pardon, the members of the Concordi elected her "prince" of the academy. In 1755, following her father's death, they renewed this position, an unusual step since the academy generally stipulated one year as the maximum term for this office.[52] During the five years of Roccati's membership, she had not once attended a private meeting of the academy, as opposed to the public assemblies that frequently included women. Her elevation as head of the academy did not alter this pattern. "[H]er modesty not allowing her to sit in assemblies," Roccati handed the gavel to the councilor of the academy who presided over the meetings *in loco Principis*. Yet she nonetheless reserved the prince's right to set the first subject of the annual meetings, choosing "whether silence or speech is more praiseworthy" in 1755.[53]

With her every action, Roccati demonstrated that she was not a libertine aristocrat who frequented the salons but a Christian noblewoman who strove to maintain decorum. Yet even her cautious and circumspect behavior did not quell critics. A revolt ensued among the younger academicians, who seceded from the academy in protest for almost a year, attempting to form their own academy. Their complaint? They did not want "a woman to sit in that

50. Conc. Ms. 163, BACR; *Leggi dell'Istituto delle Scienze*, 27 July 1751, law 15, *Annali dell'Accademia dei Concordi di Rovigo*, 1:223. This means of advertising public academies was quite standard; see Cochrane, *Tradition and Enlightenment*, 59.

51. Beccaria to Boscovich, 31 May 1762, in Heilbron, *Elements of Early Modern Physics*, 9 n. 38.

52. *Giornale dell'Accademia dei Signori Concordi*, 1697–1793, Conc. Ms. 250, 257, 262, BACR.

53. Pietropoli, *Accademia dei Concordi*, 352 n. 7; Conc. Ms. 250, 258, 260 (15 August 1754 and 21 July 1755), BACR. This procedure followed the 1739 acts of the academy, which gave authority to the eldest councilor when the prince was absent (*Gli Atti Accademici de' SS.ri Concordi di Rovigo*, Conc. Ms. 295, 10). The title of Roccati's topic is noted in Pietropoli, *Accademia dei Concordi*, 134.

presidential seat."[54] In time, the secessionists returned to the Concordi, heads between their tails, and Roccati's tenure as prince quickly ended. After 1756, she held no academy office and never again participated in the private meetings, even in absentia.

Roccati's relationship with the academy became exclusively intellectual. Each year she produced a few physics lectures, her health permitting, and read them in the public assemblies. Other academicians stayed away from this subject out of respect for her expertise. Completing a lecture on the "terracqueous and celestial globe" in 1752, Marco Antonio Campagnella noted that he would not touch on matters of natural philosophy because this was the subject of "the most erudite sig[nora] Cristina." As is apparent throughout her lectures, Roccati took the task of disseminating knowledge seriously and was profoundly grateful for the opportunity to make use of her talents: "[A]s long as I live, I will never forget it," she told her audience.[55] Ironically, her dishonor may have been the making of her academic career. While Padua might have offered her a position at the university, this honor was far from assured. One thing is certain: Roccati's opportunities to discuss physics in Rovigo increased after her father's death. No longer away from home and freed of all external commitments, Roccati now focused on the task of educating her fellow patricians.

Beginning in November 1751, Roccati gave at least two and usually three lectures annually until 1777, when she retired from her position to that of "extraordinary" (occasional) lecturer (fig. 10.3). While lectures of this sort normally do not survive, the Concordi required all members to leave a copy of each lecture with the academy secretary. In Roccati's case, fifty-one of her fifty-four recorded lectures still exist.[56] The scope of these lectures gives testimony to Roccati's thorough command of the modern philosophical systems of her time and her ability to present them in a clear and cogent fashion. Such lectures also provide valuable evidence about what Roccati and her mentors in Bologna and Padua thought physics was by the 1750s, and how they resolved (or chose not to resolve) the various religious, mathematical, and methodological problems inherent in this early modern discipline.

Through these lectures, Roccati hoped to stimulate the youth of her town to "search avidly for the beautiful, learned and distant curiosities" that the world offered. Physics was "one of the most beautiful sciences to which man could apply himself." Going beyond Aristotle's *physica,* it explained "active

54. Conc. Ms. 163, 1:312.
55. Campagnella, Le *"Iscrizioni" di Rovigo,* 111; Silv. Ms. 312, lecture 38 (n.d.), fol. 4r, BACR.
56. Silv. Ms. 312, lectures 1–51. De Vit, "Sulla vita e sulle opere," 64, provides a list of the titles.

Fig. 10.3.   Academy lecture in eighteenth-century Rovigo. Ms. 250, fol. 3, Biblioteca dell'Accademia dei Concordi, Rovigo.

principles" as well as "natural bodies."[57] In Roccati's opinion, physics embodied the best aspects of Enlightenment pedagogy—novelty, pleasure, and utility. The fundamental truths of nature could be fun, she suggested to her audience, as well as important. After following her course, one could understand everything from mechanics to acoustics, from the backyard water pump to the tingling in one's ears during a crescendo of a particularly fine symphony. With so much to offer, who could resist the lures of physics?

While offering her listeners a historical exegesis of the development of physics—something popularizers such as the abbé Nollet (whose books she owned) also did—Roccati indicated that modern physics was a truer picture of the universe than anything the ancients had offered. "Aristotle . . . was a great man but a bad physicist," she informed her listeners.[58] Quite often, the traditional philosophical education Roccati had received served as a counter-

57. *Definizione della Fisica* (15 November 1751), Silv. Ms. 312, lecture 1, fol. 1r, 2r, 3v. Quite possibly Roccati's reading of Agnesi's *Analytical Institutions* may have inspired her to direct her initial comments at the Italian "youth," as Agnesi had done in her textbook.

58. Silv. Ms. 312, lecture 43, fol. 4r. Roccati owned a 1619 edition of Aristotle's *Opera omnia*.

foil to the more novel views she introduced in each lecture. Thus, when explaining Stoic and Pythagorean views about sight in 1763, she emphasized that the "modern philosophers, such as the Cartesians and Newtonians, have a better opinion on vision." Yet she was careful to distinguish between the theories of modern philosophers, expressing a preference for those of "the immortal Newton," especially in her lecture on gravity.[59] Her final year of study in Padua had evidently tipped the balance toward the English approach to nature.

Roccati's portrayal of natural philosophy typified the opinion of the experimental community in Italy at midcentury. For many Italians whose intellectual coming-of-age belonged to the 1740s, modern natural philosophy began with Galileo and culminated in Newton and his disciples. Galileo's mechanics more than his astronomy earned him praise from Roccati—she was surely aware that Benedict XIV had allowed Galileo's *Dialogue* to remain on the Index in 1757, the year he rescinded the 1616 decree against Copernicanism. Her father may have needed permission from the local censors to acquire the 1744 Paduan edition of Galileo's *Works* for their library.[60] A careful reader of the works of the Accademia del Cimento, Roccati affirmed that it was through Galileo that Italians discovered "the true taste for philosophizing." Similarly, she was sensitive to the diminished status of Italian science within Western Europe and sought to redress this in her lectures. "[N]ot only we but the French, the English and all the ultramontanes should pay him homage." Lecture after lecture, Roccati brought her audience in touch with the state of knowledge about the natural world since the age of Galileo. Drawing upon her studies of modern attempts to synthesize and harmonize various philosophies and the journals of the Royal Society of London and the Paris Academy of Sciences, she demonstrated to the citizens of Rovigo how much the mathematization of physics and the development of a mechanical worldview had altered the picture of the universe—what she called "the great machine of the world."[61] Ironically, she did this under the banner of an academy whose chosen emblem included an image of a geocentric universe.

Roccati's physics was a product of the sort of Christianized natural philosophy that found the theology and quantitative physics of the Protestant Newton more comforting than the qualitative physics of the Catholic Descartes. Thus the presence in her library of such books as Giuseppe Valletta's *Letter in Defense of Modern Philosophy and Its Cultivators* (1691–94) may not have been a sign of complete agreement with Valletta's argument against the Jesuits that one

59. Silv. Ms. 312, lecture 29 (3 March 1763), fol. 1v; lecture 43.
60. Silv. Ms. 381, 81. On enlightened attitudes toward Galileo's works, see Ferrone, *Intellectual Roots*, 49–51; and Baldini, "L'attività scientifica," 527.
61. Silv. Ms. 312, lecture 43, fol. 6v; lecture 6 (13 May 1762), fol. 1r

could be a good Cartesian and a good Catholic. Her thesis defense, which argued strongly against skepticism, suggested that Roccati equated the laws of physics with the certain discovery of a deity. Cartesian doubt was an uncomfortable position for this sort of Catholic, and it is surely revealing that the works of Newton never appeared on the Index, unlike those of Descartes, further reinforcing the idea that the English physicist's views might be compatible with various pronouncements from Rome. While frequently admiring Descartes's ingenuity, Roccati decisively rejected his philosophy "because it does not correspond to the truth."[62] Leibniz appeared very rarely in her comments, though she evidenced appreciation of the work of Malebranche, Euler, and Wolff and of Newtonian disciples such as Samuel Clarke, all philosophers interested in the reconciliation of mathematics and moral philosophy. From this pattern, it is easy to imagine Roccati as the sort of conservative Newtonian one found in the Veneto by the 1750s, engaged in the fight against materialism and atheism.[63]

Roccati owned neither the works of Descartes nor of Newton, though she clearly had access to copies of their writings by the time she composed her physics lectures. In her second lecture at the Concordi, Roccati alerted her audience that Newton would be the fundamental point of departure for her entire course: "We will, as needed, seriously attend to certain laws learnedly prescribed by the most learned Newton, most splendid light . . . not only of England but of all Europe." In June 1752—just months before the family scandal—Roccati defined inertia, noting that it had been first observed "by the celebrated and immortal Newton, philosophy of profound and penetrating intellect." Early in the 1750s, she introduced the academy to his laws of motion and praised gravity as "one of the most necessary and most beautiful attributes that God has given to bodies" (fig. 10.4).[64] Such hints at the importance of gravity sparked the interest of the local citizenry. By the time she gave her lecture on this subject, the academy lecture hall was at full capacity.

As the decades passed, Roccati spent less time justifying Newton's fame and more time concentrating on the broad-ranging implications of his work.

62. Ibid., lecture 18 (9 March 1769), fol. 2r. This concerned Descartes's theory of the tides. See also her criticisms of his views on gravity, lecture 43, fol. 3v. Roccati owned the 1732 Rovereto edition of Valletta (Silv. Ms. 381, 75).

63. On reconciliation of mathematics and moral philosophy, see Dobbs and Jacob, *Newton and the Culture of Newtonianism,* 96–97; and Ferrone, *Intellectual Roots,* 86–87. For a general discussion of the range of approaches to natural philosophy, see Hankins, *Science and the Enlightenment.* Ferrone's comments on the Somascan father Giovanni Crivelli's *Elementi di fisica* (Venice, 1731) suggest a similar antimaterialist pattern; *Intellectual Roots,* 97–99. For more on Venetian science up to 1740, see Dooley, *Science, Politics, and Society.*

64. Silv. Ms. 312, lecture 47, fol. 3r; lecture 44, 8v; lecture 39, fols. 4r–6r; lecture 43, fol. 12r, BACR.

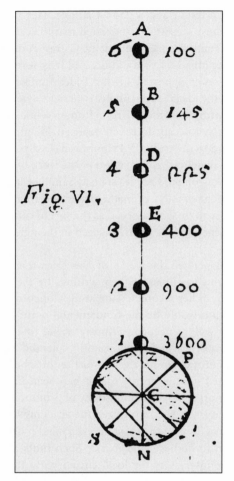

Fig. 10.4. *Lesson on Gravity* (n.d.). Roccati's demonstration of Newton's law of universal gravitation in her physics lectures at the Accademia dei Concordi. Silv. Ms. 312(43), Biblioteca dell'Accademia dei Concordi, Rovigo.

By the early 1760s, she had begun to lecture on the *Opticks*. In December 1761, she reproduced Newton's prism experiments; by March 1764, she had guided her pupils through the laws of refraction.[65] Toward the middle of this decade, she began, tentatively, to discuss Newton's ideas about subtle matter, suggest-

65. Ibid., lectures 4, 8.

ing that the elemental composition of the world might be considerably finer than most of the ancients and many moderns thought. The vast sensorium of the world consisted not in literally apprehending the brute reality of things but in understanding how much perception, in all its different forms, guided what one knew.

By 1766, Roccati had explained all five senses—predictably devoting the most room to sight and sound—and introduced listeners to the work of such Dutch physicians as Boerhaave, whose anatomies of the senses harmonized well with the physics of his countrymen Willem 's Gravesande and Musschenbroek she had first encountered in the family library. The second half of the 1760s was devoted to a detailed analysis of the four elements.[66] Here Roccati hesitated to state unequivocally that Newton's concept of ether had redefined the composition of air. In the end, she demurred: "I leave it to you, most Humane Listeners, who have better discernment than I, to decide if air can be the [first] principle of things." This sort of hesitation resurfaced in her discussion of the debate about gravity as an occult quality, to which she returned in March 1769. Invoking the objections of Euler and "after him, several other philosophers," Roccati refused to pass judgment on the suggestion that gravity was not a well worked out physical cause.[67] Most likely, she informed her audience, Newton did not err because he had been right on so many other things. Yet even Newton had his limits, she suggested. It was the combination of new ideas, experiments, and good faith that yielded most certain truths.

Like many mathematical physicists, Roccati found Newton's lack of precision in this area troubling. When she could not be precise, she did not feel empowered to speculate. The issue of certainty loomed large in her course in experimental physics, as it did in many such lectures throughout Italy. How could one know the universe? Only through God and the geometer's hand. Roccati's course—stretched out across twenty-six years—presented physics as a form of rational theology, practical and pleasing in its ordinary uses but profound in its higher meaning. In her first lecture of 1751, Roccati invoked physics

66. The structure of Roccati's physics lectures parallels Nollet's course in experimental physics, with some interesting differences. Both began with discussions of the sciences of bodies and motion, followed by such subjects as hydrostatics, mechanics, and the elemental properties of nature. But Roccati also included other subjects that reflect the influence of writers such as Musschenbroek and possibly Algarotti as well. Optics occupied a particularly important place in her curriculum. Roccati also shied away from the more nonsectarian tone of Nollet's popular physics. She openly advocated the works of Galileo and Newton and emphasized the relationship between physics and natural theology.

67. Silv. Ms. 312, lecture 14, fol. 5r; lecture 18, fols. 3v–4r, BACR. Roccati's hesitations about the full implications of Newton's particulate theories of matter accord well with the observations made by Simon Schaffer on the limits of Newtonian matter theory in the eighteenth century; see his "Natural Philosophy," in Rousseau and Porter, *Ferment of Knowledge*, 55–57

as a sublime means of knowing God. Over the years, she insisted on the role of design in the natural order, resisting the lures of materialism that tempted many of her contemporaries and sharing Newton's fears about them. After completing her lengthy and detailed tour of the Newtonian physical universe, in 1774 Roccati turned to her final subject: the creation of the world. These lectures completed the physics lessons for the Rodigini.

In her opening lecture of 10 January, Roccati conveyed the sense of accomplishment that she felt in mastering such difficult intellectual material under trying personal circumstances. Family dishonor as well as poor health had often made it impossible for her to lecture annually, at times leading Roccati to fear that she might never fulfill the duties of her office:

> Here I am again, in the presence of the Most Superior Academicians and the Most Humane Listeners, to follow the course that was hampered by those accidents that at times impede men in walking the way that they have undertaken and from which honesty also would wish that they never depart. . . . I will not stop here, in order to fully justify myself, to give you a long list of things that have made me veer off course and perhaps disappear. . . . [I]t is enough for you that I swear to you in this place that I have never neglected my obligation, neither due to a lack of esteem that I may have had for this cultivated Gathering nor due to carelessness and neglect.[68]

As Roccati suggested to her audience, undoubtedly the same divine intelligence that created order in nature had also ordained that physics be her calling. No human travails had ever swayed her from this course and subsequently, at age thirty-eight, she was ready to reveal the divine hand behind the Newtonian worldview.

Roccati most fully addressed this subject in her March 1774 lecture entitled "If the World is Made by God or by Chance." Divine creation, she reminded her audience, was a profoundly Newtonian principle, and it could be discerned in the most ordinary phenomena. To illustrate her point, Roccati recalled her student years in Bologna, where she had observed multiple dissections of the human eye in the university anatomical theater. "The structure of the eye alone . . . , if one contemplates it, can muzzle the foolish defenders of chance." To further her point, Roccati compared the world to a literary work most of her audience could recite from memory: the *Iliad*. If the world was made by chance, she affirmed, then surely the *Iliad* was merely a random collection of words?[69]

68. Silv. Ms. 312, lecture 21, 1r.
69. Ibid., lecture 21, fol. 3r; lecture 20, fol. 4v.

Such comparisons helped Roccati interpret physics and, more generally, natural philosophy in a language that her audience understood. Ordinary examples and literary analogies were standard features of popular science lessons in the eighteenth century, translating difficult concepts into a more common idiom. Yet Roccati primarily strove not to entertain her listeners but to educate and persuade them. Despite the presence of the 1740 Italian translation of Abbé Noël-Antoine de la Pluche's popular and humorous *Spectacle of Nature* in her library, she did not exactly imitate his pedagogy, though she undoubtedly approved of his physico-theology. Her physics was not a parlor game but a more sober and scholarly enterprise, replete with specific references to a wide range of scientific books consumed by eighteenth-century readers. Accomplishing these goals required an additional set of tools, tools less readily accessible in a small town and less palatable to a cultured audience. Reviewing Roccati's lectures, it is clear that she assumed no knowledge on the part of her listeners. Gradually she introduced them to the concepts, vocabulary, and instruments of physics—for instance, the definition of equilibrium and the description of "simple machines" such as the lever, balance, pulley, and inclined plane—allowing them to talk knowledgeably about such subjects as statics and mechanics.[70]

One of the fundamental skills necessary for understanding physics was geometry. As many enlightened natural philosophers noted, mathematical knowledge formed a distinct barrier between learned and popular approaches to science. "It is most false to pretend that everyone is a mathematician," wrote the marquis d'Argens to Voltaire in 1739, after reading his *Elements of Newton's Philosophy*. "Quite the contrary: among 40,000 people one finds only one of them." Even physics professors such as Poleni dispaired at the rudimentary mathematics education of many of his students in Padua and retooled his lectures accordingly. Authoritative writers such as Nollet concurred with this assessment, arguing that mathematics belonged to "a more serious study of Physics." Accordingly, his popular *Lessons in Experimental Physics* made every attempt to present a nonmathematicized physics whose proofs lay in diagrams, demonstrations, and analogies rather than in numbers. "They always presume that the majority of my Listeners are not able to understand Algebraic and Geometric expressions," wrote Nollet.[71]

Roccati's own use of geometry reflected a similar approach to the problem of popularizing physics, with one important variation. While Nollet assumed

70. Silv. Ms. 381, 91; Ms. 312, lecture 24, fol. 1r–2r.
71. D'Argens quoted in Kleinert, "La vulgarisation de la physique," 308; Nollet, *Lezioni,* 1: xii–xiii (this is the same edition used by Roccati). On the situation at Padua, see Dooley, "Science Teaching," 129.

no mathematical knowledge, Roccati attempted to offer her listeners enough basic geometry so that they could understand the rudiments of mathematics. She insisted that "geometric proofs" were crucial to experimental physics and filled her lectures with dozens of diagrams that illustrated her ideas about mass and space. Geometry became an important form of demonstrative proof in Roccati's physics. At the same time, she recognized that not all of her listeners shared her pleasure in geometry and strove to convince them of its importance to physics. Addressing the math phobias of the local elite, Roccati argued: "All the sciences have certain subjects which are murky, arid and dry. However, they will open a path to the understanding of more beautiful and delightful things, however difficult."[72] In this fashion, she traversed the difficult ground between learned and popular physics, soundly negating Algarotti's image of the female Newtonian who shunned geometry because she could not see the forest for the trees. Perhaps Roccati's insistence on the partial inclusion of this subject influenced the Accademia dei Concordi to add a chair in geometry to their Institute professorships several years later.

The problems of teaching physics in a provincial academy derived not simply from the limited science education of the local elite but also from the lack of equipment available to make physics the sort of spectacle that experimenters such as Nollet performed in the capital cities of Europe. There is absolutely no indication that Roccati had access to an experimental physics cabinet in Rovigo. Lack of equipment was a problem that plagued the teaching of physics at most European universities prior to 1750. The initial expense of stocking a cabinet with good instruments was prohibitively expensive, until governments chose to invest in this science. When Giovanni Poleni recommended that the University of Padua install a physics laboratory, his superiors responded by presenting him in 1724 with "a set of brass interlocking squares, circles, and triangles" to supplement his blackboard acrobatics. Poleni had to content himself with this mathematical toy until the image of experimental physics as a distinctive discipline finally pushed the governing board to award him a new chair in this subject, and the long-awaited physics cabinet, in 1738. By the time Roccati studied in Padua, Poleni had one of the best cabinets in Europe, containing more than two hundred machines.[73]

Prior to these encounters with experimental physics, Roccati had also enjoyed demonstrations from the collections in the Bolognese Institute for Sciences, which in 1744 divided its physics cabinet into three separate rooms, covering electricity, optics, and the motion of fluids. Two years later, Benedict XIV

---

72. Silv. Ms. 312, lecture 44, fol. 1r; lecture 41, fol. 9v, BACR.
73. Vaupaemel, "Experimental Physics," 181; Dooley, "Science Teaching," 135–38. More generally, see Heilbron, *Elements of Early Modern Physics*.

donated an instrument collection to the institute that included some fine examples of Musschenbroek's handiwork.[74] Thus, during Roccati's student days in Bologna, experimental physics was on the verge of becoming a lively discipline. While very few members of the Institute for Sciences knew how to use their machines, Laura Bassi and her husband offered private physics lessons in their home that more than compensated for these inadequacies. Finally in 1749, the famous Nollet arrived in Bologna, bringing gifts of his instruments to the institute. Roccati was among the eager witnesses of Nollet's experiments. Almost twenty years later, she recalled her first demonstration of the effects of heat and cold on iron, informing her listeners: "Many times you will see with your own eyes this experiment at the Institute for Sciences of Bologna, with the Machine donated to that celebrated place by the most famous Signor Abbé Nollet and made by his own hand."[75]

Roccati may have had aspirations to teach physics in imitation of the lessons she observed in Bologna and Padua. Yet given the difficulties of acquiring instruments and trained personnel to assist in their operation in Italy's leading intellectual centers, we can only imagine the problems of creating a decent physics cabinet in Rovigo. Roccati's adversary Maffei indicated similar frustrations in Verona. "[S]till today, and this is a shame, Verona lacks an air-pump," he wrote to a Parisian correspondent in 1736, "which makes it seem a barbarous city, without even a good microscope." In 1739, he was hot on the trail of some English prisms to replicate Algarotti's descriptions of Newton's optical experiments; in 1747, he confessed to Poleni that he would enjoy nothing better than the arrival in Verona of "those who make electrical experiments."[76] If Maffei, one of the leading lights of the Italian Republic of Letters, had difficulties acquiring instruments in a somewhat larger city, Roccati—without his range of contacts and financial resources—must have found it a virtual impossibility. And at least she had seen Nollet perform, something Maffei had only heard about through his network of correspondents.

Experimental activities were an important part of Roccati's physics course, but they appeared primarily as hypothetical entities, experiments *in potentia,* that either had been done in some other time and place—by Galileo and the Accademia del Cimento in Florence, by Boyle and Newton in London, by Nollet in Bologna—or could be done by the audience. Describing the existence of effluvia, she told her listeners: "I could demonstrate such a thing with

74. Cavazza, "L'insegnamento"; and C. de Pater, "The Textbooks of 's Gravesande and van Musschenbroek in Italy," in Maffioli and Palm, *Italian Scientists,* 233. On Laura Bassi's physics cabinet, see Cavazza, "Laura Bassi."

75. Silv. Ms. 312, lecture 27, fol. 2v, BACR. This comment occurred in the lecture on "sensory things" (especially fire), given on 12 March 1767.

76. Maffei, *Epistolario,* 2:751, 903, 1170.

many experiments but I will content myself with only a few." Such an approach not only resolved the problems of experimenting without instruments but also invoked a common pedagogical distinction between the *experimentum crucis* and the myriad observable phenomena that one might report when testing a particular hypothesis. Nollet himself announced that he provided accounts of few experiments because the goal of his lessons was "a course in experimental physics and not a course in experiences." Roccati evidently fashioned a few rough tools to complete those minimal demonstrations. In December 1761, she replicated Newton's prism experiments for her pupils in a darkened room. Yet this was an exception that only underscored the paucity of such activity in her approach to physics.[77]

For a variety of reasons, Roccati never cultivated a reputation as an experimental impresario, rearranging prisms, shifting pendula, and firing up Leiden jars. Such activities belonged to a public culture of science that demanded a certain level of investment in the materials of science and generally did not yet accept the idea of women as performers of experiments.[78] Honor as well as resources figured into this equation. The image of the popular experimenter was, in part, a libertine one, and Roccati could ill afford any further blemishes on her reputation. Mathematics appealed to her because it was a disciplined science that simultaneously tamed nature and the passions. The results of this intellectual preference created a somewhat idiosyncratic science course. Her physics existed in that nether land between the university classroom and the public assembly, a hybrid version of both worlds, neither of which she could fully inhabit.

Surely the most enduring legacy of Roccati's physics is the world of paper Newtonianism that it left behind. As Poleni had done prior to the arrival of a physics cabinet in Padua, she created a virtual laboratory out of diagrams and descriptions that emphasized pictorial vividness as a means of confirming nature's truths. Frequently she lifted her illustrations directly from the best physics textbooks of the day. Clearly Roccati understood diagrams as an important technique of persuasion, placing "proofs and geometric demonstrations before your eyes." She may have even preferred them to the experimental physics cabinet. Diagrams were a means of clarifying the geometry that she worried would overwhelm or, even worse, simply bore her listeners to death. In a discussion of the infinite divisibility of a line—a basic proof for the extension of bodies—Roccati immediately drew her audience's attention to a diagram, stating: "It is delightful and clear."[79]

77. Silv. Ms. 312, lecture 2, fol. 1r; lecture 4, fol. 2r, BACR; Nollet, *Lezioni*, 1:xx.

78. On this subject, see Schaffer, "Natural Philosophy," "Self Evidence," and "Consuming Flame." For a somewhat different approach to similar material, see Stafford, *Artful Science*.

79. Silv. Ms. 312, lecture 8, fol. 4r; lecture 48, fol. 3r, BACR.

In the absence of better resources, diagrams may have been the optimal tools for a female natural philosopher and a provincial experimenter in the mid–eighteenth century. They mediated the technical language of mathematical physics and the increased emphasis on experimentation. As Roccati promised her listeners on many occasions, "I will try as much as possible not to bore you."[80] Illustrations provided a means of making the work of analysis somewhat entertaining. They allowed her audience to visualize physics in ways that the traditional Aristotelian curriculum in natural philosophy did not, transforming an increasingly specialized discipline into a vivid and enduring account of nature's laws that gave them some appreciation of why scholars had virtually deified some English mathematician from Cambridge.

Into the Void

Roccati's lectures were not extraordinary science but instead were the product of rather extraordinary circumstances. Her unique education, the ambitions of Rovigo intellectuals, and the growing sense that knowledge was a civic responsibility all fostered her teaching of physics in a provincial Venetian town. Despite the quality of Roccati's teaching and the long tenure of her professorship, by the time of her death in 1797 she was virtually unknown, even to the Rodigini. It took almost twenty years for the Concordi to give and publish the oration commissioned for her death.[81] Her increased withdrawal from the activities of the academy after 1774, for reasons of health more than family dishonor, completed her effacement from public life. As I have suggested earlier, Roccati was never fully integrated into the Accademia dei Concordi. An uneasy academician and an absent prince, Roccati never felt that she was a member like the others. Even though she may have increased attendance at institute lectures, in the end, she was a prodigy whose luster seemed to fade with age.

Unlike Laura Bassi, Roccati did not fight to increase her privileges; her family's marginal situation after 1752 made this an impossibility. Instead, she used the opportunities that her professorship offered to put her knowledge to use in the hopes of instilling some appreciation of physics in the citizens of Rovigo. Through her teaching, Roccati played an important and exceptionally well documented role in the diffusion of scientific knowledge from centers of activity such as Bologna and Padua to the periphery of the learned world. Certainly Roccati's lectures bear witness to the fact that Newtonian physics was an important enough subject in mid-eighteenth-century Italy to be included in the pedagogical designs of a provincial academy. And it is surely significant that the members of the Concordi chose a woman to represent this particular subject.

80. Ibid., lecture 39, fol. 7r.
81. This is Grotto's *Delle lodi della dottoressa Cristina Roccati.*

In her youth, Roccati seemed poised to become one of the leading physicists in Italy. Writing to Roccati in June 1751, the Bolognese noble Marcantonio Gozzadini praised her father "who postponed that virile and unpraiseworthy policy of men to keep women away from study." Giovan Battista had had her "so well instructed that he made [her] an object of admiration to the world." [82] In Enlightenment Italy, physics *was* like a learned woman: emerging from ancient doctrines to produce the utmost novelties. Algarotti had created the template for this sort of imagery when, in his poetry about Newton's optics and in his *Newtonianism for Ladies,* he allegorized Bassi and Émilie du Châtelet as the muses of modern physics. Yet Roccati, more than any of the other learned women of her day, brought this image to its utmost fulfillment. A dedicated and even passionate Newtonian, she saw her appointment to a local academy professorship as an opportunity to disseminate knowledge. Her commitment to her subject was one of the reasons that Orazio Arrighi Landi devoted more space to Roccati than any other female natural philosopher in his *Temple of Philosophy* (1755), subtitled *The Tomb of Isaac Newton.*[83] For a brief moment, she was the muse of modern physics.

Roccati made no new discoveries, indeed hesitated even to speculate about the controversies of early modern physics as she described them to the Accademia dei Concordi. As a result, we might dismiss her as a "mediocre mind"— a passing curiosity of no particular interest—as at least one local biographer suggested in the early twentieth century. Yet to do so would be to negate the importance of disseminating scientific ideas in favor of producing them. The production of knowledge was generally not a viable option, either for female natural philosophers whose place in intellectual life was largely ceremonial, or for provincial scholars with limited access to the instrumental culture of physics. Instead they were encouraged to disseminate the learning they had mastered, to show others that the mysteries of nature were as yet accessible for the curious and diligent. The documents cannot tell us whether Roccati ever succeeded in her task: did she excite and inspire her audience about physics? She often worried that she would try their patience, suggesting that the question of scientific literacy was already a pressing issue. Whether out of idle curiosity or genuine interest, the citizens of Rovigo listened to her explain the laws of nature. She gave them some sense that they belonged to the Republic of Letters. And in that imaginary commonwealth, honor and shame were always tempered by the pursuit of knowledge—a goal that, under the right circumstances, women as well as men might seek.

---

82. Ibid., 90 (Bologna, 30 June 1751).
83. Arrighi Landini, *Il Tempio della Filosofia,* 106–10.

Primary Sources

Algarotti, Francesco. *Opere varie del Conte Francesco Algarotti*. Venice. 4 vols. 1757.

Altieri Biagi, Maria Luisa, and Bruno Basile, eds. *Scienziati del Settecento*. Milan and Naples: Riccardo Ricciardi, 1983.

Arrighi Landini, Orazio, *Il Tempio della Filosofia*. Venice, 1755.

Bandiera, Giovan Nicolò. *Trattato degli studi delle donne*. 2 vols. Venice, 1740.

Barnard, H. C., ed. and trans. *Fénelon on Education*. Cambridge: Cambridge University Press, 1966.

Doria, Paola Mattia. *Ragionamenti . . . ne' quali si dimostra la donna, in quasi che tutte le virtù più grandi, non essere all'uomo inferiore*. Frankfurt, 1716.

Fantuzzi, Giovanni. *Notizie degli scrittori bolognesi raccolte da Giovanni Fantuzzi*. 5 vols. Bologna, 1783.

Grotto, Giuseppe. *Delle lodi della dottoressa Cristina Roccati*. Venice, 1815.

Lalande, Jerome. *Voyage d'un françois en Italie, fait dans les années 1765 & 1766*. Venice, 1769.

Maffei, Scipione. *Della formazione de' fulmini*. Verona, 1747.

———. *Epistolario, 1700–1755*. Edited by Celestino Garibotto. Milan: Giuffrè, 1955.

Montagu, Lady Mary Wortley. *The Letters and Works of Lady Mary Wortley Montagu*. 2 vols. Edited by Lord Wharncliffe and W. Moy Thomas. London: Swan Sonnenschein and Co, 1893.

Nollet, Abbé. 1893. *Lezioni di fisica sperimentale*. 4 vols. Venice, 1893.

Novelle letterarie. Florence, 1750.

Poulain de la Barre, François. *The Equality of the Sexes*. Translated by Desmond M. Clark. Manchester: Manchester University Press, 1990.

Stäel, Germaine de. *Corinne; or, Italy*. Translated by Avreil H. Goldberger. New Brunswick: Rutgers University Press, 1987.

& *11* &

Going Dutch: Situating Science
in the Dutch Enlightenment

LISSA ROBERTS

Voltaire, that most witty of Enlightenment writers, had nothing but praise for the Dutch in 1737. Following a visit to the Netherlands, he wrote to one of his hosts, "I would have liked never to leave Amsterdam. Its liberty, its scholarliness, your society as a whole have completely enchanted me." With the publication of *Candide* some twenty years later, however, Voltaire presented readers with a very different image. Upon his arrival in Amsterdam, Candide witnessed only hypocrisy and intolerance. The good citizens of Amsterdam threatened him with incarceration for citing one of Newton's "rules of reasoning" while they rewarded an Anabaptist for his dissenting views by crowning him with the contents of a chamber pot.[1]

What is the truth about the Netherlands during the eighteenth century? Was it "the best of all possible worlds" or a society overrun by facile knowledge and crude commercial interests? Enlightenment authors such as Voltaire answered this question in keeping with an agenda constructed out of their own interests and values.[2] We remember that Voltaire wrote *Candide,* for example, in the wake of the great Lisbon earthquake and as part of his long sustained battle against ancien régime injustice and *l'infame.* If we want to answer this question, then, we must use the observations of foreign contemporaries with a sense of wariness. And yet to speak of the Netherlands and Dutch science as part of "Enlightenment Europe" requires that we somehow incorporate those interests and values into our analysis.

Just what key terms should we adopt and how should we employ them? A good deal of the traditional literature in intellectual history and the history of Dutch science adopts the vocabulary of the Enlightenment wholesale and valorizes those aspects of the eighteenth century that seem to celebrate ideals such as rationality and progress. Hence, for example, the focus on the Netherlands as European entrepôt for Newtonianism and the experimental method and

1. Voltaire's letter to A. F. Prévost, 16 March 1737, quoted in Vercruysse, *Voltaire et la Hollande,* 50; Voltaire, *Candide,* chap. 3.
2. Schama, "Enlightenment in the Netherlands."

the general dismissal of Dutch science during the decades following Willem 's Gravesande and Petrus (Pieter) van Musschenbroek as suffering from decline. In this prescripted analysis, the Dutch were significant precisely because and so long as they linked English and Continental centers of enlightenment by educating a generation of Newtonian do-gooders and publishing works that others dared not touch. To be Dutch according to this scenario meant not so much to engage in a complex set of culturally distinct practices, but to serve as purveyors of foreign-born knowledge, be it for the sake of truth or profit. Throughout, the emphasis is on ideas.[3]

In a discussion that contrasts the methods of intellectual and cultural history, Roger Chartier offers a fruitful challenge to this approach.

> Is it certain that the Enlightenment must be characterized exclusively or principally as a corpus of self-contained, transparent ideas or as a set of clear and distinct propositions? Should not the century's novelty be read elsewhere—in the multiple practices guided by an interest in utility and service that aimed at the management of spaces and populations and whose mechanisms (intellectual or institutional) imposed a profound reorganization of the systems of perception and of the order of the social world?[4]

Chartier poses utility as central to the Enlightenment, a claim manifestly borne out by the description of Dutch science that follows here. But this is not to say that either he or this chapter would have "utility" understood as a transcendent and univocal ideal that substantively informed all practice and, therefore, through which all actions must be interpreted and judged. Rather, it encompassed a multifaced set of discourses that were deployed in keeping with the variety of meanings and purposes with which actors at the time charged the term. Utility referred to a cluster of philosophical usages: (scientific) knowledge in the service of society and the furtherance of both secular and religious enlightenment. But it was also invoked to justify or criticize practices aimed more directly (though not always consciously) at extending institutional, cultural, or socioeconomic power and control. Finally, and perhaps most frequently for the case of the Netherlands, it provided a watchword for various forms of sociability that developed during the eighteenth century. It was under the banner of utility that more and more people sought fellowship and entertainment.

Because practice cannot be treated simply as a consequence of discourse,

3. On Dutch Newtonianism see, for example, Brunet, *Les physiciens hollandais;* De Pater, *Petrus van Musschenbroek.* On the image of Dutch scientific decline, see Snelders, "Professors, Amateurs, and Learned Societies."

4. Chartier, *Cultural Origins,* 17.

writing a cultural history of Dutch science in the eighteenth century means going beyond texts to analyze what people did in the name of science.[5] Further, because such a history traces active engagement rather than passive reception and conversion, we must abandon the myth of the Enlightenment as a monolithic system in possession of what Chartier calls a "general idiom." Active engagement filters meaning through the lens of indigenous concerns and local interests. It splinters it through translation and interpretation, requiring that we multiply the sites of our investigation in order to enrich our depiction.[6]

In this chapter, I present four sites as exemplary of the sort of practices in which various segments of the Dutch population engaged in the name of science during the long eighteenth century. They are Dutch anatomy theaters, the Leiden physics theater, the world of private lecture courses and scientific societies. These sites divide neatly into two pairs. The first turns our attention to universities and guilds, traditional institutions that proved sufficiently malleable to accommodate scientific innovation and use it for their own purposes. In both cases, as we will see, the place, structure, and presentation of science in these institutions, indeed the very fate of these institutions themselves, were ultimately determined, not by internal dynamics or the logic of ideas, but by the sociopolitical realities that swept through the Netherlands along with the revolutionary armies of the French at the end of the century.

The second pair directs us to the realm of private cultural initiative. In a society as highly urbanized as the Netherlands, it is hardly surprising to find the urge toward organized sociability expressed in nontraditional and more socially inclusive ways. Given their involvement in international travel and trade, in land reclamation and exploitation, in practices that generally required the domination and manipulation of nature, it is also not surprising that the educated and educationally interested Dutch were preoccupied with the study of nature. Following private courses and joining societies allowed those outside the traditional confines of the university to satisfy the urge to socialize in a context directed toward the purposeful unveiling of nature.

Conspicuously absent from discussion here is the popular presentation of science at Dutch carnivals and public houses, a subject I treat at length elsewhere.[7] Such an analysis broadens our focus to question the distinctions usu-

---

5. The literature on cultural history, which sports myriad definitions, is vast. A useful introduction, which includes bibliographical information for further inquiry, is Hunt's introduction, "History, Culture, and Text," in *New Cultural History.* Of more specific interest, perhaps, to historians of science, is Dear, "Cultural History of Science."

6. Chartier, "Text, Symbols, and Frenchness"; For a recent attempt to recast notions of "reception" and "conversion" into a more active sense of appropriation in which local considerations are given center stage, see Bensaude-Vincent and Abbri, *Lavoisier in European Perspective.*

7. Roberts, "Science Becomes Electric," forthcoming.

ally drawn between "official" and "popular" culture, distinctions that flow out of and reinforce more traditional, intellectual histories of the Enlightenment. For now, however, it is sufficient to examine the four sites presented here as a way of grasping how science was actively situated in the middle- and upper-class portions of Dutch society—those segments usually associated with "living the Enlightenment."

Anatomy Theaters

Although the history of anatomy theaters goes back at least to sixteenth-century Italy, Dutch anatomy theaters provide an interesting window through which to observe events and developments characteristic of the Dutch Enlightenment. Their stated purpose was "to teach, delight and to move," to teach, that is, through the twin vehicles of entertainment and appeal to the soul.[8] For what was displayed in public anatomy lessons was not so much the newest trends in anatomy as the wondrous and inviolable structures of the parallel worlds of society, morality, and nature.

An examination of the theaters' physical setting begins to make this point clear. Dutch anatomy theaters followed an architectural design first developed in Padua, but were unique in a number of telling ways. The amphitheaters were often constructed in an already standing church and, at least in the case of the University of Leiden theater, the table on which the body was placed was situated where the altar had been. Further, Dutch anatomy theaters were outfitted with a collection of skeletons and edifying art works, designed specifically to align the instructive presentation of anatomy with lessons in theologically based morality and mortality.[9] When public anatomies were not in session, the Leiden theater opened its doors as a museum of divinely sanctioned natural history and morality. Albrecht von Haller, a student of Herman Boerhaave's in the 1720s, recorded in his diary that the museum was particularly popular with country girls.

An interesting cross-section of Dutch society gathered at the public anatomy lessons that were held every winter in the theaters. Indeed, people generally attended both to see and be seen. Tickets were sold to the general public for these widely attended events—a practice that apparently predates the selling of tickets for stage performances of dramatic theaters—which enabled citizens not connected with the university or surgeons' guild to witness the anatomical spectacle while taking part in its sociocultural display. Seating was in keeping with and reinforced the reigning social hierarchy, the best seats

8. Heckscher, Rembrandt's "Anatomy," 4.
9. Lunsingh Scheurleer, "Un amphitheatre."

reserved for members of the regent class and important personages connected to the sponsoring guild or university. As the rows ascended away from center stage, the audience became increasingly plebeian. Apprentice surgeons, for example, were among those who occupied the top rungs, so far from the stage that they often could not see the lessons they were required to attend as part of their training.[10]

A festive atmosphere and general lack of decorum reigned at these public spectacles, recorded in documents such as the minutes of the Amsterdam Surgeons' Guild and reflected in regulations promulgated to monitor audience behavior.[11] No laughing or talking was permitted. Questions posed to the presiding professor in connection with the anatomy lesson were to remain decent and serious. The stealing of hearts, kidneys, livers, gallstones, and other items that were passed around by the amanuenses for closer inspection was strictly forbidden. Those who violated this edict were charged a hefty fine of six guilders. Thus public anatomies maintained at least the fiction of a learned gathering, based on the model of a university disputation. But if "proper" behavior had not been threatened by the actions of the audience, such laws would not have been considered, let alone posted.

Why were public anatomies among the most popular forms of entertainment in the Netherlands during the seventeenth and eighteenth centuries?[12] It has been suggested that a certain affinity existed between the appeal of public anatomies and public executions, which also drew large crowds and great public interest. Since cadavers of executed criminals were sometimes used for anatomy lessons, the link is more than one of analytical retrospection, and a full explanation of this point would have to take into account the very different views of life, death, and criminality that existed in early modern times. For now it is sufficient to note four things. The criminal's body provided a point of intersection between different currents of anatomical interest—it served simultaneously as the subject of investigation and display. It fed into popular medical beliefs—parts of the criminal's body were considered to have special curative qualities. It gave rise to moral considerations—the "justly deserved" punishment carried out by the executioner was augmented by the further humiliation of being publicly dissected rather than properly buried. And it raised religious sentiments—examination of the human body (even that of a crimi-

10. Hekscher, Rembrandt's "Anatomy," 32; Nuyens, "Ontleedkundig onderwijs," 75–77.

11. Nuyens, "Ontleedkundig onderwijs," 81, notes that this festive atmosphere was a regular feature of descriptions of the guild's public anatomies between 1727 and 1760.

12. Hekscher, Rembrandt's "Anatomy," 5. The University of Leiden's anatomy theater could seat four hundred spectators at a time. The theater used by the Amsterdam Surgeons' Guild had to be replaced in 1688 because it could not accommodate all those who wished to observe the spectacle.

nal) simultaneously pointed to the beautifully systematic nature of God's handiwork and displayed the "human condition" in its most visceral form.

This final religious message was prominently advertised. On the high back wall of the Amsterdam anatomy theater, for example, there appeared a poem by Casparus Barlaeus, one of the greatest Dutch poets of the seventeenth century, which included the following verse:

Qui vivi nocuere mali, post funera prosunt,
Et petit ex ipsa commoda morte salus
Exuviae sine voce docent.

[Evildoers who harm while still alive,
Prove of use after their demise.
Medicine seeks advantage, even from death.
The carcass teaches without a breath.] [13]

The poem goes on to say that God's design is evident even in the smallest portion of the human body. But it also makes clear that the fate of humanity was sealed by the characteristics of frailty and mortality. Up through the first decades of the eighteenth century, it was this Janus-faced vision of the greatness of God and the smallness of humanity that came across as the primary message of public anatomies.

Dutch criminals—like their counterparts elsewhere in Europe—were subject to myriad grotesque, public punishments, but because of the lack of centralized authority the final sentence to which they were subject varied with the locale where they were tried. In Amsterdam, for example, criminals who received capital punishment were afterward either buried, displayed in a public "gallows field," thrown in the IJsselmeer with a hundred pound weight attached to their legs (this end was generally reserved for homosexuals), or delivered to the anatomy hall. Between 1693 and 1766 some 10 percent of Amsterdam's convicts met this last fate, sent to be anatomized either because the presiding judge thought they did not deserve to be "put to their final rest" or because their particular physical condition might be of anatomical interest.[14] In Groningen, however, at least in the second half of the eighteenth century, criminals retained the final say about whether their bodies would find their way to the anatomy theater. Their carnal remains could not be dissected without their express permission, which they did not always grant, much to the distress of anatomists such as Petrus (Pieter) Camper.[15]

13. Nuyens, "Ontleedkundig onderwijs," 75.
14. Jelgersma, *Galgebergen en Galgevelden*, esp. 26–36; Spierenburg, *De Sociale Functie*.
15. I owe this information to R. P. W. Visser.

Providing cadavers for public anatomies was the responsibility of local governments, which also had jurisdiction over local hospitals and had to deal with issues such as the untimely demise of foreign residents—both potential sources of anatomy subjects. The Amsterdam City Council, for example, promised in 1673 to provide four bodies a year from the public hospital for anatomizing by the surgeons' guild. Such promises notwithstanding, finding suitable corpses remained a difficult affair. No public anatomies were held in Amsterdam in 1699 because no bodies were available. Groningen authorities solved the problem of local availability in the 1760s by having cadavers shipped up from Amsterdam. When the waterways iced over, making such transport impossible, Petrus Camper filled Groningen's anatomy theater with equally popular shows of animal dissections.[16]

Camper is known, among other things, for his contributions to the comparative anatomy of animals. Of particular interest here is the way he displayed his claims and the purpose to which he publicly put them. Camper worked in Amsterdam for a number of years, first as professor of anatomy and surgery at the Athenaeum Illustre (forerunner of the University of Amsterdam) and then also as "adjunct-surgeon of the [judicial] court" and professor of anatomy for the Amsterdam Surgeons' Guild (the first of these two positions, which he filled without pay, helped to keep him supplied with corpses for his lessons). In 1763, he accepted an invitation to become professor of theoretical medicine, anatomy, surgery, and botany at the University of Groningen. During the next ten years, he taught his courses and gave policlinical lessons in the public hospital, but also found an increasing amount of time to do research in comparative anatomy.[17]

Camper did not keep his research a private affair. Rather, announcements periodically appeared in the local newspaper inviting the public to attend one of his anatomy lessons featuring subjects such as an elephant, an orangutan, or the head of a rhinoceros. Camper could always count on playing to a full house. Where did he get his specimens and why were these public anatomies so popular? Specimens were donated from locations as far away as the Dutch East Indies and the Cape of Good Hope, sent by patrons connected either with the Dutch East India Company or the colonial apparatus of the Dutch government.[18] Camper's fortune, like that of the Netherlands generally, was inextri-

16. Tang and Wigard, *Amsterdamse gasthuizen,* 36–38; Nuyens, "Ontleedkundig onderwijs," 71; P. Camper, *Natuurkundig Verhandelingen,* 127. As Tang and Wigard discuss, a number of communities regularly relied on the public hospitals *(gasthuizen)* of Amsterdam for their cadavers (37).
17. Otterspeer, "De Aangenaamheden der Natuurlijke Historie," 7; Visser, "Camper en de Natuurlijke Historie der Dieren."
18. P. Camper, *Natuurkundig Verhandelingen,* 27–30, 130–31.

cably tied to its commercial empire. The widespread existence of botanical gardens, private animal parks, and menageries in the Netherlands simultaneously pointed to an Enlightenment preoccupation with the exotic (Dutch cabinets of natural curiosities were among the best in Europe), the urge to dominate and put nature to work (witness the Netherlands' continuous battle with the sea), and Dutch success in commanding tribute from the far-flung corners of the globe. What better symbol of the Netherlands' international dominion than the ready presence of flora and fauna from around the world?

The popularity of Camper's shows resulted from more than the sort of Dutch mirror image they provided, however. First, Camper (all 221 pounds of him by the 1770s) was a fine showman and able teacher. Further, his message had widespread appeal and marks for us a change from the moral lessons previously drawn by public anatomies. Camper studied animals as well as humans as a way to interrogate the "great chain of being." His comparative work told him that the chain was indeed continuous. He illustrated this in a special lecture he gave before the Amsterdam drawing academy (tekenacademie) in 1778 where he showed that by redrawing a few lines a cow could be transformed into a bird, a bird into a horse, and horse into a woman.[19]

Despite certain anatomical similarities, however, humans stood out for Camper and his audiences as fundamentally distinct from and superior to the animal kingdom. Unique physical and mental characteristics, particularly the ability to speak and reason, marked humanity as the crowning achievement of God's teleologically creative glory. What ultimately brought Camper and his audiences together was that they "craved to know the advantages and elevation of ourselves over all Creatures and the Orang, and on that account to glorify the Great and Divine maker of the Universe with respectful gratitude."[20] The moral weightiness and physico-theological character of previous public anatomies were hereby recast in the second half of the eighteenth century in a form that exalted God through a simultaneous celebration of human life and ability.

While all this says much, public anatomies satisfied more than the contemporary preoccupation with natural theology, concerns with human mortality and its consequent moral lessons, and desire for ribald entertainment. They also provided an income to the organization that sponsored their operation and thus served to assert the reality of a certain social and cultural hierarchy. In Amsterdam, profits (always greater if the body to be dissected was female) were used to finance lavish banquets for members of the surgeons' guild, followed by a torchlight parade. Income that remained after the praelector was

---

19. A. G. Camper, *Redenvoeringen van wylen Petrus Camper.*
20. P. Camper, "Kort berigt wegens de ontleding van verscheidene Orang Outangs," 20; cited and translated in Visser, *Zoological Work of Petrus Camper,* 109.

Fig. 11.1.   Petrus Camper, drawing from *The works of the late Professor Camper the connexion between the science of anatomy and the arts of drawing, painting, statuary . . . Illustrated with seventeen plates . . .* , translated from the Dutch by T. Cogan (London, 1794). Reproduced with permission from San Diego State University, Love Library Special Collections.

Fig. 11.2.   Cornelis Troost, *Three Governors of the Amsterdam Surgeons Guild* (1731). Reproduced with permission from the Rijksmuseum Collection, Amsterdam, the Netherlands.

paid his handsome fee went to pay for a cook, wine, tobacco, fish, lanterns, lantern carriers, and a variety of other "necessities."

These banquets and public processions emphasized the social status of those privileged enough to be guild members. Contemporary pictorial evidence reminds us further of the importance of institutional membership and status in a corporate society such as the Netherlands. Between 1603 and 1794, a series of paintings depicting anatomy lessons given by the guild's professor of anatomy were commissioned as a way to record and celebrate the functions and status of the guild. (This series includes two paintings by Rembrandt; the *Anatomy Lesson of Dr. Tulp* of 1632 and the *Anatomy Lesson of Dr. Deijman* from 1656.)[21] In keeping with this tradition, the artist Cornelis Troost was

21. For further details, see the catalog published to accompany the Amsterdam Historical Museum exhibition "De Anatomische Les van Dr. Deijman," 1994. See also, Nuyens, "Ontleedkundig onderwijs."

commissioned by the Amsterdam Surgeon's Guild in 1728 to paint *The Anatomy Lesson of Professor Willem Roell* for public display in the Amsterdam guild chamber. Three years later, he completed a group portrait of three of the guild's governors at their request. In the portrait, the three burghers seek with their gaze to engage the attendant audience they proudly assume surrounds them as they sit in front of their family crests. The center figure is clearly a busy and important man, looking up as he is from the task of recording guild business. The figure to his right nonchalantly displays his surgeon's diploma, an obvious mark of social identification.

In fact, there is more to this painting than meets the eye. The pictured diploma was falsely obtained and the three governors portrayed in the painting were accused of misusing guild funds to pay for the tableau. In 1732, they were consequently dismissed from their positions in disgrace. Why did they risk their reputations to advertise themselves in this way? In a society where position was all—one's identity and set of privileges were determined by it—such calculated risks were apparently worth taking. Relevant here is the fact that these men sought to make their reputations by publicly asserting their leading positions in the Amsterdam Surgeons' Guild. Not only were they active in a prestigious organization, they thereby accrued recognition of their status at publicly held rituals such as the yearly anatomy lessons sponsored by their guild.

The scandal that accompanied this portrait, coupled with the poor quality of the praelector Willem Roell's teaching, marked a low point in the history of the Amsterdam Surgeons' Guild.[22] The guild attempted internal reform, including the institution of fines for apprentices who did not attend anatomy lessons. For a time, thanks at least partially to the successful tenure of Camper as guild professor of anatomy 1755–61, things improved. But lethargy once again set in following Camper's departure, despite the efforts of his able successor, Andreas Bonn, and the guild showed itself to be increasingly out of touch and out of time. In addition to internal difficulties related to the poor quality of training and a surplus of surgeons in the city, guild members had also to contend with growing competition from so-called quacks.

The final solution to these difficulties was a political one, resulting from the revolutionary fervor of Dutch republicans. In 1798 the Amsterdam Surgeons' Guild, like all Dutch guilds, was disbanded, and its teaching responsibilities were taken over by a government commission for medical education. But the guild's immanent demise was already apparent to many, represented

22. Despite numerous complaints, Roell retained his position from 1727 until 1755. It was said of him that "if he was not sick, he was probably traveling or on vacation" (Nuyens, "Ontleedkundig onderwijs," 81).

for us nicely in the portrait of its last anatomy professor in 1792. The painter Adriaan de Lelie portrayed Andreas Bonn examining, not a cadaver in front of guild members, as was the pictorial custom, but a naked model before an audience at the art academy of the Felix Meritis Society!

In the final years of the guild's existence, its status waned in relation to other institutions such as Felix Meritis. Indeed, a common theme in the history of Dutch science during the second half of the eighteenth century is the replacement of traditional loci of scientific education and authority with a variety of societies, ranging from formal learned societies such as the Hollandsche Maatschappij der Wetenschappen (Dutch Society of Sciences) to a variety of local dilettante societies and reform-minded societies such as the Maatschappij tot Nut van 't Algemeen (Society for Public Welfare). This is not to say that institutions such as universities were eclipsed or replaced by societies. Rather, societies provided a space for innovation and discussion, which as we shall see, spurred cultural and intellectual developments that in turn helped spark debate, reaction, and reform within institutions throughout Dutch society.

Returning to the subject at hand, however, what does the description presented here of the world of anatomy theaters reveal about the culture of Dutch science during the Enlightenment? In addition to hinting at the rich variety of life within the walls of this institution, a pattern begins to emerge once we plot this series of details upon the matrix of social, political, and intellectual developments.[23] Among other things, we find a gradual shift in the content and deployment of the discourses of utility.

The anatomy professor Goverd Bidloo stressed the useful importance of anatomy in his inaugural lecture of 1688, describing it as the very foundation of medicine. But as evidenced by the setting, decor, and accompanying moral commentary of public anatomies in the seventeenth and early eighteenth centuries, contemporaries saw the utility of anatomy as extending far beyond medical practice and secular enlightenment. It was meant to help instill a sense of resigned modesty in its spectators. For while the efforts of doctors might prolong life by a few days, the cadaver was at once witness to the transcendence of divine creativity and warning of the ultimate fate of all humanity, already determined by original sin.[24]

When Andreas Bonn gave his inaugural lecture in 1771, the subject was still the importance of anatomy for the advancement of medical knowledge and

23. Though the institution has been totally transmuted, it still exists today. Each year, the University of Amsterdam and the *Volkskrant* (a major Dutch newspaper) jointly sponsor a public lecture on a medical topic assumed to be of wide interest; the public lecture series is entitled the "Anatomy Lesson."

24. Goverd Bidloo, introduction to *Ontleding des Menschelyken Lichaams,* cited in Rupp, "Matters of Life and Death," 276; Lunsingh Scheurleer, "Un amphitheatre," 229.

practice.[25] But as evidenced by the work of contemporaries such as Petrus Camper, a dramatic shift in the moral lesson of public anatomies had taken place by the second half of the eighteenth century. On more than one occasion, Camper admitted that he turned to the animal world for specimens to anatomize publicly because human cadavers were unavailable. But he claimed that he was thereby able to demonstrate a deep truth that contrasted with the superficiality of Linnaean natural history and set the superiority and perfectibility of humankind in sharp relief. What his audience witnessed was both science in the service of furthering knowledge and science as a justification for placing nature ever more in the service of human interests. Camper projected science, in other words, not as the herald of human frailty, but as a set of practices that celebrated humanity while serving the "public good."

As we have seen, public anatomies were also occasions for the promotion and advertisement of cultural authority and status. Anatomy professors and their colleagues, both in universities and anatomy guilds, justified their elevated positions in society by projecting themselves as the bearers of socially and spiritually useful knowledge. But, also in the name of utility, public anatomy displays spread out beyond the institutions that traditionally housed them in the latter eighteenth century. As already mentioned, Andreas Bonn took his show out of the Amsterdam surgeons' guild house and into a new arena provided by the establishment of the Felix Meritis Society. There, anatomy was related to a complex formula in which the long-standing equation of beauty and truth was bracketed by a new form of sociability. In 1791, the city council of Leiden asked the university's medical school to be of service to the city's surgeons and their students by offering public anatomy lessons in Dutch every winter. This both infringed on the traditional rights of the Leiden Surgeons' Guild and showed the university stepping beyond its self-imposed bounds of elitism by teaching in the vernacular. The professor in charge, M. S. du Pui put a rather different spin on this move in 1807, commenting that the university gave these yearly anatomy lessons free of charge so as to be "as useful as possible" to the city.[26]

In the name of utility, the university, which had been suffering a decline in enrollment and income during the second half of the century, thus sought to extend its authoritative reach to encompass not only the learned medical profession, but the traditionally separate corporation of surgeons as well. A similar story can be told of the relation between the *collegii medici* (oversight

25. Nuyens, "Ontleedkundig onderwijs," 87.
26. Molhuysen, *Bronnen tot de Geschiedenis,* 7:*92–*93. As an indication of how local differences complicate Dutch history, it should be noted that anatomy lessons in Groningen were presented in the vernacular as early as 1716.

committees composed of local and influential doctors) of cities such as Amsterdam and Groningen and the medical community writ large. Attempts to regulate medical practice invariably translated into control (often through the threat of exclusion), all done in the name of the public good.[27]

Because of the localized nature of most Dutch regulatory and corporate networks prior to the nineteenth century, such instances cannot be generalized to provide a unified picture of historical change. Even with the dismemberment of Dutch guilds in 1798 and the call for a national program of medical education reform, the pace and shape of reform differed with locale. Neither could the widespread discourse of utility fully mask the bald features of government bankruptcy and—once the locus of decision making shifted to Paris—distance. A sad note thus sounded in the chorus of pronouncements by the turn of the century, as interested educators and officials decried the state of the spaces, instruments, and collections with which they hoped to display the promise of their utilitarian visions.[28]

The Leiden Physics Theater

In January 1675, the University of Leiden granted two requests to teach experimental physics, making it the first university where such a course was taught. The chemistry professor Carolus Dematius was granted the right to incorporate an experimental physics course into the schedule of classes he taught at the university's chemistry laboratory, which had been established in 1669 under the auspices of the university's medical school. Simultaneously, the philosophy professor Burchard de Volder was not only granted the right to teach experimental physics, but the university also provided for the acquisition of instruments and a place to house and demonstrate experiments with what would become an enviable collection. Thus was the physics theater of the University of Leiden born.

Like Dutch anatomy theaters, the university physics theater was a place of demonstration and display. Unlike anatomy theaters, however, we have no records to indicate that it was overtly decorated to project a particular moral message. Neither did it house the sort of public rituals that usually attract the attention of cultural historians. Nonetheless, we can read its interior—both its furnishings and the lessons given there—as a sort of text of the institution's significance.

Strikingly, the first thing one noticed upon entering the building's foyer

27. For the growing dominance of the learned medical profession over other forms of healing in the Netherlands, see Huisman, "Gevestigden en buitenstaanders op de medische markt," esp. 121.

28. See, for example, Molhuysen, *Bronnen tot de Geschiedenis,* 7:*90.

was a large iron balance suspended from the ceiling.[29] A number of weights and instruments sat in the cabinets (some open, some closed) that lined the walls leading to the auditorium. There, one found the basic accoutrements needed for an experimental physics course: a blackboard, air pump, table, stove, and a large vice. Other demonstration instruments and models were kept in the cabinets along three walls and in the center of a room next to the auditorium. The fourth, window-filled, wall of the room provided light and offered a no-nonsense view of the alley in front of the building. From the descriptions we have of the physics theater, it is not clear whether or where demonstrations were rehearsed in the building. Certainly in the case of Willem 's Gravesande and Petrus van Musschenbroek, the two most famous professors to teach there, preparations were done at home in their own well-equipped private laboratories. In any event, as announced by the hanging balance and representative instruments that greeted students and visitors as they entered the front hall, the physics theater at the University of Leiden was a Baconian arena where nature was measured and manipulated in a concerted effort to reveal its factual truths.

This is hardly surprising once we learn that De Volder apparently got the idea to teach physics through experimental demonstration while on a visit to London, where he spent time at the Royal Society. He left no personal record of what specifically inspired him, but we can assume that the atmosphere of this community of amateurs cooperatively engaged in the direct investigation of natural and technical phenomena was extremely appealing. For neither the Royal Society nor Dutch experimental philosophers of the late seventeenth and eighteenth centuries considered the empirical pursuit of scientific "matters of fact" an end in itself. Both explicitly intended science as a way to worship God and serve the public good simultaneously. Both were guided, in other words, by the union of natural theology and social utility.[30]

This is not to say that what De Volder instituted at Leiden was a carbon copy of what he witnessed in London, nor that the curriculum's moral and philosophical messages remained constant throughout the eighteenth century. University life had its own dynamics, which made the teaching of experimental physics there a very different affair. Physics, experimental or otherwise, was taught in the faculty of philosophy, where students traditionally took their first degree. This meant that the students De Volder and his colleague Wolferd Senguerd taught were generally young—students as young as fifteen were not

---

29. The following description is drawn from an inventory taken in 1752. See "Catalogus van alle de physische instrumenten welke in het Laboratorium Physicum bevonden zijn in het jaar 1752," cited in De Clercq, "In de schaduw van 's-Gravesande," 154–55. On the symbolically rich character of the balance, see Bensaude-Vincent, "The Balance."

30. Ruestow, *Physics in Leiden*, 97; Bots, *Tussen Descartes en Darwin*. For the Royal Society, see Jacob, *Robert Boyle and the English Revolution*.

uncommon—and not necessarily interested in pursuing physics as a serious course of study. They were there largely to be entertained (their idea) and disciplined (an idea shared by their parents and the university). The professor's challenge was to see that they learned something of value along the way. It was this triangle of interests that shaped the university career of experimental physics at Leiden.[31]

An introductory course taught primarily to young, not necessarily scholarly students was bound to be superficial in terms of its scientific content. And the idea of incorporating research into the curriculum was definitely out of the question. What purposes, then, did the course serve? First, experimental physics was a popular subject that, especially in the first half of the eighteenth century with the tenure of 's Gravesande and Musschenbroek, brought students, glory and revenue to Leiden.[32] In more strictly defined pedagogical terms, it also served to engender a carefully constructed form of discipline, one directed simultaneously at training students' minds and teaching them their place in the divinely ordained worlds of nature and society.

The problem of discipline had been a pedagogical as well as social preoccupation of universities since their inception. Ramist and, later, Cartesian critics often condemned the sort of discipline that Scholasticism imparted to students, claiming that it taught them to argue incessantly rather than training their minds for the purpose of critical inquiry and deduction. While the introduction of Cartesian and Newtonian "methods" to Dutch universities have often been examined in terms of the natural philosophical/cosmological systems with which they were associated, perhaps it would be more fruitful and historically accurate to see them at least partly in light of this pedagogical tradition of seeking an appropriate and efficient method of discipline.

The very word "method" is historically significant here. Academics interested in pedagogy understood it to mean "the organization and simplification of subject matter for the sake of more facile teaching" and they spilled much ink in discussions of what constituted the proper method. When philosophers such as De Volder introduced the "experimental method" to their students, they did so in the conviction that it would make the study of physics clear, easy, and enjoyable.[33] Labeling De Volder a "Cartesian," as a number of historians

---

31. This description is also apt for similar courses taught at universities throughout Europe in the eighteenth century. See Heilbron, *Elements of Early Modern Physics*, chap. 2. Regarding students at the University of Leiden, see Schama, "Enlightenment in the Netherlands," 68: "Leiden resembled Oxford or Cambridge much more closely than a Scottish university [in the 1770s and 1780s], its student population dwindling and spending more time hunting, duelling and whoring than in the lecture theaters."

32. Ruestow, *Physics in Leiden*, 100.

33. The definition of pedagogy here is from Ruestow (ibid., 16), who refers explicitly to 's Gravesande's "experimental method" (135).

have done, can lead to great difficulties both in interpreting his natural philosophical views and explaining why he was drawn to the experimental method as a pedagogical tool if we continue to see such labels as indicating philosophical content rather than as a useful mantle of critique that one might have donned in the service of pedagogical reform.[34] In this latter view, philosophical "inconsistencies" fade to the background in favor of a more contextual understanding in which a philosophy professor is seen at work, trying to be an effective teacher and well regarded academic. A Cartesian in late-seventeenth-century Leiden can then be seen as someone who was, first and foremost, critical of the continuing attachments to Scholasticism that surrounded him and open to new methods that freed the mind from slavish attention to books in favor of directly investigation of the world. No contradiction need thus be posited between professing Cartesian allegiances and advocating pedagogical reliance on the experimental method.

An examination of lecture notes taken by one of De Volder's first students (1676–77) shows him interested in experimentally demonstrating the properties of natural elements (air, water, earth, and fire) rather than the doctrinaire truth of any particular philosophical system. Aristotle is frequently criticized, but Descartes's explanations are also called into question. If anyone stands out in a positive light it is Robert Boyle, whose clearly reported experiments were often "improved" by De Volder for presentation to his students.[35]

Judging from the reputations of 's Gravesande and Musschenbroek (both during the eighteenth century and since), the "Dutch Newtonians" who taught in the Leiden physics theater from 1724–42 and 1742–61 respectively, one might think that this perspective is less applicable to the period that begins with their tenure. In fact, the opposite is true, for rather than treating "Newtonian" as a prepackaged term with which historians might categorize these scholars, we can begin with the question of how the label "Newtonian" was used at the time. As was true for their predecessors, the professors 's Gravesande and Musschenbroek viewed experimental demonstration as a method that rendered physics both palatable and comprehensible for their students. It was, in other words, the most pedagogically effective way to discipline their field of study.[36]

If one wanted to keep abreast of the latest developments in the world of physics in the 1720s and 1730s, one had to visit 's Gravesande in his home, where

34. Heilbron, for example, refers to De Volder as a "moderate Cartesian." See Heilbron, *Elements of Early Modern Physics*, 132.

35. *Experimenta Philosophica Naturalia Auctore [Magistro] De Valdo [De Volder] Lugd. Ann. 1676.* Rienk Vermij originally uncovered this manuscript. I would like to thank him and Peter de Clercq, who kindly shared his transcription of the document with me.

36. See, for example, 's Gravesande, *Mathematical Elements*.

he kept a full stock of up-to-date experimental instruments. Visitors such as Voltaire, who was working on his own popular presentation of Newton's natural philosophy when he came to Leiden in 1736, did just that. Until 's Gravesande died and the university purchased his collection, the content of its physical cabinet had a much more modest reach. In 's Gravesande's university class, students witnessed vacuum demonstrations made with De Volder's air pump—modified to make up for its age—and scale models of machines such as mills and hoisting equipment. These practical devices could indeed serve as a vehicle for bringing the abstract propositions of Newtonian physics to life for students not well versed in mathematics. But they also inspired the same sort of entertaining fascination in their audience as did the mechanical dolls and model machines on display in shows and fairs throughout Europe.[37]

Teaching experimental physics entailed more than just keeping students entertained, of course. Musschenbroek repeatedly invoked the concept of utility to explain his predilection for this pedagogical approach. By "utility," he meant first that the vehicle of experimental demonstration most effectively facilitated the expansion of natural knowledge. But it was not knowledge for its own sake that he and his colleagues ultimately wanted to impart. Consequently Musschenbroek ordered the usefulness of various sciences in accordance with how close they brought people to the knowledge of God.[38] For like his Dutch forebears and contemporaries, Musschenbroek ultimately desired to demonstrate the teleological nature of his investigations. This, perhaps more than anything, explains what drew him and 's Gravesande to Newtonian natural philosophy. Their allegiance to Newtonianism was an effect rather than a cause of their worldviews.

In this regard, the performances given four times a week at Leiden's physics theater were not so different from those given to the public on a yearly basis in Dutch anatomy theaters. A study of nature—whether on the microcosmic level of the human body, the macrocosmic level of the heavens, or the level of any physical system in between—revealed the handiwork of God, who created and sustained the universe. In class, university orations and anatomical defenses alike, it was this lesson that stood out most clearly. Studying nature was an act of piety in the sense that it placed all natural phenomena, including human existence, in the purposeful context of divine creativity. A presentation of nature's systematic design simultaneously demonstrated the purpose inherent in God's universal plan and placed society squarely within that organized structure. Divine creativity, human modesty, and morality went hand in hand—a place for everything and everything/everybody in its place.

37. Stafford, *Artful Science.*
38. Ruestow, *Physics in Leiden,* 146; De Pater, *Musschenbroek,* 323.

With all this said, however, it would certainly be naive to assume that a curriculum built around demonstration instruments and models could escape being affected by that fact. The presence of instruments always helps to shape the complex trajectory of which they are a part. In the case of the University of Leiden's physics theater, the introduction of experimental physics in the 1670s was received as a great novelty that helped to raise the university's status and reputation. This, along with the other pedagogical considerations already discussed, turned reliance on instruments into a pedagogical imperative. If, then, the physics theater's instrument collection failed to keep pace with contemporary discoveries, innovations, and fashions, the institution risked losing its luster. By the late 1700s, the presence of a good professor was no longer sufficient to guarantee success. Indeed, the very issue of obtaining attractive professorial candidates came to rest on the state of the university's cabinet of physics.

In an interesting and revealing article, Peter de Clercq examines the upkeep and augmentation of the physics theater's instrument collection during the second half of the eighteenth century.[39] His research reveals that, at least from a quantitative perspective, the charge that physics at Leiden suffered a great decline in the decades following the deaths of 's Gravesande and Musschenbroek is overstated. In fact, the size of the instrument collection doubled between the years 1742 and 1811. As De Clercq himself admits, however, examining this issue from a qualitative perspective is much more complicated.

Two sorts of evidence are worth considering to resolve the qualitative question of how the physics theater fared in the seemingly less glorious, second half of the eighteenth century. First is the testimony of the professors regarding their inventory. Second is the relation between the instrument collection and the "growth of science" generally. By the term "growth of science," I refer to the discoveries, innovations, and fashions of the day that required the acquisition of new instruments in order to demonstrate their principles and phenomena.

As already stated, money was allocated to purchase 's Gravesande's impressive private collection when he died in 1742, making the university's physics theater the best stocked school of experimental demonstration in Europe at that time. For a number of years hence, it managed to maintain its wealth and reputation. The collection included air pumps and devices to demonstrate Newtonian physics and optics, but also model steam engines, windmills, and static electrical equipment.[40] If demonstrations of the wondrous regularities of

---

39. De Clercq, "In de schaduw van 's- Gravesande."
40. De Clercq, *Leiden Cabinet of Physics*, 8; In 1782–83, the Russian ambassador, Prince Dmitri de Gallitzin, sold his electrostatic generator—purported to be the best in Europe at that time—to the University of Leiden (ibid., 162).

nature, the spectacle of electricity, or the power of nature were not enough to provide enlightenment and entertainment, demonstrators could always turn to devices such as the magic lanterns that were also part of the collection.

The professors who followed Musschenbroek tried to carry on the tradition of entertaining and enlightening education. Jean Nicolas Sebastien Allamand purchased an electrostatic generator and nine accompanying demonstration toys, including a "thunder house" and a doll with hair from the instrument maker John Cuthbertson in 1775, and an even more dramatic (and expensive) electric machine from the collection of the Russian ambassador Prince Dmitri de Gallitzin in 1782. Allamand was also responsible for purchasing the university's first model steam engine in 1762. To ensure the greatest exposure and publicity, by 1777 it and two engines of different design had been placed in the yard of a nearby orphanage, where students and other passers-by could witness technological developments that some thought would make the windmill obsolete.[41]

Allemand's successor, Christiaan Hendrik Damen, began his tenure as professor and director of the physics theater in 1787 by ordering a thorough cleaning and replacement of worn-out instruments, which he had made largely according to 's Gravesande's fifty-year-old specifications. He saw to it that the instruments were displayed in new cabinets that highlighted their beauty and entertainment value while reflecting the "glory of the Academy." But he was generally expected to maintain the collection under his charge with an annual budget of only two hundred guilders, and he had to justify further expenditures to the university's governing body. In a letter addressed to the council in 1790, Damen employed a combined rhetoric of utility and status to plead for support: Some sixty students, including the son of the prince of Orange, depended daily on Damen and his instruments to learn the basic principles of physics. He owed it to them to teach a complete course, which depended on the availability of needed apparatus. Furthermore, the University of Leiden, "birthplace and cradle" of experimental physics, could not afford to risk its reputation by failing, in this case, to pay for the acquisition of a double-barreled air pump—a machine that could increasingly be found in competing collections throughout Europe.[42]

His request for supplementary funds was granted, but his struggle foreshadowed the difficulties that followed. Damen died in 1793 and was succeeded by the young and well-respected Pieter Nieuwland, previously lecturer in mathematics, astronomy, and navigation at the Athenaeum Illustre of Amsterdam. He died, unfortunately, after only a few months, and the search for a

41. Van der Pols, "De introductie van de stoommachine."
42. On set-up of cabinet, see De Clercq, "In de schaduw van 's-Gravesande," 168; letter quoted 149.

replacement began again in 1794. The university attempted to lure high-profile candidates such as Martinus van Marum, Jan Hendrik van Swinden, Georg Christoph Lichtenberg of Gottingen, and Simon L'Huiller of Geneva with offers of a rather substantial salary, but to no avail. Van Marum was already engaged in Haarlem, jealously nurturing the increasingly impressive instrument collection held by the Teyler's Institute. Van Swinden was happily caught in a web of activities in Amsterdam that included giving frequent demonstration lectures at the Felix Meritis Society and holding court as an enormously popular professor at the Athenaeum Illustre. And, as other universities throughout Europe became more willing to equip physics theaters of their own, Leiden could no longer count on its reputation and the call of cash to attract the candidates of their choice.[43]

Resignedly, the university chose not to appoint a professor at all, and instead hired the relatively unknown Simon Speyert van der Eyk as experimental physics lecturer in 1796. With diligence rather than an illustrious reputation to recommend him, Speyert van der Eyk actually proved the best man for the job. Promoted to the rank of professor in 1799, he served as the physics theater's bulldog during the fiscally lean years to come. As he well recognized, the challenge that faced him and the theater actually had two sources. Not only was funding scarce at the turn of the century, owing to the exigencies of revolution, war, and foreign occupation, the contemporary trends of scientific development placed enormous demands on instrument collectors who wished to stay current. In order for a course in experimental physics to retain its credibility by the turn of the century, its cabinet had to be equipped not just with the basics, but also with devices for demonstrating at least the fundamental principles of the "new" chemistry, the properties of so many recently discovered gases, and the essential phenomena of galvanic electricity.[44]

In 1807, Speyert van der Eyk wrote to the university's curatorial council and offered a sobering picture of the state of the physics theater and the instrument collection it housed.[45] The physics theater, he explained to his colleagues, was the victim of its own glorious history of pedagogical innovation. The historically constructed imperative of relying on demonstration devices meant

43. For biographical information on Nieuwland, see Snelders, *Het Gezelschap der Hollandsche Scheikundigen,* 57–62. For Van Marum's response, see letter to Leiden professor D. van Wijnpersse, cited in De Clercq, "In de schaduw van 's-Gravesande," 168, n. 39. For Van Swinden, see letter quoted in Kox, *Van Stevin tot Lorentz,* 111–12.

44. Note how the modern disciplinary boundaries between physics and chemistry had no practical meaning at the time. To further complicate matters, chemistry continued also to be taught—through a similar scheme of demonstration lectures—at the University of Leiden's medical school.

45. "Memorie van Professor Speyert van der Eyk," 1 March 1807; reproduced in Molhuysen, *Bronnen tot de Geschiedenis,* 7:90*–92*.

that course content was now dictated by what was available rather than what was of current interest. Thus, the study of hydraulics, electricity—both static and galvanic—the characteristics of gases and the principles of modern chemistry in general, steam engines, and the phenomena of light and magnetism were all seriously compromised by the incompleteness of the instrument collection. Finally, the physical plant itself needed to be replaced by a drier environment, where the collection—such as it was—could be properly conserved.

Overall, this seems a rather bleak picture with which to end a discussion of such a historically important institution. But the physics theater's cultural presence was not fully bounded by its physical condition. Even during its darker hours (from which it went on to recover later in the nineteenth century), it embodied the multifaced ideal of utility.[46] Speyert van der Eyk could thus rely on a utilitarian definition of knowledge to argue for increased funding: demonstration of scientific truth meant precisely that claims had to be physically demonstrated to pass as scientific knowledge. Further, he and his colleagues could present the theater's entire history, along with the university's and that of the Dutch population in general, as a worshipful hymn to utility.

A memoir on the University of Leiden's situation, written in 1806, began by proclaiming that the quality that best characterized the Dutch was their love of science and the arts, "especially in terms of their application to useful knowledge." This passion manifested itself not only in the disproportionate number of creative geniuses, in comparison with the rest of Europe, who graced the pages of Dutch history, but also in the attention long paid to education in the Netherlands. Even in times of difficulty and political turmoil, the Dutch did what they could to promote public instruction. William of Orange had established the University of Leiden in 1574, after all, as a way to thank the city's inhabitants for their brave assistance in vanquishing the Spanish. The gift he offered Leiden—and it was certainly not necessary that his gift be a university—was an institution meant to produce graduates capable of "governing, enlightening, and bringing honor to their country" throughout the centuries to come. The university's physics theater was explicitly intended to reflect the national character of the Dutch and help fulfill the university's mission.[47] It had

46. Speyert van der Eyk's request for relocation was granted in 1824. The collection was moved a second time in the 1850s. It was at this location that work leading to a whole series of Nobel prizes in physics by the turn of the twentieth century was carried out. On the institution's history up through the twentieth century, see De Clercq, *Leiden Cabinet of Physics*.

47. "Mémoire succinct sur la Constitution actuelle de l'Université de Leyde," 5 July 1806; reproduced in Molhuysen, *Bronnen tot de Geschiedenis*, 7:99*–106* (first quotation, 99*; second quotation, 101*). The reference to utility actually went beyond Dutch national borders. Latin was also expressly chosen as the language of instruction for the benefit of foreigners who might wish to study there. See ibid., 103*.

served this purpose for most of its history. Its caretakers, such as Speyert van der Eyk, wanted to ensure that it continued to do so.

Science Lecturers outside the University

As explicitly indicated by the book's title, Petrus van Musschenbroek wrote his *Beginselen der Natuurkunde, Beschreven ten dienste der Landgenooten* (Principles of Physics, Described as a Service to my Fellow Countrymen, 1736), as a sort of applied paean to utility. He was pleased to note that science had become a fashionable subject and hoped to put that fashion to good social use. In his introduction he commented: "Never has one met so many amateurs of Natural Science in the United Netherlands as at the present time. For this science is blossoming not only among the most scholarly, but also among a number of distinguished merchants, and people from every social rank and dignity."[48] It was especially for those outside academia that Musschenbroek wrote the *Beginselen*. The book found a ready audience, as did a number of individuals who offered experimental philosophy courses to anyone who was interested and willing to pay.

Examining the history of itinerant and private science lecturers in the eighteenth-century Netherlands takes us into a world in which the ideal of utility combined with piety, curiosity, sociability, and entrepreneurship to "advance" science in ways not usually recognized in histories that focus on ideas and elite institutions. But mixing business with pleasure, the work of the hand with that of the mind, and consideration of the here and now with the hereafter were all hallmarks of science with which the Dutch in the eighteenth century were comfortably familiar.[49] That knowledge might be produced and disseminated outside the elite circles and institutions of society was also something that rang true for a heterodox but certainly not egalitarian community such as the Netherlands.

The most famous traveling science lecturer to come to the Netherlands was John Theophile Desaguliers, who gave courses in Den Haag and Rotterdam during the early 1730s. An advertisement for his visit to Rotterdam explained that he would give lessons twice a day (once in English and once in French; he left Latin to the universities), followed by experimental demonstrations made with a host of instruments he brought from England for just that purpose. The subjects he covered included Newtonian mechanics, "applied celestial phenomena," hydrostatics, pneumatics, and Newtonian optics. An extra course of

---

48. De Pater, *Musschenbroek*, 30 (my translation).
49. See, e.g., Stewart, *Rise of Public Science*. The Musschenbroek family was a good example of this trend, combining the trade of instrument making with scholarship and demonstration. See De Clercq, *Sign of the Oriental Lamp*.

fifteen lessons in astronomy was also offered with a double promise. Those attending would receive a better education than they might otherwise get from an entire year of study. And subscription to a combination of courses could be had at a special reduced price.[50]

Desaguliers aimed his lectures at an audience not versed in mathematics but eager to glimpse the divine order of nature promised in Newton's arcane presentation. As a contemporary Dutch observer put it: "[T]hrough a series of clear experiments and observations, the entire Newtonian system was made so clear and comprehensible that people of all ranks and occupations, yes even women, attended his lectures with extraordinary eagerness, and thereby became enamored of this useful science."[51] Desaguliers's tour of the Netherlands was an enormous success.

His success could be measured in part by the flattery of imitation, with Dutch lecturers offering courses inspired by Desaguliers's example. But even before his arrival, at least one lecturer was already teaching experimental physics to interested Amsterdammers. Daniel Gabriel Fahrenheit began offering courses on subjects including mechanics, hydrostatics, optics, and chemistry in his residence on the Leidsestraat in the spring of 1718.[52] By this time, he had a growing reputation for the instruments he designed and made, a reputation he helped to establish by sending examples of his work to high-profile patrons such as Leibnitz, Christian Wolff, and after he moved to Amsterdam in 1717, Boerhaave, and 's Gravesande. His lectures, illustrated with demonstrations performed on his own instruments, further enhanced his name, and in 1719 he was asked by a group of Mennonite merchants to teach a course for them.

This Mennonite connection is not without significance. Their dissenting community proved enormously important for the propagation and popularization of science outside Dutch universities during the period under study here. Despite the Netherlands' reputation for tolerance, communities such as the Mennonites found themselves tacitly blocked from elite positions and institutions in mainstream Dutch society. De Volder, for example, had to give up his Mennonite faith and join the Dutch Reformed Church in order to obtain his professorship at Leiden. This lack of complete social integration had two significant results. First, discouraged from other avenues of advancement, Mennonites had focused their ambitions in commerce, often with great financial success. Second, they had had to find innovative outlets for their strong

<hr>

50. Van Lieburg and Snelders, *"Bevordering en Volmaking,"* 6–7.
51. Ibid., 7 (my translation).
52. Van der Star, *Fahrenheit's Letters;* Cohen and Cohen-Meester, "Fahrenheit," (2 parts). Much of Van der Star's discussion is drawn from the earlier articles by Cohen and Cohen. On imitators of Desaguliers, see for example, the discussion of Leonard Stocke, below.

interest in exploring the world of God's creation. Consequently, Mennonites were active in attending lecture courses as a means of becoming acquainted with the workings of nature, and even the education of Mennonite preachers included natural philosophy in its curriculum.[53] As we will see, they were also a moving force in the establishment of many learned societies during the second half of the eighteenth century.

Mennonites were not the only ones who found spiritual satisfaction in the investigation of nature, of course; consider the enormous popularity of Bernard Nieuwentijt's *Het regt gebruik der wereltbeschouwingen ter overtuiginge van ongodisten en ongeloovigen aangetoont* (The right manner proven of contemplating the world in order to convince atheists and infidels, 1715).[54] But they were pleased to find that Fahrenheit's courses, like those of virtually all his fellow Dutch lecturers, were flavored with a strong dose of natural theology. Over the years, their business and related social activities had taken the Mennonite community increasingly into the temporal world. Placing that world in the investigative context of physico-theology was a way of reconciling their religious predilections with what had become part of their everyday lives.

Even in this context, however, the investigation of nature was no somber affair. Alongside the laws of optics and God's wondrous construction of the human eye, for example, Fahrenheit demonstrated entertaining tricks with mirrors. Alongside the chemical relations that God had established between bodies and the medicinal uses of various substances, Fahrenheit revealed the secrets of transmuting base metals into noble ones.[55] The worldly and the otherworldly, the abstract and the mundane, were everywhere intertwined in Fahrenheit's lectures and his work in general.

Fahrenheit justified his own pursuit of natural knowledge and the construction of instruments through his faith in the teleological order of nature. At the same time, he took his units of measurement from local assayers and that handiest of beakers, the beer glass. While motivated at least partially by economic necessity (Fahrenheit had run through virtually all of his inheritance by the age of thirty-one), it was his conviction that nature was governed by consistent laws that drove him to demonstrate their natural effects. But he was also aware that his instruments' accuracy, so necessary for demonstrating that consistency, was contingent on specific details such as the provenance of the

---

53. On De Volder, see Ruestow, *Physics in Leiden*, 4, 74; on Mennonite education, Dekker, "Popularisering der Natuurwetenschap," 179.

54. See Vermij, *Secularisering en Natuurwetenschap*.

55. Van der Star, *Fahrenheit's Letters*, 9, 11; Cohen and Cohen-Meester, "Fahrenheit," 382, 383. Fahrenheit's own model of the eye, incidentally, still exists at the University of Groningen.

glass he used.[56] Fahrenheit aspired to universality while immersed in particularity. It was in this atmosphere of ambivalence that he began developing what came to be known and standardized as the Fahrenheit scale.

A number of other, less famous, lecturers used a similar recipe in their courses, combining the natural, the practical, the spiritual, and the entertaining to appeal to their audiences. Martinus Martens set up shop in Amsterdam in 1736, teaching experimental physics courses for a number of years with the express purpose of demonstrating "the utility and amusing nature of natural science" and revealing the "unending wisdom and great beneficence of the Creator." His fellow Amsterdammer Jan van den Dam constructed a planetarium to help him relate a similar message to his students. In Rotterdam, the doctor Willem de Loos taught courses to a self-styled chemical society until his death in 1771. However he advertised his work, in his lectures he focused primarily on chemical amusements such as tricks with disappearing ink.[57]

More prominent was Benjamin Bosma, who taught courses to the "good burghers" of Amsterdam from 1752 to 1790. His students included wealthy merchants, many of them Mennonites with private instrument collections from which Bosma borrowed for his lectures; doctors, also mainly Mennonites; scientific amateurs and inventors, including the instrument makers John and Jonathan Cuthbertson, who became well known themselves, particularly for the large-scale electrostatic generators they built and the natural philosophy courses that John (inspired by the work of Joseph Priestley) taught in the Netherlands during the 1770s; and a number of women, for whom Bosma organized special courses.[58]

Bosma's story is a good one for demonstrating the entrepreneurial spirit that, alongside a desire to pursue knowledge for the sake of edification and spiritual uplift, lay behind the establishment of private lecture courses. Bosma took his degree in philosophy at the University of Franeker in 1749. Upon finishing, he immediately set up shop as a private lecturer. Seeking to control and limit competition, the university senate fell back on a regulation that prohibited teaching without a special license, ordering Bosma to close down and refusing to grant him the license to practice. Bosma was thus forced to seek his fortune elsewhere. Arriving in Amsterdam in 1752, a much larger city with

56. Van der Star, *Fahrenheit's Letters*, passim. Fahrenheit repeatedly referred to his use of beer glasses to measure substances for experimental purposes and discussed the provenance of glass extensively.

57. Dekker, "Popularisering der Natuurwetenschap," 174; Lieburg and Snelders, *Bevordering en Volmaking*, 9.

58. Keyser, "Het intekenboek van Benjamin Bosma"; Hackmann, *John and Jonathan Cuthbertson*.

less restrictive regulations in these matters, he was able to begin teaching right away.

Bosma had little success in attracting students during his first three years despite the bargains and special package deals he offered. By 1757, he had cut his tuition prices in half and, perhaps more important, switched the orientation of his course offerings from logic, metaphysics, and mathematics to experimental physics. These measures did the trick. Bosma's courses began to fill up, and he was able in later years to raise his rates without suffering a decline in attendance.

It would seem that Bosma had good business sense to match his philosophical acumen. As was the practice at that time, he supplemented his income by selling wine, tea, pipes, and tobacco for consumption during his lectures. He also published a book to serve as follow-up to the courses he taught. In both his lectures and publications, he dished up a popular mix of science and religion, arguing that to be a good philosopher one must also be a Christian. The message sold well and so did his book. With such successes behind him, it is no wonder that Bosma also managed to establish an equally successful business college in 1767.

Bosma's Koopmansakademie offered a six-year program for boys who could already read and write, beginning at the age of nine. The curriculum included bookkeeping, accounting, geometry, algebra, navigation, astronomy, languages, geography, and knowledge of trade, as well as practical experience at the stock exchange and auction house. Once a week, students followed courses in natural science, mechanical engineering, commercial law, and religion. On Sundays, they studied geography of the Holy Land. Detractors questioned what a philosopher had to teach students about business, but the editor of the spectatorial paper De Koopman (The merchant) found the school important and innovative enough to dedicate an entire issue to its discussion. Further, students must have been forthcoming, as Bosma was able to hire other teachers to teach some of the courses for him.

Bosma did not mix socially or professionally with the regent elite of Amsterdam, but he was certainly a respected citizen. In 1777, he helped found the Amsterdam society Felix Meritis, an institution that would provide an alternative space to his own courses for science lectures and demonstrations as well as offering a center for bourgeois sociability in general. Bosma served as secretary to the society's governing board in its first year and opened its department of science, where he helped demonstrate electrical phenomena. But he resigned the following year for reasons that are unknown to us. Given his subsequent criticism of the society's department of commerce—he claimed that it offered the public little of practical value—one senses the sort of bitterness that might have come from helping to give birth to the competition.

Bosma was not the only lecturer forced to compete with the newly established scientific societies that were proliferating throughout the Netherlands in the last decades of the eighteenth century. It was the very success of the Rotterdam lecturer Leonard Stocke that helped shape that city's major society, the Bataafsch Genootschap der Proefondervindelijke Wijsbegeerte te Rotterdam (Batavian Society for Experimental Philosophy of Rotterdam). Stocke organized his students—mainly merchants—into an informal society with which the Bataafsch Genootschap would have to compete for social status.[59] To avoid just this competition, Lambertus Bicker, one of the Bataafsch Genootshap's founders, appealed to the city government for official recognition and the privileges that went with it. The Rotterdam society was licensed to offer public lectures in 1770 and thereafter eclipsed other private initiatives in the city.

A contemporary observer remarked that it was Stocke who was largely responsible for the taste that many Rotterdammers had for science. But just what did that taste entail? Whether speaking to the members of the Collegium Physicum Experimentale, which he organized, or to the members of the Chemisch Gezelschap (Chemical Society), which he also led following the death of Willem de Loos, Stocke's approach was similar to that of other private science lecturers of his time in the Netherlands. He turned whatever resources he had at his disposal, such as the fine collection of physical instruments assembled by the members of the Collegium Physicum Experimentale, toward the service of natural theology. Stocke opened his first course in Rotterdam in 1747, for example, with an oration on "the splendor of nature," arguing for God's unending wisdom, power, and goodness through illustrations drawn from geology, mineralogy, astronomy, and "living nature." Repeatedly in his lectures, he sought to support physico-theology in the battle against materialism, especially as represented by authors such as La Mettrie.[60]

Private science lessons thrived in the Netherlands during this period because they satisfied a number of interests simultaneously. For a bourgeois public thirsty for a mix of natural knowledge, spiritual succor, and culturally sanctioned social interaction, these courses offered a forum that was otherwise unavailable. For lecturers, these courses provided a livelihood, though one reads numerous complaints of how expensive it was to conduct demonstration exercises. Such courses also helped establish their reputations as pedagogues

59. Lieburg and Snelders, *Bevordering en Volmaking,* 10, 14.
60. Ibid., 10.

and often as doctors or instrument makers. While universities, traditional bastions of the elite, continued to conduct classes in Latin, these lecturers reached out in the vernacular to a new audience looking for edification and entertainment.

Dutch Societies

Perhaps the most salient characteristic of the Dutch Enlightenment during the second half of the eighteenth century was the widespread establishment of societies. For a growing number of people, enlightenment meant engaging in the form of sociability these organizations provided. Whatever the actuality, many stated that the primary motivation behind this movement was the principle of utility. The first volume of the Hollandsche Maatschappij der Wetenschappen's journal, for example, opened with this straightforward claim: "It is only to be of service that this society was founded." A member of the literary society Dulces ante Omnia Musae explained more generally, "Just as the flowering of arts and sciences is desirable and pleasant for every reasonable human being, so is it certain that these can nowhere be more advanced than by the founding of societies." [61] The discussion that follows does not pretend to treat the role of science in Dutch societies exhaustively. Rather, it focuses on how the practice of science in Dutch societies reflected the various ways that this claim of utility was interpreted.

The growth of societies interested in science actually goes back to the end of the seventeenth century as people began organizing themselves into informal, local societies dedicated to natural investigation, a trend that would peak in the last quarter of the eighteenth century. Beginning with the Hollandsche Maatschappij der Wetenschappen (Dutch Society of Sciences) in 1752, a few groups sought official recognition (and the privileges they hoped would go with it) from city and provincial governments. The granting of a government charter placed such societies in direct competition with the universities, which until then had exercised a monopoly over the production of official learned culture. And in some cases, as we have seen, it also placed them in competition with private lecture courses and informal societies.

The Hollandsche Maatschappij der Wetenschappen itself was at least partially an outgrowth of a number of small informal societies in the city of Haarlem. At the end of the seventeenth century, the Collegium Physicum Harlemense, an informal group of mainly Mennonite teachers and doctors, met regularly to discuss scientific matters of the day, such as local claims made

---

61. *Verhandeling, uitgegeven door de Hollandsche Maatschappij der Wetenschappen te Haarlem,* 3d ed.(1759), 1:1, quoted in Wiechmann, "Van Accademia naar Akademie," in Wiechmann and Palm, *Een elektriserend geleerde,* 60; Van den Berg, "Literary Sociability," 253.

about divining rods. This group seems to have died out quickly, but by the 1730s another society was organized in Haarlem with enough éclat to draw the attention of Musschenbroek. Members of the Natuurkundig College, also primarily Mennonites, shared an interest in astronomy, medicine, the weather, and above all, natural theology. From 1740 to 1788 they gathered in a jointly purchased building that they outfitted with an observatory, planetarium, and instrument collection.[62]

Two of the society's members were among the original members of the Hollandsche Maatschappij der Wetenschappen when it was established in Haarlem in 1752. Despite their years of experience as active society members, though, they were not placed at the organizing center of the Hollandsche Maatschappij's governance. According to the organization's structure, membership was divided between active directors, who were drawn from the elite regent class, and the more passive general membership, which was made up of men who were invited to join in recognition of their "scientific reputation" or in response to an article they published in the society's journal. Thus it was that the orientation of the society came to reflect the interests of its governing elite.

The very name of the organization is telling in this regard. Originally, the society's name was to be the Genootschap van Nederlandsche Geleerden ter bevorderinge der welvaart van hun Vaderland, opgericht door enige Heeren Directeuren resideerende te Haarlem (Society of Dutch Scholars for the promotion of their Fatherland's prosperity, established by some gentlemen directors residing in Haarlem). The Leiden professor Johan Lulofs saved his fellow society members by suggesting that "the simplest is the best. . . . Only the matters [of the society] which are useful for the Fatherland and religion deserve the name science, therefore the title 'Hollandsche Maatschappij der Wetenschappen' is all-inclusive."[63] Prosperity, utility, religion: three watchwords intended to be understood through the invocation of "science." But what did these words mean in the context of a society that represented the cultural interest of the regent class?

Perhaps the best place to look for an answer to this question is the list of topics chosen for the society's regularly offered essay competitions. These essay competitions were an important part of the society's functioning, listed in its charter as a privilege befitting the society's status. The University of Leiden had protested against granting the society any form of official status and saw to it that the universities maintained a monopoly on scholarly Latin lectures. This

62. Bierens de Haan, *Hollandsche Maatschappij,* 3; Sliggers, "Honderd jaar natuurkundige amateurs te Haarlem," 68–71.

63. Bierens de Haan, *Hollandsche Maatschappij,* 8 (my translation).

left the Hollandsche Maatschappij with vernacular discussion, the *Verhandelingen* (Proceedings), and their precious essay contests as the only marks of elite institutionality.[64]

Religion was certainly on the society's agenda. Between 1753 and 1815, prizes were awarded on two occasions (1787 and 1789) for discussions of Mendelssohn's and Kant's arguments for God's existence. Perhaps even more telling, though, was the large number of natural history topics, which reflected the continued interest in natural theology and underscored the directors' utilitarian outlook: natural knowledge was seen to be useful knowledge in that it brought people closer to an understanding and appreciation of God the creator.

Utility thus linked society members to the world beyond, but so too did it tie them firmly to earthly considerations. Among the essay questions posed were a large number of "applied" topics: what to do about the silting of rivers, how to deepen waterways and improve them for transport generally, opium as a treatment for dysentery, the utility of ventilating ships, treatment of sailors' diseases. Trade at home and in the colonies—the source of Dutch regent wealth during the golden age of the seventeenth century—also stood behind the directors' support of of utilitarian science. Yet while Haarlem suffered economically during the second half of the eighteenth century, they made no attempt to stimulate innovative technology or industry through the offices of their society. Instead they looked back to the glory days of Dutch commercial ascendance and sought to turn the energies of the Hollandsche Maatschappij to the task of recapturing them.

In 1778, a private organization was established in the same neighborhood as the Hollandsche Maatschappij that provided a battleground between this traditional, elite conception of scientific knowledge and polite sociability on one hand and a decidedly different conception of utility on the other. That organization was the Teylers Stichting (Teyler's Foundation) and the man largely responsible for precipitating this tension was Martinus van Marum. Teylers Stichting resulted from the behest of Pieter Teyler van der Hulst, a wealthy Mennonite businessman who patronized the arts and sciences throughout his adult life. In his will, he provided for the establishment of an institute with a five-man directorate that would oversee a number of charity projects and maintain two societies, the Godgeleerd Genootschap (Theological Society) and Teylers Tweede Genootschap (Teyler's Second Society), which was dedicated to the arts and sciences.[65] It is the Tweede Genootschap that concerns us here.

---

64. See De Bruijn, *Inventaris van de Prijsvragen Uitgeschreven*. According to their charter, publications as well as lectures and discussions had to be in Dutch.
65. Mijnhardt, "Veertig jaar cultuurbevordering," and *Tot Heil*.

Martinus van Marum, famous for helping to spread the "new" chemistry to the Netherlands and beyond, moved to Haarlem in 1776, where he set up practice as a physician and began giving experimental philosophy lessons in his home.[66] Almost immediately, he was made a member of the Hollandsche Maatschappij der Wetenschappen and became director of the society's cabinet of natural rarities. In 1779, he was invited to join Teylers Tweede Genootschap, and in 1784 was named director of that foundation's library and cabinet. Van Marum used these positions to promote a vision of utilitarian science quite different from the elite conception embodied in the activities of the Hollandsche Maatschappij and favored by the most influential directors with whom he had to deal at Teylers.

Both cabinets overseen by Van Marum were intended to be at the disposal of their respective society's members, and Van Marum was expected to make his knowledge and expertise similarly available. A regular part of his schedule involved giving demonstration lectures to society members. He, however, was convinced that the resources of Teylers should be turned toward experimental work and made available to a wider audience through public demonstration lectures. Throughout his tenure at Teylers, he was forced to do battle with its directors to ensure that money went to the purchase of instruments, fossil collections, and the sponsorship of lectures whose content was socially useful as he understood the term.

Van Marum's successes were intermittent. Not only did he have to struggle against the elitist views of the Teylers Society's most powerful and conservative directors. He had also to struggle with the sociopolitical and economic realities of revolutionary Europe at the turn of the century. With money in short supply, the directors refused to provide a meeting space large enough for the public lectures Van Marum wished to offer and repeatedly tightened the purse strings, making it difficult for him to purchase all the instruments he considered vital for pedagogical and research purposes. (They did, however, find resources to support the foundation's charitable activities and expand its art holdings.) When he and his colleague Adriaan van den Ende did give public lecture courses, the directors monitored their choice of subject matter and manner of presentation, pressing whenever possible for a physico-theological orientation.

No less an observer than Napoleon commented on the wasted opportunity this attitude led to.[67] But Van Marum carried on, teaching chemistry and experimental philosophy to a broad public audience when he could and lobbying constantly for a change in the directors' attitudes. Beginning in 1792, he

66. Forbes, Lefebvre, and De Bruijn, *Martinus van Marum;* Wiechmann and Palm, *Een elektriserend geleerde;* Roberts, "Science Dynamics."
67. Mijnhardt, *Tot Heil,* 335.

offered a yearly cycle of chemistry lectures, relying on Lavoisier's *Traité élémentaire de la chimie* the first year, then switching to Antoine François Fourcroy's popular textbook the following year. He introduced his lectures by justifying chemistry as useful in modern society and industry. Support for chemistry in France, after all, had helped assure military victory.[68]

In 1795, Van Marum answered the directors' cuts in his budget with a new offensive, arguing that the public—increasingly aware and critical of elite activities during this revolutionary era—felt that Teylers did not do enough that was useful or enjoyable for the residents of Haarlem. Perhaps if he could use the society's internationally renowned instrument collection for a series of public lectures, the directors would score a public relations coup. Their response was positive, so long as he promised to begin his lectures with an informal discussion of domestic chemistry (one assumes in place of a salute to chemistry in the service of revolutionary victories). He complied but introduced the opening lecture with an announcement that it was no longer tenable to lecture before such a large audience because of lack of space. Perhaps in the future, he wished aloud, the directors would see things differently and contribute more generously to the public demonstration of chemistry's social and economic utility.

Van Marum continued his public lecture series until 1797, by which point it was clear that the directors had no intentions of providing funds for a larger auditorium or more instrument purchases. Henceforth he was to lecture only for the directors, society members, and a small circle of invited guests. Public lectures under the society's auspices began again only in 1803, taught by Van Marum's friend Van den Ende. But now the directors insisted that the content be physico-theological rather than directed toward social improvement. They further insisted that the audience be limited to members and twenty guests invited from the outside. It would seem that the sort of educational reform and public-mindedness that Van Marum had so long agitated for would have to wait for the different political environment of Napoleonic occupation.

While Van Marum struggled within an established institution, others chose to organize themselves separately in a conscious effort to place scientific and technological advance in the service of the more general community. In 1784, the Genootschap van Kunsten en Wetenschappen tot Nut van 't Algemeen (Society of Arts and Sciences for Public Welfare) was established with the express goal of promoting social, moral, and educational reform. The elite societies, the Nut's organizers argued, had done little or nothing to foster en-

68. For the content of these textbooks and their relation to the chemical revolution, see Bensaude-Vincent, "View of the Chemical Revolution."

lightenment and virtue on a more general level, content to pursue and enjoy the fruits of knowledge themselves. It was thus left up to members of this new society, drawn from the less elevated and politically powerful levels of society, to help alleviate the materially deprived and morally depraved conditions in which too many of their countrymen lived.[69]

Despite the growing atmosphere of revolution in the Netherlands, the Nut was a consciously apolitical organization. The reform movement it was intended to spearhead was driven by sentiments of Christian morality, and both anti- and pro-Orangists were welcomed to its ranks. The goal of social improvement through education of the heart and mind cut across the field of political allegiance. Indeed, the dream and general ideology of the Nut's membership was that only through broad-based cooperation could society's problems be overcome. Universities with their elitist Latin culture and elite scientific societies that promoted nothing beyond their own interests, except perhaps to provide scanty entertainment and lessons in natural theology to a broader audience, could not be counted on to help regenerate either Dutch society or the economic malaise into which it had fallen. Only the kind of education—both scientific and moral—that would help produce a skilled labor force and an upright society could return the Netherlands to its former status as an international power to be reckoned with.

While the directors of the Hollandsche Maatschappij had to make hefty financial contributions to the organization, membership dues for the Nut were left at a reasonable sum so that no interested party would be left out (except Jews, who were banned from membership).[70] While most other societies either aspired to national status or were content to remain local affairs, the Nut featured a national organization and scope reinforced by local chapters. Programs for reform were devised in Amsterdam and sent out to the local chapters for approval and enactment. The local chapters also saw to it that its members were enlightened and entertained through ongoing lecture series. Judging from membership figures, the formula was a resounding success. By the beginning of the nineteenth century, more than 10,000 people belonged to the Nut. It was in the name of a significantly large constituency, then, that the Nut successfully led a drive in the early nineteenth century for national reform of elementary education.

The second half of the eighteenth century in the Netherlands has rightly

69. Mijnhardt, *Tot Heil,* 259–60 (for general background, see chap. 6, 259–94); and "Dutch Enlightenment," 220–22.

70. Nonpaying members were welcome to contribute to the society by submitting essays for discussion and publication, but only directors—carefully chosen in keeping with their social status—had a say in the society's organization. See Bierens de Haan, *Hollandsche Maatschappij,* 9.

been referred to as the "age of societies." They sprang up throughout the country in a variety of forms and in answer to a variety of interests. In major urban centers, large societies such as Rotterdam's Bataafsch Genootschap and Amsterdam's Felix Meritis provided a forum for sociability that centered on information and demonstration. Lectures on a number of scientific and cultural subjects were a regular feature, and demonstrations of new scientific claims and technologies, such as the decomposition of water and steam engines, gave members much to think and talk about. Smaller communities also joined the trend, with the focus of local societies just as likely to be literature as science. In Middelburg, a special society for the promotion of science among women was established in 1785. There, a group of thirty-nine women received instruction in science based on the popularizing publications of the abbé Nollet.[71]

A full account of the culture of Dutch scientific societies would have to cover not only all these permutations, but also the professional societies, not discussed here but organized in a conscious effort to promote the social, cultural, and professional interests of specific groups. Given constraints of space, however, only one more—quite unique—organization will be mentioned. In 1790, a small group of friends formed the Bataafsche Societeit (Batavian Society), which quickly became known as the Gezelschap der Hollandsche Scheikundigen (Society of Dutch Chemists).[72] This little coterie (only six people ever belonged) cooperated in experimental work and were actively engaged with the international world of scientific research and publication through a journal of their own and original articles submitted to the major scientific journals of Europe. They were also active locally, lecturing on the "new" chemistry to members of larger societies and participating in a number of essay competitions.

Three of the members were medical doctors (Jan R. Deiman, Nicholas Bondt, and Gerard Vrolik, who was also professor of botany in Amsterdam); one a pharmacist (Anthonie Lauwerenburg); one a lecturer in mathematics, astronomy, and navigation at the Athenaeum Illustre of Amsterdam (Pieter Nieuwland); and one a wealthy merchant (Adriaan Paets van Troostwijk). Together they gained such an international reputation for their work that many assumed they were a much larger society. Of interest here, though, is the fact that, as a group of chemists, they remained largely outside the structure and strictures of the Netherlands' official institutions of knowledge production.

71. Cohen and Cohen-Meester, "Natuurkundig Genootschap der Dames."

72. On professional societies, see Van Berkel, Van Lieburg, and Snelders, *Spiegelbeeld der wetenschap,* in which they discuss the example of "surgical sociability" (the term is Frank Huisman's) manifested in new societies that brought together surgeons and physicians. On the Batavian Society and the Society of Dutch Chemists, see Snelders, *Het gezelschap* , 136–38.

This might appear surprising at first glance, but when one considers the great distance between their active research program and the constraints of university chemistry teaching, the surprise quickly dissipates. Even at the end of the eighteenth century, academic culture placed chemistry in a subordinate position to medicine or dismissed it as a source of display and amusement in experimental physics courses. It continued to be taught, not as an independent discipline with its own research trajectory, but as a body of knowledge and practices that served the professions of medicine and pharmacy. No wonder, then, that despite plans ordered by the Amsterdam City Council to construct a chemistry laboratory for teaching purposes, none was built. Because Amsterdam lacked a medical school of its own and the Amsterdam Athenaeum's chemistry professor Dirk van Rhyn did not supplement his lectures with any experimental demonstrations, the university could not see the sense.[73] It was only much later in the nineteenth century that universities would sponsor original chemical research. For now, only the private initiatives of individuals such as the members of the Dutch Society of Chemists kept the Netherlands on the map of international science.

Conclusion

In this chapter, I have sought to do two things. The first was to describe with sufficient richness the settings and practices of four different sites where people engaged with science to imply more generally the cultural history of Dutch science during the long eighteenth century. As the reader has no doubt noticed, a certain consonance emerges in the recurring themes of natural theology, (theatrical) display, moralism, and fellowship. These were certainly among the central hallmarks of the Dutch Enlightenment, and it is hardly surprising that they should have manifested themselves in contemporary involvement with science.[74]

But what of the theme with which this chapter began? It was my second intention to say something about the role of utility in the culture of eighteenth-century Dutch science and to use that, by way of conclusion, as a vehicle for placing the Dutch experience in the broader context of "enlightened Europe." We have seen how utility served as a banner for the direction of knowledge — knowledge for the sake of secular enlightenment and spiritual edification, but also directed toward underscoring social status, stratification, and expectations. Within institutions, the ideal of utility was invoked to justify the position of individuals and as a means of lobbying for internal support. Utility was used

73. Snelders, "New Chemistry," 137, on Van Rhyn.
74. See Mijnhardt, "Dutch Enlightenment," for discussion and bibliography.

further as an argument for institutional aggrandizement. We saw, for example, the University of Leiden medical school encroaching on the traditional rights of the city's surgeons' guild with the claim of disinterested concern for the public good.

On a much wider scale, utility provided a standard for new forms of sociability. Not only was it present as an important concern behind the attendance of private science lectures from the early 1700s, it also served as an explicit rallying cry for the establishment of societies during the second half of the century. However, this is certainly not to say that its meaning was constant. For those who attended private lecture courses, utility had an introspective character. Studying nature and its ways was an edifying (and entertaining) means of finding and appreciating one's place in the divine scheme of things. The establishment of societies recast this private urge into a social movement. In accordance with their organizations' charters and manifestos, people came together in a drive for public enlightenment.

But how was that public construed and what sort of "enlightenment" was this organized sociability intended to serve? In a somewhat simplified representation of the Dutch situation, this chapter presents two contrasting models. Elite chartered societies such as the Hollandsche Maatschappij der Wetenschappen reflected the interests of their directors, the regent class, who had grown rich from the Netherlands' imperialist exploits and who dominated urban politics and society. During the eighteenth century, as the Dutch increasingly lived with the consequences of economic downturn and apparent loss of international power, directors of the Hollandsche Maatschappij steered its programs and chose topics for its essay contests with an eye toward their own intellectual and cultural cultivation and toward the revival of the sorts of transportation and trade that had fueled their wealth and power during the "golden age" of the seventeenth century.

The reforming Maatschappij tot Nut van 't Algemeen, founded late in the eighteenth century (1784), sought to institute a rather different utilitarian vision. Propelled by a strong sense of moralism, members of the Nut hoped to regenerate their society by a broad-based program of education. They believed that moral and scientific education needed to go hand in hand to give rise to a generally enlightened and industrious population. While this movement had virtually no effect on Dutch universities, which continued to hold on to the traditional ideals of corporate status for some time to come, it did lead to national reform of elementary education by the early nineteenth century.

In order to understand the case of the Netherlands in the eighteenth century, it is crucial to recall that the Nut, like all Dutch societies that were not explicitly concerned with politics, was a consciously apolitical organization.

According to its constitution, when members gathered, politics were left outside the door. Unlike the situation in France, Dutch (scientific) sociability did not lead to the construction of an arena distinct from the sphere of state authority and critical of its power—what Jürgen Habermas has labeled the "public sphere."[75] The obvious reason for this is that once the Spanish Hapsburgs were defeated, the Dutch never had to rebel against cultural decrees promulgated by a (centralizing) court or the authority of official academies. They were left to create and maintain institutions and forms of sociability more in keeping with local interests and demands. Consequently, the culture of Dutch science in the eighteenth century did not become implicated in the sort of struggles that ultimately adopted utility as the watchword for the standardization and surveillance of society.

Acknowledgments

I would like to thank Jack Spaapen, Geert Somsen, Rob Visser, Peter de Clercq, Rienk Vermij, Frank Huisman, Wijnand Mijnhardt, my colleagues from the "Ijkpunt 1800" project, and the staff of the rare book room at the University of Amsterdam central library for their help. Part of this chapter was written under a grant jointly sponsored by the Huizinga Institute for Cultural History and the Nederlandse Organisatie voor Wetenschappelijk Onderzoek, in association with the NWO priority program "Dutch Culture in European Context."

Primary Sources

Allamand, Jean Nicolas Sebastian. *Oeuvres philosophiques et mathémathiques de Mr. 's Gravesande rassemblées et publiées par J. N. S. Allamand, qui a ajouté l'histoire de la vie et des écrits de l'auteur.* 2 vols. Amsterdam, 1774.

Bidloo, Goverd. *Ontleding des Menschelyken Lichaams. Gedaan en beschreeven door Goverd Bidloo. Uitgebeeld, naar het eeven, in honderd en vijf aftekeningen, door de heer Gerard de Lairesse.* Amsterdam, 1690.

Camper, Adriaan Gilles, ed. *Redenvoeringen van wylen Petrus Camper, over de wyze, om de onderscheidene harstogten op onze wezens te verbeelden; over de verbaazende overeenkomst tusschen de viervoetige dieren, de vogelen, de visschen en den mensche en over het gedaante schoon.* Utrecht, 1792.

Camper, Petrus [Pieter]. "Kort berigt wegens de ontleding van verscheidene Orang Outangs. . . ." *Algemeene Vaderlandsche Letter-Oefeningen* 1, part 2 (1779): 18–36.

———. *Natuurkundig Verhandelingen over den Ourang-outang; en eenige andere aap-soorten. Over den rhinoseros met den dubbelen horen; en over het rendier.* Amsterdam, 1782.

Experimenta Philosophica Naturalia Auctore [Magistro] De Valdo [= De Volder] Lugd. Ann. 1676. MS Sloane 1292. British Library, London.

Gravesande, Willem 's. *Mathematical Elements of Natural Philosophy, confirmed by Experiments; or, An Introduction to Sir Isaac Newton's Philosophy.* Translated by J. Th. Desaguliers. London, 1720–21.

---

75. Habermas, *Structural Transformation of the Public Sphere.* For France, see Goodman, *Republic of Letters.*

Musschenbroek, Petrus [Pieter van]. *Beginselen der natuurkunde, Beschreven ten dienste der Landgenoten.* Leiden, 1736.

————. *Essai de physique.* Leiden, 1739.

Nieuwentyt, Bernard. *Het regt gebruik der wereltbeschouwingen ter overtuiginge van ongodisten en ongeloovingen aangetoont.* Amsterdam, 1715.

Verhandeling, uitgegeven door de Hollandsche Maatschappij der Wetenschappen te Haarlem. 3d ed. Vol. 1. 1759.

Voltaire, François-Marie Arouet de. *Candide.* Amsterdam, 1759.

12

Daedalus Hyperboreus: Baltic Natural History and Mineralogy in the Enlightenment

LISBET KOERNER

The historical literature on the sciences of the Baltic region has typically been written in terms of national origins. It has constructed erudite genealogies, in which the ethnic provenance of scientific heroes are made into consolation prizes for lesser modern nations. Such national teleologies have obscured the messier and more material histories of the connections, exchanges, and assimilations that characterized Baltic science in the eighteenth century. And it has profoundly obscured that there was *a* Baltic Enlightened science, and that this denotes a Germano-Scandinavian one.

Baltic naturalists had strong family resemblances. Typically they were Lutherans, of German or Scandinavian origins, and from modest homes. The most distinguished of them (but not their students, and thus not their recruits and successors) were educated west of the Elbe. They belonged to European-wide correspondence networks; they were in close contact with one another; they subscribed to a broad idea of the "new science"; and they were civil servants.

In this essay, I investigate the local inflections of the Enlightenment project as it was imported into the learned worlds of the Baltic. Historically, it stands sandwiched between the romanticizing and phantasmagorical linguistic sciences that characterized Baltic higher learning both in the seventeenth century (e.g., Sweden's Scythianism, Finland's Wendianism, and Poland's Sarmatianism) and in the nineteenth century (e.g., pan-Scandinavianism, pan-Slavism, and the linguistic nationalisms of peoples such as Estonians, Finns, and Sami).

At its most general, "Enlightenment" in this Baltic world means a material, universalizing, utilitarian, rational, and encyclopedic style of thought. Baltic elites and naturalists also agreed that they needed to "catch up" to the Atlantic empires, and that in this endeavor science, properly utilized, was key. Mineralogy and prospecting had a particular pride of place here. No Baltic statesman forgot how, in the seventeenth century, Sweden had financed her military empire by the sale of iron and copper (mined, often, by Calvinist confessional refugees).[1]

Typically, too, Enlightenment Baltic naturalists made their own peripheral status, their own difference and locality, an object of their investigations. At times they even celebrated their impoverished "polarity" (northernness) as enabling a differently constituted natural knowledge, which they vaguely imagined as more original, and more pure.

Stated so broadly, this complex of ideas existed from the mid–seventeenth to the later nineteenth century, and even beyond, if we categorize Nazi race science as one last stab at a "Nordic" science of nature, self, and nation. Here, I investigate the mythos of place particular to the Baltic Enlightenment, from roughly 1720 to 1800. Particularly, I look at how eighteenth-century Baltic naturalists, in studying their own regions, foreshadowed their Romantic and nationalist successors.

Since much of the Baltic region was under Russian occupation from 1945 to 1991, it is still relatively unknown in the Anglo-American world, including the scholarly. To complement other chapters in this volume, most of my examples will be drawn from the North and the East Baltic, from voyages of discovery, and from mineralogy, topography, ethnography, and anthropology. By "Baltic" I mean all countries then facing the Baltic Sea. This includes Denmark and Sweden. It incorporates Finland, which was part of Sweden until 1809, when it was ceded to Russia (it became independent in 1919). It also includes Russia, because after defeating Sweden in 1709, and capturing Ingria, Russia occupied the eastern Baltic littoral. In the easternmost part of the Gulf of Finland, she even built a new city, St. Petersburg, to replace Archangel, her ice-ridden White Sea port. "Baltic" here also means Denmark's Atlantic dependencies: Norway (ceded to Sweden in 1814, and independent in 1905); Iceland (independent in 1943); and Greenland (partially independent in 1979). It also encompasses Poland, although by 1772 that country was blocked from the Baltic seaboard, as Russia conquered Polish Livonia. Finally, it includes East Prussia, with its Baltic seaport of Königsberg, although it excludes Brandenburg, with its inland capital of Berlin.

Balticum

Before the advent of railroads, cars, and airplanes, water united and land divided. Mental maps were akin to *portolano* charts, Portuguese seafarers' route

---

1. On the Calvinist migration as one example of a technology transfer mechanism, see Schilling, "Innovation through Migration," 7–33.

maps of the late Middle Ages, which recorded every wrinkle of the coastline but nothing of the hinterlands, and which thus charted the sea's boundary rather than the perimeter of the land.

Balticum is one of Europe's three main geographic water regions, alongside the Mediterranean and the Atlantic. But in the 1700s, it profoundly differed from these other two regions. The Atlantic seaboard was Europe's center of social modernity, and its ocean link to Asia and the New World. The Mediterranean was a backwater, terrorized by slave-raiders from Tripoli (present-day Libya). But, next to the Balkans and the Eastern Mediterranean, *Balticum* was Europe's poorest, most archaic, and most ethnically complex region.

From south to north, the Baltic region divides into mixed broadleaf forest, taiga, and tundra. These ecological divides in turn explain why, during the Enlightenment, Europe's most profound social division ran through the Baltic. To the north, there were free yeomen, represented in national diets. To the south, there were enserfed peasants, sold away, even, from the land and from their families.

Northern Scandinavia's cattle farmers, lumberjacks, fur trappers, prospectors, fishermen, and reindeer herders sold furs, honey, game and fish, timber, tar and pitch, and bar-iron and copper into Europe.[2] The South Baltic's German aristocrats exported rye, malt, and barley, grown by serfs on vast demesne farms.[3]

Baltic governments included parliamentary rule (Sweden); Enlightened absolutism (Denmark and Prussia); elective monarchy (Poland); orders of knights (e.g., Kurland and Oesel); urban patriciates (e.g., Memel and Reval); and imperial autocracy (Russia). Around the Baltic Sea lived Catholics, Calvinists, Eastern Orthodox Christians, Jews, and Lutherans (including many types of Pietists). And the Baltic peoples spoke many languages: Slavic (Polish and Russian); Finno-Urgic (Finnish, Ingrian, Votian, Estonian, and Liv); Baltic (Latvian, Old Prussian, and Lithuanian); Scandinavian (Danish, Norwegian, and Swedish); Germanic (High German, *plat* German, and Yiddish); and Samish (which in turn divided into many tongues).[4]

The Baltic academies and universities of the Enlightenment formed a world of their own, and that world was German and Scandinavian. Placed by the shores of the Baltic Sea, they served the German diaspora and their patron

2. Indeed, they barely fed themselves. The last nonpolitical famine in the Baltic was in Swedish Lapland, in 1867.

3. The trade in all these products was controlled by Dutch, German, and Scandinavian merchants. On trade patterns of the eighteenth century, see Saskolskij, "Baltic Trade," 41–53; and Cieslak, "Baltic Sea-Borne Trade," 239–271.

4. South Samish, which had about ten native speakers in 1980; West Samish, divided into Lule and Norsk; and East Samish, nearly extinct today and divided into Skolt, Enare, and Kola.

military empires, and not the peasantries of their hinterlands. They taught the
future parsons and civil servants of Sweden, Denmark, Russia, and Prussia.
Only two Baltic universities were important centers of learning: Königsberg,
with Immanuel Kant (1724–1804); and Uppsala, with the botanist Carl Linnaeus (1707–78), the astronomer Anders Celsius (1701–44), the mathematician
Samuel Klingenstierna (1698–1765), the physician Nils Rosén von Rosenstein
(1706–73), and the linguist Johan Ihre (1707–80).[5] Most others were, in effect,
Lutheran seminaries—in part because professorial chairs were hereditary by
customary right.[6]

The Baltic, Slavic, and Finno-Urgic people were largely absent in this Baltic republic of learning. Speaking what Baltic Germans dismissed as "slave
languages" or "peasant speech," they lacked vernacular printed literatures.[7]
Visitation protocol shows that even basic religious literacy was weak in the
Lutheran parts of the East Baltic. (Parsons preached in German, a language
their parishioners did not understand.)[8] It is recorded that when in 1739 one
professor in natural history at the University of Åbo, Finland, was ordained
and took on a parish, his colleagues noted "with wonder and astonishment"
that he learned Finnish, merely in order to speak with his flock of farmers.[9]

By contrast, Lutheran Scandinavia and Germany had long promoted reading through catechisms and hymns, and through house visitations and parish

5. On the achievements of Uppsala scholars 1719–92, see Annerstedt, *Upsala universitets historia*, vol. 3, book 2:338–483. On the Swedish Enlightenment generally, see Frängsmyr, "Enlightenment in Sweden," 164–75.

6. The University of Copenhagen was the fief of the Bartholin family, who produced ten
professors of medicine, as well as ten in other faculties. The University of Uppsala, too, was dominated by a few intermarried clans: Celsius, Aurivillius, Benzelius-Benzelstiernas, and Rudbeckius-
Rudbeck. The Universities of Greifswald and Rostock were famously provincial. The University of
Dorpat (Tartu) was closed in 1710, when Livonia and Estonia passed from the Swedish to the
Russian empire. And the University of Vilna, which had been a Jesuit institution of great distinction
in the sixteenth century, was now in decline (see Swianiewicz, "University of Wilno," 29–46).

7. Snell, *Beschreibung der russischen Provinzen*,164: "Sklavensprache" and "Bauern-Sprache."
Snell argued serfs were not taught German because they were seen as animals ("Creatur"), and to
prevent them from escaping, by imprisoning them in a local dialect.

8. Merkel, as cited in Jennison, "Christian Garve and Garlieb Merkel," 355. Jennison describes Merkel as highly unusual, in that he was an ethnic Latvian parson in Latvia, and also notes
he tried to hide his origins. See also Rann, "Development of Estonian Literacy," 115–26. Luther's
Small Catechism had already been translated into the languages of the East Baltic in the sixteenth
century. By the seventeenth century, it was commonly printed and bound with ABC's. Yet, even
though much of the East Baltic was Lutheran, the Bible was only translated into Finnish in 1642,
Latvian in 1689, Lithuanian in 1735, and Estonian in 1739. (By contrast, it was rendered into Swedish
in 1541, Danish in 1550, and Icelandic in 1584).

9. Carleson, *Åminnelse-Tal*, 15: "med förundran och häpenhet." Typically the parsons of
Finland used translators for their sermons. Browallius became bishop of the Åbo diocese, and
spend his later life attempting to prove, in a Gothicist fashion, that Finnish was the original tongue
of mankind, and the language of Eden.

examination registers (*husförhörslängder* in Swedish). By the eighteenth century, literacy had become required for confirmation, marriage licenses, and land-holding contracts.[10] The Livonian realms, East Prussia, Denmark, and Sweden all built rural elementary schools, although Catherine II's plans for Russian village schools—on which Diderot advised her—foundered.[11] By 1800, Danish bishops insisted that even Lapland's reindeer herders and Greenland's seal hunters read Luther.

Latin and the Scandinavian vernaculars thus remained the languages of instruction at the North Baltic universities, and Latin and German at the South and East Baltic ones.[12] The scientific academies of the Baltic Enlightenment, too, were German or Scandinavian.[13] These linguistic barriers worked against non-Germanic participants in science, especially since the Muscovite aristocracy followed its European peers in preferring French.

Russian became a language of science only in Peter the Great's naval academy and schools of engineering, artillery, and medicine (opened between 1701 and 1715), and in Russia's Artillery and War School (estab. 1772), Mining Institute (estab. 1774), and Medical Surgical Academy (estab. 1798).[14] Another vernacular project remained on paper: the Polish-language schools projected by the Polish-Lithuanian Commonwealth's Commission of National Education in 1773, after Pope Clement XIV abolished the Society of Jesus, and thus higher education, in the East Baltic Catholic realms.

Admittedly, in 1767 the University of Moscow began translating parts of

10. Guttormsson, "Popular Religious Literacy," 7–35, 10.

11. Kirby, *Northern Europe,* 375; Guttormsson, "Popular Religious Literacy," 16. By 1700, eight in ten Swedes could read. Across the Baltic Sea, eight in ten Finns were illiterate.

12. North Baltic universities were Uppsala (estab. 1477), Copenhagen (estab. 1479), Åbo (estab. 1640), and Lund (estab. 1668). Norway gained its first university in 1811, and Iceland in 1911. South Baltic universities were Rostock (estab. 1419), Dorpat (or Tartu, estab. 1632), Vilna (estab. 1578), Königsberg (estab. 1544), Greifswald (estab. 1456, and under Swedish rule 1637–1815), and Moscow (estab. 1755). Of the 1,169 books published in Königsberg between 1700 and 1800, 1,017 were in German, 127 in Latin, 14 in French, 10 in Polish, and 1 in Lithuanian: see on this Gause, *Geschichte der Stadt Königsberg,* 2:241.

13. Kungliga Svenska Vetenskapsakademien (Stockholm, estab. 1739), Det Kongelige Danske Videnskabernes Selskab (Copenhagen, estab. 1742), Kongelig Norsk Videnskabers Selskap (Trondheim, estab. 1760), Finska Vetenskaps-Societeten (Helsingfors/Helsinki, estab. 1838), and the St. Petersburg Academy of Science (estab. 1724). On the last, see Schulze, "Russification of the St. Petersburg Academy," 305–335, 309. Between 1726 and 1733 six of its thirty-eight students were ethnic Russians; reflecting its polyglot uncertainty, the academy decreed it would teach in Latin 1724–34, German 1734–42, Latin 1742–46, and French from 1746 on.

14. In 1783 Catherine II also founded a St. Petersburg institute for teaching Baltic physicians, which was specifically designed as a German institute. See also Melton, "Enlightened Seigniorialism," 694–95, on how Russian was the teaching language in the Sheremetevs' school for training serfs as estate clerks. The Kantseliarskii Institut, in Ostankino outside Moscow, taught arithmetic, history, grammar, geography, catechism, and penmanship.

Diderot's encyclopedia into Russian. But after "Mineralogy," "Geography," and a handful of medical articles were published, the project foundered.[15] Russia never saw the equivalent of Christoph Friedrich Nicolai's *Allgemeine Deutsche Bibliothek* (107 vols., 1765–92), which cast itself as a periodical equivalent of Diderot's *Encyclopédie*. There were also few Slavic-language newspapers popularizing science, on the model, say, of *Königsbergsche gelehrte und politische Zeitungen* and *Preussische Archiv* (both in Königsberg), *Lärda Tidningar* (Stockholm), *Mitauische Monatschrift* (Mitau), *Neuen Nordische Miscellaneen* (Riga), or *Dansk-Norsk Økonomisk Magasin* (Copenhagen).[16] Russian economic and agricultural magazines such as *S. Peterburgskii vestnik* (St. Petersburg, 1778–81), *Ekonomicheskii magazin* (Moscow, 1780–89), and *Ezhenedelnye izvestiya* (St. Petersburg, 1788–89) were short-lived, and averaged between two and six hundred subscribers.[17] Further impeding Slavic science, the Russian crown's print monopoly (abolished in 1783) was so slovenly that schools and monasteries found it cheaper to hand copy books than to buy them.[18] As one historian has dryly noted, Russian provincial booksellers "had dozens of titles in stock."[19]

With rare exceptions, such as St. Petersburg's German-speaking Freie Ökonomische Gesellschaft, before the 1790s Russia saw few of the nodes of scientific sociability typical to the Enlightenment, such as private circulating libraries, parish libraries, reading societies, coffeehouses, spas, and salons. Nor did Russia see the scientific societies common to the Baltic Enlightenment, such as Copenhagen's Kongelige Landhusholdningsselskab and Drejer's Club, Stockholm's Patriotiska Sällskapet, Åbo's Aurora, Helsingfors's Finska Hushållningssällskapet, Danzig's Societas Physicae Experimentalis, Vilna's Societas Jabloniviana, and Riga's Livländische Gemeinnützige und Ökonomische Sozietät.[20]

Yet the most ambitious scientific institution of the Baltic Enlightenment was arguably the St. Petersburg Academy of Science. It taught mathematics and

15. Donnert, "Zur Verbreitung bürgerlicher Wissenschafts- und Gesellschaftslehren an der Universität Moskau," 25–34, 30. Diderot also planned a larger Russian encyclopedia: see on this Proust, "Diderot, L'Académie de Pétersbourg, et le projet d'une Encyclopédie Russe," 103–31.

16. The main exceptions were *St. Petersburgische Zeitung*, with its Russian twin *St. Peterburgskiye Vedomosti* (1727–1915), published by the St. Petersburg Academy of Science, and *Moskovskiye Vedomosti* (1756–1800), published by Moscow University.

17. Marker, "Russian Journals and Their Readers," 88–101, esp. 91–94.

18. The only other remaining European manuscript culture comprised the Faroe Islands, Iceland, and Greenland. On these remote Atlantic outposts, the medium only declined in the mid–nineteenth century. By comparison, in 1794 Denmark had twenty-seven printshops: see Markussen, "Development of Writing Ability," 37–63, 50.

19. Marker, "Merchandising Culture," 46–71, 59.

20. I borrow this point from Marker, ibid.

"physics," which included anatomy, chemistry, and botany, and it sported an astronomical observatory, a physics cabinet, a museum, an anatomy theater, a botanical garden, and an instrument-making workshop. The academicians included Johann Georg Gmelin (1709–55), a botanist from Tübingen who authored *Flora Siberica;* Joseph Delisle (1688–1768), a geographer from Paris; and Nikolaus Bernoulli (1695–1726) and Daniel Bernoulli (1700–1782), mathematicians from Basel. Joseph Gottlieb Koelreuter (1733–1806), whose plant breeding experiments famously foreshadowed the work of Gregor Mendel, worked there. So did the mineralogist Johann Gottlob Lehmann (1719–67), and the great mathematician Leonhard Euler (1707–83).

One indigenous academician was Mikhail Vasilievich Lomonosov (1711–65), who studied with Christian Wolff in Marburg, and who wrote on electricity, optics, fossil ice, icebergs, and the Northwest Passage. He also invented a self-recording compass and an "aerodrome machine," and started mosaic factories.[21] Yet science work by Slavs remained limited. The midcentury purges of "foreign influences" left the academy largely Russian, but also largely idle and without impact. By 1782, Euler had published more than the entire academy combined, Russians as well as foreigners.[22]

The University of Moscow, too, found rearing native scientists difficult. It was founded in 1755 with ten chairs, including chairs in chemistry, natural history, anatomy, and physics. To it were attached two grammar schools (one for nobles, one for commoners). The St. Petersburg Academy of Science similarly attached a gymnasium and even a short-lived university, since the Orthodox diocesan schools and seminaries clearly failed to prepare students. But these secular preparatory schools never rooted themselves among the indigenous Russian nobility, perhaps because the boys were expected to read Greek, Latin, German, and French.

Also, once at the academy or university, promising Russian scientists were so quickly pressured into the civil service that their scientific careers were short-changed. The demands for utility typical of the Enlightenment here worked against the possibility of sustained scientific inquiry. As for the other Slavic realm, Poland, in this orthodox Catholic country, higher science was largely defunct. Before 1750, and in the land of Copernicus, the fact that the earth revolves around the sun was never referred to in print.

21. Two other important Russians were M. I. Afonin (1739–1801), who studied at Uppsala with Linnaeus and at Paris with Antoine-Laurent Lavoisier (1743–1785), and Stepan Yakovlevich Rumovskii (1734–1812), who studied at Berlin with Euler.
22. Schulze, "Russification of the St. Petersburg Academy," 323. I here argue against the revisionist thesis of Carver, "Russia's Contributions to European Science," 389–405.

Philosophies of Science

By the mid–eighteenth century, the philosophy of science taught in Baltic universities was that of the Prussian philosopher, Christian von Wolf(f) (1679–1754). Wolff aimed to unify Lutheran orthodoxy with a Leibnizian mathematical and deductive logic.[23] The resultant metaphysics, he hoped, would philosophically ground a modern apologetic science. In his *Theologia naturalis* (2 vols., 1736–37), Wolff rejected revealed theology as an unsound basis for a science, which he defined in conventional Enlightenment terms as a critical study of nature's regularities.[24]

Wolff's natural theology was introduced into the North Baltic by two of his students, the Danish mathematician and Newtonian Jens Kraft (1720–65), and the Swedish mathematician Samuel Klingenstierna (1698–1765), who also introduced Newton's calculus, and geometrically demonstrated the possibility of constructing achromatic lenses. Both in his university lectures and in his role as tutor of the Swedish crown prince (later Gustav III), Klingenstierna dutifully drew on his erstwhile teacher to refute Voltaire's deism.

More spectacularly, Wolff inspired the Swedish mineralogist and mystic Emanuel Swedenborg (1668–1772) to seek to prove empirically the immortality of the soul. Wolff's famous assertion that "we live in the best of all possible worlds" also animated Swedenborg's and his second cousin Carl Linnaeus's botanical mysticism, and particularly how they celebrated, and even revered, sexual union as the animating principle of nature. (Linnaeus asserted that a male and female spirit jointly inhabited each plant. Swedenborg cast the plant as male and the earth as female).

Generally in the Baltic Enlightenment, natural theology was the intellectual fashion, along with its more specialized offshoots, such as astro-theology, cosmo-theology, and hydro-theology. It was typical that, when Linnaeus delivered the first public lecture of the newly founded Swedish Academy of Science in 1739, he discussed the "economy of nature" by means of "curiosities among insects," and in the form of a thematic sermon on some verses in the Book of Job.[25]

Not surprisingly, it was in this realm that English scientists gained a Baltic following. Thus Robert Boyle was mainly known for his apologetic sermons, the Boyle Lectures, and John Ray for his intricate argument from design, *The Wisdom of God Manifested in His Works of Creation* (1688). Yet to Linnaeus, at

---

23. Calinger, "Newtonian-Wolffian Confrontation," 417–35.
24. Saine, "Who's Afraid of Christian Wolff?" 107–33.
25. Linnaeus, *Märkvärdigheter uti insekterna;* Lindroth, "Two Faces of Linnaeus," 1–62, esp. 19; Sahlgren, "Linné som predikant," 40–55.

least, physico-theology did not solve the problem of theodicy. In a pamphlet of 1762 on edible native flora, *Plantae esculentae patriae,* he stated that humankind has at its disposal "an infinite larder. . . . In short, all inhabitants of the earth, the air, and the sea are at [its] service."[26] He dedicated that same pamphlet, however, to "those among the countryside Dwellers, who in famine years are forced either to take to unnatural means to survive, or to die entirely because of a severe hunger."[27] "I once heard him say," a former student of Linnaeus remembered in 1820, that if the world was made for the benefit of man, "God would have made the globe into a cheese and us to worms in the cheese."[28]

In 1765, when the Free Economic Society (Freie Ökonomische Gesellschaft) of St. Petersburg first met, its members noted for the protocol that the eighteenth century was an "economic century" *(ökonomisches Jahrhundert).*[29] As one member, Johann Georg Eisen von Schwartzenberg (1717–79), a parson originally from Franconia who settled in Livonia in 1741, wrote a friend: "You talk in favor of the home country *[Vaterland],* I will work for it, Luther did it for the heart, and Wolff for the spirit, I will do it for the stomach. Which means my Reformation will spread quicker and become more common, than those of those two."[30]

In the Baltic realms at least, the eighteenth century's many "reforms," "projects," and "improvements" of science and technology were informed throughout by the economic doctrine of cameralism. Originally developed in the Germanies in the seventeenth century by thinkers such as Johann Joachim Becher (1635–82), and without any lasting insights to its credit, cameralism had become an established academic discipline around the Baltic by the mid-1700s. Later, we will look at how this creed affected the study of agriculture and the peasantry. Here, we will look at its growth as a university science, and its influences over the manufactures.

---

26. Linnaeus, *Plantae esculentae patriae,* 4: "et outtömmeligt förrådshus . . . Kårteligen, alla jordenes, luften och hafwets inwånare äro til hennes tjenst."
   27. Ibid., 4: "til deras underrättelse, bland landets Inbyggare, som i swåra år nödsakas, at antingen gripa til onaturliga medel til sit lifs uppehälle, eller och aldeles omkomma genom en swår hunger."
   28. Ehrström, "Minnesanteckningar om Linné," 69–72, 70: "Om icke så wore (hörde jag honom en gång säga) så hade Gud kunnat göra jorden till en ost och oss till maskar deruti."
   29. Neuschäffer, "Der livländische Pastor," 120.
   30. Cited in ibid., 121: "Reden Sie nun fürs Vaterland, so wie ich dafür arbeiten will, Luther hat fürs Herz, und Wolff für den Geist gearbeitet, ich thue es für den Magen. Was gilts, meine Reformation breitet sich geschwinder aus und wird allgemeiner, als jener beiden ihre." Note that I have changed the verb tense, to better capture the breathless and informal form used by Schwartzenberg.

In 1741, the University of Uppsala established Sweden's first chair in cameralism.[31] It merged law, physics, and natural history, "combined and applied to" economics.[32] On arriving in Uppsala, its first holder, Anders Berch (1711–74), founded an "economic-mechanical theater"—a medley of handbooks, production samples, and tool models.[33] Over the next decades, further chairs were endowed at the universities of Uppsala, Åbo, and Lund in "practical economics, based on natural science" (cameralist theory, mining, agriculture, and manufactures).[34]

In Stockholm and Copenhagen, and among merchants, English mercantilist theories also flourished, and toward the end of the century freer economic doctrines such as physiocracy and Adam Smith's liberalism began to reach the Baltic. In the 1780s, a fashion spread for the *Smithianismus* taught at the University of Königsberg. Adam Smith's doctrines were also foreshadowed in the works of the obscure Swedish clergyman Anders Chydenius. His passionate orations in defense of liberty so inflamed the 1766 diet of the Swedish estates that they not only abolished censorship of the press but also allowed the towns of Ostrobothnia to export tar and pitch.

Such liberties signified the defeat of cameralist economics, which favored closely governed monopoly productions of technological novelties and former imports, such as silks. A typical expression of this ideology is the *Judenabnahmequantum* of the Berlin china factory, or the quota of porcelain each Jewish household was legally required to purchase of its unsellable wares. In a similar spirit Enlightenment reformers redesigned the Dano-Norwegian twin kingdom as a closed economy, supposedly reproducing within its borders the world at large. The resultant Danish grain monopoly meant that thousands of Norwegians starved to death in the mid-1780s.[35]

On a more positive note, the interest in economics in the Baltic world

31. See Berch, *Sätt att igenom politisk arithmetica* and *Inledning til allmänna hushålningen.* On cameralist and mercantilist theories and their relation to natural knowledge in the Swedish Enlightenment see also Hildebrand, *Kungl. Svenska Vetenskapsakademien,* 14–22; and Koerner, "Linnaeus' Floral Transplants," 144–69.

32. Berch, quoted in his entry in the *Svenskt Biografiskt Lexikon:* "kombineras och lämpas."

33. Stavenow-Hidemark, *1700-tals textil,* esp. 14–22. *Theatrum oeconomico-mechanicum,* usually referred in Uppsala to as *Theatrum Oeconomicum.* Berch's textile collection, encompassing 1,672 samples, is the second most important eighteenth-century collection after the Collection Richelieu, now kept in Bibliothèque Nationale, Paris, and consisting of 4,818 samples collected 1732–37.

34. Quoted from donation letter of the 1760 Uppsala chair from Borgström, "Två svenska akademiprogram av Linné," 111: "den practiska oeconomien, grundad på naturkunnoghet."

35. On the monopoly structures typical to the Scandinavian Enlightenment, see Ericsson, "Central Power and Local Community," 173–176, and subsequent reports in the *Scandinavian Journal of History.*

during the eighteenth century helped the development of mineralogy. Sweden was the only Baltic country with significant metal deposits.[36] Along with their European counterparts, Swedish mineralogists adhered to the chemical school of Georg Stahl, and admitted four major mineral classes in the earth's outer crust: earth and stones, metals, salts, and bituminous substances (sulfurs). But at the University of Uppsala, they also did pioneering theoretical work in mineralogy. One example is Linnaeus's mineralogical nosology and his analyses of illnesses typical of miners, which opened a new chapter on occupational disease.

In 1750, too, the University of Uppsala had received its first professorial chairs in physics (Klingenstierna) and chemistry. This last post was given to Johan Gottschalk Wallerius, a hydrologist and agricultural chemist, who in 1753 published a *Mineralogie*. He in turn was succeeded by Torbern Bergman (1735 – 84), author of the important *Physisk beskrifning öfver jordklotet* (1766). Elaborating the order in which rocks are superimposed on each other, this work inspired Abraham Gottlob Werner's (1749 –1817) famous classification of strata columns, and thus helped found historical geology.[37]

Nevertheless, Baltic mineralogy remained predominantly a practical, applied science throughout the eighteenth century. In university lectures and in mining colleges, prospecting tips, and the lore of mining, assaying, and smelting loomed large. Baltic Enlightenment mineralogists—including the young Emanuel Swedenborg—typically specialized in the location and properties of metal ores, rock salts, kaolin (for china), limestone (for quicklime), clay (for bricks), sand (for glass), and rock (for roads).

Voyages of Discovery

The Baltic Enlightenment's most prominent contribution to science lies in the realm of the voyage of discovery. These voyages began outside of the Baltic (and often outside of Europe itself). Over the century, they focused increasingly on the Baltic realms. And toward the end of the century, they became increasingly fantastic—voyages of the mind. As the century progressed, that is, Baltic naturalists turned toward their own locality and specificity as an object of their investigations. In terms of disciplines, they turned from a general natural history, to agricultural technologies and economics and to ethnography and

---

36. Nasafjäll (1634, silver), Kedkevare (1659, iron), Gällivare (1742, iron), and Stora Kopparberg (since the Middle Ages, copper).

37. "Kurze Klassifikation und Beschreibung der verschiedenen Gebirgsarten," in *Abhandlungen der Böhmischen Gesellschaft der Wissenschaften* (1796), as discussed in Hedberg, "Influence of Torbern Bergman," 186–91.

anthropology (*Völkerkunde* or *Menschenkunde*), or what they also termed "customs, dress, and manners of the inhabitants." Any and all topics made do: for example, "the Edda and the elephantiasis of Iceland."[38]

Certainly the most important Slavic science productions of the eighteenth century were the great Russian voyages of discovery across northern Eurasia and the Pacific Rim. Led by a Dane, Vitus Jonassen Bering (1681–1741), from 1728 on, these journeys involved more than ten thousand people (counting support staff).[39] The topographies, natural histories, and anthropologies, produced mostly by Scandinavian and German naturalists, followed Leibniz's famous memorandum to the Russian vice chancellor of 1716 in stressing linguistics, ethnography, and geography.[40]

One important late finding was Ivan Fedorovich Kruzenshtern's (1770–1846) discovery of the equatorial countercurrents during his 1803–5 circumnavigation of the world.[41] Other findings, such as the close observations of Northwest American peoples, are still largely unexplored. Up to the fall of the Soviet empire in 1991, many of the discoveries of the Russian Enlightenment remained state secrets, since they involved, by remote implication, army matters.

Carl Linnaeus, who trained many of the Baltic scientific travelers, even spoke of the voyage of discovery as a Swedish invention that was now copied by Joseph Banks (Australia and the Pacific), Johann Georg Gmelin (Siberia), Michel Adanson (Senegal), and José Celestino Mutis (South America).[42] As I have also noted elsewhere, Linnaeus's pride in his travelers is understandable, considering their careers. Some of them signed on as technical staff on the

38. From the subtitle of *Letters on Iceland,* by Uno von Troil (1746–1803), Sweden's archbishop and a friend of Joseph Banks and Carl Linnaeus.

39. The most important travelers were arguably Johann Georg Gmelin (1709–55), who produced the four-volume *Flora Sibirica* (1747–69), as well an extensive *Reise durch Sibirien;* D. Joh. Anton Güldenstädt (1745–81); Daniel Gottlieb Messerschmidt (1685–1735); Gerhard Friedrich Müller (1705–83); Georg Wilhelm Steller (1709–46); Johann Gottlieb Georgi (1729–1802); Johan Petter Falck (1732–71); Peter Simon Pallas (1741–1811), who authored the *Flora Rossica;* and Joseph Billings (1762–1806). The most important expeditions were the First Bering Expedition, 1724–29; the Second Bering Expedition, 1733–43; the Orenburg Expedition, 1768–74; and the Billings Expedition, 1787–94. For a stimulating general discussion of these travels, see Slezkine, "Naturalists versus Nations," 170–95.

40. Calinger, "Newtonian-Wolffian Confrontation,"417–35, 421–22. On German-Russian relations, see Robel, "Die Sibirien Expeditionen."

41. Malakhovskii, "Russian Discoveries in the Pacific," 27–47.

42. Linnaeus, letter to the Swedish Academy of Science, 75–89, 79: "ryska regeringen sänt Falck, Pallas, Gmelin till Sibirien och Tartariet, Danskarna Forsskål åt Arabien, Koenig till Tranquebar; Tyskarne Hallers till Vera Cruce, Fransoserne Adanson till Senegall, Engelsmännerne Solander till södra werlden, Spaniorerne Mutis till Mexico, Sardinien till Orienten, Holländarne Thunberg till Japan etc."

great European voyages of discovery. Some traveled on their own, typically supporting themselves as physicians to East India companies. And some were sponsored by Baltic institutions. Students of Linnaeus traveled on Cook's first and second circumnavigation of the globe (1768–71 and 1772–75), and they explored southern Africa, coastal China, Japan, Spanish South America, the Ottoman empire and the Arabian peninsula, Iceland and Lapland, and the Caucasus, Kazan, and Western Siberia.[43]

As Linnaeus and his followers saw it, such travels aimed at national self-sufficiency, by means of naturalization of foreign samples and delicacies.[44] Yet the only botanic transplant of importance in Linnaeus's lifetime, potatoes, were brought to the North Baltic by soldiers returning from the Swedish military campaigns in Pomerania in 1760 and 1761.[45] Linnaeus even conjectured it was poisonous, as it is related to the deadly nightshade. The Linnaean imaginary economy was thus a form of Edenic medicine revolving around what we (insofar as we are considering the trajectory of modernization) must regard as relative ephemera: leeches, Spanish flies, opium, rhubarb.

Yet it had high and anxious hopes. In a letter of 1746 to the Swedish Academy of Science, Linnaeus elaborates this. As we would put it, sending explorers abroad was crucial to a proper strategy of national improvement, one based on ecological diversification rather than on military conquest: "If Oaks did not grow in Sweden, and some mortal wanted to get Oaks into [the country], and they then grew here as they do today, wouldn't he serve the country more than if with the sacrifice of many thousands of people he had added a Province to Sweden?"[46]

Among Linnaeus's traveling students we find Swedes from Livonia, Lapland, Latvia, and Finland, whose childhoods already represented a diaspora of sorts. They typically prepared for their longer voyages by shorter trips within the Baltic. Indeed, Linnaeus's first professorial lecture in 1741 was "on the utility of scientific journeys within the fatherland," and he himself undertook

43. Koerner, "Linnaeus' Floral Transplants."
44. Ibid., and see also forthcoming biography on Linnaeus by Koerner, to be published by Harvard University Press.
45. Jörberg, "Nordic Countries," 375–485, 383; Utterström, "Potatisodlingen i Sverige under frihetstiden," 145–85.
46. Linnaeus, memorandum to the Academy of Science, Uppsala, 10 January 1746, in *Bref och skrifvelser*, ser. 1, 2:60: "Om Ekar intet wäxte i Swerige, och någon dödelig wille skaffa dem in, at de då fortkommmo här, såsom nu skier, tienade han landet mehra, än om han med många tusende menniskors upoffrande hade lagt en Province til Swerige." See also the similar formulation in a letter from Linnaeus to Pehr Elvius, Uppsala, 28 August 1744, ser. 1, 2:30: "Tänk om Ek ej woro bekant i Sverige, och om någon kunne skaffa Ek i Swerige, woro icke der mehr förtient af Sverige, än den provincier tillade ett rike."

five such regional explorations, through Lapland (1732), Dalarna (1734), Öland and Gotland (1741), Västergötland (1746), and Skåne (1749).[47]

By the end of the century these shorter journeys were privileged over travels further afield. As a Thuringian writer living in Livonia put it in 1794, *Heimatkunde* was "more important for the national economist *(Staatsökonomen)* than the discovery of a new fixed star."[48] The *Calcul* of "the products of a country," he went on to explain, "shows, what the country was, what it is, and what it can be." Linnaeus similarly equated indigenous flora to the dainty tidbits of abroad: "Yes here are some [plants] that can be used instead of Capers, Groceries, Coffee, Lemon juice, Sugar, Tea, and Saffron."[49]

It was typical that the China traveler Pehr Osbeck first toured Sweden's Atlantic coast (1749); that the Siberia explorer Johan Petter Falck first inventoried the Baltic island of Gotland (1760); and that the America voyager Pehr Kalm journeyed through western Sweden (1742), the mid-Baltic archipelago (1743), Finland, Ingria, Carelia, and northwest Russia (1744). The characteristic results of such travels were local "economic-technical floras."[50]

As I also discuss elsewhere, Linnaeus's student Pehr Kalm was also typical in planning a doctoral dissertation on "which of our domestic plants can be used instead of bread, and for porridge and gruel." He abandoned this project when he discovered he was unable to finish it in three days. But he then turned to another "oeconomic" subject, also inspired by his "own observations," namely, "how a [Finnish] Farmer ought to follow the example of the Chinese . . . and [use] the smallest pieces of land on his farm."[51] Kalm planned

47. Linnaeus, *Oratio, Qua peregrinationum intra patriam.*

48. Friebe, *Physisch, ökonomisch und statistische,* 1: "Die Beschaffenheit, und vorzüglich die Produkte eines Landes kennen zu lernen, ist für den Staatsökonomen wichtiger, als die Entdeckung eines neuen Fixsterns. Er zieht einen Calcul, dieser zeigt, was das Land war, was es ist, und was es seyn kann." Friebe went on to argue that a "state economist" should produce maps, topographical descriptions, and descriptions of local manufactures. On maps, see also Varep, "Maps of Estonia," 88–93.

49. Linnaeus, review of his *Plantae esculentae patriae* (1752), in *Lärda Tidningar,* 105: "Ja, här äro de, som kunna brukas i stället för Cappris, Specerier, Coffé, Citronsaft, Såcker, Thé och Safran."

50. I borrow the term from Friebe (1762–1811), *Oekonomisch-technische Flora.*

51. This story is from Hulth, "Kalm som student," quoting a letter from Pehr Kalm to Linnaeus, 18 April 1745: "derefter började iag på at arbeta om plantarum indigenarum usu oeconomico in Åbo, och tog mig egenteligen före, at drifva hvad man af våra hemväxter kunde bruka i stället för bröd, samt til gröt och välling" (47); "men sedan iag arbetar deri 3. dagar, och intet fann det så väl utarbetat som iag ville, samt at tiden til en så härlig materies rätta och tilbörliga utarbetande vore för kort, så kastade iag bort den" (47–48); "nu har jag satt mig före, at utarbeta en materia, deri jag vil visa, huru en Landtman bör sig bära åt, at han, efter Chinesernas exempel, ei har den ringaste jordlapp på sina ägor, som han ei både kan, och bör använda til sin nytta: iag vil alt af exempel bevisa huru det är möjeligt; iag skal af egna observationer bevisa altsammans" (48).

to devote eight precious pages to how the slash-and-burn corn farmers scattered across Scandinavia's taiga ought to emulate the rice growers of the Yangtze plains. This, too, he failed to write. Nonetheless, the Swedish cabinet appointed him professor in economics and natural history at the University of Åbo in 1747, at age thirty-one, and without a doctoral degree.

Lapland, with its average winter temperatures of about minus forty Celsius, its vast tundra, and its nomadic peoples, held a singular interest for Swedish voyagers and state elites, who hoped to colonize and settle the region. (Siberia played a similar role for Russia). In 1695, the Uppsala professor Olof Rudbeck the Younger had traveled to Lapland as part of an astronomical expedition to measure the sun's refraction at midsummer at the Arctic Circle. In 1711, Sweden's first scientific society, the Uppsala Royal Society of Arts and Sciences (Societetas Regia Literaria et Scientiarum, estab. 1710), sponsored an expedition by Henric Benzelius (1689–1758) to correct the measurements of 1695. Benzelius, who later become Sweden's archbishop, was also instructed to determine the specific density of the Arctic air, analyze Arctic plants for salt and sulfur, record Sami animal names, determine if kitchen vegetables grew on the tundra, and extract salt from seawater by means of cold winds and melted snow.[52] In the 1720s, the mechanic and inventor Christopher Polhem (1661–1751), together with the theologian and botanist Olof Celsius the Elder (1670–1756), drew up Lapland "experiments," to be undertaken by a "Societas Physico-Mathematica."[53]

In 1732, the Uppsala Royal Society again sponsored a Lapland voyager, Linnaeus. His tasks involved mineral prospecting, botany, zoology, ethnography, and medicine. From further afield, the Paris Academy of Science and the French royal treasury in 1736 financed an expedition to Lapland led by Pierre-Louis Moreau de Maupertuis (1698–1759). The goal was to measure the length of a degree at the Polar Circle, and thus test Newton's hypothesis—derived from his theory of gravity—that the earth was a sphere flattened at the poles.[54]

Enlightenment Lapland travelers also studied permafrost, aurora borealis, ice formation and glacier formation, the production of salt and other goods from local sources, and the North Atlantic heather moor and its relation to deforestation. Especially, they debated the supposed sinking of the Baltic

---

52. Sydow, "Vetenskapssocieteten," 138, 142–44.

53. Christopher Polhem, "Förtekning på några experimenter som på LappFiällen och i des Dahlar wore nödige att werkställas," and Olof Celsius the Elder, "Hwad en Societas Physico-mathematica här i Swerige hade at gjöra," cited in ibid., 153–54.

54. Terrall, "Representing the Earth's Shape," 218-37.

Sea (compressed by the last Ice Age's glaciers, the Scandinavian peninsula still rises).

The Question of the Peasantry

The crucial question of the Baltic Enlightenment, however, was the legal status and economic role of the peasantry. We now turn to why this was the case, and how naturalists addressed it. Particularly, we will look at how in the 1700s "the peasant" emerged as an object of scientific investigation, and how this inquiry became entangled with his status as, mostly, ethnically different from educated men.

Norwegian, Icelandic, and Swedish farmers were free.[55] But Denmark's ruling dynasty, the German Oldenburgs, allowed its commercial-minded aristocracy to extract ever more onerous services from its serfs, cottars, and tenant farmers. Indeed, Danish nobles lost the right to torture their peasants only in 1788. Prussia's ruling dynasty, the Hohenzollern, allowed its *Junkern* to rule their serfs, in return for state duties. In both Swedish and Prussian Pomerania, a German *Ritterschaft* enlarged their demesnes throughout the century. And Poland's magnates—those few families who ruled the country until its final subdivision in 1795—used their seignorial jurisdiction to plunder the serfs on their Polish, Lithuanian, and Ruthenian latifundia.

In Baltic Russia and Sweden, the German nobility, incorporated in the knightly orders of Livonia, Estonia, Lithuania, East Prussia, Courland, and Oesel, exploited their Lithuanian, Estonian, and Latvian serfs with impunity. In the infamous "Rosen Declaration" of 1739, the German Baltic nobility turned to Roman law to define their serfs as chattel. Reformers failed even to insert a clause giving serfs the legal right to own movable property such as clothes. Serfs were also auctioned—both to other nobles, and to army recruiters.[56] In the Muscovite realms, from 1721 on, merchants could buy serfs as factory slaves. More broadly, the nobility, the Eastern Orthodox church, and the imperial court all acquired the right to sell their peasants.

There was little local opposition to these slave regimes. The South and East Baltic shores were dotted with diaspora trading cities whose German patriciates carefully guarded their independence (an independence inherited from the Hanseatic League but eroded throughout the century, especially in Prussia). Neither they nor the German nobles particularly worried themselves about

---

55. In mainland Sweden, there was at most corvée service, and that only in the south, and in the west-central lands.

56. Kirby, *Northern Europe,* 378; Bartlett, "Russian Nobility," 233–44; Spekke, *History of Latvia,* 273; Wittram, *Baltische Geschichte,* 152, 153.

the subjugated native peoples of the East Baltic, peoples they still—after five hundred years—conceptualized only vaguely, as *Sklaven* (slaves), *Undeutsche* (non-Germans) or *Graue* (gray ones).[57]

It was thus highly unusual when Johann Georg Eisen von Schwarzenberg, the Germano-Livonian parson we encountered above, began agitating in 1764 for the emancipation of the serfs (although it should be noted that he published his opinions anonymously, and in German, a language not spoken by the serfs themselves). More commonly, heartland Germans and Scandinavians were attracted to the hacienda life of the East Baltic. As one contemporary noted, German prelates "flocked to Kurland as if to the East Indies." Where else could a Lutheran country parson enjoy freedom from taxes, noble status, and serfs of his own on vast parsonage glebelands?[58]

No wonder that one German pastor in Livonia began comparing his parishioners to "Algerian slaves"; and that in 1806, the rector of the University of Dorpat compared the Order of the Teutonic Knights to the Spanish conquistadors, as described in Bartolomé de las Casas's *Brief Relation of the Destruction of the Indies* (1542, and often reprinted in the Enlightenment).[59] Other writers compared them to Caribbean slaves, the Jews in ancient Egypt, and the "bound Prometheus."[60]

In 1776, the Free Economic Society of St. Petersburg posed as its annual prize question a query that in its spurious and bland "neutrality" typifies the earlier Baltic Enlightenment: "the advantages and disadvantages of serfdom." In 1778, a reform-minded German Baltic parson, proposing another prize question, naively summarized the central quandary of the Baltic Enlightenment: "Is there no way to improve the condition of the Livonian peasant without making him free *[ohne das er frei wird]*, and without diminishing the income of his lord?"[61]

The question of the peasant, then, dominated the Baltic Enlightenment. But it was debated as a question of *science* rather than politics. Indeed, it was only toward the end of the century that this issue was understood to involve

57. *Sklaven* was also the term used for New World slaves. Baltic German of the eighteenth century at times made no distinction between the two. On the general topic of the *Undeutsche,* see also Elias, "Die undeutsche Bevölkerung," 481–44.

58. On the average, glebelands were about 156 hectares. See Kirby, *Northern Europe,* 380.

59. Jürjo, "August Wilhelm Hupel," 506; Friede and Keen, *Bartolomé de las Casas,* 23. These analogies are not necessarily farfetched. See Kolchin, *Unfree Labor,* for an argument contrasting the absentee landlord ideologies of Russia to the southern planters' paternalist ideologies; discussed in the excellent article by Melton, "Enlightened Seigniorialism," 675–708.

60. Merkel, cited in Jennison, "Christian Garve and Garlieb Merkel," 352; Friebe, "Etwas über Leibeigenschaft," 744–68; Friebe, *Physisch, ökonomisch und statistische,* 71, and recurring throughout.

61. Cited in Jürjo, "August Wilhelm Hupel," 506.

social relations as well as methods of production. Previously around the Baltic Sea, enlightened and improving landlords, such as Baron Sten Bielke in Sweden, Count Hans von Borcke in Pomerania, and Count Caspar von Saldern in Holstein-Gottorp, had attempted to teach the peasantry—free and unfree alike—the methods of "scientific" agriculture: to use seed drills and scythes, plant lucerne and hay, rotate fallow land and grain crops, enclose their fields, practice animal husbandry, and settle wastelands.[62] But such attempts to "improve the condition of the peasant without making him free, and without diminishing the income of his lord" largely failed, even if Borcke could note that his Pomeranian estate was so well regarded that newcomers "offered themselves as serfs" (gaben sich unterthänig).[63]

Undaunted, other Enlightenment improvers, such as the prolific Germano-Livonian writer Wilhelm Christian Friebe (1761–1811), encouraged local farmers to try potato fields and winter crafts. The Königsberg Preussischen Archiv (1790–98) presented its noble readers with an inimitable medley of notices, now on animal diseases, new crops, and farming tools, and now on Virgil's Georgica—a perennial favorite on local estates—and the interminable genealogies of the East Prussian aristocracy.[64] Riga's Neuen Nordische Miscellaneen similarly leavened its science with family trees and histories of the German knightly orders.

Other Baltic reformers worked to make the Lutheran clergy, who themselves farmed church lands, the mediators of agricultural reform, and of the utilitarian and applied science general to the Baltic Enlightenment. Linnaeus gave expression to this program in 1740: "The Gentlemen Graduates become most all of them Parsons, spread over the entire country, mostly in the Countryside. . . . The common Man's inclination and money don't allow him to do experiments; but [he] copies everything that he sees in his Church that his Parson succeeds with."[65] Linnaeus also helped compose a pro memoria on this, submitted to the 1746 diet of estates by the chancellor of justice. In it he urged that all university students be compelled to study natural history—including the care of Spanish sheep and silk worms.[66] He demanded, too, that theology students be required to take a degree in medicine before they were admitted into the Church.

Linnaeus did not mean to teach curates pathology, physiology, or even nosology, as is evident from his summary of the art of doctoring: "This whole

62. Saldern, who was governor of the duchy and an abolitionist, directed Holstein's typically named General-, Landes-, Oeconomie-, und Verbesserungs-Directorium (estab. 1766). See Brandt, Caspar von Saldern, esp. 195.

63. H. A. Graf von Borcke, cited in Behrens, Society, Government, and the Enlightenment, 150.

64. Kozielek, "Aufgeklärtes Gedankengut," 321–47.

65. Linnaeus, "Tanckar om grunden," 422: "Ty Herrar Magistri blifwa merendels alla Prästmän, planterade öfwer hela Riket, mäst på Landsbygden. . . . En gemene Mans hog och pung tillåter inga försök at anställa; men antager alt, hwad som han ser wid sin Kyrkia lyckas för sin Präst."

66. This plea was repeated in a later memorandum, Linnaeus, "Angående Oeconomiens," 195.

science the students could easily learn . . . in eight days at the most."[67] Rather, he meant to instruct the "Gentlemen Graduates" in a neo-Hippocratic preventive medicine (typically, he called it a *diaeta naturalis*). Here he drew on his primitivist *Zivilisationskritik*, which in turn drew on ethnographic literature, and on the nostalgic ruralism inherited from Roman antiquity. In this spirit, he also sought to limit pharmacopoeias to indigenous herbal simples, and to study the plant lore and materia medica of the Baltic peasantry.[68]

Linnaeus and Nils Rosén von Rosenstein, author of a pioneering handbook in pediatrics, also wrote for the farmers' almanac to promote breast feeding and caution against wet nurses, swaddling, and drinking while pregnant. Such simultaneous moral and medical campaigns (what we today would call "public health") were common around the Baltic Sea in the Enlightenment.[69] These campaigns were humane in spirit. Linnaeus wrote with passionate sorrow of the battered children, covered with runny wounds, he saw in his practice.[70] They were also inspired by cameralist doctrines and particularly by the pro-natalism common to eighteenth-century *Balticum*. In this spirit, Sweden appointed a national health commission (in 1737), and the world's first central statistical bureau (in 1749). Headed by Pehr Wargentin (1717–83), secretary to the Swedish Academy of Science, the bureau investigated church registers epidemiologically and demographically. Again, parsons had a crucial role: collating statistics as they doctored their flocks of farmers.[71]

On East Baltic estates, German nobles, too, practiced an applied and vernacular medicine. From the 1760s on, special magazines on the problems of the *Landarzt* were published to help them treat their serfs' illnesses. These "medical ABCs" were also published in Estonian (from 1766) and Latvian (from 1768), to help deputies, or "sober, careful people, that each lord can select from his people."[72] It was typical that Eisen von Schwartzenberg, the

---

67. Linnaeus, quoted in Heckscher, "Linnés resor," 10: "Hela denna vetenskapen vore lätt lärd av de studerande. . . . högst på åtta dagars tid."

68. Hedbom, "Linnés inhemska medicinalväxter," 65–110.

69. Linnaeus, *Nutrix noverca*; Rosenstein, *Underrättelser om barn-sjukdomar*. For a general overview on Baltic Enlightenment medicine, see Huard and Wong, "Structure de la médecine russe," 125–34; Grmek, "Medical Education in Russia," 303–27; Kock, "Medical Education in Scandinavia," 263–99; Lindskog and Zetterberg, "Protokollen über das Examen Medicum," 768–70; Bogucka, "Illness and Death in a Maritime City," 91–104.

70. Linnaeus, *Nutrix noverca*, 15: "halta, puckelryggiga eller fistulösa."

71. Ischreyt, *Der Arzt als Lehrer;* Utterström, "Labour Policy and Population Thought," 272.

72. Ischreyt, "Zu den Wirkungen," 256, 251, discussing Peter Ernst Wilde, a Pomeranian doctor working in Livonia, and his Mitau publications *Landarzt* (1765–66), *Der praktische Landarzt* (1773–74), *Lühhike Öppetus* (Estonian, published in Oberpahlen 1766, and translated by Hupel), and *Latweeschu Ahrste* (Latvian, published in Oberpahlen 1768): "medicinisches ABC-Buch," "nüchternen ordentlichen Menschen, die ein jeder Herr aus seinen Leuten auswählen." See also Hubertus Neuschäffer, "Der livländischen Ökonomischen Sozietät," 337–44. In the *Baltische Wochenschrift für Landwirtschaft,* such medical handbooks were discussed within an economic frame, as were village schools.

German pastor in Livonia we encountered above, after studying medicine and cameral sciences at the University of Jena, pioneered smallpox inoculation in his parish. He also fought the practice, common among East Baltic serfs, of purposely infecting their small children—a practice justified in the mothers' eyes: "It is better, that the child dies, when it has to die, than that it first eats a lot of bread, and then dies."[73]

In his campaign for emancipation, Schwartzenberg also invoked pro-natalist arguments. For example, he warned that German landowners' refusal to allow female household serfs to marry explained the high Estonian rates of infanticide. Garlieb Merkel (1769–1850), the famous Germano-Latvian aboli-tionist, similarly claimed that the larger populations of western Europe proved serfdom inefficient. Look, he said, at the peasants emigrating to America and the Ukraine. They might have settled in Latvia.[74]

We turn now from technological improvements to legal reforms. Already in 1739, the Holstein count Hans Rantzau-Ascheberg (1693–1769) had given his tenants some freehold properties.[75] Freiherr Karl Friedrich Schoultz von Ascher-aden (1720–82), a Baltic German who had introduced civil (private) law on his Livonian estates, even argued, in the 1765 convention of the Livonian noble estate (Landtag), that demesne services be legally regulated. Proving his sin-cerity, on his own estates he had allowed the translation into Latvian of an estatewide legal document, a Bauernrecht, that granted his peasants a condi-tional hereditary tenure, regulated their labor services, and allowed them per-sonal ownership of movable property.[76]

Yet such experiments remained limited to individual estates. The youthful Catherine II's travels of 1764 in the Baltic provinces, to investigate "the question of the emancipation of the Livonian peasants," predictably came to nothing.[77] Saldern, the governor of Holstein-Gottorp, failed to abolish serfdom there, and had to content himself with creating a miniature model for enlightened abso-

73. Merkel, Die Letten vorzüglich in Liefland, 53: "Es ist besser, dass das Kind jetzt sterbe, wenn es sterben soll, als dass es erst viel Brot isst; und dann hingeht."
74. Merkel, cited in Jennison, "Christian Garve and Garlieb Merkel," 357–58; see also Die-derichs, "Garlieb Merkel," 38–83. Specifically, they—including ethnic Swedes—moved to the Kherson region in southeast Ukraine.
75. Barton, Scandinavia and the Revolutionary Era, 59.
76. Neuschäffer, "Der livländische Pastor," 127, notes Ascheraden was supported also by the Livonian governor general Count Georg von Browne (1698–1792). On the Bauernrecht, see Neus-chäffer, "Katharina II," 112; Wittram, Baltische Geschichte, 154, who also notes it was modeled on Swedish laws; see also Melton, "Enlightened Seignorialism," on such private law codes.
77. Neuschäffer, "Katharina II," gives a good overview of the question. Wittram, Baltische Geschichte, 154, notes there were eleven private codes issued between 1770 and 1816 in Courland, and four between 1789 and 1801 in Estonia. Klose, in "Die Jahrzehnte der Widervereinigung," 91, also notes that Baron Georg Wilhelm Soehlenthal, the teacher of Frederik V, pointedly liberated the serfs on the estates the king gave him.

lutism. And while the liberal Struensee government of Denmark briefly eman-
cipated the peasantry in 1771, the reform promptly failed.

Through ordinances of 1776 and 1781, Denmark at least began abandoning
common cultivation *(faellesdrift)* and natural rent and labor. Similarly, Sweden's 1757 diet of estates decreed that villagers could abandon common cultivation. But this innovation floundered, since it was made conditional on village unanimity. As an idea, land reform *(storskifte)* was embraced only in the
1780s, after a south Swedish baron, Rutger Macclean, enclosed his estates,
breaking up his corvée tenants' strip-field villages for dispersed, unified farms:
his newly prosperous farmers became the envy of neighboring estate tenants.[78]

Arguably, around the Baltic, "the peasant question" was, as an intellectual
inquiry, solved in the later Enlightenment. By 1783, even Denmark's most powerful courtier, Count Andreas Peter Bernstorff (1735–97), had came to argue
for "total reform." Having been sent in political exile on his Mecklenburg estates, he had suddenly understood how his serfs—his "miserable slaves"—
lived.[79] By 1800, even the most conservative proprietors knew their peasants
were listless and slovenly because they were bound to the manor. Enlightenment writers had long made this point, drawing on both pre-Smithian and
Smithian economic doctrines to show serf labor was less profitable than free
labor.[80] Garlieb Merkel, for example, argued that German nobles' occasional
sadistic killings of Latvian serf women was "uneconomic." (Turning ethnography on its head, other reformers likened the Baltic German nobility to "savages" and "African despots").[81]

Admittedly Eisen von Schwartzenberg failed to convince his patron, the
duke of Courland, to free experimentally some seven thousand of his serfs. His
theories on serf labor, too, were rejected by serfowners as "Donquichotterien,"
even though he had been awarded a chair in economics at the University of
Mitau and had become locally famous for having invented novel ways to dry
vegetables, mix herbal medicines, brew vodka, build wooden houses, cure
plaguey cattle, and freshen stale air.[82] But in 1766, in the St. Petersburg Eco-

78. On land reform in Scandinavia (excepting Iceland and Finland), see Tönneson, "Tenancy, Freehold, and Enclosure," 191–206.

79. Quote from Neuschäffer, "Der livländische Pastor," 134; see also Barton, *Scandinavia in
the Revolutionary Era,* 131.

80. Neuschäffer, "Carl Friedrich Frhr. von Schoultz-Ascheraden," 325–27, discusses his
physiocracy, and how in 1767, Catherine II also invited the famous French physiocrat Paul Pierre
Mercier de la Rivière (1720–94) to Russia.

81. Merkel, cited in Jennison, "Christian Garve and Garlieb Merkel," 356. As Jennison points
out (351), Merkel identified sadist serf owners by initials or place, thus shaming them (there were
only around 350 Germano-Livonian noble families). See also Merkel, *Die Letten vorzüglich in Liefland,* 153: "Wilden," "Afrikanischen Despoten."

82. Cited in Neuschäffer, "Der livländische Pastor," 132, also 142; Bartlett, "J. G. Eisen,"
95–104.

nomic Society prize question on the best agrarian legal system, only 2 of the 162 answers supported serfdom. During the same decade, Ivan Andreevich Tretyakov (1735–79), who had studied with Adam Smith, controversially lectured at the University of Moscow on the topic "From which does the state receive the greater benefit, from the slave or from the freeman."[83]

In the terms employed in the Baltic Enlightenment, "self-interest" *(Eigennutz)* and "love of humanity" *(Menschenliebe)* seamlessly merged in the case for serf emancipation.[84] The Holstein noble we encountered earlier, Rantzau-Ascheberg, furnished practical proof when he demonstrated that the rents he gathered from his new peasant leaseholds of 1739 covered his losses in labor services. Similarly, Schoultz von Ascheraden doubled his Livonian estate incomes after introducing a rule of law.[85] Christian Garve (1742–98), a Breslau professor who translated Adam's Smith's *Wealth of Nations* in 1794, explained why: Since any surplus was expropriated by the lord, "since they must deny every other desire, they seek the only pleasure left to the helpless, to rest."[86] Eisen von Schwartzenberg argued on this note that free yeomen were "the only basis" for a happy *(glückselig)* nation, since serfs, to protest their "Negro-like slavery," sabotaged their work.[87] To prove serfs behaved rationally when they were "lazy and spendthrift," Garlieb Merkel cited an old Latvian peasant: "An apparent easing off means either we get stuck right away with another tougher job, or the lords are dunging us, like their fields, so they can get more out of us later."[88]

At the same time, Enlightenment reformers noted that armies suffered from their dependence on serf soldiers—sickly and ill-willed. "If the [East Baltic] farmers became free, one could raise from among them 10,000 recruits."[89] The Prussian count Wilhelm von Schaumburg-Lippe (1724–77), who

83. See Neuschäffer, "Der livländische Pastor,"136–37; Donnert, *Russia in the Age of Enlightenment*, 88. Serfdom was adressed in lectures on anatomy, medicine, pure and applied mathematics, chemistry, and natural history.

84. Neuschäffer, "Der livländischen Ökonomischen Sozietät," 341.

85. See Degn, "Die Herzogtümer in Gesamtstaat," 227; Neuschäffer, "Carl Friedrich Frhr. von Schoultz-Ascheraden," 325. Rantzau-Ascheberg was also able to abolish his overseers, and his penal apparatus.

86. Garve, cited in Jennison, "Christian Garve and Garlieb Merkel," 349; Adam Smith, *Untersuchungen über die Natur und die Ursachen des Nationalreichtums*.

87. Donnert, *Johann Georg Eisen*, esp. 26–32. The manuscript is entitled *Beweis, dass diejenige Verfassung des Bauern, wenn selbiger seinem Herrn als ein Eigentümer von seinem Bauernhof untertan ist, der einzige Grund sei, worauf alle mögliche Glückseligkeit von aller Unvollkommenheit in demselben dafürgehalten werden könne.* The quote is from Petri, who held the same view, and here cited from Heeg, "Die Publicationen Johann Christoph Petris," 131. Petri saw ethnic Russians, as opposed to Estonians, as too primitive to be legally free.

88. Merkel, cited in Jennison, "Christian Garve and Garlieb Merkel," 357, 354. See, however, Hupel, "Über den Nationalkarakter der Russen," 36–37, where Hupel argues (unusually) both that Russian slaves worked harder than Estonian ones, and that the differential proves that the condition of slavery in itself does not explain the characteristic of laziness.

89. Eisen von Schwartzenberg, cited in Neuschäffer, "Der livländische Pastor," 125. See also Garve, *Über den Charakter der Bauern*, 155.

had introduced a "general conscription" on his estates, at least housed his serf veterans in special "colonies," and lightened their labor burdens.[90]

There was, too, the matter of the physical security of the German nobles. Here, of course, the French Revolution loomed large. Peasant riots swept repeatedly across the East Baltic, for example, in 1777, 1784, 1797, and 1802. Enlightenment ethnographers warned of the hatred of Germans in the East Baltic, and of a future when the "slavish dog" would turn into a "tiger" with "a taste for blood." Latvian parents, they noted, hushed their children into terrified silence by crying out: "Teutsche kommen!"[91]

By 1800, we've noted, even the East Prussian aristocracy understood that enserfed labor was unproductive. Still, these famously conservative *Junkern* often defended it. When, in the 1807 "revolution from above," the Prussian government abolished *Erbuntertänigkeit,* it balanced the abolition of peasants' personal bondage against the new right of nobles to freely alienate their estates. Since financial provisions were made for the redemption of all other rights of lordship, most manorial peasants were not freed properly until 1850.

Further east, reforms were even slower. In part, this was because of a land speculation bubble. By 1800, two-thirds of Livonian landlords had mortgaged their estates up to newly inflated values, values that reflected an increasingly rare situation in Europe, namely, the estate owners' life-and-death powers over their peasants. Admittedly, the Livonian *Landtag* of 1803 decided "to not treat [the peasants] as things anymore" *(nicht mehr als Sache zu behandeln).* The Estonian *Landtag* of 1816 even renounced what its members still termed "our traditional hereditary right to own the peasants." But even as emancipation crept east, the East Baltic German nobility still kept all land, retained labor services, and forbade peasants from leaving their estates.[92] While a few peasants obtained ownership of the holdings they farmed, most failed in their land claims, and also lost their rights of common, such as wood gathering, gleaning, and pasturing. Most famously of all, in the Muscovite realms, peasants remained serfs until the 1860s (only to be re-enserfed in the 1920s).

90. Neuschäffer, "Der livländische Pastor," 138.

91. Spekke, *History of Latvia,* 278; Merkel, *Die Letten vorzüglich in Liefland,* 245–46 ("sklavi-she Hund," "Tieger," "Schmack in Blut"), 37. The parents' warning also cited in Jennison, "Christian Garve and Garlieb Merkel," 354.

92. See Jennison, "Christian Garve and Garlieb Merkel," 257; Wittram, *Baltische Geschichte,* 158; Kahk, *Peasant and Lord,* 83. Wittram (155–61), notes how gradual reform was. For example, the 1802 Estonian *Landtag* forbade the sale of serfs who had land rights, but not of landless serfs. This was amended in 1804 by the head of the Estonian Order of the Knight, G. H. von Rosenthal, who was a friend of Washington and had fought as "John Rose" in the American Revolutionary War. Yet Estonian serfdom was only formally abolished in 1817. Half a century later, in 1878, only 3 percent of the land was owned by persons outside the rank of nobility. In Livonia in 1900, the average estate remained around ten thousand acres.

The Emergence of the "Peasant"

In the later Enlightenment, Baltic reformers began asking why farmers did not better themselves, despite their efforts to educate them. As we have seen, when they began to define the problem of the countryside as one of social relations, reformers appealed both to such typical Enlightenment constructs as "reason" and "observation" and to economic doctrines such as cameralism, physiocracy, and Smithianism.[93] But as the century drew to a close, and as Baltic naturalists grew increasingly disenchanted with economics as the science of fashion, they also began elaborating their explanations with peasant psychologies.

Throughout the Enlightenment, German and Scandinavian scholars had vaguely compared Finno-Ugric, Baltic, and Slavic peoples to Greenlanders, American Indians, "Orientals," "Arcadians," children, wild men, "Hottentots," and the Iroquois.[94] Now, they began developing more specific ethnographies of, for example, Latvians, Old Prussians, and Estonians. Most famously, Christian Garve wrote an ethnopsychology of the Silesian peasantry, characterizing it as *tückisch,* or willfully obstinate.[95] Christian Wilhelm Friebe wrote a similar piece on the Livonian peasantry; Eiler Hagerup, on the Danish peasantry; August Wilhelm Hupel (1737–1819), on the Estonian peasantry; and Garlieb Merkel, on the Latvian peasantry. All agreed that their objects of investigation—their parishioners, often enough—were lazy, dirty, and cunning. All, too, went on to blame this on their *Sklaverey.*

As I discussed above, these learned Germans and Scandinavians thus explained their observations in material terms, arguing that the character traits of the serfs were produced by their condition *as* serfs. To bring home the point, Garve compared their treatment with that of the Jews. Both groups, he said, were self-evidently warped and twisted; both, too, had become so through their legal disabilities.[96] Most influential among these ethnopsychologies was Garlieb Merkel's *Die Letten, vorzüglich in Liefland, . . . Ein Beytrag zur Völker*

93. Neuschäffer, "Katharina II," esp. 116, on economic doctrines and their dating in the East Baltic.

94. For Native American analogy, see "Auszug aus den Gedanken eines Ungenannten, über den bessern Wohlstand der zur Brüdergemeine gehörenden liefländischen Bauern," in Friebe, *Beyträge zur liefländischen Geschichte,* 302–8. The article argues that Moravian Brethren serfs were richer than Lutheran serfs, and that therefore serfdom per se does not explain serf poverty. For other comparisons, see Merkel, cited in Jennison, "Christian Garve and Garlieb Merkel," 355; Ischreyt, "Zu den Wirkungen," 256, again quoting Peter Ernst Wilde, the Pomeranian doctor working in Livonia ("Kinde"); and Garve, *Über den Charakter der Bauern,* 24.

95. Garve, *Über den Charakter der Bauern,* e.g., 37, and repeatedly. He defines it as "Unterart von der List" (52) and a semislave characteristic. See also Jennison, "Christian Garve and Garlieb Merkel," 347–48; and Snell, *Beschreibung der russischen Provinzen,* 175, 184 (where it is equated with "hinterlistiges").

96. Garve, introduction to *Über den Charakter der Bauern;* Behrens, *Society, Government, and the Enlightenment,* 149; Jennison, "Christian Garve and Garlieb Merkel," 344–63, esp. 346–47.

und Menschenkunde (1797), which opened prophetically, and with reference to the French Revolution: "Reason has triumphed, and the century of justice begins" *(Die Vernunft hat gesiegt und das Jahrhundert der Gerechtigkeit beginnt).*[97]

At the same time, these ethnopsychologies blossomed into more general investigations of what was becoming an increasingly alienated *Heimat.* One typical example is an anonymous pamphlet published in Riga in 1772 on the importance of *Heimatkunde,* or local ethnography and topography. Its very title foreshadows the emergent nationalisms of the smaller European peoples, even as its choice of language demonstrates the continuance of the Germano-Baltic republic of learning: *An das Lief- und Ehst-ländische Publicum.* The Courland pastor Gotthard Friedrich Stender (1714–96), author of a 1761 Latvian grammar, a 1766 collection of Latvian fables, and a 1789 Latvian dictionary, provides another example. On his gravestone, this native German speaker chose to memorialize himself by one single—albeit Latin—word: *Latwis* (Latvian).

There are many examples of such Baltic Enlightenment *Heimatkunde,* with its attendant typologies *(Nationalkarakter)* of Baltic, Finno-Ugric, and Slavic peoples. One prolific practitioner was the Lutheran pastor August Wilhelm Hupel, a native of Thuringia who studied at the University of Jena, and who took up a parsonage in an Estonian-speaking region of Livonia in 1757. Besides publishing accounts of Cossaks, Samoyeds, Tartars, and Ostyaks, he authored a four-volume topography of Livonia and Estonia (1774-87) and an assessment of the *Nationalkarakter* of Russians (1781).[98]

In a similar vein, the Germano-Baltic clergyman and noble Johann Heinrich von Jannau (1752–1821) wrote his *Geschichte der Sklaverey und Charakter der Bauern in Lief- und Ehstland: Ein Beytrag zur Verbesserung der Leibeigenschaft* (1786); the chief Lutheran prelate of Riga produced a broad description of all of Russia's people; and the rector of the cathedral school in Riga published a analogous account for the Baltic realms.[99] Other ethnographic collectors included physicians, mayors, and school inspectors. Everything was grist for the ethnographic mill: childrearing, festivals, belief systems, crafts, cooking, "the value of virginity among the Estonians and Latvians."[100]

All of these writers were self-professed *Aufklärer.* But, as Simon Schama

---

97. Wittram, *Baltische Geschichte,* 157, quotes one Baltic German noble of the times as having felt a "beschämenden Gefühl der Wahrheit" as he read Merkel's book.

98. Hupel, *Topographische Nachrichten,* and *Über den Nationalkarakter der Russen.*

99. Sonntag, *Das Russische Reich;* Snell, *Beschreibung der russischen Provinzen.*

100. Friebe, "Über der Werth der Jungfrauschaft unter Ehsten und Letten," in *Beyträge zur liefländischen Geschichte,* 279–98. Friebe discovers that the concept of female virginity does not exist.

has noted for Dutch reformers in the late Enlightenment, they rejected "a cosmopolitan, Francophone, universally applicable, rationally discerned set of natural laws, in favor of a highly particular, inward-looking, evangelical, protoromantic cult of the Fatherland."[101] In the Baltics, as opposed to the Netherlands, the celebration of the fatherland was immediately enmeshed in the question of the writers' diaspora status. Thus Hupel, a native German speaker, published as his *Heimatkunde* a compendium of Russian folk sayings, an Estonian grammar, and an *Idiotikon* of the East Baltic German dialect.[102]

No wonder that it was a German pastor's son from the East Baltic, Johann Gottfried Herder (1744–1803), who first fashioned the German Enlightenment concept of *Heimatkunde* into the Romantic cult of the European primitive, initially in his *Stimmen der Völker in Liedern* (1787), a text that included eight Estonian folk songs.[103] Indeed, Estonia so interested German literati that by 1800, a special term had been coined: *Estophilen.* By 1803, the University of Dorpat (Tartu) had established a lectureship in Estonian. Soon thereafter emerged the first scholarly treatment of the language, Johann Heinrich Rosenplänter's *Beiträge zur genauern Kenntniss der esthnischen Sprache* (20 vols., 1813–32).[104] More generally, Herder profoundly influenced later German and Scandinavian ethnographers in the Baltic. The implicit program for later research became a paraphrase of Friedrich Schlegel's famous motto for the Athenaeum: "Im Balticum müssen wir das höchste Romantische suchen."

Herder came from Königsberg, and so he knew the East Baltic region. But the appeal of *Balticum* to Scandinavian and German ethnographers was not only a matter of chance, or of a generalized Romantic sensibility. It was also a function of the relative ease with which the complex and archaic Baltic folklore fitted into a romanticizing frame. This was a region where in 1794, an old Germano-Baltic *Graf* could remember, without too much of an imaginative effort, how his father's serfs sang "songs of nature" "in a truly Homeric or Ossianic style," "without shyness, because they didn't feel they had to be embarrassed in front of me, a child."[105]

---

101. Schama, "Enlightenment in the Netherlands," 54–71.

102. Donnert, "Das russische Imperium," 94; Hupel, *Idiotikon der deutschen Sprache.*

103. Herder also dedicated a chapter of his *Ideen* (book 16, chap. 2) to the Finnish and Baltic peoples.

104. Martis, "Tartu University in the National Movement," 317–25. Baltic peasants continued to be fashionable to study throughout the nineteenth century: see, e.g., Count de Bray, *Essai critique sur l'histoire de la Livonie;* Ritchie, *Journey to St. Petersburg;* Kohl, *Die deutsch-russischen Ostseeprovinzen.*

105. Friebe, *Physisch, ökonomisch und statistische,* 284–85: "die Leute sangen ohne Scheu, weil sie sich für mich, als eiem Kinde, nicht geniren zu müssen glaubten," "Gesänge der Natur," "wirklich ein Homerischer oder Ossianischer Style."

As late as 1835, a Swedo-Finnish scholar could publish thirty-two cantos of *Kalevala,* a magnificent oral epic and pagan creation myth (in a pleasing local touch, the hero is conceived when his mother swallows a blueberry). Seventeen years later, in 1849, illiterate Finnish peasants supplied dazzled ethnographers with another eighteen cantos of this work, which is arguably the Baltic's equivalent to Homer's *Ulysses.*[106] The Estonian *Kalevi Poeg,* a related oral epic, was recorded even later, in 1857. Here was a wilderness more poetic and more novel even than that of the Orient. *A Thousand and One Nights* had already been translated between 1704 and 1708; the *Bhagavad Gita,* in 1785.

Conclusion

In this chapter I have explored the cultural valences of Baltic medicine and natural history between roughly 1720 and 1800, emphasizing throughout the manners and mechanisms scientists employed to address their geography, locality, and space. Their remoteness posed a problem that, in terms of their own life histories at least, scientists on the Baltic periphery had always to address. For centuries, people of real ability, who subscribed neither to an ethos of Nordic rusticity nor to Lutheran orthodoxy, had emigrated to find more congenial and more central sites of learning and spirituality. Olaus Magnus (1490–1557), Sweden's last Catholic archbishop and author of the most important geography of Scandinavia in the early modern period, *Historia de gentibus septentrionalibus* (1555), died in exile in Rome. Indeed, the Swedish Lutheran Reformation of 1527 so depleted science and medicine—the only university, Uppsala, closed down altogether—that when its instigator, King Gustav Vasa (1496–1560), lay on his deathbed there was not a single physician left in the realm to attend him.

The learned daughter of Gustav II Adolf, Queen Christina (1626–89), who brought René Descartes (1596–1650) to Stockholm and saw him die there, later renounced her throne, converted to Catholicism, and died as a penitent in Rome. The Danish anatomist and naturalist Niels Stensen, or Nicolaus Steno (1636–86), who famously argued that "figured stones," or what we know as fossils, were the petrified remnants of animals and plants, died as a Catholic bishop in Hamburg. Emanuel Swedenborg, who analyzed Sweden's first steam engine in his role as a member of Sweden's Board of Mines *(Bergskollegium),* and who published Sweden's first scientific magazine, *Daedalus Hyperboreus* (1716–18), died—conversing with angels—in London. Minor figures, too, sought out foreign lands. For example, Daniel Solander (1733–82), Linnaeus's

---

106. Lönnrot, *The Kalevala; or, Poems of the Kalevala District.*

student and Joseph Banks's companion on Captain Cook's first circumnavigation (1768–71), settled in London, in the pleasant role of companion and man-about-town.

Naturalists who remained in the Baltic felt compelled to explain themselves. Mårten Triewald (1691–1747), a captain in the fortification division of the Swedish army and a popularizer of Newtonian physics who helped found the Swedish Academy of Science in 1739, spoke of his return from England in terms of selfless patriotism. Linnaeus even turned to science to justify his return to Sweden from Holland in 1738. In his newfound capacity as a doctor of medicine, he diagnosed himself as suffering from a medical condition, *Gnostalgie,* and prescribed for himself a return to his ancestral lands. In doing so, he drew on a piece of Baltic scientific lore. In a period where love of *patria* was increasingly lauded, Baltic naturalists valorized nostalgia as a peculiarly Nordic disease, apt to strike Inuits and Sami.[107]

The cultural insecurities of Baltic naturalists, which bracketed love of home as at once a disease, an ethnic marker, and a virtue, are easily explainable. For ever since Hippocrates and the elder Pliny, Continental scholars have envisioned the far north as a mirror image of the far south, that is, as a repository of savagery. In the Enlightenment, this tradition was powerfully restated in the climate theories of C. L. de S. Montesquieu (1689–1755). The most important natural historian of the Enlightenment, the French *comte* Georges-Louis Leclerq de Buffon (1707–88), consequently lumped Arctic Sami and African San together as dwarfish degenerates, in a manner reminiscent of the ancients' coupling of Scythians and Ethiopians.[108]

The English naturalist Oliver Goldsmith (1728–74), in his 1774 *History of the Earth,* ranked the Sami as the lowliest of his six races. An earlier French naturalist, Jean-François Regnard (1655–1709), compared "this little animal that is called a Lapp" to apes. As the Swedish historian Gunnar Broberg has shown, Linnaeus reflected such continental conceptions when, in pioneering a global racial order in his canonical tenth edition of *Systema naturae* (1758), he placed the Sami in the freak category *Monstrosus,* and as dwarfs, alongside mythic Patagonian giants.[109]

As I have argued throughout this chapter, Baltic Enlightenment naturalists often sought to investigate their own place, and thus their own peripheral status.

107. In the 1953 Nordic conference, Svenska Samernas Riksförbund, Saami Liito, and Norske Reindriftssamers Landsförening (the Swedish, Finnish, and Norwegian Sami national associations) jointly requested that they be called Sami, and not "Lapp." (The Kola peninsula Sami, collectivized in 1927, were forbidden to attend.) See IWGIA, *Self Determination and Indigenous Peoples.*

108. Broberg, *Homo Sapiens L. Studier,* 243-44.

109. Linnaeus, Regnard, and Goldsmith are all cited in ibid., 222, 244, 245, repspectively.

I now turn to how the Arctic North as the realm of the *Monstrosus* also became a site of imagined mastery, a place, that is, where Baltic naturalists could enact, and invert, their cultural insecurities. Here the anthropology of the Sami and Samoyeds become particularly important.

On the Scandinavian peninsula, ethnography was a well-established tradition in the Enlightenment. Already in the 1670s, the Uppsala professor Olof Rudbeck the Elder complemented his philological studies of the Sami by measuring their skulls, including that of Laurentius Granius, an Ume Sami and a Bible translator. The West's first anthropological monograph on a single people is in fact on the Sami: *Lapponia* (1673), by the Uppsala law professor Johannes Schefferus. In Denmark, in 1722 Jacob Winslöw (1669–1760)—who later become professor of medicine and surgery in Paris at Jardin du Roi—made a study of the skeletons and anatomy of the Inuits of West Greenland. And in 1760, Jens Kraft (1720–65), professor at the University of Copenhagen, pioneered an ethnological survey of the world translated into German as *Die Sitten der Wilden* (The customs of savages).

One nice example of how researchers enacted the exotica of the far North is Linnaeus's 1737 commission of a full-length oil portrait of himself "on the occasion when he showed himself in a Lapp costume to display that People's dress for a Company gathered at Mr. Clifford's." [110] To the Dutch, like Linnaeus's patron George Clifford, Linnaeus's ad hoc assemblage of souvenirs from his 1732 Lapland journey made a nice display. He dangled from his belt a Sami runic calendar, a shaman's drum, a knife, birch-bark boxes, and pouches made of reindeer fur, and he dressed in the brightly colored Lübeck wools that these reindeer herders typically wore in the eighteenth century.

To the Sami themselves, however, Linnaeus in his "Lapp" costume would have cut a poor figure. His beret, which a Swedish tax collector had given him, was part of the Ume Sami women's summer clothing. His reindeer fur livery was a Torne Sami man's winter garment bought in Uppsala after his 1732 travels. His reindeer leather boots were of a type the Sami only manufactured for export. And his shaman's drum, an artifact illegal to own in Lapland itself, had been presented to him by an Uppsala professor as he packed for Holland. [111] Depicting a smiling Linnaeus wearing his Sami fedora rakishly askew, the portrait reflects the inevitable problems of veracity for a struggling explorer of the Enlightenment, forced to transform *himself* into a noble savage, since he could not afford to bring a real one with him.

110. Wiklund, "Linné och lapparna," 64, quoting from J. J. Björnståhl's travels in Holland: "vid det tillfället, då han visade sig i Lappdrägt, för att visa detta Folkets klädebonad för et hos Hr Clifford samladt Sälskap." See also Fries, "Linné i Holland," 141–55; Wachenfelt, "Ett återfunnet Lapplandsporträtt," 21–35; and Tullberg, *Linnéporträtt*.

111. Wiklund, "Linné och lapparna," 62.

Linnaeus's strategy of self-presentation, his theatrical display of all that was most exotic and foreign in his *patria,* exemplifies a key theme of Baltic natural history in the Enlightenment. It was, of course, a general Enlightenment strategy to acquire knowledge of self through the perspectival contemplation of "savage" peoples. In the case of Baltic naturalists, however, the "savages" not only inhabited but even came to define the scholars' own "fatherland." Yet Linnaeus's Sami, at least, were in some respects an imaginary community. By the 1690s, Lutheran missionaries had largely rooted out this indigenous pagan culture, Europe's last. The Lapland missions aimed at full-scale assimilation, and they modeled themselves on Continental efforts to preach to the Jews. Sami shamans were fined, tortured, imprisoned, and in one case at least, burned at the stake. By the time Olof Rudbeck the Younger arrived in Lapland in 1695, the Sami had become tourist guides of their own culture's destruction, showing sightseers their former cult places. However, as the younger Rudbeck warned Maupertuis in 1736, these guides did not like it when the visitors laughed.[112]

The alienation that German and Scandinavian naturalists experienced in relation to their polylingual and (as they saw it) wild-eyed parishioners translated into an anthropology, or theorization of their difference. In turn, this became the "factual" grounding for the smaller Baltic peoples' romanticizing constructions of their national identities in the nineteenth and twentieth century.[113] And this was so not only because Baltic Enlightenment naturalists had stored up a treasure trove of ethnographic facts. As important, the activity of collecting itself helped occasion the Romantic era. The naturalists of the Baltic Enlightenment eventually naturalized *all* social worlds, including their own, and regarded *all* local people, including their parishioners, as material resources. This process was general to the Enlightenment, of course. But it is particularly nicely exemplified in the Baltic lands, for there a Slavic/Finno-Ugric "wilderness" was also a German/Scandinavian "home."

The Enlightenment's scientific travel journals enumerated what we today term natural and human resources, and what they typically grouped together as "discoveries and comments on the economy, natural objects, antiquities, mores, and ways of living."[114] As one minor Swedish naturalist, Pehr Forsskål, explained from Asia Minor in 1762, his task was too pressing to make

112. See Sydow, "Vetenskapssocieteten," 152–53; Broberg, "Olof Rudbecks föregångare," 18. On Sami religion, see Bäckman and Krantz, *Studies in Lapp Shamanism,* 27, citing Rudbeck the Younger's undated note to Maupertius. The missionary effort was inspired especially by Daniel Djurberg, a pietist and professor of theology at the University of Uppsala.
113. Estonian nationalists also became the model for Armenian nationalists: see Balekjian, "University of Dorpat (and the Armenian National Awakening."
114. Linnaeus, *Skånska resa.*

more narrow distinctions: "People, earth and nature were new to me; the Plants were all new. I had time for nothing but collecting and viewing."[115] Within the shock of the new, all the natural historian had time for were lists and descriptions, which made no distinction between nature and culture, curiosity and custom.

The consequences of this were eventually far-reaching. Consider, for example, the first pages of Forsskål's diary (1761–63), where we meet Linnaeus's twenty-nine-year-old student as he sets sail for the Levant on 19 February 1761. The moment the ship leaves the south Swedish harbor of Helsingborg, he busies himself with observations. He determines the salinity of the seawater and investigates the nature of phosphoric light. Even as the anchor is hauled aboard, he scrapes off its seaweed and analyzes its type. He engages the seamen in "the fishing of Sea-animals,"[116] which he then bottles in spirits. He notes his successes and his failures, as with a jellyfish that promptly dissolved in the preservation. And along with such *naturalia,* he observes how the sailors man the ship, and queries them endlessly about their methods and knowledge.

As he set out across Öresund, the narrow entrance of the Baltic Sound, Forsskål threw himself into the work of science as if he were studying unknown regions and peoples. In fact, of course, he sailed in home waters, and indeed in the most crowded, well-mapped sea route in all of northern Europe. Forsskål thus exemplifies the curiosity that screened the Enlightenment observer from the community around him as much as it enabled new knowledge. Representing the social world, he also naturalized it. His sailors, the modern reader feels, are wooden figures.

In this mental distancing from home and from self, the thought-world of Forsskål, and more broadly the Baltic Enlightenment, thus neatly converges with that of a fashionable Swedish countess who in the 1770s peopled her garden with what to a visitor "seem[ed] at first sight to be real persons."

> They are painted in full colors and depict all kinds of Garden workers, occupied with digging, shoveling and such things. This, so natural in a garden, betrays the hasty eye, especially where they are glimpsed between trees or bushes. It is the last thing that one expects, that they are flat painted wooden boards, or so-called *figures postiches.*"[117]

115. Forsskål, *Resa till lycklige Arabien,* 64: "Folk, jord och natur war nytt för mig; Örterna voro alla nya. Jag hant ej mer än samla och betrakta."

116. Ibid., 11: "opfiskande af Sjö-djur."

117. A letter from one of her visitors, quoted in Lamm, *Upplysningstidens romantik,* 432: "tyckas vid första åsynen vara verkelige personer. De äro målade i fulla färger och förestälal hvarjehand Trädgårds arbetare, sysslosatte med gräfning, skyffling och dyligt. Detta, så naturligt i en trädgård, låter ögat i hast bedragas, särdeles där de röjas mellan träd eller buskar. Man väntar allraminst, att de är flata målade bräder eller så kallade figures postiches."

One sometimes wonders whether the tableau of labor that Forsskål and other Baltic naturalists of the eighteenth century paint in their descriptions of the workings of nature, and of natural workers, does not also include the naturalists themselves, fulfilling their duties as functionaries of the crown or state, or indeed whether what we see in the natural histories of the Enlightenment as forerunner of science's experimental "distance" is in fact also an allegorical play. Be that as it may, the Enlightenment's investigations of the natural world were certainly not an unambiguous, self-confident taming of the the world's wildernesses. If anything, the enlightened traveler brought wilderness home with him, or more precisely, learned to see his home as itself wild.

And this was true to a special degree in the Baltic, that northern frontier of Europe. It is thus within the realm of "Orientals," in Raymond Schwab's sense, that we should interpret the Baltic naturalists' most weird imaginary, their *tückisch* peasant, whose profoundly alien ethnopsychology was first conjured up as a response to the technological and economic failures of the earlier Baltic Enlightenment.[118] As *golems* do, he took on a life of his own. And in his manifold mitosis into a Livonian, an Estonian, a Dane, he came to signify the irreducible multiplicity of humankind.

Toward the end of his life, in the twelfth edition of *Systema naturae* (1766), Linnaeus came to describe himself as a "voyager of discovery in the fatherland" *(peregrinum in patria)*. And in his attempt at objectivity, which is to say, the assumption of distance, he even meditated on how the preconscious ordering of our sense impressions depends on the relativity of scale: "If our eyes were microscopes, every person would look like a hideous masked scarecrow."[119]

At times the travel reports in the Baltic Enlightenment call to mind the quandary experienced by a sergeant in Jaroslav Hasek's famous novel of the First World War, *The Good Soldier Schweik*. Setting out to give his orders to his sullen conscripts, the sergeant finds himself unable to describe anything at all. As a consequence of his privates' *tückisch* strategy of blank stares and mulish silences, he comes to doubt everything he thought he knew. He doubts, that is, what Maurice Merleau-Ponty has famously termed the "horizon" of thought. Similarly, the ethnographic voyage typical of the Baltic Enlightenment undermined scientists' sense of home. Intended as a method of belonging, it occasioned instead the paranoid nostalgia of Romanticism.

118. Schwab, *Oriental Renaissance.*
119. Linnaeus, *Systema naturae,* dedicatory poem to Carl Gustaf Tessin; and *Skånska resa,* 260: "voro våra ögon mikroskoper, så skulle var människa se ut som en maskerad skråpuk" ("skråpuk" means both "scarecrow," and "hideous face").

Primary Sources

Berch, Anders. *Sätt att igenom politisk arithmetica utröna länders och rikens hushåldning.* Stockholm, 1746.

———. *Inledning til allmänna hushålningen innefattande grunden til politie, oeonomie och cameral wetenskapere: Til deras tiänst, som biwista de almänne föreläsningar inrättad.* Stockholm, 1747.

Borgström, Eric Ericsson. "Två svenska akademiprogram av Linné" (1760). Edited by Telemak Fredbärj in *Svenska Linnésällskapets Årsskrift* 37–38 (1954–55): 97–114. Borgström's article was also reprinted in his commentary on Carl Linnaeus (1750, 1759).

Carleson, Carl. *Åminnelse-Tal öfver Kongl. Vetenskaps Academiens Medlem Biskopen och Pro-Cancellarien Herr Doct. Johannes Browallius, efter Kongl. Vetenskaps Acad. ns Befallning, Hållit uti Stora Riddarehus-Salen, den 18. Martii 1756.* Stockholm: Lars Salvius, 1756.

De Bray, Count. *Essai critique sur l'histoire de la Livonie suivi d'un tableau de l'état actuel de cette province.* 3 vols. Dorpat, 1817.

Friebe, Wilhelm Christian. "Etwas über Leibeigenschaft und Freiheit, sonderlich in Hinsicht auf Liefland." In *Materialien zu einer liefländischen Adelsgeschichte nach der bey der lezten dasigen Matrikul-Commission angenommenen Ordnung: Nebst andern kürzern Auffässen etc. Der nordischen Miscellaneen 15tes, 16tes und 17tes Stück,* edited by August Wilhelm Hupel. Riga: Johann Friedrich Hartknoch, 1788.

———. *Herrn W. Chr. Friebe's Beyträge zur liefländischen Geschichte aus einer neuerlichst gefundenen Handschrift: Nebst andern kürzern Auffässen etc. Der nordischen Miscellaneen 26stes Stück von August Wilhelm Hupel.* Riga: Johann Friedrich Hartknoch, 1791.

———. *Physisch, ökonomisch und statistische Bemerkungen von Lief- und Ehstland oder von den beiden Statthalterschaften Riga und Reval.* Riga: Johann Friedrich Hartknoch, 1794.

———. *Oekonomisch-technische Flora für Liefland, Ehstland und Kurland.* Riga: in der Hartmannischen Buchhandlung, 1805.

Garve, Christian. *Über den Charakter der Bauern und ihr Verhältniss gegen die Gutsherrn und gegen die Regierung: Drey Vorlesungen in der Schlesischen Oekonomischen Gesellschaft gehalten.* Breslau: Wilhelm Gottlieb Korn, 1786.

Herder, Johann Gottfried. *Ideen zur Philosophie der Geschichte der Menschheit.* Riga, 1791.

Hupel, August Wilhelm. *Topographische Nachricten von Lief- und Ehstland.* Riga: Johann Friedrich Hartknoch, 1774–87.

———. "Über den Nationalkarakter der Russen" (1781). In *Der nordischen Miscellaneen erstes Stück von August Wilhelm Hupel.* Riga: Johann Friedrich Hartknoch, 1774–87. Reprint, Hannover-Döhren: Verlag Harro v. Hirschheydt, 1970.

———. *Idiotikon der deutschen Sprache in Lief- und Ehstland.* Riga: Johann Friedrich Hartknoch, 1795.

Kohl, J. G. *Die deutsch-russischen Ostseeprovinzen.* 2 vols. Leipzig, 1841.

Letters on Iceland. . . . London: Richardson et al., 1780.

Linnaeus, Carl. "Tanckar om grunden til oeconomien genom naturkunnogheten ock physiquen." *K. Vet. Akad. Handl.* 1740.

———. *Oratio, Qua peregrinationum intra patriam asseritur necessitas, Habita Upsaliae, in Auditorio Carolino Majori MDCCXLI Octobr. XVII. Quum Medicinae Professionem Regiam & Ordinariam susciperet.* Uppsala: Typis Academ. Reg. Upsaliensis, 1741.

———. *Skånska resa, på höga öfwerhetens befallning förrätad år 1749: Med rön och anmärkningar uti Oeconomien, Naturalier, Antiquiteter, Seder, Lefnads-sätt. Med tillhörige Figurer.* Stockholm: L. Salvii, 1751.

———. *Plantae esculentae patriae, Eller Wåra Inländska Äteliga Wäxter: Med Medicinska Facultetens bifall wid Kongl. Academien i Upsala . . . Et Academiskt Snilleprof, år 1752 d. 22 Febr. på Latin, och Nu med någon Tilökning på Modersmålet framstälde Af Johan Hiorth.* Stockholm: Lars Salvii kostnad, 1752.

———. *Systema naturae per regna tria natura: secundum classes, ordines, genera, cum characteribus, differentiis, synonymis, locis.* 12th ed. Holmiae: L. Salvii, 1766.

————. "Angående Oeconomiens och landthushåldningens uphielpande genom Historiae Naturalis flitiga läsande af them, som til nästa promotion och sedan framledes tänker emottaga honores Philosophicos" (1746). In *Linné: Lefnadsteckning*, by Th. M. Fries, vol. 2. Stockholm: Fahlcrantz & Co., 1903.

————. *Bref och skrifvelser af och till Carl von Linné.* Series 1, vol. 2. 1746. Series 1, vols. 1–8, published by Th. M. Fries; series 2, vols. 1–2, published by J. M. Hulth. Printed variously in Stockholm, Uppsala, and Berlin by Aktiebolaget Ljus, Lundequistska bokhandeln, Akademiska bokhandeln, and R. Friedländer & Sohn, 1907–43.

————. *Märkvärdigheter uti insekterna* (1739). In *Carl von Linné: Fyra skrifter,* edited by Arvid Hj. Uggla, illustrated by Harald Sallberg. Stockholm: Nordiska Bibliofilsällskapet/Esselte, 1939.

————. Review of *Plantae esculentae patriae* from *Lärda Tidningar* (1752). In *Linnés disputationer: En översikt,* edited and compiled by Gustaf Drake af Hagelsrum. Nässjö: Nässjötryckeriet, 1939.

————. *Nutrix noverca* (1752). In *Valda avhandlingar av Carl von Linné,* edited by Telemak Fredbärj, vol. 4. Åbo: Svenska Linné-Sällskapet, 1947.

————. *Skånska resa år 1749.* Edited by Carl-Otto von Sydow. Stockholm: Wahlström & Widstrand, 1975. First edition published 1751.

————. Letter to the Swedish Academy of Science (1771). In Sverker Sörlin, "Apostlarnas gärning. Vetenskap och offervilja i Linné-tidevarvet," *Svenska Linnésällskapets Årsskrift,* 1990–91.

Merkel, Garlieb. *Die Letten vorzüglich in Liefland am Ende des philosophischen Jahrhunderts: Ein Beytrag zur Völker und Menschenkunde.* 2d ed. Leipzig: Heinrich Graf, 1800.

Ritchie, Leich. *A Journey to St. Petersburg and Moscow through Courland and Livonie.* London, 1836.

Rosenstein, Nils Rosén von. *Underrättelser om barn-sjukdomar och deras bote-medel: tillförene styckewis utgifne uti de små allmanachorna, nu samlade, tilökte och förbättrade.* På Kongl. Wet. Acad. Kostnad. Stockholm: Lars Salvius, 1764.

Rudbeck the Younger, Olof. *Iter Laponicum: Skissboken från resan till Lappland 1695.* II Kommentardel. Stockholm: René Coeckelbergh, 1987.

Smith, Adam. *Untersuchungen über die Natur und die Ursachen des Nationalreichtums.* Breslau, 1794.

Snell, Karl Philip Michael. *Beschreibung der russischen Provinzen an der Ostsee; oder, Zuverlässige Nachrichten sowohl von Russland überhaupt, als auch insonderheit von der natürlichen und politischen Verfassung, dem Handel, der Schiffahrt, der Lebensart, den Sitten und Gebräuchen, den Künsten und der Litteratur, dem Civil- und Militairwesen, und andern Merkwürdigkeiten von Livland, Ehstland und Ingermannland.* Jena: Akademische Buchhandlung, 1794.

Sonntag, Karl Gottlob. *Das Russische Reich, oder Merkwürdigkeiten aus der Geschichte, Geographie und Naturkunde aller der Länder, die jetzt zur Russischen Monarchie gehören.* Riga: Johann Friedrich Hartknoch, 1791.

Werner, Abraham Gottlob. "Kurze Klassification und Beschreibung der verschiedenen Gebirgsarte." In *Abhandlungen der Böhmischen Gesellschaft der Wissenschaften.* 1796.

13

The Death of Metaphysics
in Enlightened Prussia

WILLIAM CLARK

What before this time period used to be called "metaphysics" has been
rooted out root and branch, so to say, thus vanishing from the realm of
the sciences. Where can or where may one hear anything of erstwhile
ontology, rational psychology, cosmology or even erstwhile natural
theology? . . . Since science and common sense have joined hands to
work the demise of metaphysics, the strange spectacle has appeared of *a
learned folk without any metaphysics.*
Hegel, *Wissenschaf der Logik,* preface to the 1st ed. (1812)

The metaphorical death of traditional metaphysics took place in the context
of the Enlightenment. Since then, "metaphysics" has come to connote more
a pathological European illness than a science. Those enlightened European
scholars and scientists who denounced metaphysics lumped it with superstitions, such as astrology, which the Enlightenment had to combat. Many Prussian academics, however, regarded metaphysics as neither ideology nor illness.
Metaphysics was, rather, a *Wissenschaft,* a science, and one that operated by
"pure reason," free from all sensual experience, and delivered certainty about
its subject matter, which included the most important topics for all humans:
the natures of the deity, the cosmos, and the soul. The Prussian Enlightenment,
the *Aufklärung,* remained ambivalent about forsaking scientific knowledge
about these topics. It was above all academics and academicians in Prussia who
fought to save metaphysics: once the handmaiden of theology and queen of
sciences, metaphysics should not be consigned to being but the charwoman
of science, if not dismissed altogether.

In the light of other enlightenments, German or otherwise, Prussian projects to save metaphysics must have appeared at the time, to say the least,
provincial. The later prominence of Prussia should not let us forget its provinciality for a good part of the eighteenth century. Even within the Holy Roman
Empire of the German Nation(s), the specter that would be Prussia did not
begin to arise until the Seven Years' War (1756–63), a view recorded in Goethe's
autobiography, *Dichtung und Wahrheit.* It was during that war, moreover, that

one first began to call Friedrich II Friedrich the Great—albeit for militarism and not for philosophy. Goethe himself belonged to the generation before Prussia, as it were, co-opted German culture. While Hegel and Berlin belong together, the notion of Goethe in Berlin is a bit odd; indeed, this son of Frankfurt am Main felt more at home in Strasbourg (where Parisian influence was less oppressive).

In this chapter, we consider the dilemma of metaphysics as a science in enlightened Prussia and take Hegel's circumscription as a guide: German tradition clove metaphysics into "general metaphysics" as ontology and "special" metaphysics as rational psychology or pneumatology, cosmology, and rational or natural theology.[1] The parts of special metaphysics had emerged from the Cartesian reduction of all being to soul, matter, and God. For want of space, I shall mostly omit consideration of ontology and shall restrict attention to special metaphysics. There are two additional caveats. First, psychology, or pneumatology, will get shorter shrift in favor of philosophical anthropology, whose rise coincided with the demise of traditional metaphysics in the *Aufklärung*. The second caveat is that I shall not do justice to the structure of metaphysics as a demonstrative science.

Mindful of the provinciality of Brandenburg-Prussia, we shall make a journey toward the ever more peripheral: from Halle, to Berlin, to Königsberg. Halle, in the Prussian province of Sachsen-Anhalt, lay in the academic fertile crescent of the eighteenth-century Germanies, a strip containing Halle, Leipzig, Jena, and Göttingen. After considering the lay of the land in Halle, we'll move to Berlin, in Brandenburg, still in the middle of nowhere for a good part of the eighteenth century and a place that many intellectuals pass through, but few remain. The final station will be Königsberg, in Prussia proper and at the periphery of the German political space. In Halle the period 1700–50 falls into view, and here theology, Christian Wolff, and his legacy form the center of attention. In Berlin the years more or less spanning 1730–80 come into focus, for which Leonhard Euler, the Academy of Sciences, and cosmology set the agenda. In Königsberg, 1750–1800 sets the period of analysis, which is devoted to Immanuel Kant, the dissolution of rational theology and cosmology, and the "birth" of anthropology.

Wolff and the University of Halle

"The foundation of the University of Halle forms the most exceptional achievement of the absolute bureaucratic state, unmatched by ages before or

1. See Eberhard, *Kurzer Abriss;* Feder, *Logik und Metaphysik,* esp. 425. On scholastic and Jesuitical foundations, see Lewalter, *Spanisch-jesuitische und deutsch-lutherische Metaphysik,* 60–65; cf. also Heidegger, *Kant,* 8–9.

after."[2] The foundation at Halle was the inception of the "cameralist" university: the first bureaucratically well-managed university, given little of the traditional corporate privileges of a university, and run instead by ministries in Magdeburg and Berlin. During the eighteenth century, among other Prussian universities and until the foundation of the University of Berlin, Halle was the flagship university. Despite Halle's prominent role in bureaucratic modernism, it was the last university founded in the Germanies for mostly confessional reasons. The two universities serving western Prussian lands were Reformed or Calvinist. Wanting was a university to train Lutheran pastors, academics, and other potential bureaucrats for the Prussian state. So the University of Halle was founded in 1694 as Lutheran. Though religion formed an overriding aim for the foundation, it was hoped the university would be oriented on a pragmatic view of learning and the world. One of the university's first guiding lights, Christian Thomasius, seemed the right person to establish the envisaged pragmatism. Pietism, which advocated a return to revealed religion and the cultivation of piety, soon found a stronghold in Halle. "The place of activity [of the early enlightener] was Halle, and Leipzig to an extent. Most of them were touched by Halle's Pietism. They stand in stark opposition to the antecedent age and its school philosophy, while the next generation, under the lead of Wolff, returns precisely to it."[3] That was the tragedy of Halle and the early *Aufklärung:* their inception lay in a rejection of traditional metaphysics in favor of more pragmatic learning. But after the Wolff "affair," metaphysics, as the apotheosis of reason, became identified with the Prussian *Aufklärung.*

Christian Wolff was first educated in Wroclaw, which the Germans called Breslau. Here Wolff fell under syncretic influences of Jesuit and Lutheran teachers. His university education came at Jena and Leipzig, and his first academic position at the latter. Wolff received a chair as professor of mathematics at Halle in 1706, much thanks to Leibniz. At Halle Wolff and the Pietists never really got along. The Pietists' consternation about Wolff grew in 1720 when his "German Metaphysics," the *Vernünfftige Gedancken von Gott, der Welt und der Seele des Menschen,* appeared, with a rising sun on its title page, later taken as an emblem of enlightenment (see fig. 13.1). In 1721 Wolff held *Oratio de sinarum philosophia,* first published in 1726, as his final oration as prorector at Halle, whereafter by tradition the old prorector installed the new one, Joachim Lange. In this speech, Wolff intimated that Europeans, like the Chinese, might attain virtue without Christian theologians. Lange and his Pietist pals, foes of Wolff's

---

2. Bornhack, *Geschichte der Preussischen Universitätsverwaltung,* 54. On Halle, see Schrader, *Geschichte der Friedrich-Universität.*

3. Wundt, *Die deutsche Schulphilosophie,* 19. On Thomasius, see Schneiders, *Christian Thomasius.*

Fig. 13.1.   Title page to Christian Wolff's "German Metaphysics" (1720). By permission of the Universitätsbibliothek Göttingen.

Neo-Scholasticism, then sought to have him expelled from Halle and all Prussia. They succeeded in 1723. Wolff went to Marburg and became a cause célèbre of the nascent Enlightenment and, to show all things were possible in the *Aufklärung,* was endorsed even by Viennese Jesuits. In 1733, to add insult to injury, the Académie française made him the first German external member since Leibniz. When a German was celebrated in both Vienna and Paris, even Berlin took notice. Wolff was recalled to the University of Halle, where he reconciled himself with Lange the day after returning as professor in 1740. The science of German metaphysics had been reincarnated and canonized in Wolff's Latin textbooks: *Philosophia prima sive ontologia* (1730), *Cosmologia generalis* (1731), *Psychologia rationalis* (1734), and *Theologia naturalis* (1736–41).

The reincarnation of metaphysics in the early Prussian *Aufklärung* was then easily identified with the Wolffian legacy.[4]

In this chapter I am not concerned with the Wolff affair per se. My interest lies in the problems that Lange and others saw in Wolff's metaphysics, as well as its ramifications in Prussia. Theology provides the focus of this section, which has three subsections: "Minds, Machines, Miracles"; "Rational Theology"; "Revelation and Culture."

MINDS, MACHINES, MIRACLES

Mr. Wolff was teaching the system of preestablished harmony at Halle, when the king asked about this doctrine much spoken of then. A courtier responded to his majesty that, according to this doctrine, all soldiers were nothing but machines and that, should one desert, it would be a necessary consequence of its structure, so that it would be unjust to punish him, as if one might punish a machine for producing such and such a motion. The king was so furious upon hearing this view that he ordered Mr. Wolff to be driven out of Halle [and all Prussia] and to be threatened with hanging if he remained there more than twenty-four hours.[5]

Whether this anecdote gives the proximate cause of Wolff's banishment (and Mr. Wolff actually had forty-eight hours to leave), Euler nonetheless has deftly interwoven many issues surrounding the affair, showing the interconnection of the elements of special metaphysics. If the controversy between Wolff and the Pietists set the terms of the Prussian *Aufklärung* up to 1740, we can see how debates over mechanization and dehumanization, the themes of Simon Schaffer's chapter, arose in Prussia too.

Putting the Pietists' case, Lange said Wolffian metaphysics led to fatalism. He set one of Wolff's big errors in his mechanization of the world and the reduction of humans to automata. To this theme, Lange and his accomplices returned again and again. In his *Caussa Dei*, Lange said Wolff had committed the particular error of Spinoza, by believing "in absolutely mechanical fate." In *Hundert und dreyssig Fragen*, Lange saw Wolff's cosmos as a Spinozist clockwork, "a spiritual automaton." In *Bescheidene . . . Entdeckung*, he said, "The

---

4. On Pietism and Wolff, see Gierl, "Auf den Boden"; Holloran, "Banishment of Christian Wolff;" Philipp, *Das Werden der Aufklärung*, 16; on Wolff, Lange, Pietists, and Jesuits, see Wolff, *Werke*, esp. ser. 1, 10:170; Ludovici, *Sammlung*, 1:177; Ulmigena, *Send-Schreiben*, 126–46; Schöffler, *Deutscher Osten*, 194–222; Wundt, *Die deutsche Schulphilosophie*, 199–230; Hinrichs, *Preussentum und Pietismus*, 388–441; Schneiders, *Christian Wolff*; Albrecht in Wolff, *Oratio*; Bianco, "Freiheit gegen Fatalismus," esp. 113–16. About thirty-three primary writings are at the University of Göttingen Library (8° phil. III, 705–9).

5. Euler, *Lettres*, vol. 2, letter 84.

Wolffian system holds the whole world, with every thing and event in it, to be a mere machine, similar to a clock, . . . subject to [Spinozist] mechanical *fate.*" The charge of mechanical necessity and fatalism was high on Lange's list against Wolff, portrayed as more a protégé of Spinoza than of Leibniz.[6] One must wonder whether the Wolffian philosophy seemed strangely Calvinist to the officially Lutheran theology of Halle Pietists.

"Everything is thus certain and determined in advance in humans, as everywhere else, and the human spirit is a sort of spiritual automaton."[7] Wolff was at base a Leibnizean, and thus we must delve into that philosophy. Leibniz opposed the mechanical philosophy as preached by Descartes and Hobbes: their philosophy sought to do away with, among other things, "final causes," entelechies, teleology, and so on, at least from natural science. Such a Cartesian-Hobbesian mechanical philosophy rid nature of souls and spirits. Physics and cosmology became a realm of mechanical or "efficient" causes, with teleological or final causes banned to the realm of psychology or pneumatology, if not banished altogether. Seeing himself as a restorer of the ancients and school metaphysics, Leibniz held one must restore teleology, purposiveness, and even intelligence in nature.[8]

Leibniz formulated his "system of preestablished harmony" to explicate the relation of body to soul, of machines to minds, of efficient to final causes, of mechanics to teleology. Taking "clock" as a metaphor for both minds and machines: "Consider two clocks or watches perfectly synchronized. This is possible in three ways. The first consists in the mutual influence of one watch upon the other; the second consists in the care taken by the person keeping them; the third consists in their own [synchronous] exactitude . . . my hypothesis, *the way of preestablished harmony.*"[9] The first system, mutual influence, commonly attributed to Hobbes, was also in part the Cartesian system. The second system, "occasional causes," was Malebranche's, holding that minds and machines, souls and matter, may only affect one another via God, the person keeping them. Against Descartes, Malebranche, and Hobbes, Leibniz conceived the system of preestablished harmony to free minds, including the divine, from all influence over machines, thus to remove the need for constant miracles in the cosmos. The clocks that are the soul and body, mind and machine, were set in a harmony preestablished at conception. For this

6. Lange, *Modesta disquisitio,* 31–66; *Placide vindiciae,* 34, 57, 62; *Caussa Dei,* 108, 125, 360–97; *Hundert und Dreyssig Fragen,* 13–14 (sep. pag.), 8, 11, 68–69; in Ludovici, *Sammlung,* 1:14; Ulmigena, *Send-Schreiben,* esp. 3–18, 31, 65–66, 119–25.

7. Leibniz, *Schriften,* 6:131.

8. Ibid., 4:434–36, 448, 478–79; 6:114–15, 170–73, 182–83, 198–200, 229, 288, 341, 541.

9. Ibid., 4:500; see also Clark, "Scientific Revolution."

reason, Leibniz also found fault in Newton's matter theory, which held that motion was not conserved in collisions at the atomic level, entailing the necessity of regular divine intervention—miracles—to wind up the mainspring of the cosmic order.[10]

Machines and minds versus miracles was the matter. To relate the first two without miracles, Leibniz was driven to his system of preestablished harmony. In *Psychologia rationalis,* Wolff well summarized and expanded the Leibnizean position. Not counting Hobbes, Wolff reduced all systems on the commerce of body and soul to three: (1) the system of physical influx (Cartesian); (2) the system of occasional causes (Malebranchean); (3) the system of preestablished harmony (Leibnizean).[11] The first, the system of physical influx, held that mind might affect matter (and, for materialists, vice versa). Wolff objected that such an influx of mind over matter was a *qualitas occulta* and that conservation of Leibnizean "living power," *vis viva (mv²),* was also violated. Any motion induced by the mind on our body, a machine, created new *vis viva,* unless God or another spirit annihilated a like quantity of *vis viva,* which would be equivalent to system 2. The system of physical influx, if not reduced to that of occasional causes, allowed for the possibility of a *perpetuum mobile* and of the first sort: one that created new power or force. This system thus compromised the physical integrity of the cosmos and, given the problem with perpetual motion, was held impossible on metaphysical grounds.[12] Wolff held that the second system—occasional causes—needed perpetual miracles, as any intervention by God over minds or machines was a miracle. Miracles were possible but, by principles of perfection, order, and economy, should be as few as possible. Given Wolff's setup, the system of preestablished harmony won by default (the materialist system being beyond the pale), as it entailed but one initial miracle: the original harmony between body and soul. Thereafter, it explained everything naturally, needed no more miracles, and preserved the mechanical order of nature.[13]

The Leibniz-Clarke debate, which ran from 1715 to 1716, bore remarkable similarities to the Wolff-Lange debate after 1723: in view of the doctrine of preestablished harmony between minds and machines, both Leibniz and Wolff had to defend themselves from charges of fatalism. European intellectuals had distanced themselves from belief in astrological fatalism after the 1670s, but

10. See Leibniz, *Schriften,* 4:500–501; 6:136–37, 241, 352; Descartes, *Oeuvres,* 11:354–58; Newton, *Mathematical Principals,* 1:543–47, and query 31 in *Opticks,* 375–406.

11. Wolff, *Werke,* ser. 2, 6:474–75 (§ 553).

12. Ibid., §§ 578, 582. On Leibniz and *perpetuum mobile,* see Schaffer, "Show That Never Ends," 163–65; in general, Gabbey, "Mechanical Philosophy."

13. Wolff, *Werke,* ser. 2, 6:526–29 (§ 603), 556–57 (§ 623), 566 (§ 629).

Leibnizeans now seemed to have fallen into a mechanical fatalism. Leibniz made himself most clear in the long-unpublished *Théodicée,* where he held fate as one of the two great labyrinths of reason and said he accepted neither "Fatum Mahumetanum" nor "Fatum Stoicum," but only Christian fatality, "a certain destiny of all things, regulated by the prescience and providence of God." But in trying to explicate this, he turned to the fatalistic metaphors of deus ex machina and marionettes.[14]

Insofar as the system of preestablished harmony was definitive of Leibniz's final philosophy, Wolff was a Leibnizean. Adopting preestablished harmony, he held body to be a sort of automaton, and he maintained that body and soul could not per se affect one another. Wolff thought he could prove: the cosmos was one spatiotemporal causal nexus; the nature of things did not change; and nothing went out of the world by real annihilation, unless by a miracle. All things were, however, continuously conserved by God, which was a miracle (and a rather Cartesian view). Miracles were possible, but their grounds did not lie in the cosmos, and they could be explained neither from the structures of matter nor from the laws of motion.[15] Wolff's cosmos was a machine, a *horologiam automaton.* It was this notion of the cosmos as clockwork, appearing in Wolff's "German Metaphysics" of 1720, that precipitated great concern by Lange and others. Made pressing by the Pietist critique, the theo-cosmological problem for a Wolffian became how to unite belief in mechanical determinism and preestablished harmony with disbelief in astrological and mechanical fatalism? The Leibnizean-Wolffian and Prussian solution would be neither (Newtonian) to allow God to clean and repair the cosmic clockwork from time to time, nor (Cartesian) to allow the world to run purely mechanically. The Prussian solution to cosmic order lay in final causes, labeled "metaphysical" by "O" in Diderot's *Encylopédie.*[16]

Along with the system of preestablished harmony, "monadology" formed a central part of Leibnizean metaphysics. Underlying the Leibnizean-Wolffian mechanical order was a realm of organs and souls, or monads. The problem was its nature: whether it was metaphysical or noumenal, as opposed to physical or phenomenal. Though without extension, monads manifested them-

14. Leibniz, *Schriften,* 6: 29–32, 65, 135–37, 230, 319–27, 356; 7:355–59, 365, 377, 389–91, 416–19.

15. Wolff, *Werke,* ser. 1, 2:380, 478–88; ser. 2, 4:55–60, 94–97, 396–409, 437–47; 7:44, 823, 828–29, 845; 8:299–300.

16. *Encylopédie,* 2:789; Wolff, *Werke,* ser. 1, 2: §§ 556–61; foreword to vol. 7; ser. 2, 4:62–67, 103; 7:487–89; 8:509. The problem was more complex, as Wolff had to address the necessity versus contingency of laws of nature. On the medieval tradition, see Funkenstein, *Theology and the Scientific Imagination,* chap. 3. On Wolff, see Mayr, *Authority, Liberty, and Automatic Machinery,* 73–77.

selves as bodies and "this body is organic when it forms a sort of automaton or machine of nature"; "each organic body of a living being is a sort of divine machine or a natural automaton surpassing all artificial automata infinitely." The monads were psychic entities, with perception, appetite, and even memory. And every monad mirrored the entire world from its own point of view.[17] Leibniz thus underlaid his cosmology with a pneumatology; he set an intelligible order of final causes under the mechanical order of efficient causes. By fusing pneumatology and cosmology, the monadology staked a position pretty far out, even for Wolffians.

A Wolffian member of the Berlin Academy of Sciences, Formey, said Leibniz and Wolff differed in that the former stressed the mirroring and perspectivity of monads, while the latter stressed the interconnectedness and embeddedness of monads so that, given one state of the cosmos, all others in past and future follow. Formey implied Wolff's view was the more deterministic. Gottsched, also a Wolffian, said Leibniz stressed the internal as representational, a psychic power; Wolff supposedly sidestepped the issue, wishing to keep cosmology and pneumatology piously far apart.[18] This Wolffian step may have been taken under Pietist duress, as he was compelled to insulate spirit from mechanical fatalism.

Wolff held that all action was by contiguity, so action at a distance was impossible. Wolffian matter had a passive power *(inertia)* and an active power *(vis motrix)* taken as a tendency *(conatus),* so that matter was virtually always in continuous motion. Matter did not actually change its own state, owing to inertia and the resistance of contiguous bodies; moreover, the *vis motrix* of each atom or monad tended to move it in every direction at once, giving a sum zero change of state. Wolff saw Leibniz as having held the elements, as monads, to have a power continuously inducing internal changes; Wolff himself posited numerous sorts of elements with different powers: Each body had a fundamental power *(vis primitiva)* grounded in elements, qualitative and nonextended. But, as a good Leibnizean, he held that all matter was elastic, so that Leibnizean *vis viva* (mv^2) was conserved in collisions.[19] These were seen as metaphysical positions.

Wolff said mechanical principles determined bodies by figure, magnitude, place, and motion, whereas physical principles determined phenomena such as chemical mixtures in ways that as far as we could know were not reducible to mere mechanical principles. "Organic" were phenomena whose composi-

17. Leibniz, *Schriften,* 6:599, 618; cf. Baumgarten, *Metaphysica,* § 433.

18. See Formey, *Mêlanges,* 2:263–388; Gottsched, *Erste Gründe,* § 389–95.

19. Wolff, *Werke,* ser. 1, 2:368–71; ser. 2, 3: §§ 722–28; 4:112–21, 128, 140–41, 145–48, 229–30, 240, 259, 372–80.

tion depended on structure; all instruments, such as clocks, were thus organic.[20] That position might seem to reduce the organic to the mechanical,
though in a backhanded way one might argue the opposite. As we'll see, the
problem of structure and the systematic was what led Prussians to a critique of
Cartesian-Hobbesian mechanism: machines could only be seen as structural
or systematic ensembles when set within a framework, be it of "economics" or
teleology, ultimately organic or intelligent. The Leibnizean-Wolffian view saw
system or structure as the deus ex machina, the ghost or mind in the machine.
This set a teleological in a mechanical order, set final, purposive causes alongside efficient causes. But Wolff held that physics preceded teleology, which, as
an "experimental theology," supported physics.[21]

In this discussion of minds, machines, and miracles, we have been considering the Leibnizean-Wolffian monadology and the system of preestablished
harmony, which envisaged a "teleo-mechanics," a pneumato-cosmology.[22]
Like Leibniz (contra Clarke), Wolff (contra Lange et alia) risked a determinate,
mechanistic, and possibly fatalistic teleo-mechanics in order to keep theology
and miracles out of the natural world. Astrological fatalism now beyond the
pale, mechanical fatalism emerged as the issue. Lange's polemics pushed Wolff
from ontological to epistemological formulations: "This world, which exists,
we shall call the perceived world." Another critic, Ulmigena, laid his finger on
the same point, declining to see it as being held in good faith: Wolff held that
the cosmic nexus was not mechanical, but rather only rational, a nexus of
things known (nexus rerum sapiens), thus not fatal.[23] Wolff's world-as-perceived was an organic clock, a rational machine or automaton, whose soul
inhered in teleology, an intelligible structure. Cosmology was distanced from
theology, but bound up closer with pneumatology by the monadology.

RATIONAL THEOLOGY

"Natural theology is the science of God insofar as can be known without
faith."[24] "Natural" meant theology based on the natural light of reason, and,
as demonstrable, natural was rational theology. The revival of German rational
theology may be dated at least from Leibniz, whose aim was to reconcile reason
with revelation. Given scholastic terminology, this set "dogmatics" as primary:
dogmatic reason showed its conciliation with dogmatic faith, but also its independence, as it was able to arrive at the same dogmas. Here lay the potential

20. Ibid., ser. 2, 4:36–37, 180–81, 210–13, 398.
21. Ibid., ser. 1, 9: § 179; ser. 2, 1(1): §§ 85, 100–107; 3: §§ 886–943; 4:28–29.
22. On teleo-mechanics, see Lenoir, Strategy of Life.
23. Wolff, Werke, ser. 2, 4:45, 173, 224–25; Ulmigena, Send-Schreiben, 124–25.
24. Baumgarten, Metaphysica, § 800; cf. Philipp, Das Werden der Aufklärung, 18, 72.

problem with rational theology, the most influential of which in the *Aufklärung* was Wolff's, the crown of his science of metaphysics. Like his mechanic-pneumatic cosmos, Wolff's apotheosis of reason in theology found no favor with Lange and Pietists.

We shall follow Kant by dividing rational theology into onto-theology, cosmo-theology, and physico-theology.[25] These three span the field of metaphysics. Working at the hinge of logic and ontology, onto-theology claims to argue purely rationally or a priori, by mere logical analysis, in establishing God's existence as an *ens realissimum,* the absolutely supreme being. Taking up the center of special metaphysics, relating cosmology and theology, cosmo-theology argues from the notion of a given datum: the contingency of the world. Given only the existence of the cosmos as a contingent entity, cosmo-theology seeks, chiefly by the "principle of sufficient reason," to prove the existence of the deity as an *ens necessarium,* the absolutely necessary being. Moving furthest from pure reason, as a bridge from metaphysics to physics and natural history, physico-theology turns to some actual entity in our cosmos for its datum, for instance, a tulip in Candide's garden.

In volume 2 of *Theologia naturalis,* Wolff pursues onto-theology. As essentially ontological, it lies mostly beyond the scope of this essay. So we shall turn now first to cosmo-theology: volume 1 of *Theologia naturalis.* Wolff claims to prove the existence of God through the existence of the human soul, the principle of sufficient reason, and an argument from contingency: our world as a nexus in space-time of contingent entities needs some entity outside it as its cause. As self-caused *ens necessarium,* this God is argued to have the properties known by revealed theology. The deity of metaphysics has an intellect and will, is a spirit, without senses or imagination or memory, and which neither sleeps nor knows fear nor loathing nor laughter. The deity of rational theology has intuitive, nondiscursive knowledge, is possessed of foresight, including all human choices and thoughts, as well of all states of the world. Pure reason's deity cannot will the impossible, though its will is free and may will miracles; indeed, the creation of the world was a miracle. But what can be done naturally, God, adhering to principles of economy, does not work by miracles. God is lord of the world, has humans as servants, and is their just judge. Metaphysics' deity is, in short, like an absolute but enlightened Prussian monarch: "God is the absolutely supreme historian" and "is the absolutely supreme philosopher" (and, from onto-theology, God is a mathematician).[26] So we see part of Wolff's

25. Kant, *Schriften,* 2:63–163; 3:396, 482; also Leibniz, *Schriften,* 6:72, 74, 86.

26. Wolff, *Werke,* ser. 2, 7:140–41, 231, 244, 314–15, 344–45, 402, 438–42, 452–53, 757–58; cf. Leibniz, *Schriften,* 6:297: "All sages agree that chance *(hazard)* is nothing but an apparent entity, just like fortune: it is the ignorance of causes whence it arises."

rational theology, purporting to establish the truth of revealed theology by the natural light of reason.

"In the transformation to the era of the *Aufklärung*, the matter of the garden is a theological question."[27] From onto- and cosmo-theology, we turn to physico-theology, the final sort of rational theology and the metaphysics of the garden. Wolff claimed to have begun physico-theology, a claim belied by the British tradition before him. But he did make physico-theology popular by his "German Teleology," *Vernünfftige Gedancken von den Absichten der natürlichen Dinge* (1724). As teleology, physico-theology was experimental theology confirming rational theology.[28]

German physico-theology blossomed from the 1730s to the 1760s. Ahlwardt wrote a "Thunder Theology" (1745) in which lightening and thunder cried out: "Here is the only true God, who possesses infinity, the God who is alone wise, good, holy, just and all-powerful." Ahlwardt detailed the evidence, and to the atheist who said these phenomena were simply nature, he queried: "Are natural things absolutely necessary and of themselves?" Benemann wrote a "Tulip Theology" and a "Rose Theology" (1741) in which these flowers were praised in registers of types to show that "through these fair witnesses [one] may be led to knowledge of the supreme being." Denso wrote a "Grass Theology" (1750) as a poem. Fabricus outlined a "Fire Theology" (1732) and wrote a "Water Theology" (1734) in which, for instance, the number of rivers on earth was pondered in praise of the Creator. Heinsius penned a rather short "Snow Theology" (1735). The king of physico-theology, Lesser, wrote a "Stone Theology" (1735) amounting to 1,300 pages plus register. Even if humans misused them, stones allowed us to marvel at God's wisdom, despite humans' misuse. Lesser also wrote an "Insect Theology" (1740), in which tiny creatures offered lessons of divine wisdom. Finally, our king wrote a "Snail Theology" (1744), a meaty 984 pages plus register. Of a snail shell Lesser marveled, "What an infinite understanding must God not have? One can rightly conclude from the work of an artist to the mind and talent of the latter." Rathelf wrote a "Locust Theology" (1748–50), containing a natural and human history of locusts, with all attestations of their plagues. Were locusts necessary to the world? No. But they showed us the "wonderfulness" of God. Richter wrote a "Fish Theology" (1754), Rohr a "Plant Theology" (1740), Schierach a "Bee Theology" (1767), and, yes, Zorn a "Bird Theology" (1742–43).[29]

---

27. Philipp, *Das Werden der Aufklärung*, 163.
28. Wolff, *Werke*, ser. 1, 7: §§ 2–14; ser. 2, 1(1): § 85, 101, 107; see also Leibniz, *Schriften*, 6: 144–45, 165, 264, 545, 605, 621–22. In general, see Philipp, *Das Werden der Aufklärung*, 16–20.
29. Quotations from Ahlwardt, *Bronto-Theologie*, § 63; Benemann, *Die Rose*, 195–96; Lesser, *Lithotheologie*, § 616, and *Insectotheologie*, 511.

To sum up, onto-theology lay at the joint of ontology and logic. Cosmo-theology moved to the center of special metaphysics, joining cosmology and theology, by basing itself on the contingency of the cosmos and seeking to establish the existence and nature of the absolutely necessary being. At the far edge of metaphysics, joining it to natural science, lay physico-theology, which appealed to the marvelous details of nature and their contingency, whence, following cosmo-theology, arose their dependence on an external intelligible and purposive power or mind. Physico-theology sought, indeed, to establish the existence and nature of the absolutely "architectonic" being: God as ulti-mate architect-artisan. As in the case of minds and machines, teleology emerged here too as a key. Linking theology, pneumatology, and cosmology, teleology was a handmaiden to rational theology. Given impetus from Wolff's "German Teleology," physico-theology, from Ahlwardt to Zorn, embodied a sort of experimental theology confirming rational and even revealed theology.

REVELATION AND CULTURE

Insofar as Prussians under Friedrich the Great soon claimed to speak for all, the reinstatement of Wolff in 1740 seemed to seal the victory of *Aufklärung*. But Lange and the Halle Pietists had ensured that the Prussian *Aufklärung* would proceed neither atheistically nor even deistically, as the Enlightenment had at least in part in France and Britain. The reconciliation of Pietism and Wolffi-anism had far-reaching consequences for revealed theology. New currents from Britain were important too. They brought not only physico-theology but also new methods into church history and scriptural theology. A sentiment arose that enlightened church historians and nondogmatic theologians had to be impartial, had to put personal faith aside as scholars. This view would come to bear on the possibility of the miracle of revelation and the metaphysics of history.

Within this framework, a new German theological school emerged, the Neologians. The heart of their program upheld, against Deists, the possibility of the miracle of revelation while undermining its content by banishing par-ticular miracles. The neological stress on reason in history won emphasis from Wolffianism: Revealed was supernatural theology, thus miraculous.[30] Revela-tion and other miracles were metaphysical in the bad sense, since they could not be known by dint of pure reason and lay outside the natural order. With the dissolution of the mediating realm of the preternatural in the seventeenth

---

30. Wolff, *Werke,* ser. 2, 7:10. See Hinrichs, *Preussentum und Pietismus,* 441; Aner, *Die Theo-logie,* chaps. 1, 3; Frei, *Eclipse of Biblical Narrative,* 96–103, 114. Taking the University of Göttingen as a barometer, see ibid., 167–73; Heussi, *Mosheim,* chap. 4; Moeller, *Theologie in Göttingen;* Völker, *Kirchengeschichtsschreibung,* 11–13. Edited by Lessing and published anonymously, Reimarus's *Frag-mente* embodies the most rationalistic position in the *Aufklärung.*

century, moreover, the gulf between the natural and supernatural, including revelation, widened. In the name of reason, Neologism called for reading the Scriptures almost as a novel, a new "realistic" genre much limiting the supernatural. Mindful of such neological views, and in the light of the emerging bourgeois novel, it is interesting that Germans, unlike French and British, did not develop realistic fiction then, while they did develop analogous sentiments for historical writing and demanded conformity from revelation.[31]

Despite his ambivalent relation to Neologism, Semler at Halle, progeny of Lange and Wolff reconciled, was the leading theologian applying the new historical and philological methods to the Scriptures.[32] He argued that there was an external history of Scripture, which could be considered as a book like any other. Taking this point consequently, and moving to internal history, proved devastating. Hebrew tradition held its Scriptures were revealed, but as per Semler, such revelation could not be verified, so was scientifically questionable. In fact, Semler doubted the Hebrew Scriptures were revealed and thought some parts of the canon were not worthwhile.[33] As for the New Testament, it was difficult to trace the Gospels back beyond the fourth century, making it impossible to assess authenticity. Semler said one must take the Evangelists seriously as human historians. Like "cultural context" for some moderns, "historical circumstances" *(Umstände)* served as shibboleth for Semler. The atemporal meaning sought by a dogmatic theology gave way to a view of the cultural-historical embeddedness of Scripture and its authors. Opposed to dogmatic theologians, Semler's nearly anthropological theology focused on the "singular, local and characteristic of each people and time." Figurative language did not incorporate a universal message; rather, it dated the Scriptures. Semler later engaged in damage control and, trying to put space between himself and the Deists, asserted miracles were not impossible.[34]

But other works of his aimed at eliminating miracles. He claimed the Jews had brought the views of superstitious peoples from their surroundings into the Scriptures; moreover, the origins of miracles in the New Testament must be set in the fourth century, when the Romans sought to create a new mythology to Christianize barbarians. Enlightened Protestants now doubted the validity of miracles in the New Testament, as well as the reports putatively from

31. See Frei, *Eclipse of Biblical Narrative.* On the preternatural, see Daston, "Marvelous Facts."

32. For Aner, *Die Theologie,* 98–112, Semler was at best a lapsed Neologian; for Reill, *German Enlightenment,* 82, he was a leading Neologian; for Frei, *Eclipse of Biblical Narrative,* 61, the greatest one; also see Hornig, *Die Anfänge der historish-kritischen Theologie;* Span, "Auf dem Wege," 81–82.

33. Semler, *Vorbereitung,* 2: § 2; *Abhandlung,* 24, 67–77.

34. Semler, foreword to Baumgarten, *Untersuchungen theologischer Streitigkeiten,* 1(1), §§ 4–5, 9; *Neue Versuche,* 4–8; *Abhandlung,* 41–46; *Vorbereitung,* § 7; *Beantwortung.*

early church history that violated the laws of physics.[35] In his 1776 demonology, Semler claimed he was editing the work of an anonymous author. The author (Semler) claimed the New Testament could be read so that devils and demons did not appear. With some sleight of hand, he got the devils out. Much of Scripture, we hear, was figurative. Thanks to superstition, other parts were full of wonders and devils. The serpent in Genesis was not a devil; besides, serpents could not speak and the Fall did not need a devil. Demonic possession was a figment of unenlightened imaginations. In an appendix, Semler endorsed some of his own above views and traced the belief in devils to a "Jewish mythology," taken from Chaldeans, Greeks, and other pagans. Semler's theo- logical *Aufklärung*, necessitating expulsion of miracles and demons from Scrip- ture, demonized Jews for polluting revelation. Halle's new theology aimed to expunge the "base, **uncultured** [sic] manner of thinking of so many enthusi- astic Jews." Greeks and Romans were cultured, as opposed to "the actually very incapable and uncultured Jews."[36]

So the ends of Lange and Wolff reconciled, Prussian *Aufklärung* in Halle? "Realistic" and nearly "positivistic" neological views demanded a disenchant- ment of the Scriptures, now to be read like a realistic novel, where miracles and demons did not exist. This profanation of biblical tradition abetted a demonization of the Jews and allowed for the sacralization of culture. The notion of *Kultur* would be used by Semler's colleague at Halle, F. A. Wolf, to fashion a science of antiquity, for which Egyptians, Hebrews, and Persians only attained "bourgeois policing" *(bürgerliche Policirung)* or "bourgeois civi- lization," whereas Greeks and Romans attained "higher genuine culture of spirit" *(höherer eigentlicher Geistescultur)*. From the science of antiquity, Wolf argued, "Asians and Africans, as not literally cultured, but only rather civilized folks, . . . will be excluded from our borders," as well as "the Arabs, later so important," all of whose works are "left to Orientalists."[37] *Kultur,* with *Geist,* would become part of the secret new metaphysics of the Romantic era.

---

35. Semler, *Vorbereitung,* 1, § 4; *Versuch einer . . . Dämonologie,* 44–49, 54; *Neue Versuche,* 17; *Versuch einiger . . . Betrachtungen,* 2–6, 37, 40–42.

36. Semler, *Abhandlung,* 44 (quotation); *Versuch einer . . . Dämonologie,* 13–16, 26–28, 44– 49, 67–68, 74, 275–77, 300, 313–59.

37. On the above, see Wolf, *Darstellung,* 11–17; in *Encyclopädie,* 8–9, he claims the Hebrews never became a "gelehrte Nation," whereas the Greeks and Romans did. The above two works were based on Wolf's lectures in the late eighteenth century in Halle, then after 1806 in Berlin. An early cultural anthropology was *Grundriss der Geschichte der Menschheit* (1785) by Christoph Meiners— in general, see Linden, *Untersuchungen zum Anthropologiebegriff.* Though Meiners's *Grundriss* of 1785 is a racist work, it was perhaps par for the period. Beyond the pale are his articles in journals he coedited. Here he argued for the natural superiority of the "Germanic-Celtic" folks. The Ger- mans and Celts had "blue eyes and blond hair," and "the Germans surpassed all other Celts, as much in size, strength and beauty, as in the blondness of their hair and the blue and fire in their

Euler and the Berlin Academy of Sciences

The Berlin Academy of Sciences traced its origins to the Royal Brandenburg Society of Sciences, founded at Berlin in 1700, six years after the foundation of the University of Halle. The spirits behind the academy were three: Daniel E. Jablonksi, a court preacher in Berlin, Sophie Charlotte, queen of Brandenburg-Prussia, erstwhile princess electress of Hanover and pupil, correspondent, and patron of the academy's first president, Leibniz. Sophie Charlotte tried to elevate the abysmal state of arts and sciences in Berlin, though often to little avail, given the lack of cash and interest bestowed by the king. Inaugurating a modern German academic lifestyle, Leibniz began commuting between Hanover and Berlin. Sophie Charlotte's death at the age of thirty-six in 1705, Leibniz's death in 1716, and above all the lack of any real funding let the society drift into a moribund state.

Friedrich II wanted to change that on his accession to the throne in 1740. The reforms of the academy in 1743–44 and 1746 say much about Prussian *Aufklärung*. The reform of 1746 abolished the democratic-republican institutional structures and set up a bureaucratic presidency, left vacant after 1759, that Friedrich, the "first servant of the state," might more easily dominate. The notorious mechanization of the state under Friedrich II and his father arose in the context of a bureaucratization of the state, with which metaphysics was strangely related.[38]

The reform of 1743–44, preserved in that of 1746, made an intellectual change in the academy. Though Friedrich II was lukewarm about if not hostile to metaphysics, Prussia showed its provinciality here. Besides the typical sections of other European academies, the Berlin academy acquired a philosophy section, including metaphysics. On 23 May 1740, Crown Prince Friedrich II, disingenuously perhaps, wrote to Wolff, "Good Sir, every thinking and truth-loving person must treasure your newly published work. . . . I treasure it the more, since you have dedicated it to me. Philosophers should be the teachers of the world and of princes. . . . I have read and studied your works for a long time." Less than a week earlier, he had written to Voltaire: These "simple, thinking substances [monads] seem very metaphysical to me. . . . And it seems

---

eyes." See Meiners and Spittler, *Göttingisches Historisches Magazin* 8 (1787–90): 12; in general, 6 (1790): 387–450; 7(1790): 117–45, 229–30; 8(1790): 1–25, 119–21; *Neues Göttingisches Historisches Magazin* 1 (1792): 17.

38. On bureaucracy and mechanization, see Stollberg-Rilinger, *Der Staat als Maschine;* also see Rosenberg, *Bureaucracy.* On the Berlin Society and Berlin Academy, see Harnack, *Geschichte der königlich Preussischen Akademie;* McClellan, *Science Reorganized,* 68–74; Finster and van den Heuvel, *Liebniz.*

to me humanly impossible to speak about the characters and acts of the creator. . . . [I]f God had wanted to make us metaphysicians, He would have given us a higher degree of insight."[39]

Crown Prince Friedrich was laying plans to bring Voltaire and Wolff to Berlin as twin pillars of a reinvigorated academy of sciences. Voltaire had no desire to live in Berlin. Nor did Wolff, who had put in his own time in purgatory at Marburg. Wolff wanted to be at a university, which Berlin did not have. So he returned to Halle in 1740. For Berlin, the newly crowned king had to settle for Maupertuis and Euler. With Maupertuis as president of the academy after 1746, a metaphysics of teleo-mechanics would blossom in Berlin.

The article by "O" on cosmology in Diderot's *Encylopédie* is interesting here.[40] O explained that cosmology considered the general laws of the cosmos, looking for the nexus in nature, leading to a view of the cosmic author. O reviewed Wolff's *Cosmologia generalis,* then moved to Maupertuis versus Wolff on the notion of action. Maupertuis's usage, O said, reconciled mechanical and final causes, of interest to those interested in the latter, which most French were not. O then mentioned Euler's papers of 1750–51 on "least action." From a Prussian view at midcentury, O did well in moving from Wolff to Maupertuis to Euler on cosmology. Thus the following discussion centers on cosmology in three subsections: "Atoms and Monads"; "Rational Mechanics"; "Stars, Aether, Aliens."

ATOMS AND MONADS

Wolff's triumphal return to Prussia in 1740 soon met with polemics from Berlin. A Wolffian work, *Belle Wolffiene,* by Formey, secretary of the academy, did appear in 1741. In the same year, the marquise du Châtelet's *Institutions physique* also appeared. The marquise saw herself as a Leibnizean and remarked of the great change in that philosophy introduced by Wolff, who nonetheless saw her *Institutions* as the best presentation of his own metaphysics. And 1741 saw as well a critique of Voltaire's Newtonian work and a defense of Leibnizean organicism and teleology by a certain Kahle.[41] But in the same year, Euler arrived in Berlin, and then in 1745 Maupertuis. Both were interested in metaphysics, but neither enamored of Wolffianism. Indeed, Euler replaced Lange as Wolff's chief tormentor after 1740 and went after the roots of Leibnizean-Wolffian metaphysics.

Though preestablished harmony seemed the root of most religious objec-

39. Friedrich II, *Die Philosophie,* 82, 64–65.

40. *Encylopédie* (1754), 4:294–97.

41. Du Châtelet, *Institutions physiques,* 137; Wolff, *Werke,* ser. 1,10:176; Kahle, *Vergleichung der . . . Metaphysik.*

tions to the Leibnizean-Wolffian philosophy, in second place came the monadology. J. F. Müller's anti Leibniz-Wolff tract of 1745, *Die ungegründete und Idealistische Monadologie* raised the essential physical objection: How could unextended simples, in however large numbers, affect an impenetrable or even extended body? How could metaphysical substances fill a physical space? In 1746, Euler published anonymously *Gedancken von den Elementen der Körper,* an anti-Wolffian work in a similar spirit. The work was immediately recognized as bearing his signature and occasioned swift critique by Formey.[42] An anonymous essay by C. A. Körber in 1746 also took up arms against Euler's essay. Körber argued that neither Leibniz nor Wolff anywhere maintained that monads made up physical bodies or could induce motion in them. "It seems that philosophical texts are little known to the *Herrn Gegner* [Euler] and that he has looked little in them." Also after Euler went a likewise anonymous essay in 1746 by Stiebritz, who presented a Wolffian monadology in which he denied monads representational powers.[43]

The controversy on monads erupted in 1746. Euler himself recollected:

> There was a time when the dispute about monads was so lively and general that one spoke of them heatedly in all companies, even in the *corps-de-garde.* There was almost not a single lady at court who had not declared herself for or against monads. Everyone's conversation fell upon monads everywhere and no one spoke of anything else. . . . The Berlin Royal Academy took much part in this dispute and, as its wont was to propose a prize question once per year . . . , it chose for the year 1748 the question on monads. A great number of entries were received on this matter. The president, Maupertuis, nominated a commission to examine them. . . . In the end, one found those wanting to establish the existence of monads were so weak and chimerical that they would overturn all the principles of knowledge. So one decided in favor of the opposite sentiment, and awarded the prize to a piece by Mr. Justi. . . . Your Highness can easily imagine that this move by the academy terribly irritated the partisans of monads, at the head of whom was the famous Mr. Wolff.[44]

The Berlin academy prize question for 1747–48 was on the monadology. The question quickly became so heated, the academy decided that, though the question had been posed by the philosophical class, all four classes of the academy would determine the winner.[45] That helped Euler to rig the decision, so that his "tool," Johann Justi, might win. Justi's essay was published, with

---

42. Formey, *Mélanges,* 1:263–453; Euler's *Gedancken* is in *Opera,* ser. 3, 2:351–66.
43. Körber, *Gegenseitige Prüfung,* 5; Stiebritz, *Widerlegung der Gedancken,* §§ 11–17: one may see rudiments here of the later Boscovich-Kant solution.
44. Euler, *Lettres,* vol. 2, letter 125; Harnack, *Geschichte der königlich Preussischen Akademie,* 1(1):403.
45. See Harnack, *Geschichte der königlich Preussischen Akademie,* 1(1):402–3; 2:305.

others, by the academy in 1748. A review in the *Göttingische Zeitungen von gelehrten Sachen* may be taken as a good barometer of opinion outside Prussia, as Göttingen was no friend of metaphysics or Wolff. The review noted Euler's authorship of the *Gedancken* of 1746 as common knowledge. It also noted the academy had taken pains to indicate that prize-winning essays did not perforce represent their views. The review went on, "We believe many readers will find here an apology by the academy for its choice, which did not come about with the consent of all members."[46]

In 1748, the academy published Justi's essay not once but twice, in French and German.[47] Justi's strategy was a *tertium non datur* between mathematics and physics: metaphysical issues were declared void of sense or made into mathematics or physics. Simples might exist in mathematics, but "in all of nature, one sees nothing of simples" (§ 23). One might allow Leibniz his notion of monads being possessed of perception as an internal power, but that gave no basis to explicate spatial extension and motion, as Wolff had tried to do (§§ 43, 50, 57, 89–90). If monads individually filled no space, how could a collection of them do so (§ 49)? "I cannot conceal the fact that I have seen nothing in the world so poorly tied together as the doctrine of monads" (§ 62).

Besides Justi's essay, the academy published seven other essays.[48] Save Justi's, only the last two in the 1748 collection were not anonymous. The last, by Ploucquet, did not address the question; but it allowed its author to be identified as that of the fourth essay, "Primaria monadologiae capita," which weighed pros and cons, including Euler's 1746 essay, then insisted one should separate metaphysical from mathematical matters (§ 163), and tilted toward the Leibnizean view—this was the essay that the review of 1748 in *Göttingische Zeitungen* said should have won the prize instead of Justi's. The second essay, "De Elementis," defended a Newtonian physical atomism (pt. 3, § 16). The third essay, "Systema mundi," defending Leibniz's monadology, was implicitly against Wolff's (chap. 4, § 67), but also explicated a theory of physical point atoms toward its close (chap. 5, § 104), which was Euler's view. The fifth, "Essai sur . . . les monades," was against monadology, as was the seventh, "Wiederlegung der . . . Monadologie," while the sixth, "Les monades," was for a reformed monadology, but against Wolff. Of all these essays, only two supported a monadology, and none was pro-Wolff.

In 1748, an independent essay by G. A. Müller on the prize question also attacked the Leibnizean view. Müller claimed it led to idealism and could not

---

46. *Göttingische Zeitungen von gelehrten Sachen* 121 (7 Nov. 1748), 966–68.

47. Parenthetical citations refer to Justi, "Untersuchung" and "Dissertation," identical by paragraph sections.

48. From Berlin Academy of Sciences, *Dissertation,* the essays are cited by paragraph sections here.

explicate extension or solidity and that, if not as mere mathematical points, monads made no sense. An anonymous, undated essay by A. Clavius, also probably from 1748, also used the prize question to attack Leibnizean views. Two other anonymous essays of 1748, however, supported Wolff and made the true statement that Justi did not understand metaphysics.[49] As 1748 promised a bumper crop of polemics, Justi managed to respond in the same year.

Justi defended the academy's decision and responded to some of his critics, including the last two above. He again essentially reduced metaphysical issues into mathematics or physics, and claimed the Leibnizean-Wolffian monadology was an erroneous translation of mathematical principles into physical ones. He also played the provinciality card, by noting that German metaphysics was viewed ill abroad, where one laughed about monads. About du Châtelet's support of monads, he said, "Women tend more than men to marvelous and mysterious teachings."[50] Misogyny aside, the remark is interesting in view of Euler's comment above. Translated from French, it reads: "There was a time when the dispute about monads was so lively and general that one spoke of them heatedly in all companies and even in the *corps-de-garde*." The German version of the same work renders the last part of the sentence thus: "that it spread from the schools into women's salons *(Frauenzimmergesellschaften)*."[51] The French then goes on: "There was almost not a single lady at court who had not declared herself for or against monads." All this provides more grist for the view, not unknown to Leibniz scholars, that the monadology grew out of Leibniz's correspondence with women, and above all with the Hanoverians, whom he feared losing to Locke, worried as they were about the relation between their minds and bodies.

Properly peeved at Euler, Wolff wrote with a sigh to a friend in 1748:

Scholars who love real learning are disappearing everywhere and a very superficial sort of knowledge gains the upper hand, as one has mixed a so-called Newtonian philosophy with the French world of flattery. . . . Mr. Euler, who may relish his well-deserved fame in higher mathematics, now wants to use his power to dominate all sciences, [even those] which he has never pursued. . . . Hereby he damages his own reputation much . . . , for which the controversy on monads instigated by him gives evidence, above all since he elected as his tool an arrogant and foolish and so impudent sophist *(Rabulisten)*, named Justi, and thus for want of cleverness sacrificed the interests of the academy for his own passions. It is

49. Müller, *Unpartheyische Critik,* esp. 12, 20–24, 37–39; Clavius, *Bericht;* and the anonymous *Jeremaias W*** and *Sendschreiben an . . . Justi.*
50. Justi, *Nichtigkeit,* 39–42, quotation at 42.
51. Euler, *Briefe,* vol. 2, letter 125. No translator is listed.

unfortunate that Mr. President Maupertuis is a Frenchman who neither speaks German nor understands the conditions of scholars in Germany and who, in things other than mathematics, possesses no more insight than Mr. Euler, although he is more clever and politer than Mr. Euler and would hold the latter more in check if he could only read German and knew enough about the conditions of scholars in Germany. It is sad when someone wants to build on one side by tearing down what others are building on the other side.[52]

Wolff rightly saw Euler as the puppet master pulling the strings behind the scenes in Berlin. Like Lange, Euler thought the Leibnizean-Wolffian view made the human body into a machine, removing the basis of morality. If humans were like marionettes, there seemed as little right to make moral judgments as one could about clocks not working. So Euler polemicized against the Leibnizean-Wolffian pneumato-cosmology and advanced a more Cartesian system, whereby the soul sensed via the body's nerves and used the body. If God as a spirit acted on matter, then why not human souls on matter?[53]

Euler was a traditional metaphysician. He held humans knew the real essence of matter: extension, inertia, and impenetrability. Inertia was grounded in the principle of sufficient reason, while impenetrability was the ultimate essence of matter. Euler did not accept a principle of action-reaction per se and rejected Newtonian gravitational attraction as violating inertia. Eulerian impenetrability replaced Newton's action at a distance, thus setting mechanical over physical forces. A typical two-front battle emerged, as Euler opposed himself to Newtonians and to Leibnizean-Wolffians. He said the Wolffian view of matter—the continuous inclination of each body to change its state—violated inertia and made elements into souls. Matter could not have representations, according to Euler, who, more than Wolff, wanted to drive a wedge between pneumatology and cosmology. Engaging in caricature, Euler claimed the monadist view held that physical division ultimately came upon indivisible but extended elements. His position was that matter was divisible indefinitely, without necessarily arriving at extended simples. Euler's "point atoms" moved from the mathematical to the physical by acquiring inertia.[54]

Preference for point atoms over monads shows what Euler liked most in Maupertuis's principle of least action, and it was not final causes. The principle of least action reduced analyses of motion to equilibrium states (like

52. Wolff, *Werke*, ser. 1, 16:142–43.

53. On the above, see Euler, *Lettres*, vol. 1, letters 60, 80–87.

54. Ibid., letter 26; vol. 2 letters 80, 85, 122-27, 132; Euler, *Opera*, ser. 2, 1:49–51; 3:20–21, 46; 5:152–98, 250–56; ser. 3, 2:347–72. Cf. Kästner, *Anfangsgründe der höhern Mechanik*, foreword, 3, 291. See Körner, "Der Begriff," 34–47.

d'Alembert's principle) and made a theory of matter, in regard to elasticity, irrelevant. Euler's rational mechanics depended on knowing matter a priori, without experiments. (That is why Euler could not like constants of nature, since they could only be determined experimentally.) While Justi, following Anglo-Franco sentiments, seemed set on reducing metaphysics to mathematics and physics, if not to nonsense, Euler remained a good metaphysician of the Cartesian cast.

RATIONAL MECHANICS

Euler set inertia as the foundational principle of mechanics, and derived force from impenetrability. He reduced all forces to the material via impenetrability and the psychic via spirit. No Newtonian attractive force existed; and what are now termed Newton's laws were neither ascribed to Newton by Euler nor accepted by him. Mentioned by "O," Euler's papers of 1750–51, as well as others, envisaged reduction of mechanics to analysis of equilibrium states. Not adhering to a Newtonian foundation of rational mechanics, Euler's principle for statics and mechanics gave three equations: $2Mddx = \pm Pdt^2$, $2Mddy = \pm Qdt^2$, $2Mddz = \pm Rdt^2$. Reducing this, he saw the foundation of mechanics as $dv = \mu pdt/A$ (or $dds/dt^2 = \mu p/A$, where μ is used as a proportionality constant, A as mass, v as velocity, p as *vis*). Euler claimed this as a necessary truth, giving him a rational mechanics metaphysically founded, the basis of his rational system of the world.[55]

In 1756–58, the absent president of the academy, Maupertuis, cast doubts on the necessity of the laws of mechanics, thus on the possibility of rational mechanics as a metaphysical science. He took "necessity" as meaning essentially mathematical necessity and argued that, given the conditions of experience or experiment, laws or principles of nature could not be necessary. The only seeming nonmathematical, metaphysical sense he would give "necessity" involved final causes or teleology, thus ultimately theology.[56] As with Wolffian rational theology in Halle, teleology emerged then in Berlin as central to the articulation of rational mechanics, or cosmology, as a metaphysical science.

Following on Maupertuis's essays, the Berlin academy prize question for 1758, repeated in 1759–60 as no one won, was: "Are the laws of statics and mechanics necessary or contingent?" No one won, and only seven entries survive in the academy's archive. In their nice variation and anonymity, these essays give a good barometer of the sense of such a metaphysical question then, in regard to rational mechanics. I refer to the essays by the last few digits of

---

55. Euler, *Lettres*, vol. 1, letters 19–24, 71-73, 77-79; vol. 2, letter 121; *Opera*, ser. 2, 3:66–67, 81–82; 5:90, 111, 118, 152–93; ser. 3, 1:18–51. See Hankins, "Reception of Newton's Second Law."
56. Maupertuis, "Examen philosophique," esp. 422–24.

their identifying archival numbers, 529 to 535. Written in Latin are 531, 533, 534, 535; in French, 529, 532; in German, 530.[57] Of the three great Berlin prize questions concerning metaphysics around midcentury, that on monads was the first, and this on the possibility of rational mechanics was the second.

Essay 529 sees dynamics as giving contingent laws, while phoronomics has necessary ones, since it abstracts from physical force and uses only point atoms. Essay 532 sees the laws of statics and mechanics as necessary, but not in all possible worlds. Maupertuis, according to 532, extended mechanics to include teleology; but 532 does not think the academy expects this profundity. Mechanics, for 532 following d'Alembert, is founded on pressures, not forces; and now following Euler, 532 sets the central formula of statics-mechanics as $pdt = dc$. Essay 535 sets as principles: inertia, action-reaction, $2mv = Pt$, and $2mdv = Pdt$, with P as *vis*. Analyzing the latter, 535 follows Euler's claims on necessity. Essay 534 does not answer the question, but rather transposes it, moving from analysis of mathematical to mechanical to physical points. Mathematical points in motion produce phoronomics, which has necessary laws; mass makes these into mechanical points. Necessary principles are those, such as inertia, that can be derived from the nature of such points without experience. The crucial step comes in moving to physical, extended points that bear the ideal mechanical ones, and the question is: In what sense are the laws of the latter true of the former, the knowledge of which depends on sensation? Essay 531 begins with a definition of "necessity," tying it to essences, then reducing that of body to extension, *realitas*, and inertia, using the latter contra Wolff to preclude self-alteration of states. Then 531 moves to statics founded on equilibrium, and claims its principles are necessary. Leibnizean *vis viva* is valid, but with such principles as least action, we leave rational mechanics or metaphysics, since such principles are not absolutely but only hypothetically necessary, as experience is involved. Essay 533 is more philosophical, convinced that the laws of mechanics, interpreted as relating to physical body, are not necessary: mechanical qua physical principles can be neither reduced to nor deduced from mathematics. 533 is puzzled that this metaphysical question has been posed by the mathematics section of the academy (which seems to be a mistake). This essay gives a history of positions on the laws of motion, much of which reduces to a discussion of monads and the Leibniz-Clarke debate; it rejects the monadists' active powers in body and argues that, seen physically, mechanics depends on the structure of matter, which cannot be shown to be necessary.

57. Berlin Brandenburgische Akademie der Wissenschaften, I-M529 to I-M535. Pre-1762, there is no way to tell whether the current holdings are complete or not. Essay 529 has a "No. 3" in different ink, perhaps by the academy; 531 has a "No. 7," likewise perhaps by the academy; 535 has a "D," probably by the academy. A better historian would have figured out who wrote these. The famous "O" was of course d'Alembert.

So we see a range of views from entries 529 to 535 for the Berlin academy's prize question for 1758 and 1759. The laws of mechanics did not seem to set Berlin abuzz that way that monads had. And, while resolution had largely risen against monads in Prussia, at least in view of the number of essays against them, that no one won this prize might indicate irresolution had set in about what was meant by "necessary," "rational," "mechanics," and above all by "metaphysics." Most saw no problems with statics. Mechanics was the problem and had come to have two parts: phoronomics and dynamics. Phoronomics was easily mastered by analytic geometry and statics; but dynamics involved collisions and so seemed to entail a theory of matter, about whose status everyone was now uncertain. Could there be a metaphysics of matter?

Before the 1750s, wishing a rational mechanics preserving conservation of *vis viva* and continuity in collisions, Leibnizeans had posited matter as elastic. Wolff and Euler were united in adhering to the idea of continuity, which meant opposition to Newton. Newtonians were happy to do away with continuity, conservation, and *vis viva*. Euler's option offered point atoms and principles, above all least action, saving continuity and making the structure of matter irrelevant. By the 1760s, monads might no longer have been in fashion, but Euler's point atoms, good for phoronomics, did not seem enough for dynamics. Metaphysicians disputed with mathematicians, philosophers with physicists. The result would be a "phenomenological" and even positivistic solution by physicists.

In and after the 1760s, the two most widely used mathematics-mechanics textbooks in the Germanies were by Kästner, professor in Göttingen, and by Karsten, professor in Rostock, and later in Halle. Karsten said that although we have no experience of perfectly hard bodies, he would not rule them out a priori via appeal to continuity as a metaphysical principle. Rules of collision must consider inelastic and elastic bodies. Kästner in fact posited inelastic bodies, rejected a principle of continuity for collisions, and saw the point of the principle of least action to lie in its application to elastic and inelastic bodies.[58] Kästner and Karsten, prey to a now fashionable Newtonianism, were closer to Maupertuis's view than to Euler's: What we know of matter is grounded empirically in phenomena. Principles of mechanics are judged positivistically by the results obtained, not by intrinsic warrant. This view made rational mechanics, as a metaphysical science, essentially impossible.

Let us return to the erstwhile president of the academy whose essays in the 1750s had provoked the second great Berlin prize question on metaphysics and

---

58. Karsten, *Lehrbegriff*, vol. 4, §§ 228–46, 266–73; Kästner, *Anfangsgründe der höhern Mechanik*, 293–94, 351–85.

helped pave the way for the acceptance of Newtonianism. As we have seen, Maupertuis only found necessity in physics insofar as teleology and final causes were at work. What remained of metaphysical science was but cosmo-theology: Maupertius's principle of least action was a proof of God's existence, though he sometimes also argued the opposite, deriving cosmological principles from theology.[59] In any case, Maupertuis's least action promised to make teleology more than a handmaiden of mechanics. As Berlin was a purgatory of controversies in the mid–eighteenth century, a dispute about least action emerged. In the controversy, debate revolved around not only who—Wolff or Maupertuis—had coined the relevant sense of "action," but also around how to measure "action." However one measured it, Maupertuis's principle held that, in transmission of light and in collision of bodies, nature always worked to minimize action.

Euler's papers of 1750–51 gave the controversy its decisive turn: the principle of least action became one of maxima and minima for a path or a system of bodies. Euler approximated paths or states of a system by taking initial and final positions.[60] The teleological moment switched from economics (minima) to systematics (paths and states), returning to the Leibnizean view that mechanics, as a study of efficient causes, could not explain the possibility of the cosmos as a structure or system, seen by Leibnizeans as involving final causes. The principle of least action, as a principle of system, showed how to see the mind in the machine without the miraculous: teleology.

That makes Maupertuis's view of physico-theology interesting: He made fun of it. He thought it degrading to deal in a theology of details. Instead, he would have one pursue cosmo-theology, the only part of metaphysics scientifically feasible. Reimarus produced two popular works in physico-theology and asked, What are details? He said that Maupertuis misunderstood the point.[61] Physico-theology meant that one saw the garden as a metaphysical matter where no detail was too small to reflect God. But at midcentury in Berlin, while the central status of teleology was maintained, metaphysics seemed to be collapsing at the president's hands, despite Euler, to its great center in cosmo-theology.

To illuminate the tie of teleology and cosmology, let us look at someone who longed to be in the purgatory of controversies that most sought to leave. Johann Lambert arrived in Berlin to stay in 1764, becoming a member of the

59. Maupertuis, *Essai,* 9, 18, 39, 75; *Oeuvres,* 4:21.

60. See Calinger, "Newtonian-Wolffian Confrontation"; Polonoff, *Force;* Pulte, *Das Prinzip;* Terrall, "Culture of Science." The papers are in Euler, *Opera,* ser. 2, vol. 5.

61. Reimarus, *Abhandlung,* 208–43, esp. 222; Maupertuis, *Essai,* 9, 18, 39, 75; *Oeuvres,* 4:21; "Examen philosophique."

academy in 1765, a year before Euler left. Lambert saw a problem with all constructions of matter, since science rested on sensation and it was possible there were matters we could not sense. Dynamics was contingent on instrumental verification, thus it was "phenomenological," a word he coined. Mechanics concerned only the world as phenomenon, making experiment and measurement essential to knowledge of nature.[62] Lambert's positivism on rational mechanics seems to have gone beyond that of even Kästner and Karsten in the 1760s.

But, being in Berlin, Lambert argued for what was implicit in Euler's construal of Maupertuis's cosmo-theology: as opposed to mechanics founded only on efficient causes, cosmology emerged by the introduction of teleology, or final causes, grounding the possibility of a system.[63] Lambert took the Newtonian as opposed to Leibnizean solution to the system: gravity, rather than monads, was the systematic moment in the cosmos. Nonetheless, a would-be servant to the Prussian crown, Lambert held that teleology helped establish the structural aspect of natural laws and aided in their discovery. Like mechanics, teleology was a constructive part of natural philosophy and underlay the possibility of a system of the world. The "nature" of Prussia remained teleo-mechanical.

Stars, Aether, Aliens

We have first considered atoms and monads as the ultimately small: the microcosm. Then we have considered whether laws or principles, as rational mechanics, founded a metaphysical science of nature: the system. We now complete the metaphysical dimensions of the cosmos by considering the ultimately large or distant: the macrocosm. As dogmatic metaphysicians, Leibniz and Wolff held no other worlds existed, since ours was the best possible.[64] The notion of possible worlds, in a sense other than spatially distant, remained too metaphysical for most. More typical were considerations about the extent of the cosmos, what filled it, the possible number of stars, and whether they or their planets were inhabited.

Director of the Berlin observatory and member of the academy, Johann Bode held space to be probably infinite, with the cosmos as finite and God beyond it. Whether or not such assertions in modern cosmology are grounded in anything other than metaphysics, they were not in the eighteenth century.[65]

62. Lambert, *Schriften*, 1:462–63, 48–92; cf. Kästner, *Anfangsgründe der höhern Mechanik*, foreword; Karsten, *Lehrbegriff der gesammten Mathematik*, vol. 4, foreword.

63. Lambert, *Briefe*, viii–ix, xxi, 61, 80. See Biermann, "Lambert und die berliner Akademie."

64. Leibniz, *Schriften*, 6:107, 210; Wolff, *Werke*, ser. 1, 6: § 80; 8: § 37; ser. 2, 2:319–26, 332; 7: 133–34, 328, 358, 401–2, 780–82.

65. Bode, *Betrachtungen*, 185–87; cf. *Anleitung*, 669. See Alfvém, "Cosmology."

The post-Cartesian cosmos had become, at the least, indefinite. For many in the enlightened Germanies, the indefinite cosmic space was filled by aether, the cosmic mediator between advocates of gravity and fans of monads. Aether theory, like physico-theology, served as a linkage point between theology, pneumatology, and cosmology. Aether instantiated the system of the world.

Euler was the great aether theorist in the Germanies. Posit of an enlightened aether there typically followed in his footsteps, even if not taking his aim of doing away with Newtonian action at a distance and Leibnizean monads. Whether following Euler or not, German aether theorists usually used aether as a space-filling entity, often the bearer of light and sometimes of gravity. Aether worked by mechanical principles, so was of no relevance to a new sort of doctrine of astral influx. By midcentury, monads, as the principle of system, found little favor, as pneumatology and cosmology had been cloven by aether.[66]

The Enlightenment mechanized and disenchanted the cosmos. In the post-astrological, clockwork cosmos, there could be no necessity in a particular number of planets, and in any case, the magic number of seven had not survived the seventeenth century. Planets beyond Saturn in our solar system had been mooted by the 1750s, so construction of "Uranus" in the early 1780s proceeded with ease. Debate quickly revolved around the cometary versus planetary status of the newly observed body. After Herschel's first glimpse in March 1781, a number of astronomers, including Bode in Berlin, had advanced a planetary view by September, though by early 1782 it had not been generally endorsed. Kästner in Göttingen, for example, initially proceeded cautiously in discussing the matter.[67]

Interesting for enlightened cosmology was the naming of the new planet. Though some wanted to give it such silly names as George (after the king of England), Bode claimed to be the one to have suggested the name "Uranus," the mythical father of Saturn.[68] Bode also suggested a symbol for the new planet developed along the lines of traditional astrological symbols (see fig. 13.2). Mindful of the concurrent changes in chemistry in regard to the tradition of naming, cosmology seems to have become at ease with its mythical roots, insofar as aware of them. That the first new planet in millennia could be so quickly recognized, then given the name "Uranus," with a symbol from the astrological tradition, shows how dead astrology really was. Cosmology had achieved control of its mystical, metaphysical past.

Prussian *Aufklärung* made aether and stars less spooky, naturalizing,

66. See Euler, *Opera*, ser. 3, 1:3–15, 112ff., 134ff., 149ff.; *Lettres*, vol. 1, letters 17 and following, letters 54 and following. On German aethers, cf. Clark, "German Physics Textbooks"; Wise, "German Concepts of Force"; on astrology, see Clark, "Der Untergang der Astrologie."
67. Kästner, *Anfangsgründe der Astronomie*, 168–72; also Wurm, *Geschichte*, 9–17.
68. Bode, *Von den neu entdekten Planeten*, 87–95.

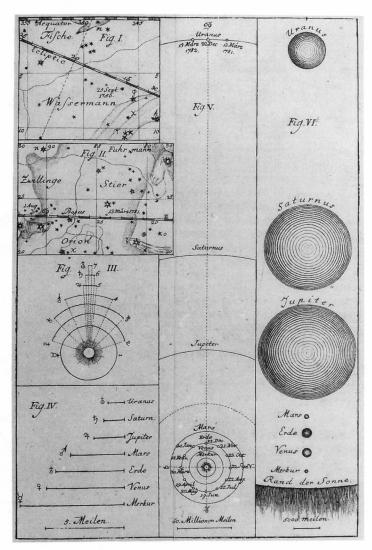

Fig. 13.2.   Illustration of the integration of Uranus into the known solar system from Johann Elert von Bode's *Von den neu endeckten Planten* (1784). By permission of the Universitätsbibliothek Göttingen.

mechanizing, dispiriting, disenchanting them. How oddly other, then, the views about aliens. While belief in extraterrestrial life still seems exceptional in the Baroque, it becomes the rule among intellectuals in the *Aufklärung*.[69] It is almost as if the mental energy once devoted to the pneumato-cosmology of judicial astrology moves to a "cosmo-anthropology." And like its cousin, physico-theology, cosmo-anthropology becomes part of popular science.

Near midcentury, one might write poems about aliens or speculate of them in works on comets.[70] In a popular work, later made more academic, Kindermann formulates a cosmic anthropology. His Venutians are not lazier than "Indians," as he relates, but owing to the heat, rather more spiritual than corporeal. They know no Fall, as Earthlings had; they speak Hebrew and have no sects. Mercurians have even better minds and senses than Venutians, and their bodies are nearly wholly fluid, though because of their very short diurnal cycle they probably do not study astronomy. Mercurians live in forests with huge trees for shade; indeed, they live in trees. If you object that only savages live in forests, and never civilized peoples, Kindermann refers you to Adam and his kin: Where did they live? On Mars things are bad: They don't study astronomy. They swear and curse and so probably have a lot of lawyers. Martians fight too much. Think of those who live in cold climes, such as Russians and Swedes, as opposed to Indians. The former fight a lot, so it makes sense that Martians do too. The other planets get peoples, but comets do not, since they are erstwhile planets that have experienced a Last Judgment and lost their souls.[71]

An equally famous cosmo-anthropology comes from Krüger. His Mercurians are black and think differently than "we" do. They are superstitious and foolish. It's a madhouse. Venutians are more educated. Since the heat is less, they are more human in form and less mad than Mercurians. There is no grass on Venus, but wine-producing plants exist and no one needs to work. But Venus has a dark side. Lunarians are bad philosophers. They think teleologically. Witches and magicians abound. Martians inhabit cities on big cliffs. Everyone is mean. Men brandish swords and drink blood. Kids are naked and villages are freqently burned. But Jovians are cool. They live on air and water, and no one ever gets sick. Some live to be a thousand years old. And Jovians

69. Dick, Plurality of Worlds; Guthke, *Der Mythos der Neuzeit*. See Wolff, *Werke*, ser. 1, 7: § 30; Leibniz, *Schriften*, 6:114; also Fontenelle, *Dialogen über die Mehrheit der Welten*, with commentary by Bode.

70. As in Mylius, "Lehrgedicht von den . . . Cometen"; J.M.R., *Gedancken*, 20; the anonymous *Sendschreiben an einen Freund*, 11, and *Curiöses Gespräch*, 21–22; also Heyn, *Versuch einer Betrachtung über die Cometen*, 19–25.

71. Kindermann, *Reise in Gedancken*, 50–55, 61–65, 70–74, 82, 84, 206.

like science. Saturnians live in caves since it's so cold. Stars have inhabited planets, about which we know little. . . .[72]

Such views were not restricted to popularizers. In an otherwise technical work, the later professor and member of the Berlin Academy of Sciences, Ernst Fischer, explained that the inhabitants of Uranus and Saturn get little light, so their senses must be better than ours. Saturnians have more time to penetrate the secrets of nature and may therefore have reached a higher state than we; but Mercurians must be several steps lower than we, and have a shorter life, with little time to study nature, and coarser senses needing much stimulation. And cometary inhabitants must be more noble than planetary dwellers. The most important Prussian astronomer, Bode, thought planets were inhabited and probably comets as well. Even the sun might be populated, and other solar systems too. Bode held thinking beings would be better the further from the center of the galaxy, since the rotation was less.[73]

The midcentury *Aufklärung* coveted cosmic disenchantment, contemplating the death of metaphysics or collapsing it to cosmo-theology, if not to cosmo-anthropology. Cosmology was freed of astrology and pneumatology: Beyond the aspect of system as the mind in the machine, the cosmos was rid of the deus ex machina and atoms rid of souls. But in banning monads, miracles, and fate, the Prussian *Aufklärung* held firm to teleology. The enlightened Lambert spoke well when he said all parts of the cosmos must have life by a teleological principle of plenitude. "I agree with you entirely that those who doubt or even deny the existence of inhabitants of other planets have been left far behind and are so very restricted in their understanding. . . . Most know only a few villages, their birthplace."[74] Enlightened Prussians were cosmopolitans. Cosmo-anthropology and physico-theology were the twin progeny of popular metaphysics as teleology. Aliens ruled. "Philosophical anthropology has become cosmology and cosmology philosophical anthropology."[75]

Kant and the Heavenly City

"Kant was not destined for easy victories and perhaps it was just as well that his academic aspirations were not fulfilled at the time [the 1750s and 1760s], and that he remained the stout-hearted provincial in Königsberg rather than a fallen Icarus in Berlin."[76] A year before Wolff died in 1754, Maupertuis

72. Krüger, *Träume*, 97–120.
73. Fischer, *Betrachtung über de Kometen*, 4–6, 15–16; Bode, *Anleitung*, 642–47, 662–66; and *Betrachtungen*, 105–18, 166, 174–75.
74. Lambert, *Briefe*, 106, also x, 62–68, 80.
75. Guthke, *Der Mythos*, 236.
76. Polonoff, *Force*, 123.

absented himself from Berlin and died himself in 1759. Euler left Berlin in 1766 and returned to the Academy of Sciences in St. Petersburg, another brainchild of Leibniz. In 1764, the year Lambert arrived in Berlin, the University of Königsberg and the ministry in Berlin, embarrassed that the "world famous" Immanuel Kant was still but a lecturer, offered him the vacant professorship in eloquence and poetry. Though such an offer might seem strange to us, it was a typical early modern practice in the Germanies, where professorships were still treated essentially as sinecures, and where subjects in the faculty of arts and sciences were still effectively treated as a unified corpus.

Lecturer Kant, however, showing modern sentiments, turned the offer down. He wanted the professorship of logic and metaphysics, for the latter was his great love in life, to which he intended to be true. Kant likewise remained true to Königsberg. When Wolff was driven from Halle in 1723, J. J. Lange replaced him as professor of mathematics and philosophy. Between 1763 and 1765, given the junior Lange's advanced age, then death, the ministry in Berlin went so far as to appoint Kant to Halle in 1763–65, in effect, giving him Wolff's first chair. But Lecturer Kant did not go to Halle, even though, like Semler, he had grown up in the traditions of Wolffianism and Pietism. Indeed, for most of his career, Kant used the textbook of a Wolffian from Halle, Baumgarten's *Metaphysica,* which begins: "§ 1. Metaphysics is the science of the first principles of human cognition. § 2. . . . To metaphysics are referred ontology, cosmology, psychology and natural theology." Kant stayed in peripheral Königsberg, became professor of logic and metaphysics in 1770, and tried to rescue but ended up terrorizing his great love in life.

In the provinces, Kant mulled over problems, such as monads and fate, left as passé by the academicians in the capitals. After his "critical" and "transcendental" turn of 1781—these being Kant's terms of art to distinguish his metaphysics from the "dogmatic" and "transcendent" of Leibniz, Wolff, and Euler—Kant's path put him at odds with the phenomenological and positivistic turn emerging among the new physicists: Kästner, Karsten, and Lambert. This final section, on the Kantian attempt to rescue and reform metaphysics, has three subsections: "Cosmology," "Theology," and "Anthropology."

Cosmology

In his first work (1747), Kant declared himself a Leibnizean by adhering to a principle of continuity to resolve the *vis viva* controversy, then passé.[77] In *Monadologia physica* of 1756, on an issue also passé, he adopted Eulerian point atoms, with some big wrinkles: Kant transmuted monads into a "dialectic of

---

77. Kant, *Schriften,* 1:21–28, 37, 181.

powers." His physical monads filled space by a sphere of repulsive power, whence Eulerian impenetrability was derived. But if bodies were to have a finite volume, instead of self-repelling themselves to infinity, attractive power must complement repulsive. Kant posited Newtonian attraction for that necessary opposition. This theory of conflicting powers preserved the Leibnizean emphasis on monads as active principles, and accommodated the Wolffian notion that monads had a motive power tending to move them in all directions at once, appearing now as a tendency to fill space by self-repulsion, which, thanks to attraction, gave bodies a finite volume. From the dialectic of powers, perfect elasticity followed, since in collisions the spheres of equilibrated powers would be compressed until the repulsive forces of the monads overcame the compressing force. Moreover, one might hold that Eulerian point atoms never touched in collisions, since the repulsive sphere prohibited that by rising asymptotically as distance went to zero. The theory thus preserved continuity and allowed conservation of *vis viva* (mv^2).[78]

After the first critique, *Kritik der reinen Vernunft* (1781), Kant took the question of the divisibility to and composition from atoms or monads to be the "second antinomy" of dogmatic metaphysics: a dilemma that neither pro nor contra could resolve. His critical metaphysics, in place of the dogmatic, held one could not decide the question of whether atoms or monads in the strict sense existed. In *Metaphysische Anfangsgründe* of 1786, however, Kant attempted a critical metaphysics of matter. Here he did not advance much in dynamics over the essay of 1756, as a letter of 1792 shows he knew. The difference in 1786 was the transcendental turn grounding the construction of matter on his critically metaphysical or transcendental principle of "the possibility of experience." Abandoning the approach of the 1756 work as dogmatically metaphysical, Kant in effect took the view of the new physicists—Kästner, Karsten, Lambert—and asked: What are the necessary conditions underlying the possibility of sensing matter? Kant's physical monadology of 1756, insofar as it survived in the work of 1786, traded Leibnizean monadology for the dialectic of powers, which was now taken as the answer to Kant's new transcendental question.

"Give me matter alone and I shall build you a world," said Kant, moving from the microcosm in his first work (1747) to the macrocosm in his second work (1755), where he set a mechanistic cosmogony in place of Moses' creation story. Did this remove the basis of physico-theology? The philosopher of Königsberg was obscure here. Kant did people the planets in his 1755 work. As opposed to Kindermann's, Kant's aliens, like Fischer's, improved the further

78. Ibid., 473–87; cf. Boscovich, *Philosophia naturalis*.

they got from the center of the solar system. But he also stressed the need to keep theology out of natural science, so the cosmos was allowed to work mechanically, at least macroscopically. In 1755, Kant exposited not only a solar system mechanically born and working, but also a system of solar systems wrapped up in systems, perhaps in systems of systems. . . . It seems that a misunderstanding of Thomas Wright's work on the Milky Way gave him the impetus for his theory. By the principle of continuity, Kant argued not only for higher-order systems, or galaxies, above solar systems, but also for the less dramatic but important point that other planets probably existed beyond Saturn.[79]

Kant extended the systematic unity of the cosmos not only in space but also in time. "Creation is never at end. It had once a beginning, but will never end." Centers expanded creation outward, while themselves later ultimately collapsing. "The developed world finds itself limited between the ruins of the destroyed and the chaos of undeveloped Nature." The theory of 1755 spoke only of a cosmos without bounds, while that of 1747 had seen multiple worlds in other dimensions as at least metaphysically possible, though not likely.[80] The critical Kant, however, would come to see the question of the extent of the cosmos in space-time to be the "first antinomy" of dogmatic metaphysics: another dilemma that neither pro nor con could resolve. Critical metaphysics held that "of the cosmos as a whole we can say nothing." The vogue of cosmoanthropology would also become suspect. The tendency of Kant's thinking was to exorcize spirits from matter, to separate pneumatology and cosmology. He followed Euler in seeing matter as necessarily dispirited: "All natural science rests on the proposition: matter has no representations." As it seems to have been Wolff's fate to be defamed, Kant claimed that it was Wolff who reduced power to the power of perception or representation, leading to Spinozism![81]

About the aether, Kant might have had other tendencies. In *Kritik der reinen Vernunft,* he argued that the "principle of the possibility of experience"—his critical, transcendental substitute for Leibniz's dogmatic and transcendent principle of sufficient reason—entailed the necessity of a causal order in nature, amounting to a more or less mechanical one. Kant's *Metaphysische Anfangsgründe der Naturwissenschaften* of 1786 fleshed out the metaphysical foundations of this mechanical order, with Leibnizean organics replaced by a dynamics or dialectic of powers, discussed above, which explicated the

79. Kant, *Schriften,* 1:222–23, 229 (quotation), 230, 247–58, 333–34, 351–66; see Schaffer, "Phoenix of Nature."
80. Kant, *Schriften,* 1:22–25, 256–57, 308–22; quotations at 314, 319.
81. Ibid., 3:294–303, 340, 354–62, 355 (first quotation); 4:476–78, 499, 508, 523–25, 554–55; 11:376–77, 395–96; 28(1):441, 449 (second), 511–12.

possibility of sensing matter. In his unpublished work *(Opus postumum)*, he envisaged the final step as a metaphysics of experimental physics. In this work, in which Kästner and Karsten were criticized, Kant wanted to supply the missing transcendental basis to the positivistic experimental physics. The *Opus postumum* used the aether as a psychophysical posit, and in essence articulated the dialectic of subject-object in experimentation.

This was to be part of the new metaphysics he envisaged. We have seen that, for the critical Kant of the transcendental turn, attempts to decide both the extent of the cosmos in space-time and the divisibility to and composition from atoms or monads constituted the first and second antinomies of dogmatic metaphysics. On these two great cosmological issues—the microcosm and macrocosm—he now denied metaphysical science knowledge. The mechanical constitution of the cosmos—the system of the world—and its ramifications for the question of fate versus freedom constituted the "third antinomy" of dogmatic metaphysics for him: a third dilemma that neither pro nor contra could resolve.[82] Around this antinomy, the first Leibnizean labyrinth of reason, revolved theology, cosmology, and pneumatology. In the case of this third dilemma of metaphysics, however, he argued that both pro and con were correct, but of different spheres or worlds. Mechanics treated of the sensible, phenomenal world, and morals of the intelligible, noumenal world. Having served as the theme of his inaugural dissertation of 1770, these two kingdoms or worlds, which Leibniz had sought to unite in his teleo-mechanics, were now cast asunder. And instead of being the handmaiden of theology in answering the great cosmic questions of life, metaphysics seemed now rather more stuck in the phenomenal world, as a servant of positive science, establishing the possibility of rational knowledge through experimentation. But Kant made a most interesting remark in the late work of 1786: "An absolute vacuum and absolute hardness are in natural science what blind chance and blind fate are in metaphysical science."[83] Even critical metaphysics could rest assured pietistically that neither fate nor fortune existed when blind.

THEOLOGY

There had been other critiques of Wolff's rational theology and the tradition it inspired. But Kant's critique spelled its death knell philosophically and historically. His critique of rational theology appeared in a work of 1763 and then in the *Kritik der reinen Vernunft* (2d ed., 1787).[84] In these works, he inaugurated a modern style of metaproof: to prove that the existence of a thing, in this case

82. Ibid., 3:308–13, 362–77.
83. Ibid., 4:532.
84. Ibid., 2:63–164; 3:314–21, 378–81, 392–426.

God, could not be proven. The style of reasoning was different from skepticism as well as from arguments by reductio ad absurdam.

Kant separated critical from dogmatic metaphysics, and both from mathematics. Critical metaphysics set limits to reason, and its positive results came only by the principle of the possibility of experience, as the basis of a new transcendental logic. Unlike formal and transcendental logic, mathematics engaged in proofs of existence, contained "synthetic a priori" truths. Mathematics, Kant held, contained such truths since it had access to a priori "intuitions" or apprehensions of space and time. Unlike mathematical truths, a "direct synthetic [existential] proposition from concepts [alone] is a *Dogma.*" A metaphysical dogmatics of reason was to be replaced with a critique of it. As we have seen, Kant argued that rational theology reduced to onto-theology, cosmo-theology, and physico-theology. Following Leibniz and Baumgarten, he argued that cosmo-theology and onto-theology were interrelated and his strategy was to expose a vicious circle.[85]

The ontological proof claimed that the existence of an absolutely supreme being followed from articulated conception alone. Kant cut the Gordian knot by arguing that "existence" was not a predicate, so that existential or "synthetic" propositions depended on data—pure intuitions or sensible experience—beyond the analysis of concepts. Insofar as mathematics could not offer a base, onto-theology must depend on cosmo-theology for its a priori data. So Kant turned to cosmo-theology, which argued from the contingency of intramundane beings and relations, as its sole datum, to the absolute necessity of an extramundane deity. Kant argued that there was no way to move from the cosmos, as a series of contingent beings and relations to anything outside that series. He called this the "fourth antinomy" of dogmatic metaphysics, which completed the list and formed a bridge to cosmology per se.[86] And should one concede the existence of an extramundane being, there was no way to establish whether this was the deity of revelation, as opposed to a world soul or one deity among many, unless cosmo-theology appealed circularly and viciously to the absolutely supreme being of onto-theology.

Kant thought he had established a vicious circle between onto- and cosmo-theology, as well as unveiling the dogmatic presuppositions of each. So he saw no consequence in Maupertuis's theological application of the principle of least action, which Kant saw as an overly abstract physico-theology or an overly detailed cosmo-theology.[87] If taken as the latter, then it fell under the fourth antinomy of dogmatic metaphysics; if taken as the former, it would

85. Ibid., 3:482; see Henrich, *Der ontologische Gottesbeweis,* 45, 65–66, 137.
86. Kant, *Schriften,* 3:314–21, 378–81.
87. Ibid., 2:116–37; 3:413–20; 5:436–42.

stand or fall with physico-theology. And physico-theology, as we have seen, remained central to enlightened Prussian metaphysics and would remain so for Kant, though no longer as a constructive part of science. While Wolff saw physico-theology, qua teleology, as an experimental theology or metaphysics, for Kant it would become part of a critical metaphysics of experiment and observation. As part of dogmatic metaphysics, however, physico-theology could never prove anything about any deity, Kant argued, since its evidential force fell with cosmo-theology. Moreover, he took issue with the basic presumption underlying physico- and cosmo-theology: That order might not arise from mechanical causes. His *Allgemeine Naturgeschichte* of 1755 set out to argue that the cosmic order could have arisen from mechanical causes alone. While culminating the Leibnizean-Wolffian project to remove the miraculous divine hand even from creation, the young Kant expressed Pietistic admiration nonetheless for the spirit of physico-theology.

Perhaps having lost his Pietistic faith, in his later years Kant advanced against revealed theology. We have discussed above the appearance of Neologism, fusing Pietism with Wolffian rationalism, and trying to undercut Deism by naturalizing particular miracles. Kant tried for a time to make way for Lange along with Wolff. In arguing in the first edition of *Kritik der reinen Vernunft* that reason could not prove the existence of a deity, he was in effect arguing the Pietists' case contra Wolff; moreover, as Kant noted in the preface to the second edition, he had shown reason its limits in order better to secure the rights of faith. But by the 1790s, reason got in the way of faith, and Kant seemed to be heading to a radically Neological position, denying the possibility of the miracle of revelation. He claimed that governments and churches liked ancient miracles, but not modern ones. Churches were sectarian, since they depended on revelation, always particular and miraculous. "One can say with reason that 'the Kingdom of God has arrived to us' only when the principle of a gradual transition from belief in Churches to a general religion of Reason, and thus to a (divine) ethical state on earth, obtains general and also *public* roots somewhere." The rationality of claims to revelation could never be admitted. "For if God would really speak to a human, the latter could never know that it was God who spoke."[88]

Belief that the deity spoke at a certain time to a certain people was opposed by Kantian *Aufklärung,* bent to remove the miraculous, as dogmatically metaphysical, from history. If not even a revealed theology remained possible, did any? The Cartesian instauration of modern metaphysics had depended on the possibility of rational theology, especially onto-theology. The German tradition

88. Ibid., 6:83–89, 102–115, 122 (first quotation); 7:63 (second).

of Leibniz and Wolff set cosmo-theology alongside onto-theology as jewels in the crown of metaphysics, queen of the sciences and handmaiden of revealed theology. Wolff and the *Aufklärung* added physico-theology, as the final jewel in the trinity of rational theology, the loftiest human science. In the course of the eighteenth century, a cosmo-anthropology had emerged too, which with physico-theology formed a sort of popular metaphysics. If, save Euler, the climate of opinion in enlightened Berlin by midcentury was casting dark clouds over the field of academic metaphysics, at least cosmo-theology seemed safe, given shelter by Maupertuis. Allowing physico-theology and cosmo-anthropology as enlightened and pious but essentially nonscientific pastimes, Kant's attack on cosmo-theology seems then the crux of his attack on transcendent metaphysics—and one might wonder whether that was essentially his attack on the principle of sufficient reason as dogmatic metaphysics.[89] With the transcendent labeled "illusion," was the science of metaphysics now but the charwoman of physical science?

ANTHROPOLOGY

> The field of philosophy . . . may be reduced to four questions: (1) What can I know? (2) What should I do? (3) What dare I hope? (4) What is human *(Was ist der Mensch)*? Metaphysics answers the first question, morals the second, religion the third, and anthropology the fourth. But one could really take all of these as anthropology, since the first three questions relate to the fourth.[90]

A Prussian view of the Enlightenment: Kant saved metaphysics by making it anthropology?

Nature and Freedom

Kant's first critique, *Kritik der reinen Vernunft,* turned general metaphysics as ontology into a "transcendental logic" and, specifically, into a "transcendental analytics," followed by a "transcendental dialectics" on special metaphysics. The first part of the critique concerned the question, What is an object? Kant transformed that question into a new one: What is the objective structure of the world? He argued from his new transcendental, critical principle of the possibility of experience: if (intersubjective) experience of a world is to be at

---

89. Leibniz, for example, begins with the cosmo-theological proof in *Monadologie* and moves toward the onto-theological; his argument depends on the principle of sufficient reason, and he claims that without this principle "we would never be able to prove God's existence" (*Théodicée,* § 44); see Leibniz, *Schriften,* 6:127, 612–14. From the Middle Ages up to the Enlightenment, cosmo-theology seems the clear favorite for academics.

90. Kant, *Schriften,* 9:25.

all possible, what are the conditions on the structure of the world? The medieval concern with possible worlds, revived by Leibniz, became with Kant a concern with the conditions underlying the possibility of the experience of a world, taken as its objective structure, that is, nature. Probably formulated in view of of Pietist fears of fatalism, as noted, Wolff's world-as-perceived had already moved toward phenomenalism. An Anglo-American tradition tends to see this as a turn from ontology to espistemology, while a Continental tradition speaks rather of a turn to philosophical anthropology: the metaphysics of nature rests on a metaphysics of human (inter)subjectivity.[91]

The second critique, *Kritik der praktischen Vernunft,* answers the question, What is a person? The condition of the possibility for recognizing a person is Kant's "categorical imperative": A person is an entity that is an end-in-itself, never a means. The second critique aims also to establish the actuality of freedom, given the necessary nexus of the cosmos as a seemingly fatalistic mechanical order. As always in enlightened Prussia, it involves the reconciliation of Lange and Wolff. The third antinomy in the first critique set the stage for this reconciliation. Indeed, the Leibnizean kingdoms of nature and grace, the worlds of the sensible and the intelligible, are what Kant reformulates in his first and second critiques. The second critique defines the "person" as that which inhabits an intelligible world beyond the sensible and so partakes of the kingdom of grace, called a sphere of freedom as duty by Kant, a Prussian professorial civil servant. In this critique and in subsequent works on morals, he articulates a moral philosophy, perhaps the first, that applies to all rational beings. All such beings must be seen as persons. The moral philosophy or theory of the person following from this and its implications also includes angels and aliens, automata and apes, insofar as the latter could become rational. If there is a heavenly city in the Enlightenment, it is Kant's universal kingdom of rational (free) beings.

Kant envisaged two metaphysical sciences: a science of nature and a science of freedom, that is, of objects and persons—but not of humans. The first critique prepared a metaphysical science of nature, set out in *Metaphysische Anfangsgründe der Naturwissenschaften* and *Opus postumum.* The second critique prepared a metaphysical science of persons, set out in *Grundlegung zur*

---

91. As cited above, see Wolff, *Werke,* ser. 2, 4:45, 173, 224–25. The argument about Kant's transformation of metaphysics into anthropology has been well known since Heidegger and more recently through Foucault's work: see Heidegger, *Kant,* esp. §§ 36–38 (*Holzwege,* 73–110, esp. 89–91, 96–98, makes this turn Cartesian); Foucault, *Les mots et les choses,* 351–54. Foucault's view of Kant was long essentially Heidegger's, and it is no accident that Foucault's "complementary thesis" at the Sorbonne was a translation and commentary on Kant's *Anthropologie in pragmatischer Hinsicht.* Cf. Macey, *Lives of Michel Foucault,* 88–90, 111; Eribon, *Michel Foucault,* 110–15. In the late 1970s, a critical turn emerged, as Foucault developed interest in Kant's essay "Was ist *Aufklärung?*" Cf. Kelly, *Critique and Power.* On empirical anthropology, see Linden, *Untersuchungen;* Pickerodt, *Georg Forster;* Riedel, "Historizismus und Kritizismus."

Metaphysik der Sitten and *Die Metaphysik der Sitten*. But the third critique, *Kritik der Urteilskraft*, laid no such metaphysical foundation for a science of humans. For the third critique really answered the question, What is a human?

Aesthetics and Teleology

Kritik der Urteilskraft has as its three chief parts analyses of the beautiful, the sublime, and the purposive; or, to collapse these, of aesthetics and teleology. At the close of the introduction to the third critique, Kant tells us that *Kunst* mediates between nature and freedom, that is, between objects and persons.[92] By *Kunst*, he means art and techniques, or once more, aesthetics and teleology. The artistic-technical as the aesthetic-teleological mediates between nature and freedom. Art and aesthetics provide the link to morals, while technics and teleology provide the link to nature. Between the realms of objects and persons exists the realm of humans, fusing persons with bodies, and so possessed of and by aesthetics and teleology. Properly understood, the third critique is philosophical anthropology.

The science of aesthetics grew in the Wolffian school. From Wolff's world-as-perceived arose the notion that aesthetics, a theory of human sensibility or subjectivity, lay at the base of our knowledge. Besides writing an influential Wolffian textbook on metaphysics, Alexander Baumgarten had also coined the term "aesthetics" in its modern sense, a word seen as typical of German metaphysics by those outside and long resisted abroad.[93] Baumgarten's *Aesthetica* begins: "§ 1. Aesthetics . . . is the science of sensitive cognition." Aesthetics used "taste" as a central concept. Taste had been the most subjective of the senses and, from antiquity through the Renaissance, had had nothing to do with discourses on art and the beautiful, which had been set rather around oral and visual, ultimately mathematical, harmonies. Up to the Baroque, the sphere of art and the beautiful had remained the opposite of the aesthetic: the noetic. This sphere was akin to those of the true and the good. "The emergence of taste [as a theoretical term] constituted the aesthetic subject" in the post-Cartesian era. Unlike angels and automata, humans were now constituted as aesthetic subjects. Wolff's world-as-experienced became, after Baumgarten and Kant, an aesthetic object. The emergence of the notion "culture" was part of the same historical process: the human world became an aesthetic object, one whose essence lay in experience and expression.[94]

92. Kant, *Schriften*, 5:198.

93. Ibid., 50 n. B35: "The Germans are the only ones who use the word 'aesthetics' to designate what others call the critique of taste." The article on "aesthetics" in the *Oxford English Dictionary* is instructive on English resistance to it.

94. Bäumler, *Das Irrationalitätsproblem*, 2. In the eighteenth century, emphasis on taste, or *Geschmack*, in education shifted to *Kultur* and *Bildung*. Note too, Kant, *Schriften*, 15(2):569: "The sphere of women's sciences is characterized only by taste. Its teaching [text breaks off]."

Kant pursued aesthetics in the narrow sense in the third critique. But he had also opened the first critique with a "transcendental aesthetics." Here he claimed space-time to be the aesthetic base of human cognition, though by implication not necessarily that of other rational beings. In this sense, the first critique opened in an anthropological key; indeed, as a critique, instead of a doctrine or dogma, it was per se a philosophical anthropology. "Critique" as a concept and method came to Kant, not so much from the philological tradition, but rather from discourses on art and taste—in other words, from what becomes aesthetics. Critique is necessary as a method in disciplines for which, in scholastic terms, "dogmas" are not possible.[95] Kant displaced the metaphysical foundations of knowledge into a transcendental aesthetics, that is, a critique of human subjectivity.

The anthropological motif reappeared as well at the close of the first critique, as Kant recast science into the teleological realm definitive of the (non-angelic) human condition. Elsewhere he noted that Leibniz's *Monadologie* was "in itself a correct Platonic conception of the world, insofar as it is not an object of the senses, but rather . . . an object of the understanding, which however underlies the appearances of the senses." What did that mean? The "Architektonik" (B860–79) of Kant's first critique argued that each science was, in effect, a Leibnizean organic ensemble. "By 'architectonic' I mean the art of system. Since systematic unity is that which makes normal knowledge first into a science, i.e. a system out of a mere aggregate, and so is architectonic the doctrine of the scientific in our knowledge altogether."[96] He then equated the mechanical with aggregates, and the organic or purposive with the architectonic or systematic. The monadology was a correct theory of the systematic, teleological unity that human knowledge must have in order to be scientific. Kant remade the monadology into a philosophy of science.

Besides aesthetics, Kant's third critique treated of teleology, the key to enlightened Prussian and Kantian metaphysics. Kant defined the "organic" or the "organized" as an entity in which parts and whole mutually cause or reflect one another. This he tied as well to the notion of a system—as opposed to a machine—as a teleological unity.[97] In the third critique, he argued that in order for sciences of nature to be at all possible for us, we must presume an intelligible order in nature. We must presume, for example, that nature is undergirded by a purposive order that gives us such principles as the law of

95. See in general Cassirer, *Die Philosophie der Aufklärung,* chap. 7.
96. Kant, *Schriften,* 4:507; 3:538.
97. Ibid., 5:372–75; also 381–84. Much of this was made most clear, alas, in the "first version" of the introduction to the third critique, not yet available in the *Akademie-Ausgabe;* it is in Kant, *Kritik,* however.

economy *(lex parsimonia):* nature never does with more what it can do with less, that is, entities are not to be multiplied without necessity. Without that principle and others, science as a system of knowledge would not be possible for us. Such principles of economy and system presumed a purposive or teleological order as a regulative ideal we posited for research.[98] The order of scientific knowledge, as the systematic, was neither mechanical nor rational; it was, rather, architectonic. For a science of nature to be possible, we must view nature as though it were an artistic-technical work. Kant made physico-theology and teleology into a philosophical anthropology of science: we were architects of the system.

Kant's first critique declared the old metaphysical sciences impostors. Ontology he transmuted into a transcendental analytics in the first chief part of this critique. In the second chief part of this work, he declared pneumatology, cosmology, and theology to be illusions, a dialectic of reason. By critique of these pseudosciences, he wanted to set metaphysical reason on the path of science in regard to the deity, the cosmos, and the soul. Yet he declared these dialectical, dogmatic pseudosciences to be inevitable lures of reason. As a Leibnizean-Wolffian, making the order of things mechanical and the order of persons rational in the first two critiques, he made the order of humans cultural in the third: artistic-technical or aesthetic-teleological. Kantian humans were embodied rational beings condemned to pursue culture, as art and science, with a natural disposition to metaphysics, ultimately part of anthropology.[99] From his dogmatic slumber, Kant awoke to the anthropological sleep in the apotheosis of culture.

Kant, Aufklärung, Kultur

The life-history of Immanuel Kant is difficult to describe, since he had neither life nor history. He lived a mechanically ordered, almost abstract bachelor life. . . . I don't think the great clock of the cathedral there [in Königsberg] dispatched its daily work with less passion and more regularity than its fellow citizen Immanuel Kant. Getting up, drinking coffee, writing, lecturing, eating, taking a walk, everything had its fixed time, and the neighbors knew precisely that the clock must be at three-thirty if Immanuel Kant, in grey great-coat and Spanish cane in hand, stepped out his door to walk down the small Lindenallee. . . . But if Immanuel Kant, this great destroyer in the realm of ideas, by far exceeded Maximilian Robespierre in terrorism, he still has with the latter many similarities. . . . Indeed, to the highest degree, the character of the provincial revealed

98. Kant, *Schriften,* 2:427–43; 7:89; 8:89–106, 157–84; 9:311–20, 377–436.
99. Ibid., 4:362, a troubling passage.

itself in both—nature determined them to weigh coffee and sugar, but fate willed that they weigh other things, and laid the one a king and the other a God on the scale.[100]

Immanuel Kant had a small, skinny, and frail body, but he was never once seriously ill, till his death neared. Over the course of fifty years, he arose every morning at five o'clock, awakened by Lampe, his servant for forty years, with the simple, military call: "It's time" *(Es ist Zeit)*. Kant drank for breakfast two or three cups of weak tea and smoked a pipe of tobacco. After the morning lectures, in his early years, he would often go to a café where he would play billiards now and again, and drink a cup of tea. Besides his tea, Kant drank usually only water and wine. He ate a big lunch and tended to put mustard on everything at every meal, as he was convinced that mustard aided memory. His favorite foods were cod, peas, turnips, caviar, and Göttingen sausages. He took a walk of an hour or more every evening between 4 and 7 P.M. The only recorded case of him missing his walk was when he was reading Rousseau's *Émile*. Kant was what is politely known as thrifty, as his students could testify, few of whom he ever allowed into his classes for free and from whom, or anyone else for that matter, he did not appreciate contradiction. He refused to lend money to poor Fichte, who found his lectures sleepy; and, though being without spouse or children, Kant left at death an academically immense estate of about 20,000 taler. A proponent of the French Revolution, even when it was passé, he never left Königsberg, so never visited the capital city of the king whose name he associated with *Aufklärung* and whose century he said it was. Kant once went twenty-five years without visiting his sister, although she lived in the same town as he. "Kant was namely a *misogynist.*"[101]

"Our anthropology can be read by everyone, even by ladies at their *toilette,* since it is so entertaining." In his more well known lectures, Kant wrote that his anthropology was for "the reading public."[102] While Kant was interested in theological and cosmological issues, he seems to have been less interested in the pneumatological. One of his few works on pneumatology, *Träume eines Geistersehers, erläutert durch Träume der Metaphysik* (1766), is the strangest thing he ever wrote. What filled this void in the Kantian corpus was an interest in anthropology, philosophical and empirical. He was one of the first German professors to lecture on a subject he called *Anthropologie*. Indeed, from the late

---

100. Heine, *Werke,* 4:123–24.
101. Based on Gross, *Immanuel Kant,* 54–56, 79, 163, 184–93, 198, 225; Schwarz, *Immanuel Kant;* Rink, *Ansichten,* 109, 135; Kuhrke, *Kant,* 82–83; Metzger, *Aesserungen über Kant,* 8–9, 10 (quotation), 13, 15–16, 35, 42; Vorländer, *Immanuel Kant;* Schultz, *Immanuel Kant;* and Benninghoven, *Immanuel Kant.*
102. Kant, *Menschenkunde,* 6; *Schriften* 7:121.

1760s onward, for about thirty years, he lectured on anthropology in the winter semesters. The two passages cited above show the elision in the Enlightenment from women to the public. Along with cosmo-anthropology and physico-theology, does anthropology herald the collapse of metaphysics into the non-science of enlightened popular culture?

"Human reason has the special fate in one species of its knowledge: that it is pestered by questions which it cannot shrug off, since they are set by the nature of reason itself, but which it also cannot answer, since they surpass all ability of human reason." So began the preface to the first edition of Kant's first critique. It went on to relate that this special fate of human reason (spared angels and automata) was called "metaphysics" and that it had been a "battle-ground of endless controversies." Metaphysics was not only a science, she was "the *Queen* of all Sciences. . . . Now the fashionable tone of the time has brought it to showing her all contempt, and the matron laments, forsaken and forlorn." These metaphors of metaphysics as erstwhile queen of sciences were dropped from the second edition. In the first edition, royal metaphors had set an underlying political critique: "Our age is actually the age of critique, which subjects all to it. Religion by its sanctity and legislation by its majesty usually wish to withdraw themselves from that." But Reason objects.[103]

These remarks, which Kant also left out of the second edition, show that the first critique was meant as *Aufklärung* of a queen, which was just what he said on the two previous pages, where she was "the mother of chaos and night," awaiting "re-creation and *Aufklärung.*" Kant wanted a constitutional monarchy, whereby a criticized, nondogmatic, enlightened metaphysics was restored as philosopher queen of science. In the preface to the second edition, he exchanged the royal metaphor for "experimentation." Critique was analo-gized to an experiment in natural science, in a hypothetico-deductive sense. The dogmatic queen would be set on the critical path by an experiment. So had metaphysical critique become depoliticized by 1787, signaling the end of *Aufklärung* in Prussia?

"We are *cultured* to a high degree by art and science. We are *civilized* to the point of overtaxing by all sorts of social courtesy and propriety"—so Kant's now famous saying from 1784, which Elias used as an epitome of the German opposition of *Kultur* and civilization.[104] We need not dwell on the fact that this

103. On the above, Kant, *Schriften*, 4:7–9 (Avii–xii).
104. Elias, *Über den Prozess der Zivilisation*, 1:8; the original is in Kant, *Schriften*, 8:26. Meiners's cultural anthropology, discussed in note 37 above, was historico-ethnological. By con-trast, "philosophical anthropologies" studied humanity in terms of the natural and the spiritual: not in regard to nature and culture or society, but rather in regard to body and soul or *Geist*. That indicates the crucial role of the notion of *Geist* in the emergence of the notion of *Kultur,* linking anthropology with pneumatology.

opposition may not have been as clear as Elias would have it. Like others, Kant often conflated *Kultur, Zivilisation,* and *Aufklärung.* In his other famous essay of 1784, "What is *Aufklärung?*" Kant began by contrasting the animal with the enlightened existence. As appropriate to Prussian concerns, the motif of the animal became displaced in the essay by the motif of the mechanical. As the Wolffian clockwork cosmos seems to have been condensed into Kant as the Prussian automaton or mechanical man, the essay on *Aufklärung* ended with the sociopolitical insight that a human is more than a machine. *Kultur* and *Aufklärung* defined the truly human condition of society, as opposed to the animal or mechanical. Elsewhere Kant queried, "What is the natural condition of a person? The highest culture. What is the condition wherein that is possible? Bourgeois society." It was after all in such a society, as opposed to the academic or bureaucratic, that the "cosmopolitan" view, as he called it, was possible for philosophy. "On this view, philosophy is the science of the relations of all knowledge to the essential ends of human reason *(teleologia rationis humanae).* . . . Metaphysics is the perfection of all *culture* [in view] of human reason." [105]

The perfection of Prussian culture for Kant meant transcendental foundations for two sciences (of nature and freedom) and a critique of special metaphysics, which had concerned the three beings of interest to all humanity: the deity, the soul, the cosmos. The second part of the first critique, the "Transcendental Dialectic," recast this metaphysics into a rational reconstruction of the history of unreason. What had been metaphysical sciences in Halle into the 1760s, having largely then collapsed to cosmo-theology in Berlin, became a monument to metaphysical illness. As a Robespierre presiding over the Terror of the modern Prussian soul, Kant must be charged with a fitting crime. Driven now like Fichte to turn logic or *Wissenschaftslehre* into metaphysics, Hegel in the preface to *Wissenschaft der Logik* found the appropriate indictment: Kant had sided with positive science and common sense.

Conclusion

When he was once speaking with Goethe on the nature of tragedy, Napoleon said that the moderns distinguished themselves from the ancients in that we no longer have a fate . . . and, in place of ancient fate, politics has arisen *[La politique est la fatalité].* This must be used then as the new fate for tragedy, as the irresistible force of circumstances to which individuality must cede. [106]

105. Kant, *Schriften,* 15(2): 885, 3:542, 549; see also 3:522, 542–50; 9:24–25.
106. Hegel, *Werke,* 12:339.

A number of things characterize the Enlightenment. One of them is the death of traditional metaphysics. Another is the birth of modern politics. Somehow bound with those is a critique of the notion of fate and the end of classical tragedy. That is why so many attempts to overcome the Enlightenment have turned not only to a rejection of modern politics, and with it the public sphere, have turned not only to a restoration of metaphysics, disguised often as the modern litany on the death of metaphysics, but have also returned to a habilitation of *fatum,* destiny, the miraculous, and, if not astrology, then teleology.[107]

Conceived in the reconciliation of Lange and Wolff, Prussian *Aufklärung,* as a state religion, though rational, could not be political. "In Germany, philosophical genius wanders further than anywhere else, nothing stopping it, and the very absence of a political career, so tragic for the masses, gives still more liberty to thinkers."[108] If the advent of modern politics and the novel were hallmarks of the French and British Enlightenments, then the Prussian *Aufklärung* showed its provinciality here. For, even in the slow death of metaphysics, wrestling with the reconciliation of nature and freedom, Prussian *Aufklärung* led not to politics and the novel but rather to *Anthropologie* and *Kultur.*

"An . . . essential appearance of the Modern Era lies in the event that art moves into the horizon of aesthetics. That means: the work of art becomes an object of experience, and thus art counts as an expression of human life. . . . [Another] modern appearance arises in that human activity is comprehended and fulfilled as culture."[109] The nature of Prussia became a cultural product, an artistic-technical object to be experienced and expressed. The apotheosis of culture at the close of the *Aufklärung* emerged not only in Halle with Semler's and Wolf's antimetaphysical, historico-critical theology and philology, but also in Königsberg with Kantian *Aesthetik* and *Anthropologie.* Nietzsche remarked that we would never be free of God so long as we still believed in Grammar. Let us say too: We shall never be free of Metaphysics so long as we still believe in Culture.

Like Leibniz's monadology, Kant's heavenly city has been the object of satire. In *Götzen-Dämmerung,* Nietzsche wrote: "'Who is the perfect human?' The bureaucrat. 'Whose philosophy provides the supreme formula for the bureaucrat?' Kant's." His theory of the person, founded on a realm of duty, may

107. On the miraculous, see Brunschwig, *La crise de l'état,* 217–69.
108. Staël, *De l'Allemagne,* 1:137; cf. Marx, "Zur Kritik," 381–87.
109. Heidegger, *Holzwege,* 73. This essay (from 1938) seems aimed against Cassirer.

be seen as a legitimation of the Prussian bureaucracy. The enlightened Prussian state was a teleo-mechanical automaton, an artistic-technical system governed by angelic and automatic bureaucrats. Kant's heavenly city was the metaphysical foundation of the new circle of knowledge from which no one could escape, the irresistible force to which Prussians had to cede: *la bureaucratie est la fatalité*. Friedrich II said as much in his dictum: The king is first servant of the state. This too was the secret of Hegel's master-servant thesis: The civil servant as bureaucratic master. In Romantic Prussia, Leibnizean-Kantian metaphysics, as a phenomenolgy of spirit *(Geist)*, lived on in *Anthropologie, Kultur und die Staatsdienst*.

From the unfinished work of the civil servant philosopher of the heavenly city:

> The transc. philos. is the science of the forms by which one produces a whole in intuition and an object in thought synthetically according to principles.

> NB. Domestic: the dried fruit from HE Lehmann will not be given to the cook for keeping, but rather behind *the stove* in my dining room in a sealed sack or in 2 sacks which are often shaken . . .

> The subjective comes before the objective in intuition. The consciousness of oneself before the exterior and external thing.

> *Forma dat Esse rei.*

> There is in our well-water a styptic taste that I notice when gargling.

> The sack with the dried fruit behind the stove . . .

> Lampe hung my coat my nightshirt in the dining room behind the stove so that after the meal it could be put on while warm so that it would not be put on while cold. The cook like a mad one reproached Lampe that he leave her in peace and that she would not obey him as though he were Master *(Herr)* of the house. But she is the one who wants to play Master.[110]

In January 1802, Kant had to fire Lampe, his servant of forty years, who had a fondness for drinking that led him to indulge in the master's store. Lampe had also gotten married behind the bachelor Kant's back. Shaken by

110. Kant, *Schriften*, 21:112, 114, 121.

Fig. 13.3.   Kant mixing mustard to improve his memory.
The sketch is by Friedrich Hagemann, drawn in 1801 after
a visit to Kant the previous year. Reproduced in K. H.
Classen's *Kant-Bildnisse* (1924). By permission of the
Universitätsbibliothek Göttingen.

having to remove his servant, while going slowly senile, as his *Opus postumum*
above shows, so mixing more and more mustard to preserve his memory, as
figure 13.3 shows, and perhaps meaning his servant's wife, rather than the man
himself, the great philosophical anthropologist of the heavenly city penned, as
the anecdote goes, one of his "memory notes" *(Gedächtniszettel)* to himself as

a tool of countermemory, the better not to forget to remember: "The name
Lampe must become now completely forgotten." [111]

Acknowledgments

In addition to all those involved in this volume, I would like to thank Alix
Cooper, Martin Gierl, John Holloran, and André Wakefield.

Primary Sources

Ahlwardt, Peter. *Bronto-Theologie.* Greifswald and Leipzig, 1745.
Baumgarten, Alexander. *Aesthetica.* 2 vols. in 1. Frankfurt am Main, 1750–58.
————. *Metaphysica.* 7th ed. Halle, 1779. Reprint, Hildesheim: Olms, 1982. First edition published 1739.
[Benemann, Joahnn Christian]. *Die Tulpe zum Ruhm ihres Schöpfers.* Dresden and Leipzig, 1741.
[————]. *Die Rose zum Ruhm ihres Schöpfers.* Leipzig, 1741.
Berlin Academy of Sciences. *Dissertation qui a ramporté le prix par l'Academie royale des sciences et belles lettres sur le système des monades avec les pieces qui ont concouru.* Berlin, 1748.
Bode, Johann Elert. *Von den neu entdeckten Planeten.* Berlin, 1784.
————. *Allgemeine Betrachtungen über das Weltgebäude.* Berlin, 1804.
————, trans and comm. *Anleitung zur Kenntniss des gestirnten Himmel.* 4th ed. Berlin and Leipzig, 1778.
Boscovich, Roger Joseph, S. J. *Philosophia naturalis theoria redacta ad unicam legem virium in natura exitentiam.* Vienna, 1763. Translated by J. M. Child as *A Theory of Natural Philosophy* (Cambridge: Harvard Univerity Press, 1966). First edition published 1758.
Châtelet, marquise du (Gabrielle-Émilie le Tonnelier de Breteuil). *Institutions physiques.* 2d ed. 1742. Reprinted in Wolff, *Gesammelte Werke,* ser. 3, vol. 28. First edition published 1741.
[Clavius, Andreas]. *Bericht von dem gefahrlichen Vorurtheile worin die Lehre von den Elementen der Körper zu diesen Zeiten gerathen ist.* N.p., n.d.
Curiöses Gespräch zwichen einem Astronomo und einem einfältigen thüringischen Bauer.... N.p., 1744.
Denso, Johann David. *Beweis der Gottheit aus dem Grase.* Amsterdam, 1750.
Descartes, René. *Oeuvres.* 2d ed. 11 vols. Paris: Vrin, 1982–91.
Eberhard, Johann August. *Kurzer Abriss der Metaphysik mit Rücksicht auf den gegenwärtigen Zustand der Philosophie.* Halle, 1794.
Euler, Leonhard. *Lettres à une Princesse d'Allemagne.* 3 vols. 2d ed. Paris, 1787–89. First edition published 1768-74.
————. *Briefe an eine deutsche Prinzessin.* Leipzig, Riga, St. Petersburg, 1769–73. Reprint, Brunswick: Vieweg, 1986.
————. *Opera Omnia.* Multiple vols. in 3 series. Geneva: Birkhäuser, 1912–.
Fabricius, Joahnn Albert. *Pyrotheologie....* Hamburg, 1732.
————. *Hydrotheologie....* Hamburg, 1734.
Feder, Johann Georg H. *Logik und Metaphysik.* 6th ed. Göttingen, 1786.
Fischer, Ernst G. *Betrachtung über die Kometen....* Berlin, 1789.

111. Wasianski (1804) in Schwarz, *Immanuel Kant,* 326; on the firing of Lampe, 319–21. Cf. Weinrich, "Warum will Kant." Wasianski had Kant's *Gedächtniszettel* and so on, so the anecdote is doubtless based in truth. On Hagemann's drawing, see Classen, *Kant-Bildnisse,* 25–27; Kuhrke, *Kant,* 82–83. Cf. Foucault, *Dits et écrits,* 2:153: "De toute façon, il s'agit de faire de l'histoire un usage qui l'affranchisse à jamais du modèle, à la fois métaphysique et anthropologique, de la mémoire. Il s'agit de faire de l'histoire une contre-mémoire."

Fontenelle, Bernard le Bovier de. *Dialogen über die Mehrheit der Welten.* Berlin, 1780. Translated with commentary by J. E. Bode as *Entretiens sur la pluralité des mondes* (Paris, 1686).

Formey, Jean H. S. *Mêlanges philosophiques.* 2 vols. Leiden, 1754.

Friedrich II, king of Prussia. *Friedrich der Grosse und die Philosophie: Texte und Dokumente.* Stuttgart: Reclam, 1986.

Gottsched, Johann Christoph. *Erste Gründe der gesammten Weltweisheit.* 2 vols. Leipzig, 1762. Reprinted in Wolff, *Gesammelte Werke*, ser. 3, vol. 22, 1–2.

Gross, Felix, ed. *Immanuel Kant: Sein Leben in Darstellungen von Zeitgenossen.* Berlin: Deutsche Bibliothek, 1912.

Hegel, Georg W. F. *Werke.* 20 vols. 1832-45. Reprint, Frankfurt am Main: Suhrkamp, 1971.

Heine, Heinrich. *Werke.* 4 vols. Edited by C. Siegrist. 1968. Reprint, Frankfurt am Main: Insel, 1994.

Heinsius, Balthasar. *Chionotheolgie. . . .* Züllichau, 1735.

Heyn, Johann. *Versuch einer Betrachtung über die Cometen. . . .* Berlin and Leipzig, 1742.

Jeremais W** Bürger und Meister zu G** erläutert dem S. T. Herrn Rath Justi eins unds andere, das ihm zu schwer ist in den Anfangs-Lehren der Metaphysik. . . . Leipzig, 1748.

Justi, Johann Heinrich Gottlob von. "Untersuchung der Lehre von den Monaden und einfachen Dingen. . . ." In Berlin Academy of Sciences, *Dissertation*, 1–52. 1748.

———. "Dissertation sur les monads." In Berlin Academy of Sciences, *Dissertation*, 53–110. 1748.

———. *Die Nichtigkeit aller Einwürfe und unhöflichen Anfälle, welche wider seine Untersuchung der Lehre von den Monaden und einfachen Dingen zum Vorschein gekommen sind. . . .* Frankfurt and Leipzig, 1748.

J.M.R. *Mutmassliche Gedancken von dem Ursprung, Wesen, Lauff und Bedeutung des Cometen. . . .* Frankfurt and Leipzig, 1744.

Kahle, Ludwig M. *Vergleichung der Leibnitzischen und Newtonischen Metaphysik . . . dem . . . Voltaire entgegen gesetzt.* Göttingen, 1741.

Kant, Immanuel. *Menschenkunde oder philosophische Anthropologie.* 2d ed. Edited by F. C. Starke. Quedinburg and Leipzig, 1839. Reprint, Hildesheim: Olms, 1976.

———. *Gesammelte Schriften.* Multiple vols. Berlin: Reimar und de Gruyter, 1902–.

———. *Kritik der Urteilskraft.* Edited by W. Weischedel. Frankfurt am Main: Suhrkamp, 1974.

Karsten, Wenceslaus Johann Gustav. *Lehrbegriff der gesammten Mathematik.* 8 vols. Greifswald, 1767–77.

Kästner, Abraham G. *Anfangsgründe der höhern Mechanik.* Göttingen, 1766.

———. *Anfangsgründe der Astronomie.* 4th ed. Göttingen, 1794. First edition published 1759.

[Kindermann, Eberhard C.]. *Reise in Gedancken durch die eröffneten allgemeinen Himmels-Kugeln. . . .* Rudolstadt, 1739.

———. *Vollständige Astronomie . . . , Vormahls unter dem Titul: Reise in Gedancken. . . .* Rudolstadt, 1744.

[Körber, Christian A.]. *Gegenseitige Prüfung der Gedancken von den Elementen der Körper.* Frankfurt and Leipzig, 1746.

Krüger, Johann Gottlob. *Träume.* Halle, 1785. First edition published 1754.

Lambert, Johann H. *Cosmologische Briefe über die Einrichtung des Weltgebäudes.* Augsburg, 1761.

———. *Philosophische Schriften.* 9 vols. Hildesheim: Olms, 1965–68.

Lange, Joachim. *Modesta disquisitio novi philosophiae systematis. . . .* Halle, 1723. Reprinted in Wolff, *Gesammelte Werke*, ser. 3, vol. 23.

———. *Placide vindiciae modeste disquisitionis. . . .* Halle, 1723.

———. *Bescheidene und ausführliche Entdeckung der falschen und schädlichen Philosophie in dem Wolffianischen Systemate Metaphysico. . . .* Halle, 1724.

———. *Amerkungen über des Herrn . . . Wolffens Metaphysicam . . . Nebst Wolffens gründlichen Antwort.* Cassel, 1724.

———. *Ausführliche Recension, der wider die WolfianischeMetaphysic auf 9 Universitäten und anderwärtig edirten sämmtlichen 26 Schriften . . .* Halle, 1725.

———. *Caussa Dei et religionis naturalis adversus atheismum.* . . . Halle, 1727. Reprinted in Wolff, *Gesammelte Werke,* ser. 3, vol. 12.
———. *Philosophische Fragen aus der neuen Mechanischen Morale.* . . . Halle, 1734.
———. *Hundert und dreyssig Fragen aus der neuen Mechanischen Philosophie.* . . . Halle, 1734.
Leibniz, Gottfried Wilhelm. *Die philosophischen Schriften.* 7 vols. Edited by C. J. Gerhardt. 1875–90. Reprint, Hildesheim: Olms, 1978.
———. *Das Neuste von China. Novissima Sinica,* Deutsche China-Gesellschaft, Schriftenreihe, 2. Edited and translated by H. G. Nesselrath and H. Reinbothe. Cologne, 1979. First edition published 1697.
Lesser, Friedrich C. *Lithotheologie.* . . . Hamburg, 1735.
———. *Insectotheologie.* . . . 2d ed. Frankfurt and Leipzig, 1740.
———. *Testaceo-Theologia.* . . . Leipzig, 1744.
Ludovici, Carl G., ed. *Sammlung und Auszüge der sämmtlichen Streitschriften wegen der Wolffischen Philosophie.* 2 vols. Leipzig, 1737–38. Reprinted in Wolff, *Gesammelte Werke,* ser. 3, vol. 2.
Marx, Karl. "Zur Kritik der Hegelischen Rechtsphilosophie" (1844). In vol. 1 of *Werke von Karl Marx und Friedrich Engels.* 40 vols. with registers. Berlin: Dietz, 1957-90.
Maupertuis, Pierre L. M. de. *Essai de Cosmologie.* Berlin, 1750.
———. *Oeuvres.* 4 vols. Lyon, 1756.
———. "Examen philosophique de la preuve de l'Existence de Dieu employée dans l'essai de Cosmologie" (1756). *Historie de l'Académie Royale des Sciences et Belles-Lettres de Berlin* 12 (1758): 389–424.
Meiners, Christoph. *Grundriss der Geschichte der Menschheit.* Lemgo, 1785.
———. *Über wahre, unzeitige, und falsche Aufklärung.* . . . Hanover, 1794.
Meiners, Christoph, and Ludwig T. Spittler, eds. *Göttingisches Historisches Magazin.* Hanover, 1787–90.
———. *Neues Göttingisches Historisches Magazin.* Hanover, 1791–94.
[Metzger, Johann Daniel]. *Aeusserungen über Kant, seinen Charakter und seine Meinungen.* [Königsberg], 1804.
Müller, Gerhard A. *Unparteyische Critik der Leibnizischen Monadologie.* Jena, 1748.
Müller, Jacob F. *Die ungegründete und Idealistische Monadologie.* . . . Frankfurt am Main, 1745.
Mylius, Christlob. "Lehrgedicht von den Bewohnern der Kometen." *Belustigungen des Verstandes und des Witzes,* May 1744, 383–92.
Newton, Isaac. *Mathematical Principles of Natural Philosophy.* . . . 3d ed. 2 vols. Translated by Andrew Motte. 1729. Revised by Florian Cajori. Berkeley and Los Angeles: University of California Press, 1934. First edition published 1687.
———. *Opticks.* 1704. Reprint of 4th ed., New York: Dover, 1979.
Rathelf, Ernst L. *Akridotheologie.* . . . 2 vols. Hanover, 1748–50.
[Reimarus, Hermann Samuel]. *Fragmente des Wolfenbüttelschen Unbenannten . . . Bekanntgemacht von G. E. Lessing.* Berlin, 1788.
———. *Abhandlung von den vornehmsten Wahrheiten der natürlichen Religion . . .* 6th ed. Hamburg, 1791. First edition published 1754.
———. *Allgemeine Betrachtungen über die Triebe der Thieren.* . . . 4th ed. Hamburg, 1798. First edition published 1760.
Richter, Johann Gottfried. *Ichthyotheologie.* . . . Leipzig, 1754.
Rink, Friedrich T. *Ansichten aus Immanuel Kants Leben.* Königsberg, 1805.
Rohr, Julius Bernhard, von. *Phyto-Theologie.* . . . Frankfurt and Leipzig, 1740.
Schierach, Adolf. *Melitto-Theologia.* . . . Dresden, 1767.
Schwarz, Hermann, ed. *Immanuel Kant: Ein Lebensbild nach Darstellungen der Zeitgenossen Borowski, Jachmann, Wasianski.* 2d ed. Halle: Peter, 1907.
Semler, Johann Salomo. *Vorbereitung zur theologischer Hermeneutik.* 4 vols. Halle, 1760–69.
———. Foreword to *Untersuchungen theologischer Streitigkeiten,* by S. J. Baumgarten, vol. 1. Halle, 1762.

————. *Versuch einiger moralischen Betrachtungen über die vielen Wundercuren und Mirackel in den ältern Zeiten.* . . . Halle, 1767.

————. *Abhandlung von freier Untersuchung des Canon.* . . . Halle, 1771.

————. *Versuch einer biblischen Dämonologie.* Halle, 1776.

————. *Beantwortung der Fragmente eines Ungenannten insbesondere vom Zweck Jesu und seiner Jünger.* Halle, 1779.

————. *Neue Versuche die Kirchenhistorie der ersten Jahrhunderte aufzuklären.* Leipzig, 1788.

Sendschreiben an einen Freund in H. von Beschaffenheit, Bedeutung, Wirckung und vernünftigen Betrachtung der Cometen. . . . N.p., [1744?].

Sendschreiben an . . . Herrn Justi . . . von der Wiederbringung der Monaden in das Reich der Wirklichkeit. Liegnitz, 1748.

Staël, Germaine de. *De l'Allemagne.* 2 vols. 1813. Reprint, Paris: Flammarion, 1968.

[Stiebritz, Johann Friedrich]. *Widerlegung der Gedancken von den Elementen der Cörper.* Frankfurt and Leipzig, 1746.

Ulmigena, Eusebius [Johann F. Bertram]. *Send-Schreiben an einem guten Freund von den Wolfischen Fato, Das ist von der Mechanischen Verknüpfung und Nothwendigkeit aller Dinge.* . . . Bremen, 1737.

Walpurger, Johann Gottlieb. *Cosmologische Betrachtungen derer wichtigsten Wunder und Wahrheiten im Reiche der Natur.* . . . 4 vols. in 2 parts. Chemnitz, 1748–52.

Wolf, Friedrich A. *Encyclopädie der Philologie.* Edited by G. H. Stockman. Leipzig, 1831.

————. *Darstellung der Alterthumswissenschaft.* . . . Edited by S. F. W. Hoffmann. Leipzig, 1833.

Wolff, Christian Freiherr von. *De differentia nexus rerum sapientis et fatalis necessitatis.* . . . Halle, 1724.

————. *Monitum ad commentationem luculentam de differentia.* . . . Halle, 1724.

————. *Gesammelte Werke.* Multiple vols. in 3 series. Hildesheim: Olms, 1962–.

————. *Oratio de Sinarum philosophia practica: Rede über die praktische Philosophie der Chinesen.* Edited and translated by M. Albrecht. Hamburg: Meiner, 1985.

Wurm, Johann Friedrich. *Geschichte des neuen Planeten Uranus.* Gotha, 1791.

Zorn, Johann Heinrich. *Petinotheologie.* . . . 2 vols. Pappenheim and Schwabach, 1742–43.

Departures

14

Inner History; or, How to End Enlightenment

NICHOLAS JARDINE

It has become a commonplace that European enlightenment ended with a transformation in historical discourse and consciousness—indeed, as we shall see, such views can be traced right back to the end of the Enlightenment itself. I address particular claims of this sort by Michel Foucault in *The Order of Things* (1966) and Reinhart Koselleck in *Futures Past* (1979). At the outset, I shall consider some ways in which the development of natural history and *Naturphilosophie* in the German lands appears to undermine Koselleck's and Foucault's theses about the new historicism. I shall argue, however, that a core of their claims survives these challenges, and can indeed be strengthened and sharpened into a new Koselleck—Foucault thesis.

Koselleck and Foucault were out to trace the genealogy of our historical consciousness, and that is also my concern. Following them, I identify the new historicism that ended the Enlightenment with the narration of "inner histories"—of life, of *Geist,* of culture—underlying the diverse and fragmented phenomena of the natural and human worlds. Much of this new historicism is unthinkable for us; but I shall argue that a part of it, especially evident in Romantic histories of the disciplines, having to do with the production of novelty, remains at once valid and deeply problematic for us. I illustrate this aspect of the new historicism through a comparison of Enlightened and Romantic histories of natural history.

Koselleck, Foucault, and Historicity

The animal, whose great threat or radical strangeness had been left suspended and as it were disarmed at the end of the Middle Ages, or at least at the end of the Renaissance, discovers fantastic new powers in the Nineteenth Century. In the interval, Classical Nature had given precedence to vegetable values—since the plant bears upon its visible form the overt mark of every possible order; with all its forms on display, from stem to seed, from root to fruit, with all its secrets generously made visible, the vegetable kingdom formed a pure and transparent object for thought as tabulation.

Michel Foucault, *The Order of Things*

Koselleck tells how Jan Comenius in *Das Labyrinth der Welt und das Paradies des Herzens* (1623) portrayed historians as carriers of telescopes, like trumpets pointing backward; and a generation later, Samuel Butler declared that "if Travellers are allowed to Lye in recompense of the great Pains they have taken to bring home strange Stories from foreign Parts, there is no reason why *Antiquaries* should not be allow'd the Priviledge, who are but Travellers in *Time.*"[1] When the Enlightenment ends these images become incongruous. The human past ceases to be a surveyable territory of whose occupants histories are compiled, on a par with the natural histories of the heavens and earth. For, as Droysen was later famously to observe, "beyond histories there is History." And the time of this new History is no longer a distance into the past; rather, Koselleck claims, it becomes a historical time, a time in which genuine novelties occur, in which new values, new thoughts, new forms of life, new types of persons become possible.[2] It is the lifetime, the course of education, the cultural growth of humanity.

At the present level of simplification and abstraction, Foucault's end-of-Enlightenment thesis exactly complements Koselleck's. Where Koselleck charts the moves from human histories to human History, Foucault announces the displacement of the manifold natural histories by the History of Nature.[3] For Foucault as for Koselleck, at the end of the eighteenth century chronology is replaced by historical time, that is, time in which genuine innovation, not merely actualization of preexisting possibilities, can occur. For Foucault the new locus of historical development is not on the surface, in the realm of visible events, but in the depths, in the internal functions and activities that constitute the lives of organic beings, just as for Koselleck the new locus is the life and culture of humanity that underlies the actions and events of history.

Soon we shall have occasion to look at some of the specificities and nuances of Koselleck's and Foucault's theses. But my simplified versions will do for the moment—for they suffice to bring out some obvious difficulties.

Against these theses we may level a first objection based on Europe-wide features of human and natural history. Koselleck and Foucault point to the appearance in the nineteenth century of new principles of historical unity. Human histories are absorbed into History, the life of mankind—variously construed as the progressive concretization of reason, as the unfolding of *Geist* in

1. Comenius, *Das Labyrinth der Welt,* 105; Butler, *Characters and Passages,* 270.
2. Koselleck, *Futures Past:* "Modernity and the Places of History" (3–20), "Historia Magistra Vitae" (21–38), "Perspective and Temporality" (130–55). The quotation from J. G. Droysen's *Historik* is on 28.
3. Foucault , "Labour, Life, and Language," chap. 8 in *Order of Things.*

the cultures of nations, as the creative reflection of Nature upon herself, etc. Natural histories are likewise absorbed into the History of Life on Earth—variously construed as the explication of a primordial ideal, as a sequence of epochs of creation, as a single vast genealogy responding to changing conditions of existence, and so on. *But,* and it is an enormous *but,* in the opening decades of the nineteenth century it is not grand, unifying schemes, but local histories that proliferate. Let us take just a couple of examples. In the German lands, we see a plethora of histories of particular towns, estates, institutions, and disciplines, often paying homage to Justus Möser's *Osnabrückische Geschichte.*[4] Or consider the history of the Earth—in the work of William Smith in England, and of Georges Cuvier and Alexandre Brongniart in France, the emphasis is on reconstructions of Earth histories on a scale that is unprecedentedly local, both chronologically and spatially.[5] Both civil and natural historians of the 1810s and 1820s habitually mock the grand historical speculations of their enlightened predecessors—thus the conjectural histories of mankind of the Scottish Enlightenment are regularly denounced along with grand theories of the Earth in the style of Buffon, Werner, and Hutton, as relics from the despotic age of systems.[6] Indeed, it seems one might go further, arguing for an inverse Koselleck—Foucault thesis, for a shift from Enlightened fascination with grand historical schemes to Romantic celebration—in the manner of Friedrich Schlegel—of *Eigentlaümlichkeit,* of fragmentation, partisanship, and irreconcilable perspectives in all fields of history.[7] For future reference, let us call this the "locality objection."

A second objection is posed by the development of natural history in the German lands. According to Koselleck, the unification of the many human histories into History entailed the separation of history proper from natural history.[8] (Koselleck even appears to suggest that this recognition of natural history as an autonomous discipline was new—but that is clearly a mistake, for a perfectly good civil/natural historical division is to be found in Pliny, even at a stretch in Aristotle.) Far from separating human from natural history, it is one of the defining characteristics of the *Naturphilosophie* of the opening decades of the nineteenth century to integrate them. As Henrich Steffens

---

4. On the reception of the *Osnabrückische Geschichte,* see Hempel, "Justus Mösers Wirkung"; Knudsen, "Justus Möser" and *Justus Möser and the German Enlightenment,* chap. 7. See Friedrich Meinecke's classic *Historism,* chap. 8, for a more *volkisch* reading.

5. See, for example, Rudwick, "Cuvier and Brongniart."

6. Indeed, Cuvier himself denounced grand speculative systems inadequately founded on *expérience:* see Cuvier, *Rapport historique,* passim.

7. For a moving celebration of Romantic *Eigentümlichkeit,* see Koerner, *Caspar David Friedrich,* chap. 4.

8. Koselleck, "Historia Magistra Vitae," in *Futures Past.*

declared: "Do you want to know nature? Turn your glance inwards, and you will be granted the privilege of beholding nature's stages of development in the stages of your spiritual education. Do you want to know yourself? Seek in nature. Her works are those of the selfsame spirit."[9] For Steffens, as for such *Naturphilosophen* as Schelling, Lorenz Oken, C. G. Carus, G. H. von Schubert, and C. J. H. Windischmann, the history of nature and human history are manifestations of one and the same inner history, the dialectical play of spirit.[10] For future reference, let us call this the "objection from *Geist.*"

The third objection impinges on Foucault alone, and arises from his choice of Cuvier's comparative anatomy as the ideal type for the new life-centered episteme. Cuvier, so the objection goes, is far from original in his program for a comparative anatomy in which function has priority over structure, and structure over visible form.[11] Vicq d'Azyr, Blumenbach, Kaspar Friedrich Wolff, even the Stagirite, appear to be worthy precursors—as Cuvier, keen to legitimate his enterprise, did not fail to admit.[12] Further, and more seriously for Foucault's thesis, the approach that really is new in early-nineteenth-century comparative anatomy is not Cuvier's neo-Aristotelian functionalism, but the so-called philosophical anatomy, or morphology, typified in the work of Goethe, C. G. Carus, Etienne Geoffroy Saint-Hilaire, and Richard Owen. The new morphology, however, is concentrated on structural plans and their transformations, with relatively little regard to the inner activities and functions of life.[13] Let us call this the "anti-Cuvier objection."

What can be salvaged? Almost everything, I think.

First, let us take the issue of locality. Foucault's grandiloquence—"tearing aside the veil to reveal the mute and invisible violence that devours living beings in the darkness," and so on[14]—is indeed suggestive of vast and sublime historical schemes; but there is no explicit commitment to unifying histories of life on Earth. Nor, in fact, is there a real difficulty here. For one may well suppose, as Cuvier and Brongniart, for example, explicitly did, that the big inaccessible history is just the sum of all the local, but sometimes more accessible ones.

In Koselleck's case, the objection, while readily met, serves to highlight an aspect of the new historicism that should give pause for uneasy thought.

---

9. Steffens, *Alt und Neu,* 2:102.

10. For an insightful survey of *naturphilosophischen* histories, see Engelhardt, "Historical Consciousness."

11. For objections to Foucault's claims for the epochal significance of Cuvier, see the discussion following Foucault, "La situation de Cuvier."

12. Cuvier, *Rapport historique.*

13. On the new morphology, see Russell's classic *Form and Function,* chaps. 4–8; Rehbock, *Philosophical Naturalists;* Nyhart, *Biology Takes Form,* chaps. 1–2.

14. Foucault, *Order of Things,* 278.

Koselleck himself does in fact have much to say about the ways in which the new historical consciousness recognized the local, perspectival, and partisan nature of all historical perception; indeed, he points to the tension between this recognition and the wish for a cosmopolitan, *unparteiliche* history of the education of mankind that arises in the works of Friedrich Schlegel and his contemporaries.[15] But, perhaps because he takes it to be too obvious to need stating, he says nothing about the way in which the great nineteenth-century philosophies of history cope with the dilemma. However, to the extent that his concern, like my concern with post-Enlightenment historicism, has to do with the genealogy of our historical consciousness, something does need to be said on this score. For the grand philosophies of history can be defined, functionally as it were, as so many resolutions of this dilemma—as so many ways of explaining how, appearances to the contrary, the many local histories conditioned by time, place, and partisanship, are aspects of one big story—the return of *Geist* to itself, the life of humanity, the realization of the end of ends, the dialectic of freedom and necessity, and so on. But we, of course, are supposed to be too sophisticated to believe such tales.

The objection from *Geist* is likewise readily met. Against Koselleck, indeed, it hardly counts as an objection. For the *Naturphilosophen,* the inner identity of human history with the history of nature as manifestations of the odyssey of the spirit by no means collapses these disciplines into one another. If anything, it enhances the distinction of human history, for the activity of the spirit in human history is privileged as the supreme manifestation of spirit—in Schelling's terminology, as *Geist* acting at its highest *Potenz.* As for Foucault's thesis, only its extension, not its intention is affected. After his Heideggerian enthusiasm in the late 1950s, Foucault fixed his gaze on France. Thus he associates the new "inner history" of life with the new French *biologie* of plants and animals. But the cosmos of the *Naturphilosophen* is hylozoic. As C. G. Carus declares: "If once we have recognized nature as being in the process of endless inner linkage, then we must at the same time consider it as the absolute living thing, from whose primordial life are derived the appearances of life of each particular living thing."[16] In such a German world, Foucault's thesis predicts an inner history for everything, quite correctly as it turns out.

What of the third challenge, the anti-Cuvier objection? Part of this can be dealt with briskly. From Aristotle onward, there had, indeed, been many who gave priority to vital activities and functions, considering structures and organs as means to functional ends. However, Cuvier goes much further than this, treating the functions of life as independent of particular organs and

---

15. Koselleck, "Perspective and Temporality," in *Futures Past;* and "Standortbindung und Zeitlichkeit."

16. Carus, "Grundzüge allgemeiner Naturbetrachtung," 85.

structures, since they can be effectively realized and performed in indefinitely many radically different ways—a point that Foucault duly emphasises.[17] With the possible exception of his near-contemporary mentor and sentimental friend at Stuttgart, Carl Friedrich Kielmeyer, it is hard to think of precursors in this.[18] The challenge to Foucault's thesis from the rise of structural morphology is a bit more serious. But even here, I think, Foucault's basic insight remains intact. The trouble arises from all his black Romantic images—his talk of abysses and obscure verticalities, of violence and struggle in the velvety darkness. Note that for the morphologists as for the functional anatomists, the emphasis was on inner structure, and that inner structure was seen as the historical product of a plan in accordance with dynamic laws. So the essential features of the new biological episteme—secret histories determining regular transformations of structures determining changes in outward and visible forms—remain in place.[19] But the register has shifted—the secrets of the morphologists are not the warm, dark secrets of the life force, but the primordial archetypes and ideas. To turn from the functional anatomy of Cuvier and Bichat to the ideal morphology of Oken, C. G. Carus, and Meckel is, to use the language of Schelling's *Bruno,* itself a reflection on the new historicities, to turn from the history of the dark inward flame of the *Lebenskraft,* to that of the blinding white flame of the *logos.*

A New Koselleck—Foucault Thesis

All visible objects, man, are but as pasteboard masks. But in each event—in the living act, in the undoubted deed—there, some unknown but still reasoning thing puts forth the mouldings of its features from behind the uneasy mask. If man will strike, strike through the mask!

Herman Melville, *Moby Dick*

17. Foucault, *Order of Things,* 264–65, 271; and "La situation de Cuvier," 73, in response to objections.

18. On Kielmeyer in relation to Cuvier, see Kohlbrugge, "G. Cuvier und K. F. Kielmeyer"; Kuhn, "Der naturwissenschaftliche Unterricht"; Schumacher, "Karl Friedrich Kielmeyer." Cuvier studied with Kielmeyer at the Hohen Karlsschule at Stuttgart and was later sent notes of Kielmeyer's lectures on comparative zoology by Christoph Heinrich Pfaff. However, only a plan of Kielmeyer's lectures is known to be extant; and this, while showing parallels in the ordering of topics with Cuvier's *Leçons sur l'anatomie comparée,* gives little specific indication of Kielmeyer's doctrines. As in Kielmeyer's *Ueber die Verhältnisse der organischen Kräfte,* of 1793, vital forces and functions are evidently treated as prior to structure and radically diverse anatomical plans are related to the predominance of different vital forces. But this does not differentiate Kielmeyer and Cuvier from other comparative anatomists of the period, for example, Blumenbach, whose course of comparative anatomy as set out in his *Handbuch der vergleichende Anatomie* of 1804 has a similar agenda.

19. For example, Michael Hagner has argued that it is not the focus on inner processes per se, but the commitment to the generation of structure by dynamic laws of inner development that distinguishes nineteenth-century teratology from eighteenth-century natural histories of monstrosity: see Hagner, "Vom Naturalienkabinett zur Embriologie," esp. 102ff.

In *The German Enlightenment and the Rise of Historicism,* Peter Hanns Reill foregrounds a dissonance within German Enlightenment historiography—on the one hand there is the concern with the global history of human culture, on the other, the increasing appreciation of the historical individuality of epochs, nations, and cultures, and likewise of historians' standpoints and affiliations.[20] Symptomatic of this is the clash between universal, providential interpretations of the Scriptures, and readings of them as local Hebrew myths and chronicles—a confrontation epitomized in the desperate attempts by the universal historian and sacred critic Johann Salomo Semler to preserve a core of universally significant canonical Scriptures in the face of the historicist interpretations of Lessing's *Reimarus Fragmente* and of his own earlier works.[21]

An analogous tension can be discerned in the natural history of the second half of the eighteenth century. On the one hand, this is the time of quest for standardized descriptions and unified systems of classification.[22] At the same time, however, natural historians and their publics are ever more fascinated by the rare, the monstrous, the singular, and the exotic. The irreducible differences between local floras and faunas and their conditions of existence are increasingly realized, as is the diversity of structural plans and modes of generation and development of plants and animals.[23]

It is this tension between appreciation of singularity and locality and the quest for universal laws and systems that provides the problematic of the new historicities described by Koselleck and Foucault. These new historicities can be seen as modes of reconciliation of the demands of the local and the global, the partial and the impartial. Thus, for example, in a marvelous recent historical reading of Schelling's early philosophy, S. R. Morgan has shown in detail how the clash between universal and historicist readings of the Scriptures provides the underlying problematic of Schelling's history of divinity in the world, itself a *fons et origo* of the new nineteenth-century historicities.[24]

A revised Koselleck—Foucault thesis should, I suggest, bring this tension into the open. In natural as in human affairs, the new historicities postulate

20. Reill, *German Enlightenment,* chap. 8.

21. For accounts of the new "sacred criticism" and the Semler/Lessing confrontation see Hornig, *Die Anfänge der historisch-kritischen Theologie,* and Frei, *Eclipse of Biblical Narrative,* chap. 6. On the local/universal tension in histories of human nature see Frasca-Spada, chap. 7 above.

22. See, for example, Daudin, *De Linné à Jussieu;* Lesch, "Systematics and the Geometrical Spirit."

23. On later eighteenth-century appreciation of the singular and the exotic, see Stafford, *Voyage into Substance,* intro. and passim; Charlton, *New Images of the Natural in France,* chaps. 3–5. On monstrosity and attempts to tabulate and order it, see Tort, *L'ordre et les monstres,* and Hagner, "Vom Naturalienkabinett zur Embriologie." On recognition of diversity in natural history, see Browne, *Secular Ark,* and Larson, *Interpreting Nature,* chaps. 3–4.

24. Morgan, *Palingenesis of Ancient Wisdom.*

inner histories, hidden dynamisms to unify the dissonant local phenomena—such are the secret histories of development of life and culture, of transformation of archetypes and ideas, of the odyssey of the spirit. Only thus can the new historicities underwrite unified narratives while doing justice to the fragmentation and individuality of the human and natural worlds. Moreover, and this, I shall suggest, is the really crucial claim common to Koselleck and Foucault, these latent histories are, to use another Schellingian term, *productive* histories. That is, they are histories in which there is radical innovation—in which new forms of living beings, new types of community, new values and new ideas, new selves become possible.

Genealogies of the Present

In history's mysteries vast,
The present's as strange as the past,
But before you condemn,
Remember—*pro tem*—
You also are one of the cast.

from the *Penguin Book of Limericks*

Both Koselleck and Foucault present post-Enlightenment historicism as a discourse in which we still participate. But their uses of that genealogy could hardly be more different. Koselleck is friendly toward the new historicism precisely for its reconciliation of partisanship with objectivity, evidently supposing, with Friedrich Schlegel, that the objectivity appropriate to history is obtainable through reflection on one's own historically conditioned standpoint.[25] And he welcomes what he sees as the new historicity's proper balance of individual freedom and historical determination, quoting with approval Marx's famous line: "Men make their own history, but they do not do so freely, not under conditions of their own choosing, but rather under circumstances which directly confront them, and which are historically given and transmitted."[26] Foucault, in stark contrast, views the new historicity with implacable hostility, for it inaugurated human subjectivity as a basic category of history, just as it inaugurated life as the defining category of biology. And Foucault, notoriously, looked forward to the day when human subjectivity would be erased from all disciplines. Doubtless he would have regarded Koselleck's endorsement of human responsibilities in history and of historians as distressing symptoms of, rather than critical comments on, post-Enlightenment historicism. However, apart from exposing the ways in which the new human subjec-

25. F. Schlegel, "Ueber die neuere Geschichte," 125–45; cited in Koselleck, *Futures Past*, 147.
26. Koselleck, *Futures Past*, 208.

tivity laid human history open to invasion by psychology, sociology, and linguistics, Foucault offers no specific critique of modernist history, contenting himself with lurid presentiments of its imminent demise in the bonfire of the human sciences.[27]

Foucault's grounds for dismissal of post-Enlightenment historicism are, I believe, entirely misguided. Consider the grand schemes of nineteenth-century latent history—of *Geist,* of reason, of *Kultur*/culture, of *Bildung,* of the life of humanity, of the will to power—the schemes that reconcile historical individuality with narrative unity, historically conditioned partiality with scientific impartiality, the many historical times with the one all-consuming time of History. It is arguable that few of these grand schemes do in fact privilege human subjectivity. But even if one accepts with Foucault and Heidegger that such systems, by virtue of their claim to represent what is objective, invoke an illegitimate subject—object divide, that is surely not what rules them out for us.[28] Rather, they are unavailable to us because the very question—What manifests itself in history?—has lost its sense. There is for us literally *no thing* whose story can unite the many local and contingent histories into a single unified narrative—no big *Ich* behind all the little *ich*s, no *Geist* behind the diversity of *mentalités,* no master Will behind the many little wills, no global Culture behind the multiplicity of local cultures.

What a contrast with the history of living beings! Whatever our doubts about the neo-Darwinian synthesis, we can no longer conceive the *absence* of some such grand history to unify the countless local episodes of life on Earth. Indeed, to the extent that a universal history of mankind does remain thinkable for us, it is precisely through a reintegration of human history into natural history: the climatic-geographical history of Thomas Buckle, the degenerationist tragedy of Spengler, the geohistory of Fernand Braudel, and so on. Of course, most of us reject these "scientific" grand narratives as well. But we reject them as being contingently false—as drawing false analogies between organisms and human collectivities; as involving a false determinism of human action, and so on—not as fantasies beyond the bounds of sense. (In making such claims about "our" attitudes I feel obscurely uneasy. To salve my conscience let me quote a wonderful anecdote from Didier Eribon's *Michel Foucault:* "Godard even said in an interview that it was against people like 'the Reverend Father Foucault' that he wanted to make films. 'If I don't particularly like Foucault, it is because of his saying, "At such and such a period they thought . . . " That's fine with me, but how can we be so sure? That is exactly

27. Foucault , *Order of Things,* 10, 367–73.
28. Foucault, "Man and His Doubles," in ibid.; Heidegger, "Age of the World System," in *Question Concerning Technology,* 115–54.

why we try to make films; to prevent future Foucaults from presumptuously saying things like that.'")[29]

In a fundamental respect the problematic of history now is like that ascribed by Peter Hanns Reill to the end of the eighteenth century. We simply cannot see how it is possible to achieve unified and objective historical accounts while doing full justice to the locality and individuality of all human affairs, including the writing of history. Indeed, in one absolutely crucial respect the collapse of post-Enlightenment historicism has left us worse off than the late Enlightenment historians. For our historical narratives are choked with the residues of historicism and Romanticism, with dead organic metaphors—of growth and decay, of proliferation and dissemination, of flowerings and witherings, of developments and evolutions.[30] "Culture" itself is such a dead metaphor: as Raymond Williams remarks, "[Culture] had meant, primarily, the 'tendency of natural growth,' and then, by analogy, a process of human training. But this latter use, which had usually been a culture *of* something, was changed . . . to *culture* as such, a thing in itself."[31] So, alas, it seems that the very cultural history through which we currently try to escape the grand narratives of development is rooted in a suspect organicism. We can no longer conceive the historian's task as, in Michelet's words, "the resuscitation of the original life in the innermost parts of the organism"—yet we simply do not know how to write off such vitalistic vocabulary.[32] It is, dare I say it, ineradicable.

Was Foucault actually right, even if for the wrong reasons? Is there really nothing we can salvage from the new historicism with which Koselleck and Foucault end the Enlightenment?

Novelty and the Disciplines

Just as in its progressive formation the organic life of mankind passes through all the stages of organic life generally, so science, as product of the development of the spiritual force of knowledge in mankind, presents a like sequence in its formation.

D. G. Kieser, "Eröffnungsrede"

29. Cited in Eribon, *Michel Foucault,* 156.
30. The roles of organic imagery in Romantic writing and theory are explored in the classic Abrams, *The Mirror and the Lamp.*
31. Williams, *Culture and Society,* xvii; for fuller accounts, see Kroeber and Kluckhohn, *Culture,* and Williams, "Culture," in *Keywords.* The trajectory of the French *culture* is parallel. German *Cultur-Kultur,* an eighteenth-century derivation from the French, was similarly hypostatized: see Wittram, *Das Interesse an der Geschichte,* 40–43.
32. Michelet, as cited in Langlois and Seignobos, *Introduction to the Study of History,* 297; for entrancing reflections on Michelet's vitalism, see Barthes, *Michelet,* 27–38, especially "Goethe-as-Dog" and "History-as-Plant."

The component of the new post-Enlightenment historicism that survives the debacle of grand philosophies of history, and that we can hope to build on for our own purposes, has to do with the productivity of historical time, which generates real novelties rather than, as Foucault puts it, "merely providing a means of traversing the discretely predetermined table of possible variations."[33] Note that Koselleck and Foucault cannot but endorse this aspect of the new historicism, for it is instantiated in their very claims about the appearance of the new historicism itself. Foucault, usually so quick on the reflective turn, is oddly silent on this score. But Koselleck shows himself well aware of it, going to considerable lengths to create an ideologically acceptable historical setting. Thus he locates the new awareness of the unprecedented and unpredictable in history both as a response to the shock of the French Revolution, and as an aspect of secularization, of the opening of the future as providential models of human history lost their hold.[34]

Romantic "inner" histories of the disciplines show especially clear-cut recognition of radical historical innovation.[35] A fine example of such an inner disciplinary history (and, incidentally, one that contains a striking prefiguration of Foucault's end-of-Enlightenment thesis) is *Beurtheilung aller Systeme in der Zoologie nach ihrer Entwiklungsfolge von Aristoteles bis auf die gegenwärtige Zeit,* published in 1811 by the conservator of the zoology cabinets at the Royal Bavarian Academy of Sciences, Johann Baptist von Spix, best known for his essay in idealist morphology, *Cephalogenesis* (1815).[36] As specimens of the enlightened disciplinary history against which Spix reacts I have chosen two works on the history of natural history: the "Préface Istorike sur l'état ancien & actuel de la Botanike," with which Michel Adanson opens his *Familles des plantes* (1763); and the "Introductory Discourse on the Rise and Progress of Natural History," with which James Edward Smith opens the first, 1791, volume of the transactions of the new Linnean Society of London.[37]

These works are superficially comparable: each covers the entire span of natural history from antiquity to the present; each is specifically concerned with the development of systematic orderings of living beings; and each openly

33. Foucault , *Order of Things,* 275.
34. Koselleck, "Space of Experience" and "Horizon of Expectation," in *Futures Past,* 267–80.
35. The following remarks owe much to accounts of the thematic and narrative forms of Romantic disciplinary history by Lucien Braun and Dietrich von Engelhardt. See Braun, "La vision romantique de l'histoire de la philosophie," chap. 6 in *Histoire de l'histoire de la philosophie;* and Engelhardt, "Romantische Naturforschung," in *Historisches Bewusstsein,* 103–57.
36. Spix, *Beurtheilung aller Systeme.* On J. B. von Spix, see the entry by F. Ratsel in *Allgemeine deutsche Biographie* (Leipzig, 1893), 53:231–32.
37. Adanson, *Familles des plantes* (note how Adanson, concerned to sweep away all artificiality, uses phonetic spelling); Smith, "Introductory Discourse" (on J. E. Smith and the Linnean Society, see Gage, *History of the Linnean Society*).

uses history to legitimate its author's views about the basis and structure of an ideal system. At the levels of narrative forms and metaphysical presuppositions, however, Spix's epic history of zoology diverges *toto coelo* from Adanson's and Smith's chronicles of natural historical progress and enlightenment. Adanson declares his aim thus:

> My object is to trace the plan of all the works of botany that have been, with a view of giving a general knowledge of this science, both in developing the foundations, and in establishing systematic methods; and of those who have treated some part of it with success, and in a manner which could serve as a model, to compare their systems and thereby to put the public in a position to judge for itself their various degrees of goodness, certitude, utility and facility.[38]

He proceeds to describe all the known orderings of the plant kingdom from Aristotle to his own day, providing a chart in which all the various systems are ranked according to the number of natural groupings incorporated in them.[39] The great Joseph Pitton de Tournefort, whose work Adanson proposes to emulate, comes out top.

Adanson views progress in botany as the product of extensive observation, meticulous descriptions, and sound methods. It is aided by the "insight and profound erudition" of such men as Tournefort, but frustrated by the vanity and jealousy of botanists, by national prejudice, and above all, by the pursuit of artificial methods.[40] Smith's "Introductory Discourse" is more cunning and sophisticated. Forswearing conjecture, he speculates in the manner of the Scottish conjectural historians on the origins of natural history in the state of nature; and with frequent declarations of the importance of impartiality in botany, he denounces French botany and praises English contributions.[41] Smith shows far more appreciation than Adanson of the changes that have occurred, of the gradual liberation of natural history from medicine, of the impact of physiological discoveries on the disciplines. And his account of the causes of stasis and progress is more elaborate. To Adanson's envy and national prejudice as obstacles, Smith adds war and the "*ignis fatuus* of theory." To Adanson's sound method and meticulous observation (here represented by Linnaeus) as causes of progress, Smith adds enlightened patronage, influx of specimens through commerce and expeditions, and laudable emulation excited by read-

38. Adanson, *Familles des plantes,* iii–iv.
39. Ibid., lxxxix–cii.
40. On Tournefort's virtues, ibid., xxx, xcix, cii, and passim; on "causes qui ont areté le progrès de la Botanike," cxlix and following.
41. Smith, "Introductory Discourse," 1–5, 31ff., 50ff.

ing histories such as his own.[42] But the underlying structures of the two works are close, and closely in accord with Foucault's classical episteme. In both, it is precisely a universal *taxinomia* that figures as the ultimate goal of natural history—a goal that Adanson promises to realize later in his book, and that Smith sees as attainable through extended application of Linnaeus's methods. And this taxonomic structure repeats itself, "is doubled," Foucault would say, in their histories, for both offer taxonomies of botanists and their systems, manifestly so in Adanson's case, thinly veiled by urbane oratory and irony in the case of Smith.

When we turn to Spix, we find ourselves in utterly different territory. After versifying on the "teeming crowds" and "thousands of budding blossoms" that met his eyes when he surveyed earlier zoological writings, Spix sounds off with a fine holistic and *naturphilosophisch* fanfare:

> I soon became aware that in the history of zoology as in the realm of animals themselves, there rules not chance and arbitrariness, but lawful necessity, and that as in time generally, so in the developmental sequence of previous systems no element from its structure can be removed. Likewise, I realized that the appearance of the individual systems precisely depends on the contemporary changes in the world, and that therefore the history of zoology must as far as possible be presented as rooted in the general history of the world. With pleasure I grasped this thread, which in this labyrinth offered the nature of the thing itself, and thus followed from Aristotle to Linnaeus and Cuvier the growth of a tree, which, rooting in antiquity, united itself into a stem and burgeoned in many green branches striving toward a single blossom.[43]

For Spix the history of zoology is a fragment of world history, "the ocean in which *Weltweisheit* mirrors itself." Accordingly, Spix ties the epochs of zoology in with epochs of cultural history—the mythical Arcadian age, the ages of Greece and Rome, the Gothic and Arabic ages, the Reformation. The narrative is by no means a straightforward tale of progress. In Greek antiquity all the germs *(Keime)* of natural history were there, but inarticulate and *gestaltlos*. Thus in Aristotle's works, presentiments of the perfect system appeared as he dealt with the souls, the inner structures, and the outer appearances of animals. Subsequent ages were marked by the conflicts and resolutions of two basic drives, one toward enumeration and description of the outward and

---

42. Smith, "Introductory Discourse," 5 (*ignis fatuus*), 20 (war), passim (prejudice), 7ff. (observation), 20ff. (patronage), 40ff. (travel), 49 (emulation).

43. Spix, *Beurtheilung aller Systeme*, vii–viii.

visible, one toward explorations of inwardness. Roman natural history, represented by Pliny, manifested the outward drive in extreme form; then, in the Gothic era, Christianity turned men's thoughts inward, away from the visible world. Natural history was only apparently frustrated by this, for at the Reformation the inward and outward drives were reconciled under the stimulus of European discovery and commerce, to rejuvenate zoology at the hands of Gessner in Germany (which for Spix evidently includes Switzerland), Wotton in England, and Aldrovandi in Italy. The next great synthesis in natural history came with the rise of comparative anatomy, culminating in the work of Cuvier, "who brings order to the chaos of rhapsodic work in anatomy." Cuvier provides Spix with another of his threads: "a thread, thin yet complete, to help us through the darkness and labyrinth of the hidden organs of animal bodies." (Foucault *avant la lettre!*) A second *Verinnerlichung* will give perfect form to the rudiments found in Aristotle. To yield full knowledge of the animal within and without, a *Seelenlehre*, a natural history of the animal soul, is needed; and thus the perfect natural system will be revealed. This is the task Spix sets for himself.[44]

In Spix's history, as in the many other Romantic histories of natural philosophy, natural history, and medicine, the Enlightenment model of progress and civilization through adherence to rational methods is decisively rejected. There is a progress of sorts as successive ages and cultures return to the origins, the primordial *Keime* of natural history, reflecting on and articulating them in ever greater depth. But this progress is local and discontinuous. Superstition and partisanship, opponents of Enlightenment, now appear at the pole of inner feeling in its dialectic with outer reason, the dialectic that underlies the growth of all cultures and disciplines. And the development of the discipline is conceived, not as the ever more effective pursuit of assigned goals, but rather as a creative and genial process in which entirely new fields of inquiry—the study of inner structure and function at the hands of Vicq d'Azyr and Cuvier, the history of the animal soul at the hands of Spix himself—become possible at particular times and places.

It is indeed startling to see in what unmistakably Foucaultian terms the *naturphilosophischen* historians describe the genesis of their own age. For Spix, the shift of gaze from the outside to the innards of animals, which culminates in Cuvier's work, marks a "wholly new direction in natural history." For Schelling, Carl Friedrich Kielmeyer, Cuvier's teacher, has inaugurated "a wholly new epoch of natural history," in which the whole system of living beings can be derived from the dialectical play of inner forces. And Georg Kieser celebrates

44. Ibid., 4, 14–28, 28–57, 57–72, 130–31, 134–35, 147.

the new *Naturphilosophie,* which has brought to light "the dependence of the diversity of form in the realms of the plant and animal world on each other and on inner forces."[45]

At the most fundamental level, appearances of new disciplines and new disciplinary horizons are seen by the Romantic historians as manifestations or "concretizations" of the spirit or life of the cosmos. This, of course, belongs to that aspect of the new historicism that is no longer an option for us. However, there is a wonderful variety in the agencies through which *Geist* contrives to expand the horizons of human inquirers.

Genius is frequently the immediate source of innovation. Alongside the traditional *topoi* of genius as aping of divine creation *(Dei creationis simia),* or as exemplary performance, the Romantics invoked other types: the original genius, innovating by giving voice to, "concretizing," the primordial ideas on which all human culture rests; genius as spontaneous feeling overflowing the bounds of reason or as titanic force breaking the shackles of conventions; the vegetative genius, in intuitive, even unconscious communion with nature; genius as a formative drive like the vital force that, on epigenetic accounts of organic generation, creates structured adults from undifferentiated germs; or, most common of all, as in Spix's account, the *genius loci,* or representative genius, embodying, often unwittingly, the best tendencies of a place, a culture, an age.[46]

Reflection on national and popular tales, traditions, and practices is seen as another major source of insights beyond the bounds of current knowledge, for the seeds of all philosophy of nature are supposedly latent in folk wisdom and the myths of the ancients.[47] Accordingly we find, especially in the monumental Romantic historians of medicine, Windischmann, Leupoldt, Damerow, Isensee, and Quitzmann, an extraordinary, often obsessively Paracelso-centric, emphasis on folk medicine and medical practice, rather than theory, as sources of innovation and renovation.[48]

45. Ibid., 123ff.; Schelling, *Sämmtliche Werke,* 2:565; Kieser, "Eröffnungsrede," 45.

46. On Romantic conceptions of genius, see Abrams, "The Psychology of Literary Innovation: Unconscious Genius and Organic Growth," chap. 8 in *The Mirror and the Lamp;* Schmidt, *Die Geschichte des Genie-Gedankens.* On genius and epigenesis see Müller-Sievers, *Self-Generation,* 63ff. Woodmansee, *Author, Art, and the Market,* chap. 2, effectively links the construction of genius as exemplary originality with the campaign for adequate recognition in copyright laws of authors' ownership of the products of their labors.

47. Johann Salomo Schweigger, for example, held that the latest discoveries concerning electricity and polarity had been known to the most ancient sages, and that fragments of this primordial wisdom were preserved in mythological form: see Schweigger, *Ueber die älteste Physik.* On Schweigger see Snelders, "J. S. C. Schweigger."

48. The basic work on Romantic medical historiography remains Seemen, *Zur Kenntnis der Medizinhistorie;* see also Risse, "Historicism in medical history"; and Engelhardt, "Historisches Bewusstsein."

Experimentation is another fundamental form of innovation. Romantic *Naturphilosophie* is still often portrayed as armchair speculation, contemptuous of observation and experiment. Quite to the contrary, as Brigitte Lohff and Michael Dettelbach have shown in masterly recent studies, the *Naturphiloso-phen* and their allies were opposed not to experiment per se, but to analytical experimentation in the French manner; and they were active in their promotion of new styles of experimentation that would bring men into more direct contact with nature through their aesthetic and synthetic faculties.[49] Alexander von Humboldt and, following him, Johann Wilhelm Ritter, reported Frankensteinian galvanic experiments on their own bodies, designed to put them directly in touch with the secret powers of nature, to short-circuit, so to speak, the processes of analysis and ratiocination.[50] In the histories of physics and chemistry composed by Ritter, Oersted, and others, experiment is, indeed, presented as a primary means of access to new scenes of inquiry.[51]

Last, but not least, for the Romantic historians the study of the history of the disciplines yields glimpses beyond the current bounds of sense. Thus Spix derives his presentiments of the perfect natural system from his history of all previous articulations of the primordial *Keime* of natural history intuited by the ancients. Thus Oersted is able to claim of his history of chemistry that "the greatest possible independence from the limitations of the age will surely be the result of this work."[52] In using disciplinary history to achieve dramatic glimpses of "new forms of thought," "new positivities," even "a new culture" beyond our current episteme, Foucault operates in a strictly Romantic manner.[53]

Conclusion

To summarize: I started by using German *Naturphilosophie* to challenge the claims of Koselleck and Foucault about the new historicism that marks the end of the Enlightenment. I then argued that the core of their claims survives, in-

49. Lohff, *Die Suche nach Wissenschaftlichkeit*, and Dettelbach, *Romanticism and Administra-tion;* see also Trumpler, *Questioning Nature*, and Henderson, "Novalis's Idea of 'Experimental-philosophie.'"

50. Humboldt, *Versuche über gereizte Muskel- und Nervenfaser*. On Humboldt's experiments, see Dettelbach, *Romanticism and Administration*, chap. 2; Ritter, *Physisch-chemische Abhandlungen* (vol. 1 contains his lectures on galvanism); Poppe, *Johann Wilhelm Ritter*. An informative, if hostile, account of Ritter's autoexperimental studies of the effects of electrical stimulations of the sense organs is in Du Bois-Reymond, *Untersuchungen über thierische Electricität*, 340–58; see also Wetzels, *Johann Wilhelm Ritter*, 97–107, and Trumpler, *Questioning Nature*, chap. 5.

51. See, e.g., Ritter, *Die Physik als Kunst;* Oersted, *Der Geist in der Natur*.

52. Spix, *Beurtheilung aller Systeme*, 145ff.; Oersted, *Der Geist in der Natur*, 172.

53. Foucault, *Order of Things*, 342, 372–73, 384–87. For other indications of Foucault's un-confessed Romanticism, see Habermas, "Critique of Reason."

deed is strengthened by, the challenge. Part of this new historicism, that which postulates inner spiritual histories underlying the diversities and contingencies of human history, is a legacy we cannot but refuse, if indeed we have not already long ago done so. But the other part, that which concerns radical innovations, the appearance in history of utterly new possibilities, remains with us. I concluded with some reflections on this theme of novelty in Romantic disciplinary histories, indicating the richness of their accounts of the formation of new scenes of inquiry and new agendas.

With Foucault I believe this issue of novelty to be the central issue of all disciplinary history.[54] As readers of *1066 and All That* will know, the world divides into Cavaliers, who are Wrong but Romantic, and Roundheads, who are Right but Repulsive. At least in this respect, in his archaeology of disciplines Foucault achieved the impossible feat of being Right *and* Romantic. But true novelty—unfated, unforced, locally situated, genuinely contingent novelty— is deeply problematic for us. It is one thing to acknowledge that disciplinary history is shot through and through with novelty, contingency, and locality, quite another to write in accordance with that recognition.

Thomas De Quincey, elaborating on Ehregott Andreas Christoph Wasianski's memoir of Kant's old age, tells how a couple of years before his death Kant dismissed Lampe, his man-servant for forty years, for "great irregularities and habitual neglects." Shortly afterward, De Quincey reports (adopting the persona of Wasianski): "I met with what struck me as an affecting instance of Kant's yearning after his old good-for-nothing servant in his memorandum-book: other people record what they wish to remember; but Kant had here recorded what he was to forget. 'Mem.—February 1802, the name of Lampe must now be remembered no more.'"[55] For "Lampe," a discerning eye may read "metaphysics." Faced with radical novelty and brute contingency in history, we become nostalgic for the grand metaphysical philosophies of his-

54. As I argue in detail in Jardine, *Scenes of Inquiry.*
55. *Collected Writings of De Quincey,* 4:355–56; Wasianski's own version reads, "Ein sonderbares Phänomen von Kants Schwäche war folgendes. Gewöhnlich schreibt man sich auf, was man nicht vergessen will; aber Kant schreibt in sein Büchelchen: der Name Lampe muss nun völlig vergessen werden" (Wasianski, *Immanuel Kant,* 326).
There is precedent for this in the Holy Writ. "Remember what Amalek did to you on your journey, after you left Egypt—how, undeterred by fear of God, he surprised you on the march, when you were famished and weary, and cut down all the stragglers in your rear. Therefore, when the Lord your God grants you safety from all your enemies around you, in the land that the Lord your God is giving you as a hereditary portion, you shall blot out the memory of Amalek from under heaven. Do not forget!" (Deuteronomy 25:17–19). Peter Lipton informs me that when Jewish scribes need to warm up their quills at the start of a session, they have the problem that whatever they write might have sacred status and so should not be scribbled on the odd piece of parchment. The traditional solution is that they write the name of Amalek and then immediately blot it out, thus warming up and performing a good deed at the same time.

tory, whether Enlightened or Romantic, that once promised to control and unify the blooming buzzing confusion. And even as we contest the metaphysical categories of grand narrative histories—progress, Enlightenment, culture, what "we" believe (remember Godard's indictment of Foucault)—we give them new currency.

Primary Sources

Adanson, Michel. *Familles des plantes.* 1763. Facsimile with an introduction by F. A. Stafleu, edited by J. Cramer and H. K. Swann. Codicote, 1966.

Butler, Samuel. *Characters and Passages from Notebooks.* Edited by A. R. Waller. Cambridge, 1908.

Carus, C. G. "Grundzüge allgemeiner Naturbetrachtung." In *Zur Morphologie,* edited by Goethe, 2:84–95. N.p., 1823.

Comenius, Jan. *Das Labyrinth der Welt und das Paradies des Herzens.* 1623. Reprint, Lucerne, 1970.

Cuvier, Georges. *Rapport historique sur les progrès des sciences naturelles depuis 1789, et sur leur état actuel.* Paris, 1809.

De Quincey, Thomas. *The Collected Writings of Thomas De Quincey.* New and enlarged edition by D. Masson. Edinburgh, 1890.

Du Bois-Reymond, E. H. *Untersuchungen über thierische Electricität.* Vol. 1. Berlin, 1848.

Humboldt, Alexander von. *Versuche über gereizte Muskel- und Nervenfaser, nebst Vermuthungen über den chemischen Prozess des Lebens in der Thier- und Pflanzenwelt.* 2 vols. Berlin and Posen, 1797–98.

Kielmeyer, Carl Friedrich von. *Ueber die Verhaltnisse der organischen Kräfte.* Stuttgart, 1793.

Kieser, D. G. "Eröffnungsrede." In *Amtlicher Bericht über die Versammlung deutscher Naturforscher und Ärzte zu Jena im September 1836,* 42–47. Weimar, 1837.

Möser, Justus. *Osnabrückische Geschichte.* Osnabrück, 1768.

Oersted, H. *Der Geist in der Natur.* In *Gesammelte Schriften,* translated by K. L. Kannegiesser, vol. 1. Leipzig, 1850.

Ritter, Johann Wilhelm. *Physisch-chemische Abhandlungen in chronologischen Ordnung.* 3 vols. Leipzig, 1806.

———. *Die Physik als Kunst: Ein Versuch, die Tendenz der Physik aus ihrer Geschichte zu deuten.* Munich, 1806.

———. *Fragmente aus dem Nachlass eines jungen Physikers: Ein Taschenbuch für Freunde der Natur.* 2 vols. Heidelberg, 1810.

Schelling, F. W. J. *Sämmtliche Werke.* Edited by K. F. A. von Schelling. Stuttgart, 1856–61.

Schlegel, Friedrich. "Ueber die neuere Geschichte: Erste Vorlesung" (1810–11). In *Kritische Friedrich—Schlegel—Ausgabe,* edited by E. Behler, vol. 7. Munich, 1966.

Schweigger, Johann Salomo. *Ueber die älteste Physik und den Ursprung des Heidenthums aus einer missverstandenen Naturweisheit.* Nuremberg, 1821.

Smith, James Edward. "Introductory Discourse on the Rise and Progress of Natural History." *Transactions of the Linnean Society* 1 (1791): 1–55.

Spix, Johann Baptist von. *Beurtheilung aller Systeme in der Zoologie nach ihrer Entwiklungsfolge von Aristoteles bis auf die gegenwärtige Zeit.* Nuremberg, 1811.

Steffens, Henrich. *Alt und Neu.* Breslau, 1821.

Wasianski, E. A. C. *Immanuel Kant in seinen letzten Lebensjahren* (1804). In H. Schwartz, *Immanuel Kant: Ein Lebensbild nach Darstellungen der Zeitgenossen.* 2d ed. Halle, 1907.

❧ *15* ❧

Afterword: The Ethos of Enlightenment

Lorraine Daston

Not "What is Enlightenment?" but "Who are the enlightened, and how did they get that way?"—that is the question raised most urgently by these essays about the enlightened sciences. The enlightened turn out to be a large and motley crew: Baltic Lutherans, French artillery engineers, English instrument makers and their barometer-mad clientele, sentimental novelists and their lachrymose readers, whole streetfuls of Parisian engravers of natural history books, Dutch patricians and country girls, monster mongers and automata acolytes, political arithmeticians who never mastered the infinitesimal calculus and *filosofesse* who did, and even (Hume help us) Prussian metaphysicians. What did Cristina Roccati, Newtonian prodigy of the Veneto, have in common with Johann Maelzel, impresario of the Turkish chess-player automaton? Or the enthusiastic spectators who attended Pieter Camper's anatomical demonstrations in Groningen with the Berlin academicians who wrangled over the physico-theological demonstrations of Christian Wolff and Leonhard Euler? Or the readers of the sentimental novel *David Simple* with the Baltic landlords who promoted mineralogy and challenged serfdom?

The most obvious answer is a dedication to utility. But as Lissa Roberts points out in the context of the Netherlands, "utility" was an elastic term, which in eighteenth-century usage expanded to cover a range of meanings. The crass utility of promoting power and profit was certainly among them: Lisbet Koerner notes that Baltic opponents of serfdom coolly argued in terms of agricultural profits as well as humanity. Andrea Rusnock seconds the Foucauldian thesis that the efforts of political arithmeticians to quantify populations promoted the "biopower" of states. Jan Golinski suggests that barometers probably contributed more to the fortunes of London instrument makers than to those of meteorology. Emma Spary shows how French naturalists exploited aristocratic patronage. Simon Schaffer discerns sinister implications in "a division between subjects that could be automated and those reserved for reason." But professions of public (as opposed to private) utility were rife among the enlightened, and at least some of the authors in this volume are cautiously inclined to take them at their word. Rusnock contends that the political arithmeticians "should not be summarily dismissed" when they claimed to act in the interest of "relieving misery and promoting happiness through the application of reasoned knowledge." Paula Findlen apparently sees no reason to doubt Roccati's pious commitment to educating the academicians of Rovigo in Newtonian physics, after her father's disgrace sequestered her from the wider world of learning. While Roberts asserts that the Dutch sometimes invoked utility "to justify or criticize practices aimed more directly (though not always consciously) at extending institutional, cultural, or socioeconomic power and control," she notes that the term also referred to efforts to educate and entertain the public.

The enlightened commitment to public utility permits of further subdivisions. On the one hand, there were material "improvements," which in these essays take the form of agricultural reforms, gathering statistics on salubrious and insalubrious locales, domesticating exotic commodities like tea and silk, the mathematization of practical ballistics, using barometers to predict the weather, and inventions like the Jaquard loom. Here suspicion returns, announced by a flurry of raised-eyebrow quotation marks: just how "public" was the "utility" of enclosing the commons and improving the accuracy of big guns? Who counted it an "improvement" to replace the skilled silk weavers of Lyons with an automatic loom that could allegedly be operated by "a horse, an ox or an ass"? It was cold comfort to the captive monsters in Peter I's collection that their presence (alive or stuffed) dispelled dark fears of demons. Eighteenth-century neo-Hippocratics, supported by statistics, proclaimed that fens and swamps were unhealthy, but draining them could be even more dangerous for the inhabitants, as Goethe's Faust discovered when he played improving engineer. Distrust of the enlightened—of their tendency to conflate their own narrow interests with the dictates of universal nature, and to substitute rationalization for rationality—is at least as old as the Romantics, and was famously elaborated by Marx. It is still alive and kicking in this volume. With only a few exceptions, the authors of these essays regard the material claims of the enlightened to serve the commonweal with a skepticism that ranges from mild to vehement.

On the other hand, public utility might refer to edification in the form of physico-theology, anthropomorphized natural history, public demonstrations of the wonders of anatomy and electricity, or even lectures on the intricacies of the new analysis. Under the spell of natural history, boorish French aristocrats came to prefer the delights of the garden to the cruder and crueler pleasures of the hunt; Dutch gentry and burghers "sought fellowship and entertainment" in anatomy theaters and public science lectures; northern Italian notables met regularly (not without some grumbling) to hear a woman explain Newtonian physics; aspiring engineers derived a new sense of self-worth and group identity by vigorously competing with one another in mathematics and technical drawing. In these examples, the devotion of the enlightened to public utility recalls the religious associations that lie at the root of the word "enlightenment." Originally, to be filled with light was to be filled with the light of God, a meaning still vivid in the Italian cognate of the enlightened, the *illuminati*. Within this etymological lineage, the enlightened and the enthusiasts ought to have been first cousins. This is why some seventeenth-century readers of Descartes interpreted his appeal to *lumière naturelle* as a form of enthusiasm, a personal claim to divine inspiration. But of course by the early eighteenth century, enlightenment and enthusiasm had become polar opposites: the alleged light of reason (wielded by the established ecclesiastical authorities in both Catholic and Protestant lands, not just by Deists) countered the alleged light of God (wielded by dissenting sects, including the Cévennes prophets, the Jansenists, the Quakers, and the Pietists). What exactly was the peculiar light of the enlightened, which now burned at the altar of public utility rather than that of private inspiration?

First and foremost, it was a sociable light. It was not the inner light of mystical vision, but rather the outer light of letters, lectures, treatises, memoirs, novels, journals, and conversations. This was not so much a distinction between the private and the public as between the intuitive and the discursive. Enthusiasm could be infectiously communicated to fellow believers, and its displays were notoriously public: "[T]he inspiring Disease imparts it self by insensible Transpiration," according to Shaftesbury.[1] Enlightenment could be and was sturdily resisted (as the enlightened never ceased to complain) by the vast majority of the population, and many of its most characteristic activities were carried out in the solitude and intimacy of one's own chamber. But whereas enthusiasm caught fire from a blinding, undeniable intuition that admitted neither elaboration nor rebuttal, enlightenment was kindled by argument, explanation, demonstration, and discussion within a network of interlocutors. Sociability need not imply amicability. Wolff and Euler were bitter adversaries, as William Clark shows, but the very fact of their adversity, governed by the rules of academic polemics, was the precondition for much of their writing, which developed in point-counterpoint to each other's positions. Metaphysics may have been the science of "'pure reason,' free from all sensual experience," but even Kant, isolated in remote Königsberg, required at least an imaginary debate with Hume to launch himself into the first *Kritik*. The reading and conversation circles described by Marina Frasca-Spada were sometimes face-to-face gatherings, considerably more cordial than academic confrontations in Berlin or Halle. But the structure, if not the affect, of these interactions was remarkably similar: views were developed, propounded, and

1. Anthony, earl of Shaftesbury, *A Letter Concerning Enthusiasm*, 69.

criticized within reactive contexts—in conversations, correspondences, disputes, and above all, reviews.

Much of Enlightenment intellectual life, including that of the sciences, resembled a great echo chamber, in which not a word was whispered but that multiple voices took it up, repeating and reshaping the sound. The commissions of the academies, the reviews of the literary journals, the gossip of the salons, the letters of the learned—the enlightened were endlessly chatting with one another. Of course almost no form of learning has lacked sociability of one kind or another: even Saint Jerome in his Bethlehem retreat kept up a correspondence of sorts; the Scholastic disputation honed wits in the thrust and parry of stylized argument; humanists cast their positions in elegant dialogues. What was distinctive about enlightened sociability was its ceaseless give-and-take, its ping-pong rhythm of brief exchanges: the repartee of the salons, the reviews and rebuttals of the journals, the memoirs and commission reports of the academies, the correspondence shuttling back and forth along the thickening web of trade routes, the public lectures and natural history treatises that at once formed and were formed by the tastes of their audiences. The net result was an intricate minuet of constant mutual adjustment. The adjustments were by no means always toward greater harmony; criticism was the moving spirit of Enlightenment chatter—allegorically portrayed in one late-seventeenth-century frontispiece as a "a venerable matron, grave, moderate," armed with compass, protractor, and other "exact measures," crowned with a classical laurel wreath but otherwise in contemporary dress, and wearing thick spectacles.[2] But even criticism implied responsiveness, just as sympathy could imply distance. In the most famous Enlightenment analysis of sympathy, Adam Smith described it as the irresistible impulse to close the gap between disparate states of soul: "[A]s nature teaches the spectators to assume the circumstances of the person principally concerned, so she teaches this last in some measure to assume those of the spectators. As they are continually considering what they themselves would feel, if they actually were the sufferers, so he is constantly led to imagine in what manner he would be affected if he was only one of the spectators of his own situation."[3]

The seesaw rhythm of such reciprocal responses suggests the movements of conversation. But if the conversation was prototypical of enlightened sociability, it was not typical in any numerical sense. The vast majority of the interactions documented in these essays were encounters across time and space, which occurred through books, journals, letters, and (in the case of scientific

2. Leti, *Critique historique*, vol. 1, frontispiece and 4r.
3. Smith, *Theory of Moral Sentiments*, 22.

commodities like barometers, automata, monsters, and public lectures) advertisements. Even the face-to-face encounters of academies and lectures were regulated by architecture and decorum: the steeply banked seats of the anatomy theaters, the seating arrangements of the academies, the nook-and-cranny cupboards and guidebook etiquette of the museums, the crowd-managing ceremonies of the spectacular demonstration. A 1727 treatise on museology gave minute instructions on everything from furniture (cabinets should be twice as high as they are wide, be painted white, and face southeast; stuffed birds go on the top shelf and crocodiles should be hung from the ceiling) to visitor comportment (have clean hands, bring a microscope, don't stare at anything too long, and follow the guide obediently).[4] These sites did not promote the fluid banter of the salon, though they allowed for more structured (and more hierarchical) exchanges. But more often, enlightened sociability imitated the allegro-paced patterns of talking in the andante media of reading and writing. For many locales without a critical mass of naturalists, political arithmeticians, barometricians, or Richardson readers to form a circle or academy, this was an imposed necessity. Ever more specialized tastes in learning made for ever more dispersed soul mates. But even residents of the metropolises of the Republic of Letters, like Paris or Amsterdam, addressed themselves to far-flung readers and correspondents. Considerations of the international book market may explain the former, but not the latter. The enlightened preferred intimacy at a distance. Clark remarks upon perhaps the most extreme expression of this peculiar combined craving for community and remoteness among the enlightened: "While belief in extraterrestrial life may still count as exceptional in the seventeenth century, it became the rule among intellectuals in the *Aufklärung*." Somewhere on Jupiter, or perhaps Saturn, sagacious naturalists awaited a letter from their earthling colleagues.

The enlightened, whether resident in London or Königsberg, Paris or St. Petersburg, if not on Saturn, imagined themselves as comrades in a common undertaking, the material and moral improvement of the human estate. But these essays suggest that they might be better understood as united by a common sensibility, that of being themselves improved. On the face of it, there seems little to connect the efforts of the political arithmeticians with the proprietors of automata, the Paris analysts with the Prussian metaphysicians. If, however, we instead focus on how the enlightened got that way, certain commonalities do emerge. First, the very notion of belonging to a loose international confederacy of the enlightened, which superseded identities grounded in confession, nationality, estate, profession, and even family, requires scrutiny.

4. Neickelius, *Museographia*, 421–24, 455–57.

Jesuits and Freemasons supply suggestive precedents, but there was nothing loose about the organization of these notoriously top-down, ritualized, and in principle all-male confraternities. The *res publica literaria* of early modern humanists, though it more closely resembled the open-textured networks of the enlightened in its structure and aims, had admitted members on the basis of certain professional accomplishments: the mastery of classical languages at a minimum, supplemented by a penchant for learned tourism and collections of antiquities and naturalia. What is novel about the enlightened is that they recognized one another by shared personae and values rather than by mastery of any particular subject matter or allegiance to any particular creed or corporate body.

Second, to be an enlightened student of nature (including human nature) implied an intense preoccupation with some domain of phenomena, which could be anything from the *moeurs* of insects to mortality rates to the mathematical analysis of the vibrating string to grain prices. What mattered was not so much the object as the intensity of the commitment. The elective affinities bound those who shared one or another commitment into a group that commanded strong and even supreme loyalty, so that resources of time, money, and labor were willingly spent in pursuit of an aim outsiders deemed trivial or even mad. These commitments had little to do with professionalization. The analysts at the Paris Académie des Sciences (and the metaphysicians at the Berlin Akademie der Wissenschaften whom they presumably disdained, on Mary Terrall's account) might have counted as professionals, as might artillery engineers and university professors like Samuel Soemmerring, Wolff, or Kant. But it is much more difficult to make such claims about the physicians who plied political arithmetic, the gentry who took barometer measurements twice a day, or the readers of Locke and Richardson. Yet all of these people belonged to ranks of the self-consciously enlightened, not only by the standards of the essays in this volume, but by the standards of the Enlightenment itself. Part of the basis for that identity was that some aspect of the natural or moral world had become, to use the Freudian term, cathected for them, to the point where they centered their identity on the object of passion.

What kind of person makes regular barometer readings or *Clarissa* the nucleus of a self? These essays engage more in intellectual and cultural than in social and economic history, so they furnish only a few scattered clues, and these do not all point in the same direction. Although leisure and means were often prerequisites for such pursuits, as in the case of Golinski's well-heeled barometer readers and Spary's aristocratic devotees of natural history, many of the enlightened appear to have come from less comfortable backgrounds. Koerner, Roberts, and Alder remark upon the modest origins of enlightened

Baltic scholars, Dutch members of popular scientific societies, and French engineers, respectively. Apparently, the enlightened were recruited from the ranks of the middling as well as the upper classes (insofar as even these vague rubrics can be applied to the very different societies that made up eighteenth-century Europe). This still would exclude the great bulk of the population, but it was a more broadly based recruitment than for any prior elite—in terms of religion, nationality, education, class, and gender. So there is little specificity concerning the identity of the enlightened to be extracted here. At least with respect to the standard categories of analysis as to how identities congeal and cohere, the enlightened seem to have been self-made.

But self-made according to some common recipes: the references to one or another kind of "discipline" scattered through these essays furnish some clues about how one became enlightened. Golinski for example remarks that "[t]he the significance of these activities [systematic barometric observations] lies more in the routinization of observational practices than in the achievement of a naturalistic understanding of the weather"; Alder describes the strict discipline of engineering education; Roberts suggests that both Cartesian and Newtonian systems as taught in Dutch universities served "at least partly in this pedagogical tradition of seeking an appropriate method of discipline." Of course there is discipline and there is discipline. Schaffer understands something rather more nasty by that word when he writes: "Soldiers, engineers, prisoners, and workers were subjected to rather literal forms of enlightened discipline. The social order actively engineered their mechanization, and thus those who did this engineering could represent this condition as natural." Between the often brutal, externally imposed discipline of the eighteenth-century atelier or barracks and the subtle, internalized discipline of rising at dawn over decades to take a barometer measurement yawns a chasm. But a bridge also spans the two extremes: one form of discipline could substitute for the other; both external and internal discipline aimed to secure social order by producing orderly individuals. Self-discipline could and was claimed to be a form of self-governance, which made the externally imposed discipline of force, whether by master or by monarch, superfluous. Hence the cultivation of self-discipline was part of the preoccupation of Enlightenment intellectuals with autonomy, in spite (or perhaps because) of their continued dependence on patrons and printers.

Self-discipline, however ubiquitous, still underspecifies the collective persona of the enlightened. Jesuits, Ramists, Calvinists had all preached and practiced various forms of bodily and intellectual discipline. These essays cry out for a more fine-grained analysis of exactly wherein enlightened discipline consisted—its distinctive bodily gestures, intellectual habits, and emotional reflexes—and of its implications for the sciences cultivated by means of those gestures, habits, and reflexes. How, for example, did the long drilling in technical drawing train the eye to take in a landscape at a glance, and to collapse automatically three-dimensional objects into two? What were the consequences of this aesthetic education for civil and military architecture, for the mapping of kingdoms, for the values of rationalized efficiency so typical of French engineers since the eighteenth century? What mental tics did the eighteenth-century practices of mathematical analysis foster (in contrast to the methods of both seventeenth- and nineteenth-century analysts), and how were these ways of decomposing mathematical problems translated into ways of solving administrative problems by high-level old regime bureaucrats?[5] How did novels and treatises on human nature school their readers in the exquisite arts of introspection, and the plumbing of covert motives behind overt conduct? How did the ubiquitous pursuit of natural theology train the intellect to match minute observations about everything from caterpillars to monsters to stars with assessments of optimalization: optimal match of form to function, optimal yield of effect to minimum expenditure of resources? Only when we have detailed answers to questions like these will we understand what it meant to be enlightened.

Who were the unenlightened? They were those whose lack of self-discipline necessitated, according to the enlightened, the imposition of external discipline. They were the ignorant and the barbarous; they were peasants, old women, and children—"lazy, dirty, and cunning," as the Baltic reformers moaned. They could also be prosperous gentlemen with large libraries, so long as they were credulous and suffered from a "love of the marvelous." They were hyper-imaginative *convulsionnaires* in either Jansenist enclaves or mesmeric tubs. They were enslaved to fear and superstition. They were the unimproved. To serve "public utility" meant to embark upon a program of improvement, first and foremost self-improvement. This was the core of the ethos of the Enlightenment. To be educated and public-spirited did not suffice; one also had to learn to think, feel, read, and write in certain ways. In the systematic study of nature during the eighteenth century, the most characteristic form of thinking, feeling, reading, and writing in an enlightened way was probably natural theology. On the basis of sheer prevalence, natural theology in all its myriad forms emerges as *the* enlightened science in these essays. But the implications of its predominance receive little systematic attention: historians of the Enlightenment sciences have yet to fill the historiographic gap between the Boyle lectures and the Bridgewater treatises.

5. Eric Brian has tried to draw these connections in his *La mesure de l'État*.

Although these essays do not undertake to fill this gap, they do point to the importance of the task. Michael Hagner claims that even in the limiting case of monsters, the challenge for eighteenth-century anatomists was not to naturalize the phenomena—this had already been accomplished more than a century earlier. Rather, the goal was one of "integration, incorporation, and domestication" within a natural order endowed not only with regularity but also "beauty and usefulness." Spary remarks that "enlightenment" and "nature" were prescriptive terms, "providing standards for behavior that extended vastly outside the realm of polite society, unlike manners or erudition." "Qui est," as Joseph de Maistre once notoriously snarled, "cette Dame, la nature?" How did nature become the bedrock of justification not only for family relationships and sexual conduct (her specialized realm of authority from roughly the thirteenth through the mid–seventeenth centuries in Latin Christendom), but for almost everything, from societies to sentiment, by the end of the eighteenth century? It was, so to speak, as natural for Lawrence Sterne's parson Yorick to appeal to the authority of nature in affairs of the heart as it was for the drafters of the "Droits de l'Homme" and the originators of the metric system—appeals that would have been as unintelligible to a fourteenth-century scholastic like Duns Scotus as they were to become to a nineteenth-century liberal like John Stuart Mill, albeit for different reasons. Duns Scotus would have reached directly to God; Mill would have insisted that human arrangements were freestanding. Neither would have granted the Enlightenment assumption that nature was the ultimate source of the good, the true, and the beautiful. This assumption represents an anthropological mutation of the first order. In natural theology, ubiquitous companion of the enlightened sciences, nature could be described in fastidious empirical detail, submitted to strict mathematical regularities, and nonetheless be pumped full of moral and aesthetic standards without once (or only once) having recourse to primary causes. Like the enlightened themselves, natural theology aimed at safe autonomy, independence without disorder.

Nicholas Jardine argues that "[t]here is for us literally *no thing* whose story can unite the many local and contingent histories into a single unified narrative." For the enlightened, nature was the principle that unified all narratives—the history of human society, as much as the history of the earth and stars, was a narrative about nature. Insofar as we still believe such all-encompassing narratives, in cosmology and evolutionary theory, they remain about nature. The enlightened sciences were central to the project of the Enlightenment not because of their achievements in theory and application, but because they claimed to reveal the true identity of nature, heroine of all narratives. It was here that the enlightened staked their highest claims to public utility. Their cultural authority derived chiefly from nature's authority, not from their ability to command, predict, or even explain nature. However triumphal a moment in the history of mechanics, Newton's *Principia* would have granted no global authority to nature or to natural philosophers in matters of morals and politics without Richard Bentley's Boyle lectures, the final queries to the *Opticks,* and the further diligent efforts of eighteenth-century physico-theologists to extract symbolic meaning and value from the inverse square law, planetary orbits, and the distribution of stars. The very quaintness of their accounts—"Tulip Theology," "Snail Theology," "Fire Theology," and so on—should compel closer reading by historians alert to the wierdness, the non-self-evidence of the Enlightenment. Who knows what darkness lurks in the hearts of the enlightened?

Primary Sources

Leti, G. *Critique historique, politique, morale, economique, & comique sur les lotteries.* 2 vols. Amsterdam: Theodore Boetman, 1697.

Neickelius, C. F. *Museographia; oder, Einleitung zum rechten Begriff und nützlicher Anlegung der Museorum oder Raritäten-Kammern.* Leipzig and Breslau: Michael Hubert, 1727.

Shaftesbury, Anthony, earl of. *A Letter Concerning Enthusiasm.* London: J. Morphew, 1708.

Smith, Adam. *The Theory of Moral Sentiments.* Edited by D. D. Raphael and A. L. Macfie. Oxford: Oxford University Press, 1976. First edition published 1759.

BIBLIOGRAPHY

Abrams, M. H. A. *The Mirror and the Lamp: Romantic Theory and the Critical Tradition.*
Oxford: Oxford University Press, 1957.

Adams, William Howard. *The French Garden, 1500–1800.* London: Scolar Press, 1979.

Adorno, Theodor. *Aesthetische Theorie.* Frankfurt am Main: Suhrkamp, 1970.

———. *The Stars Come Down to Earth.* Edited by Stephen Crook. London: Routledge, 1994.

The Age of Elegance: Paintings from the Rijksmuseum in Amsterdam, 1700–1800. Weesp: Uitgeverij Waanders, 1995.

Agnew, Jean-Christophe. *Worlds Apart: The Market and the Theater in Anglo-American Thought, 1550–1750.* Cambridge: Cambridge University Press, 1986.

Albury, W. R. "Halley and the Barometer." In *Standing on the Shoulders of Giants: A Longer View of Newton and Halley,* edited by Norman J. W. Thrower, 220–27. Berkeley and Los Angeles: University of California Press, 1990.

Alder, Ken. *Engineering the Revolution: Arms, Enlightenment, and the Making of Modern France, 1763–1815.* Princeton: Princeton University Press, 1997.

———. "Stepson of the Enlightenment: The Duc du Châtelet, the Colonel Who 'Caused' the French Revolution." *Eighteenth-Century Studies,* in press (1998).

Alfvén, Hannes. "Cosmology: Myth or Science?" In *Cosmology, History, and Theology,* edited by Wolfgang Yourgrau and Allen Breck, 1–14. New York: Plenum, 1977.

Allen, David E. *The Naturalist in Britain: A Social History.* Harmondsworth: Penguin Books, 1976.

Alma mater studiorum: La presenza femminile dal XVIII al XX secolo. Bologna: CLUEB, 1988.

Altick, Richard. *The Shows of London.* Cambridge: Harvard University Press, Belknap Press, 1978.

Altieri Biagi, Maria Luisa, and Bruno Basile, eds. *Scienziati del Settecento.* Milan and Naples: Riccardo Ricciardi, 1983.

Amies, Marion. "Amusing and Instructive Conversations: The Literary Genre and Its Relevance to Home Education." *History of Education* 14 (1985): 87–99.

Aner, Karl. *Die Theologie der Lessingzeit.* Halle: Niemeyer, 1929.

Annerstedt, Claes. *Upsala universitets historia.* 10 vols. Uppsala and Stockholm: various publishers, 1877–1914.

Apostolidès, Jean-Marie. *Le roi-machine: Spectacle et politique au temps de Louis XIV.* Paris: Minuit, 1981.

Artelt, Walter. "Die anatomisch-pathologischen Sammlungen Berlins im 18. Jahrhundert." *Klinische Wochenschrift* 15 (1936): 96–99.

Artz, Frederick B. *The Development of Technical Education in France, 1500–1850.* Cambridge: MIT Press, 1966.

Ashworth, William J. "System of Terror: Samuel Bentham, Accountability, and Dockyard Reform during the Napoelonic Wars." *Social History* 23 (1998): 63–79.

Atran, Scott, et al., eds. *Histoire du concept d'espèce dans les sciences de la vie.* Paris: Fondation Singer-Polignac, 1985.

Aumüller, Gerhard. "Zur Geschichte der Anatomischen Institute von Kassel und Mainz." *Medizinhistorisches Journal* 5 (1970): 59–80, 145–61, 268–88.

Azouvi, François. "Sens et fonction épistémologiques de la critique du magnétisme animal par les Académies." *Revue de l'Histoire des Sciences* 29 (1976): 123–42.

Baasner, Rainer. *Das Lob der Sternkunst: Astronomie in der Deuschten Aufklärung.* Abhandlungen der Akademie der Wissenschaften in Göttingen, Math.-physik. Kl., 3d ser., no. 40. Göttingen: Vandenhoeck und Ruprecht, 1987.

Babbage, Charles. *Passages from the Life of a Philosopher.* London: Longman, 1864.

Bäckman, Louise, and Åke Hult Krantz. *Studies in Lapp Shamanism.* Stockholm University's Studies in Comparative Religion, no. 16. Stockholm: Almqvist & Wiksell, 1978.

Baecque, Antoine de. *La caricature révolutionnaire.* Paris: Presses du C.N.R.S., 1988.

Baehr, Erhard, ed. *Was ist Aufklärung? Thesen und Definitionen.* Stuttgart: Reclam, 1974.

————. "Kant, Mendelssohn, and the Problem of Enlightenment from Above." *Eighteenth-Century Life* 8 (1982): 1–12.

Baier, Annette. *A Progress of Sentiments.* Cambridge: Harvard University Press, 1991.

Baker, Keith Michael. "Les débuts de Condorcet au secrétariat de l'Académie des Sciences, 1773–1776." *Revue de l'Histoire des Sciences et de Leurs Applications* 20 (1967): 229–80.

————. *Condorcet: From Natural Philosophy to Social Mathematics.* Chicago: University of Chicago Press, 1975.

————. "Scientism at the End of the Old Regime: Reflections on a Theme of Professor Charles Gillispie." *Minerva* 25 (1987): 21–34.

————. *Inventing the French Revolution: Essays on French Political Culture in the Eighteenth Century.* Cambridge: Cambridge University Press, 1990.

————. "Defining the Public Sphere in Eighteenth Century France." In *Habermas and the Public Sphere,* edited by Craig Calhoun, 181–211. Cambridge: MIT Press, 1992.

————, ed. *The Political Culture of the Old Regime.* Oxford: Pergamon Press, 1987.

Baldini, Ugo. "L'attività scientifica nel primo Settecento." In *Scienze e tecnica nella cultura e nella società dal Rinascimento a oggi,* 467–529. Vol. 3 of *Storia d'Italia: Annali.* Turin: Einaudi, 1980.

Balekjian, Wahi H. "The University of Dorpat (Tartu) and the Armenian National Awakening in the Nineteenth Century." In *National Movements in the Baltic Countries during the Nineteenth Century,* edited by Aleksander Loit, 327–36. Stockholm: Center for Baltic Studies, 1985.

Banfield, Edwin. *Barometer Makers and Retailers, 1660–1900.* Trowbridge, Wiltshire: Baros Books, 1991.

Barker, G. A. "David Simple: The Novel of Sensibility in Embryo." *Modern Language Studies* 12 (1982): 69–80.

Barker-Benfield, G. J. *The Culture of Sensibility: Sex and Society in Eighteenth-Century Britain.* Chicago: University of Chicago Press, 1992.

Barthes, Roland. *Sade, Fourier, Loyola.* Paris: Seuil, 1971.

————. *Michelet.* Translated by Richard Howard. Oxford: Basil Blackwell, 1987.

Bartlett, Roger. "J. G. Eisen: Minor Writings." *Journal of Baltic Studies* 21 (1990): 95–104.

————. "The Russian Nobility and the Baltic German Nobility in the Eighteenth Century." *Cahiers du Monde Russe et Soviétique* 34 (1993): 233–44.

Barton, H. Arnold. *Scandinavia and the Revolutionary Era, 1766–1815.* Minneapolis: University of Minnesota Press, 1986.

Battestin, M. C., and R. R. Battestin. *Henry Fielding: A Life.* London and New York: Routledge, 1989.

Battestin, M. C., and C. T. Probyn. *The Correspondence of Henry and Sarah Fielding.* Oxford: Clarendon Press, 1993.

Bäumler, Alfred. *Das Irrationalitätsproblem in der Ästhetik und Logik des 18. Jahrhundert bis zur Kritik der Urtheilskraft.* 2d ed. Darmstadt: Wissenschaftliche Buchgesellschaft, 1967. First edition published 1923.

Beaune, Jean-Claude. "The Classical Age of Automata." In *Fragments for a History of the Human Body,* edited by Michel Feher, Ramona Naddaff, and Nadia Tazi, 3: 431–80. New York: Zone Books, 1989.

————, et al., eds. *Buffon 88: Actes du Colloque international pour le bicentenaire de la mort de Buffon, Paris, Montbard, Dijon, 14–22 juin 1988.* Paris: Vrin; Lyon: Institut Interdisciplinaire d'Études Épistémologiques, 1992.

Bechler, Rosemary. "'Trial by what is contrary': Samuel Richardson and Christian Dialectic." In *Samuel Richardson: Passion and Prudence,* edited by Valerie Grosvenor Myer, 93–113. London: Vision Press, 1986.

Becker, Carl. *The Heavenly City of the Eighteenth-Century Philosophers.* 1932. Reprint, New Haven: Yale University Press, 1960.

Bedini, Silvio. "Automata in the History of Technology." *Technology and Culture* 5 (1964): 24–42.

Behrens, C. B. A. *Society, Government, and the Enlightenment: The Experiences of Eighteenth-Century France and Prussia.* New York: Harper and Row, 1985.

Beiser, Frederick. *Enlightenment, Revolution, and Romanticism: The Genesis of Modern German Political Thought, 1790–1800* Cambridge: Harvard University Press, 1992.

Bell, David A. *Lawyers and Citizens: The Making of Political Elite in Old Regime France.* New York: Oxford University Press, 1994.

Bender, John. "A New History of the Enlightenment?" *Eighteenth-Century Life* 16 (1992): 1–20.

Benedict, Barbara M. "The 'Curious Attitude' in Eighteenth-Century Britain: Observing and Owning." *Eighteenth-Century Life* 14 (1990): 59–98.

Benhamou, Reed. "From *Curiosité* to *Utilité*: The Automaton in Eighteenth-Century France," *Studies in Eighteenth-Century Culture* 17 (1987): 91–105.

Benjamin, Walter. *Illuminations.* Edited by Hannah Arendt. New York: Schocken, 1968.

Bennett, James A. *The Mathematical Science of Christopher Wren.* Cambridge: Cambridge University Press, 1982.

Benninghoven, Friedrich. *Immanuel Kant: Leben, Umwelt, Werk: Ausstellung . . . zur Wiederkehr von Kants Geburtstag.* Berlin-Dahlem: Geheimes Staatsarchiv, Preussischer Kulturbesitz, 1974.

Bensaude-Vincent, Bernadette. "A View of the Chemical Revolution through Contemporary Textbooks: Lavoisier, Fourcroy, and Chaptal." *British Journal for the History of Science* 23 (1990): 435–60.

———. "The Balance: Between Chemistry and Politics." In *The Chemical Revolution: Context and Practices,* edited by Lissa Roberts, 217–37. Special issue of *The Eighteenth Century: Theory and Interpretation* 33 (1992).

———. *Lavoisier: Mémoires d'une Révolution.* Paris: Flammarion, 1993.

Bensaude-Vincent, Bernadette, and Fernando Abbri, eds. *Lavoisier in European Perspective: Negotiating a New Language for Chemistry.* Canton, Mass.: Science History Publications, 1995.

Berg, W. van den. "Literary Sociability in the Netherlands, 1750–1840." In *The Dutch Republic in the Eighteenth Century: Decline, Enlightenment, and Revolution,* edited by Margaret C. Jacob and Wijnand Mijnhardt, 253–69. Ithaca: Cornell University Press, 1992.

Bergeron, Louis, and Guy Chaussinand-Nogaret. *Les "masses de granit": Cent mille notables du Premier Empire.* Paris: Editions de l'École des Hautes Études en Sciences Sociales, 1979.

Berkel, Klaas van, M. J. van Lieburg, and H. A. M. Snelders. *Spiegelbeeld der wetenschap: Het genootschap ter bevording van natuur-, genees- en heelkunde, 1790–1990.* Rotterdam: Erasmus Publishing, 1991.

Bernal, J. D. *Science and Industry in the Nineteenth Century.* 1953. Reprint, Bloomington: Indiana University Press, 1969.

Bernal, Martin. *Black Athena: The Afroasiatic Roots of Classical Civilization.* Vol. 1, *The Fabrication of Ancient Greece, 1785–1985.* New Brunswick: Rutgers University Press, 1987.

Bertaud, Jean-Paul. *The Army of the French Revolution: From Citizen-Soldiers to Instrument of Power.* Translated by R. R. Palmer. Princeton, NJ: Princeton University Press, 1988.

Bianco, Bruno. "Freiheit gegen Fatalismus: Zu Joachim Langes Kritik an Wolff." In *Kant als Herausforderung an die Gegenwart,* edited by Norbert Hinske, 111–55. Freiburg: Alber, 1989.

Bien, David. "Military Education in Eighteenth-Century France: Technical and Non-Technical Determinants." In *Science, Technology, and Warfare: Third Military History Symposium,* edited by Monte D. Wright and Lawrence J. Paszek. Washington, D.C.: General Publishing Office, 1969.

————. "La réaction aristocratique avant 1789: L'exemple de l'armée." *Annales: Economies, Sociétés, Civilisations* 29 (1974): 23–48, 505–34.

————. "The Army in the French Revolution: Reform, Reaction, and Revolution." *Past and Present* 85 (1979): 68–98.

Bierens de Haan, J. A. *De Hollandsche Maatschappij de Wetenschappen, 1752–1952.* Haarlem: Tjeek Willink, 1952.

Biermann, K. R. "J.-H. Lambert und die berliner Akademie der Wissenschaften." In *Colloque international et interdisciplinaire Jean-Henri Lambert,* 115–26. Université de Haute-Alsace, Centre National de la Recherche Scientifique. Paris: Ophrys, 1979.

Birtsch, G. "Die Berliner Mittwochsgesellschaft." In *Über den Prozess der Aufklärung in Deutschland im 18. Jahrhundert,* edited by H. E. Bödeker and Ulrich Herrmann, 94–112. Göttingen: Vandenhoek und Ruprecht, 1987.

Blanchard, Anne. *Les ingénieurs du "roy" de Louis XIV à Louis XVI.* Montpellier: Dehan, 1979.

Blumenberg, Hans. *The Legitimacy of the Modern Age.* 2d ed. Translated by Robert M. Wallace. Cambridge: MIT Press, 1983.

Bödeker, Hans Erich, Georg Iggers, Jonathan Knudsen, and Peter Reill, eds. *Aufklärung und Geschichte: Studien zur deutschen Geschichtswissenschaft im 18. Jahrhundert.* Veröffentlichungen des Max-Planck-Instituts für Geschichte, no. 81. Göttingen: Vandenhoeck und Ruprecht, 1986.

Bodinier, Gilbert. *Les officiers de l'armée royale combattants de la Guerre d'Indépendance des États-Unis de Yorktown à l'an II.* Vincennes: Service Historique de l'Armée de Terre, 1983.

Bogdan, Robert. *Freak Show: Presenting Human Oddities for Amusement and Profit.* Chicago: University of Chicago Press, 1988.

Bogucka, Maria. "Illness and Death in a Maritime City: Gdansk in the Seventeenth Century." *American Neptune* 51 (1991): 91–104.

Bornhack, Conrad. *Geschichte der preussischen Universitätsverwaltung bis 1810.* Berlin: Reimer, 1900.

Bos, H. J. M. "Mathematics and Rational Mechanics." In *The Ferment of Knowledge: Studies in the Historiography of Eighteenth-Century Science,* edited by George Rousseau and Roy Porter, 327–55. Cambridge: Cambridge University Press, 1980.

Bots, J. *Tussen Descartes en Darwin: Geloof en Natuurwetenschap in de achttiende eeuw in Nederland.* Assen: van Gorcum, 1971

Boudri, J. Christiaan. *Het mechanische van de mechanica: Het krachtbegrip tussen metafysica en mechanica van Newton tot Lagrange.* Delft: Eburon, 1994.

Bourde, André. *Agronomie et agronomes en France au XVIII siècle.* 3 vols. Paris: S.E.V.P.E.N., 1967

Bourguet, Marie-Noëlle. *Déchiffrer la France: La statistique départementale à l'époque napoléonienne.* Paris: Éditions des Archives Contemporaines, 1988.

Box, M. A. *The Suasive Art of David Hume.* Princeton: Princeton University Press, 1990.

Bradley, Margaret. "Scientific Education versus Military Training: The Influence of Napoleon Bonaparte on the École Polytechnique." *Annals of Science* 32 (1975): 415–49.

Brandt, Otto. *Caspar von Saldern und die nordeuropäische Politik in Zeitalter Katharinas II.* Erlangen: von Palm und Ekke, 1932.

Braudy, Leo. *Narrative Form in History and Fiction: Hume, Fielding, and Gibbon.* Princeton: Princeton University Press, 1970.

————. "Fanny Hill and Materialism." *Eighteenth-Century Studies* 4 (1970): 21–40.

Braun, Lucien. *Histoire de l'histoire de la philosophie.* Paris: Ophrys, 1973.

Brewer, John. "Commercialization and Politics." In *The Birth of a Consumer Society: The Commercialization of Eighteenth-Century England,* edited by Neil McKendrick, John Brewer, and J. H. Plumb, 197–262. Bloomington: Indiana University Press, 1982.

————. *The Sinews of Power: War, Money, and the English State, 1688–1783.* New York: Knopf, 1988.

————. "The Eighteenth-Century British State." In *An Imperial State at War,* edited by Lawrence Stone, 52–71. London: Routledge, 1994.

Brewer, John, and Roy Porter, eds. *Consumption and the World of Goods.* London: Routledge, 1992.

Brian, Eric. *La mesure de l'état: Administrateurs et géomètres au XVIIIe siècle*. Paris: Albin Michel, 1994.

Brissenden, Robert Francis. Introduction to *Clarissa: Preface, Hints of Prefaces, and Postscript*, by Samuel Richardson, i–xi. The Augustan Reprint Society, publication no. 103. Los Angeles: University of California, William Andrews Clark Memorial Library, 1964.

———. *Virtue in Distress: Studies in the Novel of Sentiment from Richardson to De Sade*. London: Macmillan, 1974.

Britten, F. J. *Old Clocks and Watches and Their Makers*. London: Spon, 1932.

Brittnacher, Hans Richard. "Der böse Blick des Physiognomen: Lavaters Ästhetik der Deformation." In *Der "falsche" Körper: Beiträge zu einer Geschichte der Monstrositäten*, edited by Michael Hagner, 127–46. Göttingen: Wallstein, 1995.

Broberg, Gunnar. *Homo Sapiens L. Studier i Carl von Linnés naturuppfattning och människolära*. Studies and sources published by the Swedish History of Science Society. Motala: Borgströms, 1975.

———. "Olof Rudbecks föregångare." In *Iter Lapponicum: Skissboken från resan till Lappland 1695*, by Olof Rudbeck (the younger), 11–21. II Kommentardel. Stockholm: René Coeckelbergh, 1987.

Brockliss, L. W. B. *French Higher Education in the Seventeenth and Eighteenth Centuries: A Cultural History*. Oxford: Clarendon Press, 1987.

———. "The Medico-Religious Universe of an Early Eighteenth-Century Parisian Doctor: The Case of Philippe Hecquet." In *The Medical Revolution of the Seventeenth Century*, edited by Roger French and Andrew Wear, 191–221. Cambridge: Cambridge University Press, 1989.

Browne, Janet. *The Secular Ark: Studies in the History of Biogeography*. New Haven: Yale University Press, 1983.

Bruijn, J. G. de, ed. *Inventaris van de Prijsvragen Uitgeschreven door de Hollandsche Maatschappij der Wetenschappen, 1753–1917*. Haarlem and Groningen: Tjeenk Willink, 1977.

Brunet, Pierre. *Les physiciens hollandais et la methode éxpérimentale en France au XVIIIe siècle*. Paris: Blanchard, 1926.

———. *Étude historique sur le principe de la moindre action*. Paris: Hermann, 1938.

Brunschwig, Henri. *La crise de l'état prussien à la fin du XVIIIe siècle et la genèse de la mentalité romantique*. Paris: Presses Universitaires de France, 1947.

———. *Enlightenment and Romanticism in Eighteenth-Century Prussia*. Chicago: University of Chicago Press, 1974.

Bryson, Gladys. *Man and Society: The Scottish Inquiry of the Eighteenth Century*. Princeton: Princeton University Press, 1945.

Bryson, Norman. *Word and Image: French Painting of the Ancien Régime*. Cambridge: Cambridge University Press, 1981.

Buchdahl, Gerd. *The Image of Newton and Locke in the Age of Reason*. London: Sheed and Ward, 1961.

———. *Metaphysics and the Philosophy of Science: The Classical Origins, Descartes to Kant*. 1969. Reprint, Lanham, Md: University of America Press, 1988.

Buck, Peter. "Seventeenth-Century Political Arithmetic: Civil Strife and Vital Statistics." *Isis* 68(1977): 67–84.

———. "People Who Counted: Political Arithmetic in the Eighteenth Century." *Isis* 73 (1982): 28–45.

Burchell, Graham. "Peculiar Interests: Civil Society and Governing the System of Natural Liberty." In *The Foucault Effect: Studies in Governmentality*, edited by Graham Burchell, Colin Gordon, and Peter Miller, 119–50. Chicago: University of Chicago Press, 1991.

Burchell, Graham, Colin Gordon, and Peter Miller, eds. *The Foucault Effect: Studies in Governmentality*. Chicago: University of Chicago Press, 1991.

Butler, Stella, R. H. Nuttall, and Olivia Brown. *The Social History of the Microscope*. Cambridge: Whipple Museum of the History of Science, 1991.

Butterfield, Herbert. *The Origins of Modern Science*. 2d ed. London: Bell, 1957. First edition published 1949.

Calhoun, Craig, ed. *Habermas and the Public Sphere*. Cambridge: MIT Press, 1992.

Calinger, Ronald. "The Newtonian-Wolffian Confrontation in the St. Petersburg Academy of Science, 1725–1746." *Cahiers d'Histoire Mondiale* 11 (1968): 417–35.

———. "The Newtonian-Wolffian Controversy, 1740–1759." *Journal of the History of Ideas* 30 (1969): 319–30.

Callot, Émile. *La philosophie de la vie au XVIIIe siècle, etudiée chez Fontenelle, Montesquieu, Maupertuis, La Mettrie, Diderot, d'Holbach, Linné*. Paris: M. Rivière, 1965.

Campe, Rüdiger, and Manfred Schneider, eds. *Geschichten der Physiognomik: Text, Bild, Wissen*. Freiburg: Rombach, 1996.

Canguilhem, Georges. "La monstruosité et les monstrueux" (1965). In *La connaissance de la vie*, 171–84. Paris: Vrin, 1992.

———. "The Death of Man or Exhaustion of the Cogito?" (1967). In *The Cambridge Companion to Foucault*, edited by Gary Gutting, 71–91. Cambridge: Cambridge University Press, 1994.

Capp, Bernard. *Astrology and the Popular Press: English Almanacs, 1500–1800*. London: Faber and Faber, 1979.

Carroll, Charles. *The Great Chess Automaton*. New York: Dover, 1975.

Carver, J. Scott. "A Reconsideration of Eighteenth-Century Russia's Contributions to European Science." *Canadian-American Slavic Studies* 14 (1980): 389–405.

Casini, Paolo. *Newton e la conscienza europea*. Bologna: Il Mulino, 1983.

———. "D'Alembert, l'économie des principes, et la 'métaphysique des sciences.'" In *Jean d'Alembert, savant et philosophe: Portrait à plusieurs voix*, edited by Monique Emery and Pierre Monzani. Paris: Éditions des Archives Contemporaines, 1989.

Cassedy, James H. "Medicine and the Rise of Statistics." In *Medicine in Seventeenth-Century England*, edited by Allen G. Debus, 283–312. Berkeley and Los Angeles: University of California Press, 1974.

Cassirer, Ernst. *Die Philosophie der Aufklärung*. Tübingen: Mohr, 1932.

———. *The Philosophy of the Enlightenment*. Translated by F. C. A. Koelln and J. P. Pettegrove. Princeton: Princeton University Press, 1951.

———. *Die Philosophie der Aufklärung* 3d ed. Tübingen: Mohr, 1973.

———. *An Essay on Man*. New Haven: Yale University Press, 1944.

Castle, Terry. "The Female Thermometer." *Representations*, no. 17 (1987): 1–27.

———. *The Female Thermometer*. Oxford: Oxford University Press, 1995.

Cavazza, Marta. *Settecento inquieto: Alle origini dell'Istituto dell Scienze di Bologna*. Bologna: Il Mulino, 1990.

———. "L'insegnamento delle scienze sperimentali nell'Istituto delle Scienze di Bologna." In *Le università e le scienze: Prospettive storiche e attuali*, edited by Giuliano Pancaldi, 155–168. Bologna: CIS, 1993.

———. "Laura Bassi e il suo gabinetto di fisica sperimentale: realtà e mito." *Nuncius* 10 (1995): 715–53.

Ceranski, Beate. *"Und sie fürchtet sich vor niemanden": Die Physikerin Laura Bassi, 1711–1778*. Geschichte und Geschlechter, no. 17. Frankfurt am Main: Campus Verlag, 1996.

Cessi, Ugo. "Una dottoressa rodigina del secolo XVIII: Nuove notizie e documenti intorno a Cristina Roccati." *Ateneo Veneto* 24, no. 1, fasc.1 (1901): 43–76.

Chapuis, Alfred, and Edmond Droz. *Automata*. London: Batsford, 1958.

Charlton, D. G. *New Images of the Natural in France: A Study in European Cultural History, 1750–1800*. Cambridge: Cambridge University Press, 1984.

Chartier, Roger. "Text, Symbols, and Frenchness," *Journal of Modern History* 57 (1985): 682–95.

———. *The Cultural Uses of Print in Early Modern France*. Princeton: Princeton University Press, 1987.

———. *Cultural History: Between Practices and Representations*. Translated by Lydia G. Cochrane. Cambridge: Polity Press, 1993.

———. *The Cultural Origins of the French Revolution*. Durham: Duke University Press, 1991.

———. *The Order of Books*. Cambridge: Polity Press, 1994.

———. *Forms and Meanings: Texts, Performances, and Audiences from Codex to Computer*. Philadelphia: University of Pennsylvania Press, 1995.

Chastel, André. "Le Baroque et la mort." In *Fables, Formes, Figures*, 1:205–26. Paris: Flammarion, 1978.

Chaussinand-Nogaret, Guy. *The French Nobility in the Eighteenth Century: From Feudalism to Enlightenment.* Translated by William Doyle. Cambridge: Cambridge University Press, 1985.

————, et al. *Histoire des élites en France du XVIe au XXe siècle: L'honneur, le mérite, l'argent.* Paris: Tallandier, 1991.

Childs, John. *Armies and Warfare in Europe, 1648–1789.* Manchester: Manchester University Press, 1982.

Chisick, Harvey. *The Limits of Reform in the Enlightenment: Attitudes towards the Education of the Lower Classes in Eighteenth-Century France.* Princeton: Princeton University Press, 1981.

Choulant, Ludwig. *Geschichte und Bibliographie der anatomischen Abbildung.* Leipzig: Weigel, 1852.

Christensen, Dan. *Det modern Projekt: Teknik og Kultur i Danmark-Norge, 1750–(1814)–1850.* Copenhagen: Gyldendal, 1996.

Christensen, Jerome. *Practicing Enlightenment: Hume and the Formation of a Literary Career.* Madison: University of Wisconsin Press, 1987.

Christie, J. R. R. "The Origins and Development of the Scottish Scientific Community, 1680–1760." *History of Science* 12 (1974): 122–41.

————. "Ether and the Science of Chemistry, 1740–1790." In *Conceptions of Ether: Studies in the History of Ether Theories, 1740–1900*, edited by G. N. Cantor and M. J. S. Hodge, 86–110. Cambridge: Cambridge University Press, 1981.

Cieslak, Edmund. "Aspects of Baltic Sea-borne Trade in the Eighteenth Century: The Trade Relations between Sweden, Poland, Russia, and Prussia," *Journal of European Economic History* 12 (1983): 239–71.

Clark, Owen E. "The Contributions of J. F. Meckel, the Younger, to the Science of Teratology." *Journal of the History of Medicine* 24 (1969): 310–22.

Clark, William. "The Scientific Revolution in the German Nations." In *The Scientific Revolution in National Context*, edited by Roy Porter and Mikuláš Teich, 90–114. Cambridge: Cambridge University Press, 1992.

————. "German Physics Textbooks in the *Goethezeit.*" *History of Science* 35 (1997): 219–39, 295–363.

————. "Der Untergang der Astrologie in der deutschen Barockzeit." In *Christentum in Europa im 17. Jahrhundert*, edited by H. Lehmann and A.-C. Trepp. Göttingen: Vandenhoeck, forthcoming.

Classen, Karl Heinz, ed. *Kant-Bildnisse.* Königsberg: Gräfe und Unzer, 1924.

Clercq, Peter de. *The Leiden Cabinet of Physics.* Leiden: Museum Boerhaave, 1989.

————. "In de schaduw van 's-Gravesande: Het Leids Physisch Kabinet in de tweede helft van de 18e eeuw." In *Het Instrument in de Wetenschap*, edited by Marian Fournier and Bert Theunissen, 149–73. Special issue of *Tijdschrift voor de Geschiedenis der Geneeskunde, Natuurwetenschappen, Wiskunde, en Techniek* 10 (1987).

————. *At the Sign of the Oriental Lamp: The Musschenbroek Workshop in Leiden, 1660–1750.* Rotterdam: Erasmus, 1997.

Clingham, Greg, ed. *New Light on Boswell: Critical and Historical Essays on the Occasion of the Bicentenary of "The Life of Johnson."* Cambridge: Cambridge University Press, 1991.

Cochrane, Eric. *Tradition and Enlightenment in the Tuscan Academies, 1691–1800.* Chicago: University of Chicago Press, 1961.

Cohen, Ernst, and W. A. T. Cohen-Meester. "Daniel Gabriel Fahrenheit." *Chemisch Weekblad* 33, no. 24 (1936): 374–93; 34, no. 45 (1937): 727–30.

————. "Het Natuurkundig Genootschap der Dames, 1785–1787." *Chemisch Weekblad* 39 (1942): 242–46.

Cohen, I. Bernard. *Benjamin Franklin's Experiments.* Cambridge: Harvard University Press, 1941.

————. *Franklin and Newton: An Inquiry into Speculative Newtonian Experimental Science and Franklin's Work in Electricity as an Example Thereof.* Philadelphia: American Philosophical Society, 1956.

Cooper, Carolyn. "The Portsmouth System of Manufacture." *Technology and Culture* 25 (1984): 182–225.

Cope, Kevin L., ed. *Compendious Conversations: The Method of Dialogue in the Early Enlightenment.* Frankfurt am Main: Peter Lang, 1992.

Craig, Edward. *The Mind of God and the Works of Man.* Oxford: Clarendon Press, 1987.

Crampe-Casnabet, Michèle. "A Sampling of Eighteenth-Century Philosophy." In *A History of Women: Renaissance and Enlightenment Paradoxes,* edited by Natalie Zemon Davis and Arlette Farge, 315–47. Cambridge: Harvard University Press, 1993.

Crawforth, M. A. "Evidence from Trade Cards for the Scientific Instrument Industry." *Annals of Science* 42 (1985): 453–554.

Crocker, Lester G. *An Age of Crisis: Man and World in Eighteenth-Century French Thought.* Baltimore: Johns Hopkins University Press, 1959.

———. "Interpreting the Enlightenment: A Political Approach," *Journal of the History of Ideas* 46 (1985): 211–30.

———. Introduction to *The Blackwell Companion to the Enlightenment,* edited by John W. Yolton, 1–9. Oxford: Basil Blackwell, 1991.

Crow, Thomas E. *Painters and Public Life in Eighteenth-Century Paris.* New Haven: Yale University Press, 1985.

Cunningham, Andrew. "Medicine to Calm the Mind: Boerhaave's Medical System, and Why It Was Adopted in Edinburgh." In *The Medical Enlightenment of the Eighteenth Century,* edited by Andrew Cunningham and Roger French, 40–66. Cambridge: Cambridge University Press, 1990.

Cunningham, Andrew, and Nicholas Jardine, eds. *Romanticism and the Sciences.* Cambridge: Cambridge University Press, 1990.

Darnton, Robert. *Mesmerism and the End of the Enlightenment in France.* Cambridge: Harvard University Press, 1968.

———. "In Search of the Enlightenment: Recent Attempts to Create a Social History of Ideas." *Journal of Modern History* 43 (1971): 113–32.

———. *The Business of Enlightenment: A Publishing History of the "Encyclopédie," 1775–1800.* Cambridge: Harvard University Press, Belknap Press, 1979.

———. *The Literary Underground of the Old Regime.* Cambridge: Harvard University Press, 1982.

———. "The Literary Revolution of 1789." *Studies in Eighteenth-Century Culture* 21 (1991): 3–26.

Daston, Lorraine J. "The Physicalist Tradition in Early Nineteenth-Century French Geometry." *Studies in History and Philosophy of Science* 17 (1986): 269–95.

———. *Classical Probability in the Enlightenment.* Princeton: Princeton University Press, 1988.

———. "The Ideal and Reality of the Republic of Letters in the Enlightenment." *Science in Context* 4 (1991): 367–86.

———. "Marvelous Facts and Miraculous Evidence in Early Modern Europe." *Critical Inquiry* 18 (1991): 93–124.

———. "The Naturalized Female Intellect." *Science in Context* 5 (1992): 209–35.

———. "Objectivity and the Escape from Perspective." *Social Studies of Science* 22 (1992): 597–618.

———. "Enlightenment Calculation." *Critical Inquiry* 21 (1994): 182–202.

———. "Neugierde als Empfindung und Epistemologie." In *Macrocosmos und Microcosmos: Die Welt in der Stube, zur Geschichte des Sammelns, 1450–1800,* edited by Andreas Grote, 35–59. Opladen: Leske und Budrich, 1994.

Daston, Lorraine, and Peter Galison. "The Image of Objectivity." *Representations,* no. 40 (1992): 81–128.

Daudin, H. *De Linné à Jussieu: Méthodes de classification et idée de série en botanique et en zoologie.* Paris: F. Alcan, 1926.

Daumas, Maurice. *Scientific Instruments of the Seventeenth and Eighteenth Centuries and Their Makers.* London: Batsford, 1972.

Dawson, Virginia P. *Nature's Enigma: The Problem of the Polyp in the Letters of Bonnet, Trembley, and Réaumur.* Philadelphia: American Philosophical Society, 1987.

Day, Charles R. *Education for the Industrial World: The Écoles d'Arts et Métiers and the Rise of French Industrial Engineering.* Cambridge: MIT Press, 1987.

De Besaucèle, Berthe. *Les cartésiens d'Italie: Recherches sur l'influence de la philosophie de Descartes dans l'évolution de la pensée italienne aux XVII et XVIII siècle.* Paris: A. Picard, 1920.

De Certeau, Michel. *The Practice of Everyday Life.* Translated by Steven Rendall. Berkeley and Los Angeles: University of California Press, 1984.

De Vit, Vincenzo. "Sulla vita e sulle opere della Dottoressa Cristina Roccati Rodigina." In *Opusculi letterati editi e inediti,* 47–73. Milan: Boniardi-Pogliani, 1883.

Dear, Peter. "The Cultural History of Science: An Overview with Reflections." *Science, Technology, and Human Values* 20 (1995): 150–70.

———. "A Mechanical Microcosm: Bodily Passions, Good Manners, and Cartesian Mechanism." In *Science Incarnate: Historical Embodiments of Natural Knowledge,* edited by Christopher Lawrence and Steven Shapin, 51–82. Chicago: University of Chicago Press, 1998.

Degn, Christian. "Die Herzogtümer in Gesamtstaat, 1773–1830." In *Die Herzogtümer in Gesamtstaat, 1721–1830,* by Olaf Klose and Christian Degn, 163–427. Vol. 6 of *Geschichte Schleswig-Holstein,* edited by Olaf Klose. Neumünster: Karl Wachholtz Verlag, 1960.

Dekker, T. "De Popularisering der Natuurwetenschap in Nederland in de Achttiende eeuw." *Geloof en Wetenschap: Orgaan van de Christelijke Vereniging van Natuur- en Geneeskundigen in Nederland* 53 (1955): 173–88.

Delaporte, François. *Nature's Second Kingdom: Explorations of Vegetality in the Eighteenth Century.* Translated by Arthur Goldhammer. Cambridge: MIT Press, 1982.

Dening, Greg. *Performances.* Melbourne: Melbourne University Press, 1996.

Dettelbach, Michael. "Romanticism and Administration: Mining, Galvanism, and Oversight in Alexander von Humboldt's Global Physics." Ph.D. diss., Cambridge University, 1992.

———. "Humboldtian Science." In *Cultures of Natural History,* edited by Nicholas Jardine, J. A. Secord, and E. C. Spary, 287–304. Cambridge: Cambridge University Press, 1996.

Dick, Steven J. *Plurality of Worlds: The Origins of the Extraterrestrial Life Debate from Democritus to Kant.* Cambridge: Cambridge University Press, 1982.

Diederichs, H. "Garlieb Merkel als Bekämpfer der Leibeigenschaft." *Baltische Monatsschrift* 19 (1870): 38–83.

Dobbs, B. J. T., and Margaret C. Jacob. *Newton and the Culture of Newtonianism.* Atlantic Highlands: Humanities Press, 1995.

Donnelly, Michael. "On Foucault's Uses of the Notion 'Biopower.'" In *Michel Foucault Philosopher,* translated by Timothy J. Armstrong, 199–203. New York: Harvester, 1992.

Donnert, Erich. *Johann Georg Eisen, 1717–1779: Ein Vorkämpfer der Bauernbefreiung in Russland.* Leipzig: Koehler und Amelang, 1978.

———. "Zur Verbreitung bürgerlicher Wissenschafts- und Gesellschaftslehren an der Universität Moskau in der zweiten Hälfte des 18. Jh." *Jahrbuch für Geschichte der Sozialistischen Länder Europas* 23 (1979): 25–34.

———. "Das russische Imperium im Urteil des Deutsch-Baltischen Aufklärungsschriftstellers August Wilhelm Hupel im 18. und beginnenden 19. Jahrhundert." *Journal of Baltic Studies* 15 (1984): 91–96.

———. *Russia in the Age of Enlightenment.* Translated by Alison Wightman and Alistair Wightman. Leipzig: Edition Leipzig, 1986.

Dooley, Brendan. "Science Teaching as a Career at Padua in the Early Eighteenth Century: The Case of Giovanni Poleni." *History of Universities* 4 (1984): 115–51.

———. *Science, Politics, and Society in Eighteenth-Century Italy: The "Giornale de' Letterati" and Its World.* New York: Garland, 1991.

Dotzler, Bernhard, Peter Gendolla, and Jorgen Schäfer. *Maschinen-Menschen: Eine Bibliographie.* Frankfurt am Main: Lang, 1992.

Doyon, André, and Lucien Liaigre, 1992. *Jacques Vaucanson, mécanicien de génie.* Paris: Presses Universitaires de France, 1992.

Drux, Rudolf. *Marionette Mensch: Ein Metaphernkomplex und sein Kontext von Hoffmann bis Büchner.* Munich: Wilhelm Fink, 1986.

Duffy, Christopher. *The Fortress in the Age of Vauban and Frederick the Great, 1660–1789.* London: Routledge and Kegan Paul, 1985.

———. *The Military Experience in the Age of Reason.* London: Routledge, 1987.

Duris, Pascal. *Linné et la France, 1780–1850.* Geneva: Droz, 1993.

Durkheim, Emile. *Professional Ethics and Civic Morals.* Translated by Cornelia Brookfield. London: Routledge and Kegan Paul, 1957. Original edition published 1898.

Duveen, Denis I., and Roger Hahn. "Laplace's Succession to Bézout's Post of Examinateur des Elèves de l'Artillerie." *Isis* 48 (1957): 416–27.

Eagleton, Terry. *The Rape of Clarissa.* Oxford: Basil Blackwell, 1982.

Eaves, T. C. Duncan. "The Harlowe Family by Joseph Highmore: A Note on the Illustration of Richardson's Clarissa." *Huntington Library Quarterly* 7 (1943): 89–96.

———. "Graphic Illustrations of the Novels of Samuel Richardson." *Huntington Library Quarterly* 14 (1951): 349–83.

Eisenstadt, Peter. "The Weather and Weather Forecasting in Colonial America." Ph.D. diss., New York University, 1990.

Ehrard, Jean. *L'Idée de la nature en France dans la première moitié du XVIIIe siècle.* 1963. Reprint, Paris: Albin Michel, 1994.

———. "La main du travailleur, la plume du philosophe." *Milieux* 19/20 (1984–85): 47–53.

Ehrström, Anders. "Minnesanteckningar om Linné av kapellanen i Kronoby (Österbotten, Finland), meddelade av Robert Ehrström." *Svenska Linnésällskapets Årsskrift* 29 (1946): 69–72.

Elena, Alberto. "'In lode della filosofessa di Bologna': An Introduction to Laura Bassi." *Isis* 82 (1991): 71–79.

Eley, Geoffrey. "Nazism, Politics, and the Image of the Past: Thoughts on the West German Historikerstreit." *Past and Present* 121 (1988): 171–208.

Elias, Norbert. *Über den Prozess der Zivilisation.* 2d ed. 2 vols. Frankfurt am Main: Suhrkamp, 1976. First edition published 1939.

———. *The Civilising Process.* Translated by Edmund Jephcott. Oxford: Basil Blackwell, 1994.

Elias, O.-H. "Die undeutsche Bevölkerung im Riga des 18. Jahrhunderts." *Jahrbücher für Geschichte Osteuropas* 14 (1966): 481–84.

Ellis, F. H. *Sentimental Comedy: Theory and Practice.* Cambridge: Cambridge University Press, 1991.

Ellis, Harold A. "Genealogy, History, and Aristocratic Reaction in Early Eighteenth-Century France: The Case of Henri de Boulainvilliers." *Journal of Modern History* 58 (1986): 414–51.

Ellrich, Robert. "Modes of Discourse and the Language of Sexual Reference in Eighteenth-Century French Fiction." In *'Tis Nature's Fault: Unauthorized Sexuality during the Enlightenment,* edited by Robert McCubbin, 217–28. Cambridge: Cambridge University Press, 1985.

Engelhardt, Dietrich von. *Historisches Bewusstsein in der Naturwissenschaft von der Aufklärung bis zum Positivismus.* Munich: Alber, 1979.

———. "Historisches Bewusstsein in der Medizin der Romantik." In *Bausteine zur Medizingeschichte: Festschrift für Heinrich Schipperges zum 65 Geburtstag,* edited by Eduard Seidler and Heine Schott, 25–35. Stuttgart: Steiner Verlag, 1984.

———. "Historical Consciousness in German Romantic Naturforschung." Translated by Christine Salazar. In *Romanticism and the Sciences,* edited by Andrew Cunningham and Nicholas Jardine, 55–68. Cambridge: Cambridge University Press, 1990.

Enke, Ulrike. "Soemmerrings erste Professor am Collegium Carolinum zu Kassel." In *Samuel Thomas Soemmerring in Kassel, 1779–1784: Beiträge zur Wissenschaftsgeschichte der Goethezeit,* edited by Manfred Wenzel, 9:75–141. Stuttgart and New York: Fischer, 1995.

Epstein, Klaus. *The Genesis of German Conservatism.* Princeton: Princeton University Press, 1966.

Eribon, Didier. *Michel Foucault.* Translated by Betsy Wing. Cambridge: Harvard University Press, 1991.

Ericsson, Birgitta. "Central Power and Local Community: Joint Nordic Research on the Granting of 'Privilegia' to Industrial Enterprises in Scandinavia during the Eighteenth Century." *Scandinavian Journal of History* 7 (1982): 173–76.

Ertman, Thomas. *The Birth of the Leviathan: Building States and Regimes in Medieval and Early Modern Europe.* Cambridge: Cambridge University Press, 1997.

Ewald, François. "Insurance and Risk." In *The Foucault Effect: Studies in Governmentality,* edited by Graham Burchell, Colin Gordon, and Peter Miller, 197–210. Chicago: University of Chicago Press, 1991.

Faber, Marion, ed. *Der Schachautomat des Baron von Kempelen.* Dortmund: Harenberg, 1983.

Fabian, Bernard, Wilhelm Schmidt-Biggemann, and Rudolf Vierhaus, eds. *Deutschlands kulturelle Entfaltung: Die Neubestimmung des Menschen.* Studien zum Achtzehnten Jahrhundert, nos. 2/3. Munich: Kraus, 1980.

Feldman, Theodore S. "Late Enlightenment Meteorology." In *The Quantifying Spirit in the Eighteenth Century,* edited by Tore Frängsmyr, J. L. Heilbron, and Robin E. Rider, 143–77. Berkeley and Los Angeles: University of California Press, 1990.

Fellmann, Emil A. *Leonhard Euler.* Hamburg: Rowohlt, 1995.

Ferrone, Vincenzo. "Les mécanismes de formation des élites de la Maison de Savoie: Recrutement et sélection dans les écoles militaires du Piémont au XVIIIe siècle." *Paedagogica Historica* 30 (1994): 341–70.

———. *The Intellectual Roots of the Italian Enlightenment: Newtonian Science, Religion, and Politics in the Early Eighteenth Century.* Translated by Sue Brotherton. Atlantic Highlands: Humanities Press, 1995.

Fiedler, Leslie. *Freaks: Myths and Images of the Secret Self.* New York: Simon and Schuster, 1978.

Figlio, Karl M. "Theories of Perception and the Physiology of Mind in the Late Eighteenth Century." *History of Science* 13 (1975): 177–212.

Filho, W. B., and C. Comte. "La formalisation de la dynamique par Lagrange, 1736–1813." In *Sciences à l'époque de la Révolution française,* edited by Roshdi Rashed, 329–82. Paris: Albert Blanchard, 1988

Findlen, Paula. "Science as a Career in Enlightenment Italy: The Strategies of Laura Bassi." *Isis* 83 (1993): 441–69.

———. *Possessing Nature: Museums, Collecting, and Scientific Culture in Early Modern Italy.* Berkeley and Los Angeles: University of California Press, 1994.

———. "Translating the New Science: Women and the Circulation of Knowledge in Enlightenment Italy." *Configurations* 2 (1995): 167–206.

Finster, Reinhard, and Gerd van den Heuvel. *Gottfried Wilhelm Leibniz.* Hamburg: Rowohlt, 1990.

Flesher, Mary Mosher. "Repetitive Order and the Human Walking Apparatus: Prussian Military Science versus the Webers' Locomotion Research." *Annals of Science* 54 (1997): 463–87.

Forbes, R. J., E. Lefebvre, and J. G. de Bruijn, eds. *Martinus van Marum, Life and Work.* 6 vols. Leiden: Nordhoff, 1969–76.

Forsskål, Pehr. *Resa till lycklige Arabien: Petrus Forsskåls dagbok, 1761–1763.* Uppsala: Almquist & Wiksell, 1950.

Foucault, Michel. *Les mots et les choses: Une archéologie des sciences humaines.* Paris: Gallimard, 1966.

———. *The Order of Things: An Archeology of the Human Sciences.* London: Tavistock Press, 1970.

———. "La situation de Cuvier dans l'histoire de la biologie." *Revue d'Histoire des Sciences et de Leurs Applications* 23 (1970): 71–92.

———. "Nietzsche, la généalogie, l'histoire." In *Hommage à Jean Hyppolite,* edited by Suzanne Bachelard et al., 145–72. Paris: Presses Universitaires de France, 1971.

———. *Surveiller et punir.* Paris: Gallimard, 1975.

———. *Discipline and Punish: The Birth of the Prison.* Translated by Alan Sheridan. New York: Random House, 1979.

————. *The History of Sexuality.* Vol. 1, *An Introduction.* Translated by Robert Hurley. New York: Vintage Books, 1980.

————. "The Politics of Health in the Eighteenth Century." In *Power/Knowledge: Selected Interviews and Other Writings, 1972–1977,* edited by Colin Gordon, translated by Colin Gordon, Leo Marshall, John Mepham, and Kate Soper, 166–82. New York: Pantheon Books, 1980.

————. *Power/Knowledge: Selected Interviews and Other Writings, 1972–1977.* Edited by Colin Gordon, translated by Colin Gordon, Leo Marshall, John Mepham, and Kate Soper. New York: Pantheon Books, 1980.

————. "What is Enlightenment?" In *The Foucault Reader,* edited by Paul Rabinow, 32–50. New York: Pantheon Books, 1984.

————. *The Foucault Reader.* Edited by Paul Rabinow. New York: Pantheon Books, 1984.

————. *Resumé des cours, 1970–1982.* Paris: Julliard, 1989.

————. *Dits et écrits.* Edited by Daniel Defert, François Ewald, and Jacques Lagrange. 4 vols. Paris: Gallimard, 1994.

Fox, Christopher, Roy Porter, and Robert Wokler, eds. *Inventing Human Science: Eighteenth-Century Domains.* Berkeley and Los Angeles: University of California Press, 1995.

Fox, Robert. "The Rise and Fall of Laplacian Physics," *Historical Studies in the Physical Sciences* 4 (1976): 89–136.

Fox, Robert, and George Weisz, eds. *The Organization of Science and Technology in France, 1808–1914.* Cambridge: Cambridge University Press, 1980.

France, Peter. *Politeness and Its Discontents: Problems in French Classical Culture.* Cambridge: Cambridge University Press, 1992.

Frängsmyr, Tore. "The Enlightenment in Sweden." In *The Enlightenment in National Context,* edited by Roy Porter and Mikuláš Teich, 164–75. Cambridge: Cambridge University Press, 1981.

Frängsmyr, Tore, J. L. Heilbron, and Robin E. Rider, eds. *The Quantifying Spirit in the Eighteenth Century.* Berkeley and Los Angeles: University of California Press, 1990.

Frasca-Spada, Marina. *Space and the Self in Hume's "Treatise."* Cambridge: Cambridge University Press, 1998.

Fraser, Craig. "J. L. Lagrange's Early Contributions to the Principles and Methods of Mechanics." *Archive for the History of Exact Sciences* 28 (1983): 197–241.

————. "D'Alembert's Principle: The Original Formulation and Application in Jean D'Alembert's *Traité de dynamique* (1743)." *Centaurus* 28 (1985): 31–61, 145–69.

Frei, Hans. W. *The Eclipse of Biblical Narrative: A Study of Eighteenth and Nineteenth Century Hermeneutics.* New Haven: Yale University Press, 1974.

French, Anne, ed. *John Joseph Merlin: The Ingenious Mechanick.* London: Greater London Council, 1985.

Friede, Juan, and Benjamin Keen. *Bartolomé de las Casas in History: Toward an Understanding of the Man and His Work.* De Kalb: Northern Illinois University Press, 1971.

Friedman, Michael. *Kant and the Exact Sciences.* Cambridge: Harvard University Press, 1992.

Fries, Robert E. "Linné i Holland." *Svenska Linnésällskapets Årsskrift* 2 (1919): 141–55.

Frolow, M. *Le problème d'Euler.* St. Petersburg: Trenke et Fusnot, 1884.

Fryer, David M., and John C. Marshall. "The Motives of Jacques de Vaucanson." *Technology and Culture* 20 (1979): 257–69.

Funkenstein, Amos. *Theology and the Scientific Imagination from the Middle Ages to the Scientific Revolution.* Princeton: Princeton University Press, 1986.

Gabbey, Alan. "The Mechanical Philosophy and Its Problems." In *Change and Progress in Modern Science,* edited by Joseph Pitt, 9–84. Dordrecht: Reidel, 1985.

————. "Cudworth, More, and the Mechanical Analogy." In *Philosophy, Science, and Religion in England, 1640–1700,* edited by Richard Kroll, Richard Ashcraft, and Perez Zagorin, 109–27. Cambridge: Cambridge University Press, 1992.

Gage, Andrew Thomas. *A History of the Linnean Society of London.* London: Linnean Society, 1938.

Garrett, Don. *Cognition and Commitment in Hume's Philosophy.* Oxford: Oxford University Press, 1997.

Gascoigne, John. "Mathematics and Meritocracy: The Emergence of the Cambridge Mathematical Tripos." *Social Studies of Science* 14 (1984): 547–84.

———. *Joseph Banks and the English Enlightenment: Useful Knowledge and Polite Culture.* Cambridge: Cambridge University Press, 1994.

Gasking, Elisabeth. *Investigations into Generation, 1651–1828.* London: Hutchinson, 1967.

Gause, Fritz. *Die Geschichte der Stadt Königsberg im Preussen.* Vol 2, *Von der Königskrönung bis zum Ausbruch des Ersten Weltkrieges.* Köln: Böhlau Verlag, 1968.

Gawlick, Günther, and Lothar Kreimendahl. *Hume in der deutschen Aufklärung: Umrisse einer Rezeptionsgeschichte.* Stuttgart: Frommann-Holzboog, 1987.

Gay, Peter. *The Enlightenment: An Interpretation.* Vol. 1, *The Rise of Modern Paganism.* Vol.2, *The Science of Freedom.* London: Weidenfeld and Nicholson; New York: Knopf, 1967–69.

Gelfand, Toby. "Empiricism and Eighteenth Century French Surgery." *Bulletin of the History of Medicine* 44 (1970): 40–53.

Gendolla, Peter. *Die lebenden Maschinen: Zur Geschichte der Maschinenmenschen bei Jean Paul, E. T. A. Hoffmann, und Villiers de l'Isle Adam.* Marburg: Guttandin und Hoppe, 1980.

Giedion, Siegfried. *Mechanization Takes Command.* New York: Norton, 1969.

Gierl, Martin. "Vom Wühlen der Aufklärung im Gedärme." *Volkskunde in Niedersachsen,* 1988, 3–10.

———. "Auf den Boden des Streits: Die Pietismuskontroverse und die deutsche Frühaufklärung. Eine kommunikationsgeschichtliche Untersuchung."Ph.D. diss., University of Göttingen, 1994.

Giglioni, Guido. "Automata Compared: Boyle, Leibniz, and the Debate on the Notion of Life and Mind." *British Journal for the History of Philosophy* 3 (1995): 249–78.

Gillispie, Charles C. "The Natural History of Industry." *Isis* 48 (1957): 398–407.

———. *Lazare Carnot, Savant.* Princeton: Princeton University Press, 1971.

———. *Science and Polity in France at the End of the Old Regime.* Princeton: Princeton University Press, 1981.

———. *The Edge of Objectivity.* 1960. Reprint, Princeton: Princeton University Press, 1990.

Gillmor, C. Stewart. *Coulomb and the Evolution of Physics and Engineering in Eighteenth-Century France.* Princeton: Princeton University Press, 1971.

Glass, Bentley, et al., eds. *Forerunners of Darwin, 1745–1859.* Baltimore: Johns Hopkins University Press, 1959.

Goldgar, Anne. *Impolite Learning: Conduct and Community in the Republic of Letters, 1680–1750.* New Haven: Yale University Press, 1995.

Golinski, Jan. *Science as Public Culture: Chemistry and Enlightenment in Britain, 1760–1820.* Cambridge: Cambridge University Press, 1992.

Goodison, Nicholas. *English Barometers, 1680–1860: A History of Domestic Barometers and Their Makers and Retailers.* Woodbridge, Suffolk: Antique Collectors' Club, 1977.

Goodman, Dena. "Enlightenment Salons: The Convergence of Female and Philosophic Ambitions." *Eighteenth-Century Studies* 22 (1988–89): 329–50.

———. "Governing the Republic of Letters: The Politics of Culture in the French Enlightenment." *History of European Ideas* (1991): 183–99.

———. "Public Sphere and Private Life: Toward a Synthesis of Current Historiographical Approaches to the Old Regime." *History and Theory* 31 (1992): 1–20.

———. *The Republic of Letters. A Cultural History of the French Enlightenment.* Ithaca: Cornell University Press, 1994.

Gordon, Daniel. "'Public Opinion' and the Civilising Process in France: The Example of Morellet." *Eighteenth-Century Studies* 22 (1988–89): 302–28.

———. *Citizens without Sovereignty: Equality and Sociability in French Thought, 1670–1789.* Princeton: Princeton University Press, 1994.

Gowing, Lawrence. "Hogarth, Hayman, and Vauxhall Decorations." *Burlington Magazine* 95 (1953): 4–17.

Grasse, Pierre. *La vie et l'oeuvre de Réaumur, 1683–1757.* Paris: Presses Universitaires de France, 1962.

Gray, Richard. "Sign and Sein: The Physiognomikstreit and the Dispute over the Semiotic

Constitution of Bourgeois Individuality." *Deutsche Vierteljahresschrift für Literaturwissenschaft und Geistesgeschichte* 66 (1992): 300–29.

———. "Aufklärung und Anti-Aufklärung: Wissenschaftlichkeit und Zeichenbegriff in Lavaters 'Physiognomik.'" In *Das Antlitz Gottes im Antlitz des Menschen: Zugänge zu Johann Kaspar Lavater*, edited by Karl Pestalozzi and Horst Weigelt, 166–78. Göttingen: Vandenhoeck und Ruprecht, 1994.

Greenberg, John L. "Mathematical Physics in Eighteenth-Century France." *Isis* 77 (1986): 59–78.

———. *The Problem of the Earth's Shape from Newton to Clairaut: The Rise of Mathematical Science in Eighteenth-Century Paris and the Fall of "Normal" Science*. Cambridge: Cambridge University Press, 1995.

Grmek, Mirko. "The History of Medical Education in Russia." In *The History of Medical Education*, edited by C. D. O'Malley, 303–27. Cambridge: Cambridge University Press, 1970.

Guerlac, Henry. "Some Historical Assumptions of the History of Science." In *Scientific Change*, edited by A. C. Crombie, 797–812. London: Heinemann, 1963.

———. *Essays in the History of Modern Science*. Baltimore: Johns Hopkins University Press, 1977.

———. *Lavoisier—the Crucial Year: The Background and Origin of His First Experiments on Combustion in 1772*. 1961. Reprint, New York: Gordon and Breach, 1990.

Guéroult, Marcel. *Dynamique et métaphysique leibniziennes*. Paris: Les Belles Lettres, 1934.

Guillerme, André. "Network: Birth of a Category in Engineering Thought during the French Restoration." *History and Technology* 8 (1992): 151–66.

Guitton, Edouard. *Jacques Delille, 1738–1813, et le poème de la nature en France de 1750 à 1820*. Paris: Klincksieck, 1974.

Guthke, Karl S. *Der Mythos der Neuzeit: Das Thema der Mehrheit der Welten in der Literatur und Geistesgeschichte von der kopernikanischen Wende bis zur Science Fiction*. Bern: Franck, 1983.

Guttormsson, Loftur. "The Development of Popular Religious Literacy in the Seventeenth and Eighteenth Centuries." *Scandinavian Journal of History* 15 (1990): 7–35.

Habermas, Jürgen. "The Critique of Reason as an Unmasking of the Human Sciences: Michel Foucault." In *The Philosophical Discourse of Modernity: Twelve Lectures*, translated by F. Lawrence, 238–65. Cambridge: Polity Press, 1987.

———. *The Structural Transformation of the Public Sphere: An Inquiry into a Category of Bourgeois Society*. Translated by Thomas Burger. Cambridge: MIT Press, 1989.

———. *The New Conservatism*. Edited by Sherry Weber Nicholson. Cambridge: Polity Press, 1989.

———. *Strukturwandel der Öffentlichkeit: Untersuchungen zu einer Kategorie der bürgerlichen Gesellschaft*. 2d ed. Frankfurt am Main: Suhrkamp, 1990. First edition published 1962.

Hacking, Ian. *The Emergence of Probability: A Philosophical Study of Early Ideas about Probability, Induction, and Statistical Inference*. Cambridge: Cambridge University Press, 1975.

———. *The Taming of Chance*. Cambridge: Cambridge University Press, 1990.

Hackmann, W. D. *John and Jonathan Cuthbertson: The Invention and Development of the Eighteenth Century Electrical Plate*. Leiden: Rijksmuseum voor de Geschiedenis der Natuurwetenschappen, 1973.

Hagner, Michael. "Vom Naturalienkabinett zur Embriologie: Wandlungen des Monströsen und die Ordnung des Lebens." In *Der "falsche" Körper: Beiträge zu einer Geschichte der Monstrositäten*, edited by Michael Hagner, 73–107. Göttingen: Wallstein, 1995.

———. *Homo cerebralis: Der Wandel vom Seelenorgan zum Gehirn*. Berlin: Berlin Verlag, 1997.

———. "Monstrositäten in gelehrten Räumen." In *Alexander Polzin: Abgetrieben*, edited by Sander L. Gilman, 11–30. Göttingen: Wallstein, 1997.

———, ed. *Der "falsche" Körper: Beiträge zu einer Geschichte der Monstrositäten*. Göttingen: Wallstein, 1995.

Hahn, Roger. *The Anatomy of a Scientific Institution: The Paris Academy of Sciences, 1666–1803*. Berkeley and Los Angeles: University of California Press, 1971.

————. "L'enseignement scientifique aux écoles militaires et d'artillerie." In *Écoles techniques et militaires au XVIIIe siècle,* edited by Roger Hahn and René Taton, 513–45. Paris: Hermann, 1986.

Hald, Anders. *The History of Probability and Statistics and Their Applications before 1750.* New York: John Wiley and Sons, 1990.

Hall, A. Rupert. *Ballistics in the Seventeenth Century: A Study in the Relations of Science and War with Reference Principally to England.* Cambridge: Cambridge University Press, 1952.

————. *The Revolution in Science, 1500–1750.* 2d ed. London: Longman, 1983. First edition published 1954 under the title *The Scientific Revolution.*

Hampson, Norman. *The Enlightenment.* Harmondsworth: Penguin Books, 1968.

Hampton, T. *Writing from History: The Rhetoric of Exemplarity in Renaissance Literature.* Ithaca: Cornell University Press, 1990.

Hankins, Thomas L. "The Reception of Newton's Second Law of Motion in the Eighteenth Century." *Archives Internationales d'Histoire des Sciences* 20 (1967): 43–65.

————. *Jean d'Alembert: Science and the Enlightenment.* Oxford: Clarendon Press, 1970.

————. *Science and the Enlightenment.* Cambridge: Cambridge University Press, 1985.

Hankins, Thomas L., and Robert J. Silverman. *Instruments and the Imagination.* Princeton: Princeton University Press, 1995.

Hannaway, Caroline. "Medicine, Public Welfare, and the State in Eighteenth-Century France: The Société Royale de Médecine of Paris, 1776–1793." Ph. D. diss., Johns Hopkins University, 1974.

————. "Vicq d'Azyr, Anatomy, and a Vision of Medicine." In *French Medical Culture in the Nineteenth Century,* edited by Ann La Berge and Mordechai Feingold, 280–95. Wellcome Institute Series in the History of Medicine. Amsterdam and Atlanta: Rodopi, 1994.

Harding, Rosamond. *The Metronome and Its Precursors.* Henley-on-Thames: Gresham Books, 1983.

Harnack, Adolf von. *Geschichte der königlich Preussischen Akademie der Wissenschaften zu Berlin,* 3 vols. in 4 parts. Berlin: Reichsdruckerei, 1900.

Harth, Erica. *Cartesian Women: Versions and Subversions of Rational Discourse in the Old Regime.* Ithaca: Cornell University Press, 1992.

Haskell, Thomas L. "The New Aristocracy." *New York Review of Books* 44 (4 December 1997): 47–53.

Hauser, Arnold. *The Social History of Art.* 4 vols. 1951. Reprint, London: Routledge and Kegan Paul, 1962.

Hawthorn, Geoffrey. *Enlightenment and Despair: A History of Social Theory.* 2d ed. Cambridge: Cambridge University Press, 1987.

Hazard, Paul. *The European Mind, 1680–1715.* Harmondsworth: Penguin Books, 1973.

————. *European Thought in the Eighteenth Century.* Harmondsworth: Penguin Books, 1965.

Heckmann, Herbert. *Die andere Schöpfung: Geschichte der frühen Automaten in Wirklichkeit und Dichtung.* Frankfurt am Main: Umschau, 1982.

Heckscher, Eli F. "Linnés resor—den ekonomiska bakgrunden." *Svenska Linnésällskapets Årsskrift* 25 (1942): 1–11.

Heckscher, William S. *Rembrandt's "Anatomy of Dr. Nicolaas Tulp": An Iconological Study.* New York: New York University Press, 1958.

Hedberg, Hollis D. "The Influence of Torbern Bergman, 1735–1784, on Stratigraphy: A Résumé." In *Toward a History of Geology,* edited by Cecil J. Schneer, 186–91. Cambridge: MIT Press, 1969.

Hedbom, Karl. "Linnés inhemska medicinalväxter," *Svenska Linnésällskapets Årsskrift* 2 (1919): 65–110.

Heeg, Jürgen. "Die Publicationen Johann Christoph Petris, 1762–1851, über Estland, Livland, und Russland." *Journal of Baltic Studies* 16 (1985): 128–37.

Heidegger, Martin. *The Question Concerning Technology and Other Essays.* Translated by William Lovitt. New York: Harper and Row, 1977.

————. *Holzwege.* 6th ed. Frankfurt am Main: Klostermann, 1980. First edition published 1950.

————. *Kant und das Problem der Metaphysik.* 5th ed. Frankfurt am Main: Klostermann, 1991. First edition published 1929.

Heilbron, John L. *Electricity in the Seventeenth and Eighteenth Centuries: A Study of Early Modern Physics.* Berkeley and Los Angeles: University of California Press, 1979.

————. *Elements of Early Modern Physics.* Berkeley and Los Angeles: University of California Press, 1982.

————. *Weighing Imponderables and Other Quantitative Science around 1800.* Berkeley and Los Angeles: University of California Press, 1993.

Heim, Roger, ed. *Buffon.* Paris: Muséum National d'Histoire Naturelle, 1954.

Heinemann, Käthe. "Aus der Blütezeit der Medizin am Collegium illustre Carolinum zu Kassel." *Zeitschrift des Vereins für Hessische Geschichte und Landeskunde* 71 (1960): 85–96.

Hempel, E. "Justus Mösers Wirkung auf seine Zeitgenossen und auf die deutsche Geschichtsschreibung." *Osnabrücker Mitteilungen* 54 (1933): 1–76.

Henderson, F. R. "Novalis's Idea of 'Experimentalphilosophie': A Study of Romantic Science in Its Context." Ph.D. diss., University of London, 1995.

Heninger, S. K. *A Handbook of Renaissance Meteorology, with Particular Reference to Elizabethan and Jacobean Literature.* Durham: Duke University Press, 1960.

Henrich, Dieter. *Der ontologische Gottesbeweis: Sein Problem und seine Geschichte in der Neuzeit.* Tübingen: Mohr, 1960.

Heussi, Karl. *Johann Lorenz Mosheim: Ein Beitrag zur Kirchengeschichte des 18. Jahrhunderts.* Tübingen: Mohr, 1906.

Hildebrand, Bengt. *Kungl. Svenska Vetenskapsakademien: Förhistoria, grundläggning, och första organisation.* Stockholm: Svenska Vetenskapsakademien, 1939.

Hildenbrock, Aglaja. *Das andere Ich: künstlicher Mensch und Doppelgänger in der deutsch- und englischsprachen Literatur.* Tübingen: Stauffenburg, 1986.

Hilken, T. J. N. *Engineering at Cambridge University, 1783–1965.* Cambridge: Cambridge University Press, 1967.

Hill, Emita. "Materialism and Monsters in 'Le rêve de d'Alembert.'" *Diderot Studies* 10 (1968): 67–93.

————. "The Role of 'le Monstre' in Diderot's Thought." *Studies on Voltaire and the Eighteenth Century* 97 (1972): 149–261.

Hilson, J. C. "Hume: The Historian as Man of Feeling." In *Augustan Worlds,* edited by J. C. Hilson, M. M. B. Jones, and J. R. Watson, 205–22. New York: Barnes and Noble, 1978.

Hinchman, Lewis. *Hegel's Critique of the Enlightenment.* Gainesville: University of Florida Press, 1984.

Hinrichs, Carl. *Preussentum und Pietismus. Die Pietismus in Brandenburg-Preussen als religiössoziale Reformbewegung.* Göttingen: Vandenhoeck und Ruprecht, 1971.

Hinske, Norbert, ed. *Was ist Aufklärung? Beiträge aus der Berlinischen Monatsschrift.* Darmstadt: Wissenschaftliche Buchgesellschaft, 1977.

————, ed. *Halle: Aufklärung und Pietismus.* Wolfenbütteler Studien zur Aufklärung, no. 15 (Zentren der Aufklärung, no. 1). Heidelberg: Schneider, 1989.

————, ed. *Kant als Herausforderung an die Gegenwart.* Freiburg: Alber, 1989

Holländer, Eugen. *Wunder, Wundergeburt, und Wundergestalt.* Stuttgart: Enke, 1921.

Holloran, John. "The Banishment of Christian Wolff: A Study of German Philosophy, Theology, and High Politics in the Age of Enlightenment." Ph.D. diss., University of Virginia, forthcoming.

Holmes, Frederic L. *Lavoisier and the Chemistry of Life.* Madison: University of Wisconsin Press, 1985.

Horkheimer, Max. "The End of Reason." *Studies in Philosophy and Social Sciences* 9 (1941): 366–88.

————. *Between Philosophy and Social Science: Selected Early Writings.* Cambridge: MIT Press, 1993.

————. "Reason against Itself: Some Remarks on Enlightenment" (1946). In *What is Enlightenment? Eighteenth-Century Answers and Twentieth-Century Questions,* edited by James Schmidt, 359–67. Berkeley and Los Angeles: University of California Press, 1996.

Horkheimer, Max, and Theodor Adorno. *Dialektik der Aufklärung*. 1944. Reprint, Frankfurt am Main: Fischer, 1988.

———, *Dialektik der Aufklärung*. Amsterdam: Querido, 1947.

———, *Dialectic of Enlightenment*. Translated by John Cumming. London: Allen Lane, 1973.

Hornig, Gottfried. *Die Anfänge der historisch-kritischen Theologie: Johann Salomo Semlers Schriftenverständnis*. Forschungen zur Systematischen Theologie und Religionsphilosophie, no. 8. Göttingen: Vandenhoeck und Ruprecht, 1961.

Houlding, J. A. *Fit for Service: The Training of the British Army, 1715–1795*. Oxford: Clarendon Press, 1981.

Huard, Pierra A., and Ming Wong. "Structure de la médecine russe au siècle des lumiéres." *Concours Méd.* 87 (1965): 125–34.

Hughes, Basil Perronet. *Firepower: Weapons Effectiveness on the Battlefield, 1630–1850*. New York: Scribner's, 1974.

Huisman, Frank. "Gevestigden en buitenstaanders op de medische markt: De marginalisering van reizende meesters in achttiende-eeuws Groningen." In *Grenzen van genezing: Gezondheid, ziekte, en genezen in Nederland, zestiende tot begin twintigste eeuw*, edited by Willem de Blecourt, Willem Frijhoff, and Marijke Gijswijt-Hofstra, 115–54. Hilversum: Verloren, 1993.

Hulme, Peter, and Ludmilla Jordanova, eds. *The Enlightenment and Its Shadows*. London: Routledge, 1990.

Hulth, J. M. "Kalm som student i Uppsala och lärjunge till Linné åren 1741–1747." *Svenska Linnésällskapets Årsskrift* 7 (1924): 39–49.

Hunt, Lynn, ed. *The New Cultural History*. Berkeley and Los Angeles: University of California Press, 1989.

Ignatieff, Michael. *A Just Measure of Pain: The Penitentiary in the Industrial Revolution*. London: Macmillan, 1978.

International Working Group for Indigenous Affairs (Oslo and Copenhagen groups). *Self Determination and Indigenous Peoples: Sami Rights and Northern Perspectives*. Copenhagen: IWGIA, 1987.

Ischreyt, Heinz. "Zu den Wirkungen von Tissots Schrift 'Avis au Peuple sur sa Santé' in Nordosteuropa." In *Wissenschaftspolitik in Mittel- und Ost-Europa. Wissenschaftliche Gesellschaften, Akademien, und Hochschulen im 18. und beginnenden 19. Jahrhundert*, edited by Erik Amburger, Michal Cieslaj, and Laszlo Sziklay, 247–58. Berlin: Verlag Ulrich Camen, 1976.

Ischreyt, Irene and Heinz Ischreyt. *Der Arzt als Lehrer: Populärmedizinische Publizistik in Liv-Est- und Kurland als Beitrag zur volkstumlichen Aufklärung im 18. Jahrhundert*. Luneberg: Verlag Norostdeutsches Kulturwerk, 1990.

Isherwood, Robert M. *Farce and Fantasy: Popular Entertainment in Eighteenth-Century Paris*. Oxford: Oxford University Press, 1986.

Jacob, James R. *Robert Boyle and the English Revolution*. New York: B. Franklin, 1977.

Jacob, Margaret C. *The Cultural Meaning of the Scientific Revolution*. Philadelphia: Temple University Press, 1987.

———. "Scientific Culture in the Early English Enlightenment: Mechanisms, Industry, and Gentlemanly Facts." In *Anticipations of the Enlightenment in England, France, and Germany*, edited by Alan Charles Kors and Paul J. Korshin, 134–64. Philadelphia: University of Pennsylvania Press, 1987.

———. "The Enlightenment Redefined: The Formation of Modern Civil Society." *Social Research* 58 (1991): 475–95.

———. *Living the Enlightenment: Freemasonry and Politics in Eighteenth-Century Europe*. New York and Oxford: Oxford University Press, 1991.

———. "The Materialist World of Pornography." In *The Invention of Pornography: Obscenity and the Origins of Modernity, 1500–1800*, edited by Lynn Hunt, 157–202. New York: Zone Books, 1993.

———. *Scientific Culture and the Making of the Industrial West*. Oxford: Oxford University Press, 1997.

Jahn, Ilse, and Konrad Senglaub. *Carl von Linné*. Leipzig: Teubner, 1987.

Jardine, Nicholas. *The Scenes of Inquiry: On the Reality of Questions in the Sciences*. Oxford: Oxford University Press, 1991.

Jardine, Nicholas, and E. C. Spary. "The Natures of Cultural History." In *Cultures of Natural History*, edited by Nicholas Jardine, J. A. Secord, and E. C. Spary, 3–13. Cambridge: Cambridge University Press, 1996.

Jardine, Nicholas, J. A. Secord, and E. C. Spary, eds. *Cultures of Natural History*. Cambridge: Cambridge University Press, 1996.

Jaynes, Julian. "The Problem of Animate Motion in the Seventeenth Century." *Journal of the History of Ideas* 31 (1970): 219–34.

Jelgersma, H. G. *Galgebergen en Galgevelden*. Zutphen: Walburg Pers, 1978.

Jennison, Earl W., Jr. "Christian Garve and Garlieb Merkel: Two Theorists of Peasant Emancipation during the Ages of Enlightenment and Revolution." *Journal of Baltic Studies* 4 (1973): 344–63.

Jones, Colin. "Bourgeois Revolution Revivified: 1789 and Social Change." In *Rewriting the French Revolution*, edited by Colin Lucas. Oxford: Clarendon Press, 1991.

Jones, Peter. *Philosophy and the Novel*. Oxford: Clarendon Press, 1975.

———, ed. *The "Science of Man" in the Scottish Enlightenment: Hume, Reid, and Their Contemporaries*. Edinburgh: Edinburgh University Press, 1989.

Jörberg, Lennart. "The Nordic Countries, 1850–1914." In *The Fontana Economic History of Europe*, edited by Carlo Cipolla, vol. 4, part 2, 375–485. New York: Barnes and Noble, 1976.

Jordanova, Ludmilla. *Sexual Visions: Images of Gender in Science and Medicine between the Eighteenth and Twentieth Centuries*. Madison: University of Wisconsin Press, 1989.

———. "Gender and the Historiography of Science." *British Journal for the History of Science* 26 (1993): 469–83.

Jordanova, Ludmilla, ed. *Languages of Nature: Critical Essays on Science and Literature*. London: Free Association Books, 1986.

Jürjo, Indrek. "August Wilhelm Hupel als Repräsentant der baltischen Aufklärung." *Jahrbücher für Geschichte Osteuropas* 39 (1991): 495–513.

Kahk, Juhan. *Peasant and Lord in the Process of Transition from Feudalism to Capitalism in the Baltics: An Attempt of Interdisciplinary History*. Tallinn: Eesti Raamat, 1982.

Keller, Evelyn Fox. *Reflections on Gender and Science*. New Haven: Yale University Press, 1985.

Kelly, Michael, ed. *Critique and Power: Recasting the Foucault/Habermas Debate*. Cambridge: MIT Press, 1994.

Kemp Smith, Norman. *The Philosophy of David Hume: A Critical Study of Its Origin and Central Doctrines*. London: Macmillan, 1941.

Kennedy, Emmet. *A Cultural History of the French Revolution*. New Haven: Yale University Press, 1989.

Kennett, Lee. *The French Armies in the Seven Years' War: A Study in Military Organization and Administration*. Durham: Duke University Press, 1967.

Kernan, Alvin. *Printing Technology, Letters, and Samuel Johnson*. Princeton: Princeton University Press, 1987.

Keymer, Tom. *Richardson's "Clarissa" and the Eighteenth-Century Reader*. Cambridge: Cambridge University Press, 1992.

Keyser, Marja. "Het intekenboek van Benjamin Bosma: Natuurwetenschappelijk en wijsgerig onderwijs te Amsterdam, 1752–1790: Een Verkenning," *Jaarverslagen Koninklijk Oudheidkundig Genootschap*, 1986, 65–81.

Kirby, David. *Northern Europe in the Early Modern Period: The Baltic World, 1492–1772*. London: Longman, 1990.

Kleinert, Andreas. "La vulgarisation de la physique au siècle des lumières." *Francia: Forschungen zur Westeuropäischen Geschichte* 10 (1982): 303–12.

Klose, Olaf. "Die Jahrzehnte der Widervereinigung, 1721–1773." In *Die Herzogtümer in Gesamtstaat, 1721–1830*, by Olaf Klose and Christian Degn, 1–159. Vol 6 of *Geschichte Schleswig-Holstein*, edited by Olaf Klose. Neumünster: Karl Wachholtz Verlag, 1960.

Knudsen, J. B. "Justus Möser: Local History as Cosmopolitan History." In *Aufklärung und Geschichte: Studien zur deutschen Geschichtswissenschaft im 18. Jahrhundert*, edited by H. E.

Bödeker, G. G. Iggers, J. B. Knudsen, and P. H. Reill, 324–43. Göttingen: Vandenhoeck und Ruprecht, 1986.

———. *Justus Möser and the German Enlightenment.* Cambridge: Cambridge University Press, 1987.

Kock, Wolfram. "Medical Education in Scandinavia since 1600." In *The History of Medical Education,* edited by C. D. O'Malley, 263–99. Cambridge: Cambridge University Press, 1970.

Koepp, Cynthia J. "The Alphabetical Order: Work in Diderot's *Encyclopédie.*" In *Work in France: Representations, Meaning, Organization, and Practice,* edited by Steven Laurence Kaplan and Cynthia J. Koepp, 229–57. Ithaca: Cornell University Press, 1986.

Koerner, J. L. *Caspar David Friedrich and the Subject of Landscape.* London: Reaktion Books, 1990.

Koerner, Lisbet. "Linnaeus' Floral Transplants." *Representations* no. 47 (1994): 144–69.

Kohlbrugge, J. H. F. "G. Cuvier und K. F. Kielmeyer." *Biologisches Centralblatt* 32 (1912): 291–95.

Kolchin, Peter. *Unfree Labor: American Slavery and Russian Serfdom.* Cambridge: Harvard University Press, Belknap Press, 1987.

Körner, Theodor. "Der Begriff des materiellen Punktes in der Mechanik des achtzehnten Jahrhunderts." *Bibliotheca Mathematica,* 3d ser., 5 (1904): 15–62.

Kors, Alan C. *D'Holbach's Coterie: An Enlightenment in Paris.* Princeton: Princeton University Press, 1976.

Koselleck, Reinhart. "Standortbindung und Zeitlichkeit. Ein Beitrag zur historiographischen Erschliessung der geschichtlischen Welt." In *Objectivität und Parteilichkeit in der Geschichtswissenschaft,* edited by Reinhart. Koselleck, W. J. Mommsen, and Jörn Rüsen, 17–36. Munich: Deutscher Taschenbuch-Verlag, 1977.

———. *Futures Past: On the Semantics of Historical Time.* Translated by Keith Tribe. Cambridge: Harvard University Press, 1985.

———. *Critique and Crisis: Enlightenment and the Pathogenesis of Modern Society.* Oxford: Berg, 1988.

Kox, A. J., ed. *Van Stevin tot Lorentz.* Amsterdam: Uitgeveeerij Bert Bakker, 1990.

Koyré, Alexandre. "Commentary." In *Scientific Change,* edited by A. C. Crombie, 847–57. London: Heinemann, 1963.

———. *De la mystique à la science: Cours, conferences, et documents, 1922–1962,* edited by Pietro Redondi. Paris: École des Hautes Études en Sciences Sociales, 1986.

Kozielek, Gerard. "Aufgeklärtes Gedankengut in der Tätigkeit der deutschen Gesellschaft in Königsberg." In *Wissenschaftspolitik,* 321– 47. OECD-Konferenz, Paris 1963. Bonn: Standige Konferenz der Kultusminister der Landes in der Bundesrepublik Deutschland, 1964.

Kranakis, Eda. "The Social Determinants of Engineering Practice: A Comparative View of France and America in the Nineteenth Century." *Social Studies of Science* 19 (1989): 5–70.

———. *Constructing a Bridge: An Exploration of Engineering Culture, Design, and Research in Nineteenth-Century France and America.* Cambridge: MIT Press, 1997.

Kreager, Philip. "New Light on Graunt." *Population Studies* 42 (1988): 129– 40.

———. "Early Modern Population Theory: A Reassessment." *Population and Development Review* 17 (1991): 207–27.

———. "Quand une population est-elle une nation? Quand une nation est-elle un état? La démographie et l'émergence d'un dilemme moderne, 1770–1870." *Population* 6 (1992): 1639–56.

———. "Histories of Demography: A Review Article." *Population Studies* 47 (1993): 519–39.

Kroeber, A. L., and Clyde Kluckhohn. *Culture: A Critical Review of Concepts and Definitions.* Papers of the Peabody Museum of American Archeology and Ethnology, vol. 47. Cambridge: Harvard University Press, 1952.

Kuhn, D. "Der naturwissenschaftliche Unterricht an der Hohen Karlsschule." *Medizinhistorisches Journal* 11 (1976): 315–34.

Kuhn, Thomas S. "The Function of Dogma in Scientific Research." In *Scientific Change,* edited by A. C. Crombie, 347–69. London: Heinemann, 1963.

———. "Mathematical versus Experimental Traditions in the Development of Physical Science" (1975). In *The Essential Tension,* 31–65. Chicago: University of Chicago Press, 1977.

Kuhrke, Walter. *Kant und seine Umgebung.* Königsberg: Gräfe und Unzer, 1924.

La Balme, Patricia, ed. *Beyond Their Sex: Learned Women of the European Past.* New York: New York University Press, 1980.

La Vopa, Anthony J. "Conceiving a Public: Ideas and Society in Eighteenth-Century Europe." *Journal of Modern History* 64 (1992): 79–116.

Laissus, Yves. "Les cabinets d'histoire naturelle." In *Enseignement et diffusion des sciences en France au XVIII siècle,* edited by René Taton, 342–84. Paris: Hermann, 1964.

Lamm, Martin. *Upplysningstidens romantik.* 2 vols. Stockholm: Gebers, 1918.

Landes, David. *Revolution in Time.* Cambridge: Harvard University Press, 1983.

Landes, Joan. *Women and the Public Sphere in the Age of the French Revolution.* Ithaca: Cornell University Press, 1988.

Langins, Janis. "Sur l'enseignement et les examens à l'École Polytechnique sous le Directoire: A propos d'une lettre inédite de Laplace." *Revue d'Histoire des Sciences* 40 (1987): 145–77.

Langlois, C. V., and C. Seignobos. *Introduction to the Study of History.* Translated by G. G. Berry. New York: Henry Holt, 1898.

Laqueur, Thomas. *Making Sex: Body and Gender from the Greeks to Freud.* Cambridge: Harvard University Press, 1990.

Larson, James L. *Interpreting Nature: The Science of Living Form from Linnaeus to Kant.* Baltimore: Johns Hopkins University Press, 1994.

Latour, Bruno. *Science in Action.* Milton Keynes: Open University Press, 1987.

———. *We Have Never Been Modern.* New York: Harvester Wheatsheaf, 1993.

Lauerma, Matti. *L'artillerie de campagne française pendant les guerres de la Révolution: Évolution de l'organisation et de la tactique.* Annales Academiae Scientiarum Fennicae, ser. B, vol. 96. Helsinki, 1956.

Lawrence, Christopher. "The Nervous System and Society in the Scottish Enlightenment." In *Natural Order: Historical Studies of Scientific Culture,* edited by Barry Barnes and Steven Shapin, 19–40. Beverly Hills and London: Sage Publications, 1979.

Lazzaro, Alfonso. *Rovigo nel Settecento.* Rovigo: Il Polesine Fascista, 1936.

Le Puillon de Boblaye, Théodor. *Esquisse historique sur les Écoles d'Artillerie pour servir à l'histoire de l'École d'Application de l'Artillerie et du Génie.* Metz: Rousseau-Pallez, 1858.

Lefkowitz, Mary R., and Guy M. Rogers, eds. *Black Athena Revisited.* Chapel Hill: University of North Carolina Press, 1996.

Leith, James A. *The Idea of Art as Propaganda in France, 1750–1799.* Toronto: University of Toronto Press, 1965.

Lénardon, Dante. *Index du Journal Encyclopédique, 1756–1793.* Geneva: Slatkine Reprints, 1976.

Lenoir, Timothy. *The Strategy of Life: Teleology and Mechanics in Nineteenth-Century Biology.* 1982. Reprint, Chicago: University of Chicago Press, 1989.

Lepenies, Wolf. *Das Ende der Naturgeschichte: Wandel kultureller Selbstverständlichkeiten in den Wissenschaften des 18. und 19. Jahrhunderts.* Munich: Hanser Verlag, 1976.

Lesch, John E. "Systematics and the Geometrical Spirit." In *The Quantifying Spirit in the Eighteenth Century,* edited by Tore Frängsmyr, J. L. Heilbron, and Robin E. Rider, 73–111. Berkeley and Los Angeles: University of California Press, 1990.

Levi-Bruhl, Lucien. "L'orientation de la pensée philosophique de David Hume." *Revue de Metaphysique et de Morale* 17 (1909): 595–619.

Lewalter, Ernst. *Spanisch-jesuitische und deutsch-lutheranishce Metaphysik des 17. Jahrhunderts,* Ibero-Amerikanische Studien, no. 4. Hamburg: Ibero-amerikanishes Institut, 1935.

Lieburg, M. J. van, and H. A. M. Snelders. *"De Bevordering en Volmaking der Proefondervindelijke Wijsbegeerte": De rol van het Bataafsch Genootschap te Rotterdam in de geschiedenis van de natuurwetenschappen, geneeskunde en techniek, 1769–1988,* Amsterdam: Rodopi, 1989.

Linden, Mareta. "Untersuchungen zum Anthropologiebegriff des 18. Jahrhunderts." Inaugural diss., University of Trier, 1975.

Lindqvist, Svante. "Labs in the Woods: The Quantification of Technology during the Late Enlightenment." In *The Quantifying Spirit in the Eighteenth Century,* edited by Tore Frängsmyr, J. L. Heilbron, and Robin E. Rider, 291–314. Berkeley and Los Angeles: University of California Press, 1990.

Lindroth, Sten. "The Two Faces of Linnaeus." In *Linnaeus: The Man and His Work,* edited by Tore Frängsmyr, 1–62. Berkeley and Los Angeles: University of California Press, 1983.

Lindskog, Bengt I., and B. I. Zetterberg. "Aus Protokollen über das Examen Medicum und Practicum im 18. Jahrhundert an der Universität Lund, Schweden." In *Verhandlungen von der 20 Internationale Kongress von der Geschichte der Medizin,* 768–70. Hildesheim: Olms, 1968.

Linebaugh, Peter. *The London Hanged.* Harmondsworth: Penguin Books, 1991.

Lively, Jack. "The Europe of the Enlightenment." *History of European Ideas* 1 (1981): 91–102.

Livingston, Donald W. *Hume's Philosophy of Common Life.* Chicago: University of Chicago Press, 1984.

Logan, Gabriella Berti. "The Desire to Contribute: An Eighteenth-Century Italian Woman of Science." *American Historical Review* 99 (1994): 785–812.

Lohff, B. *Die Suche nach Wissenschaftlichkeit der Physiologie in der Zeit der Romantik.* Stuttgart: G. Fischer, 1990.

Lönnrot, Elias. *The Kalevala; or, Poems of the Kalevala District.* Translated by Francis Peabody Magoun, Jr. Cambridge: Harvard University Press, 1963.

Lougee, Carolyn C. *"Le Paradis des Femmes": Women, Salons, and Social Stratification in Seventeenth-Century France.* Princeton: Princeton University Press, 1976.

Louis, Eleonora. "Der beredte Leib." In *Die Beredsamkeit des Leibes: Zur Körpersprache in der Kunst,* edited by Ilsebill Barta Fliedl and Christoph Geissmar, 113–54. Salzburg: Residenz, 1992.

Lovejoy, Arthur O. *Essays in the History of Ideas.* 1949. Reprint, New York: Capricorn Books, 1960.

Lucas, Colin. "Nobles, Bourgeois, and the Origins of the French Revolution." *Past and Present* 60 (1973): 84–126.

Lukina, Tatjana A. "Caspar Friedrich Wolff und die Petersburger Akademie der Wissenschaften." *Acta Historica Leopoldina* (Festschrift for Georg Uschmann), 9 (1975): 411–25.

Lundgreen, Peter. "Engineering Education in Europe and the U.S.A., 1750–1930: The Rise to Dominance of School Culture and the Engineering Profession." *Annals of Science* 47 (1990): 33–75.

Lunsingh Scheurleer, Th. H. "Un amphithéatre d'anatomie moralisée." In *Leiden University in the Seventeenth Century: An Exchange of Learning,* edited by Th. H. Lunsingh Scheurleer and G. H. M. Posthumus Meyjes, 217–69. Leiden: Leiden University Press, 1975.

Luyendijk-Elshout, Antonie M. "Death Enlightened: A Study of Frederik Ruysch." *Journal of the American Medical Association* 212 (1970): 121–26.

Lynn, John A. "The *Trace Italienne* and the Growth of Armies: The French Case." *Journal of Military History* 55 (1991): 297–330.

McClellan, James E. *Science Reorganized: Scientific Societies in the Eighteenth Century.* New York: Columbia University Press, 1985.

Macey, David. *The Lives of Michel Foucault: A Biography.* New York: Vintage, 1993.

McKendrick, Neil. "Josiah Wedgwood and Factory Discipline." *Historical Journal* 4 (1961): 30–55.

McKendrick, Neil, John Brewer, and J. H. Plumb. *The Birth of a Consumer Society: The Commercialization of Eighteenth-Century England.* Bloomington: Indiana University Press, 1982.

McKenzie, A. T. "The Systematic Scrutiny of Passion in Johnson's Rambler." *Eighteenth Century Studies* 20, no. 2 (1986): 129–52.

MacKenzie, Donald. *Inventing Accuracy: A Historical Sociology of Nuclear Missile Guidance.* Cambridge: MIT Press, 1990.

McKeon, Michael. *The Origins of the English Novel, 1600–1740.* Baltimore: Johns Hopkins University Press, 1987.

MacLean, Kenneth. *John Locke and English Literature of the Eighteenth Century.* New Haven: Yale University Press, 1936.

McNeill, William H. *The Pursuit of Power: Technology, Armed Force, and Society since A.D. 1000.* Chicago: University of Chicago Press, 1982.

Maffioli, C. S., and L. C. Palm, eds. *Italian Scientists in the Low Countries in the Seventeenth and Eighteenth Centuries.* Amsterdam: Rodopi, 1989.

Magli, Patrizia. "The Face and the Soul." In *Fragments for a History of the Human Body*, edited by Michel Feher, Ramona Naddaff, and Nadia Tazi, 2:87–127. New York: Zone Books, 1989.

Malakhovskii, K. V. "Russian Discoveries in the Pacific Ocean from the Seventeenth to the Nineteenth Century." *Soviet Studies in History* 21 (1983): 27–47.

Malson, Lucien. *Les enfants sauvages: Mythe et réalité.* Paris: Union Générale d'Editions, 1964.

Mandelstam, Osip. *Selected Poems.* Edited by Clarence Brown and W. S. Merwin. Harmondsworth: Penguin Books, 1977.

Manning, Susan. "'This Philosophical Melancholy': Style and Self in Boswell and Hume." In *New Light on Boswell: Critical and Historical Essays on the Occasion of the Bicentenary of "The Life of Johnson,"* edited by Greg Clingham, 126–40. Cambridge: Cambridge University Press, 1991.

Manzoni, Claudio. *I cartesiani italiani, 1660–1760.* Udine: Edizione "La Nuova Base," 1984.

Marcuse, Herbert. *Der Eindimensionale Mensch: Studien zur Ideologie der fortgeschrittenen Industriegesellschaft.* Frankfurt am Main: Suhrkamp, 1967.

Marion, Michel. *Recherches sur les bibliothèques privées à Paris au milieu du XVIIIe siècle, 1750–1759.* Paris: Bibliothèque Nationale, 1978.

Marker, Gary. "Merchandising Culture: The Market for Books in Late Eighteenth-Century Russia." *Eighteenth-Century Life* 8 (1982): 46–71.

——. "Russian Journals and Their Readers in the Late Eighteenth Century." *Oxford Slavonic Papers* 19 (1986): 88–101.

Markus, Thomas. *Buildings and Power: Freedom and Control in the Origin of Modern Building Types.* London: Routledge, 1993.

Markusson, Ingrid. "The Development of Writing Ability in the Nordic Countries in the Eighteenth and Nineteenth Centuries." *Scandinavian Journal of History* 15 (1990): 37–63.

Martis, Ela. "The Role of Tartu University in the National Movement." In Conference on Baltic Studies in Scandinavia, *National Movements in the Baltic Countries during the Nineteenth Century,* 317–25. Stockholm: Center for Baltic Studies at the University of Stockholm, distributed by Almquvist & Wiskell International, 1985.

Marx, Karl. *Grundrisse.* Edited by Martin Nicolaus. Harmondsworth: Penguin Books, 1973.

——. *Capital.* Vol. 1. Edited by Ernest Mandel. Harmondsworth: Penguin Books, 1976. Original edition published 1867.

Maschietto, Francesco Ludovico. *Elena Cornaro Piscopia, 1646–1684, prima donna laureata nel mondo.* Padua: Antenore, 1978.

Mattelart, Armand. *L'invention de la communication.* Paris: La Découverte, 1994.

Maugain, Gabriel. *Étude sur l'evolution intellectuelle de l'Italie de 1657–1750 environ.* Paris: Hachette, 1909.

Mayr, Otto. *Authority, Liberty, and Automatic Machinery in Early Modern Europe.* Baltimore: Johns Hopkins University Press, 1986.

Maza, Sarah. *Private Lives and Public Affairs: The Causes Célèbres of Prerevolutionary France.* Berkeley and Los Angeles: University of California Press, 1994.

Mazzetti, Adriano, and Enrico Zerbinati, eds. *Le "iscrizioni di Rovigo" delineate da Marco Antonio Campagnella: Contributi per la storia di Rovigo nel periodo veneziano.* Trieste: LINT, 1986.

Mazzolini, Renato G., ed. *Non-Verbal Communication in Science prior to 1900.* Florence: Leo S. Olschki, 1993.

Meinecke, Friedrich. *Historism: The Rise of a New Historical Outlook.* Translated by J. E. Anderson. London: Routledge and Kegan Paul, 1972.

Melton, Edgar. "Enlightened Seignorialism and Its Dilemmas in Serf Russia, 1750–1830." *Journal of Modern History* 62 (1990): 675–708.

Merrick, Jeffrey. "Royal Bees: The Gender Politics of the Beehive in Early Modern Europe." *Studies in Eighteenth-Century Culture* 18 (1988): 7–37.

Mezger, Werner. *Hofnarren im Mittelalter: Vom tieferen Sinn eines seltsamen Amts.* Konstanz: Universitätsverlag, 1981.

Middleton, W. E. Knowles. *The History of the Barometer.* Baltimore: Johns Hopkins University Press, 1964.

Mijnhardt, Wijnand W. "Veertig jaar cultuurbevordering: Teylers Stichting, 1778–1815." In *"Teyler," 1778–1978: Studies en bijdragen over Teylers Stichting naar aanleiding van het tweede eeuw feest,* 58–111. Haarlem: Schuyt, 1978.

———. *Tot Heil van 't Menschdom: Culturele genootschappen in Nederland, 1750–1815.* Amsterdam: Rodopi, 1988.

———. "The Dutch Enlightenment: Humanism, Nationalism, and Decline." In *The Dutch Republic in the Eighteenth Century: Decline, Enlightenment, and Revolution,* edited by Margaret C. Jacob and Wijnand Mijnhardt, 197–223. Ithaca: Cornell University Press, 1992.

Mild, Warren. *Joseph Highmore of Holborn Row.* London: Phyllis Mild, 1990.

Miller, David Philip, and Peter Hanns Reill, eds. *Visions of Empire: Voyages, Botany, and Representations of Nature.* Cambridge: Cambridge University Press, 1996.

Moeller, Bernd, ed. *Theologie in Göttingen: Eine Vorlesungsreihe.* Göttinger Universitätsschriften, ser. A, vol. 1. Göttingen: Vandenhoeck und Ruprecht, 1987.

Molhuysen, P. C. *Bronnen tot de Geschiedenis der Leidsche Universiteit.* Vol. 7. 's Gravenhage: Martinus Nijhoff, 1924.

Moravia, Sergio. "The Enlightenment and the Sciences of Man." *History of Science* 18 (1980): 247–68.

Morgan, S. R. "The Palingenesis of Ancient Wisdom and the Kingdom of God: An Historical Interpretation of Schelling's Earliest Philosophy." Ph.D. diss., Cambridge University, 1995.

Mornet, Daniel. *Le sentiment de la nature en France de Jean-Jacques Rousseau à Bernardin de Saint-Pierre: Essai sur les Rapports de la littérature et des moeurs.* Paris: Hachette, 1907.

———. "Les Enseignements des bibliothèques privées, 1750–1780." *Revue de l'Histoire Littéraire de la France* 17 (1910): 449–96.

———. *Les sciences de la nature en France, au XVIIIe siècle: Un chapitre de l'histoire des idées.* Paris: A. Colin, 1911.

———. *Les origines intellectuelles de la Révolution française.* 1933. Reprint, Paris: Colin, 1967.

Mortier, Roland. "Variations on the Dialogue in the French Enlightenment." *Studies in Eighteenth-Century Culture* 16 (1986): 225–40.

Morton, Alan. "Concepts of Power: Natural Philosophy and the Uses of Machines in Mid-Eighteenth-Century London." *British Journal for the History of Science* 28 (1995): 63–78.

Moscoso, Javier. "Vollkommene Monstren und unheilvolle Gestalten: Zur Naturalisierung der Monstrosität im 18. Jahrhundert." In *Der "falsche" Körper: Beiträge zu einer Geschichte der Monstrositäten,* edited by Michael Hagner, 56–72. Göttingen: Wallstein, 1995.

Mossner, Ernest Campbell. *The Forgotten Hume: Le Bon David.* New York: Columbia University Press, 1943.

———. "The Continental Reception of Hume's Treatise, 1739–1741." *Mind* 45 (1947): 31–43.

———. *The Life of David Hume.* Edinburgh: Thomas Nelson and Sons, 1954.

Mousnier, Roland. *Les hiérarchies sociales de 1450 à nos jours.* Paris: Presses Universitaires de France, 1969.

Mullan, John. *Sentiment and Sociability: The Language of Feeling in the Eighteenth Century.* Oxford: Clarendon Press, 1988.

Müller-Dietz, Heinz E. "Anatomische Präparate in der Petersburger 'Kunstkammer.'" *Zentralblatt für Allgemeine Pathologie und Pathologische Anatomie* 135 (1989): 757–67.

Müller-Sievers, Helmut. *Epigenesis: Naturphilosophie im Sprachdenken Wilhelm von Humboldts.* Paderborn: Schöningh, 1993.

———. *Self-Generation: Biology, Philosophy, and Literature around 1800.* Stanford: Stanford University Press, 1997.

Myers, S. H. *The Bluestocking Circle: Women, Friendship, and the Life of the Mind in Eighteenth-Century England.* Oxford: Clarendon Press, 1990.

Mykkänen, Juri. "'To Methodize and Regulate them': William Petty's Governmental Science of Statistics." *History of the Human Sciences* 7 (1994): 65–88.

Needham, Joseph. "Poverties and Triumphs of the Scientific Tradition." In *Scientific Change,* edited by A. C. Crombie, 117–53. London: Heinemann, 1963.

Nef, John U. *War and Human Progress: An Essay on the Rise of Industrial Civilization.* Cambridge: Harvard University Press, 1950.

Neuschäffer, Hubertus. "Der livländische Pastor und Kameralist Johann Georg Eisen von Schwarzenberg: Ein deutscher Vertreter der Aufklärung in Russland zu Beginn der zweiten Hälfte des 18. Jahrhunderts." In *Russland und Deutschland,* edited by Uwe Liszkowski, 120–43. Stuttgart: Ernst Klett Verlag, 1974.

————. "Die Anfänge der livländischen Ökonomischen Sozietät, 1792–1939." *Journal of Baltic Studies* 10 (1979): 337–44.

————. "Carl Friedrich Frhr. von Schoultz-Ascheraden: Ein Beitrag zum Forschungsproblem der Agrarreformen im Ostseeraum des 18. Jahrhunderts." *Journal of Baltic Studies* 12 (1981): 318–32.

————. "Katharina II und die Agrarfrage in den baltischen Provinzen." *Journal of Baltic Studies* 14 (1983): 109–20.

Neverov, Oleg. "'His Majesty's Cabinet' and Peter I's Kunstkammer." In *The Origins of Museums: The Cabinets of Curiosity in Sixteenth- and Seventeenth-Century England,* edited by Oliver Impey and Arthur MacGregor, 54–61. Oxford: Clarendon Press, 1985.

Nietzhammer, Lutz. *Posthistoire: Has History Come to an End?* London: Verso, 1992.

Norberg, Kathryn. "The Libertine Whore: Prostitution in French Pornography from Margot to Juliette." In *The Invention of Pornography: Obscenity and the Origins of Modernity, 1500–1800,* edited by Lynn Hunt, 225–52. New York: Zone Books, 1993.

Nuyens, B. W. Th. "Het ontleedkundig onderwijs en de geschilderde Anatomische lessen van het Chirurgijns Gilde te Amsterdam, in de jaren 1550–1798." *Jaarverslag Koninklijk Outheidkundig Genootschap,* 1927, 45–90.

Nyhart, Lynn K. *Biology Takes Form: Animal Morphology in the German Universities, 1800–1900.* Chicago: University of Chicago Press, 1995.

Oehler-Klein, Sigrid. "Anatomie und Kunstgeschichte: Soemmerrings Rede 'Über die Schönheit der antiken Kinderköpfe' vor der Société des Antiqués in Kassel, 1779." In *Samuel Thomas Soemmerring in Kassel, 1779–1784: Beiträge zur Wissenschaftsgeschichte der Goethezeit.* Soemmerring-Forschungen, edited by Manfred Wenzel, 9:189–225. Stuttgart and New York: Fischer, 1995.

Olmi, Giuseppe, 1993. "From the Marvellous to the Commonplace: Notes on Natural History Museums, 16th–18th Centuries." In *Non-Verbal Comunication in Science prior to 1900,* edited by Renato G. Mazzolini, 235–78. Florence: Leo S. Olschki, 1993.

Ophir, Adi, and Steven Shapin. "The Place of Knowledge: A Methodological Survey." *Science in Context* 4 (1991): 3–21.

Orcel, M. J. "Daubenton, 1716–1799: Organisateur du Cabinet d'Histoire Naturelle et créateur de l'enseignement de la minéralogie au Jardin du Roi, puis au Muséum." *84e Congrès des Sociétés Savantes, Dijon, 1959,* 41–62.

Ord-Hume, Arthur W. J. G. *Clockwork Music.* London: George Allen and Unwin, 1973.

Otterspeer, W. "De Aangenaamheden der Natuurlijke Historie: Leven en Werk van Petrus Camper." In *Petrus Camper, 1722–1789: Onderzoeker van Nature,* edited by J. Schuller tot Peursum-Meijer and W. R. H. Koops, 5–22. Groningen: Universiteitsmuseum Groningen, 1989.

Outram, Dorinda. *The Body and the French Revolution: Sex, Class, and Political Culture.* New Haven: Yale University Press, 1989.

————. *The Enlightenment.* New Approaches to European History. Cambridge: Cambridge University Press, 1995.

Park, Katharine, and Lorraine Daston. "Unnatural Conceptions: The Study of Monsters in Sixteenth and Seventeenth-Century France and England." *Past and Present* 92 (1981): 20–54.

————. *Wonders and the Order of Nature, 1150–1750.* New York: Zone Books, 1998.

Pasquino, Pasquale. "Theatrum Politicum: The Genealogy of Capital—Police and the State of Prosperity." In *The Foucault Effect: Studies in Governmentality,* edited by Graham Burchell, Colin Gordon, and Peter Miller, 105–18. Chicago: University of Chicago Press, 1991.

Pater, C. de. *Petrus van Musschenbroek, 1692–1761, een Newtoniaans natuuronderzoeker.* Utrecht: Elinkwijk, 1979.

Pearson, Karl. *The History of Statistics in the Seventeenth and Eighteenth Centuries against the Changing Background of Intellectual, Scientific, and Religious Thought.* Edited by E. S. Pearson. London: Charles Griffin, 1978.

Perkin, Harold. *The Third Revolution: Professional Elites in the Modern World.* London: Routledge, 1996.

Pesante, Maria Luisa. "An Impartial Actor: The Private and the Public Sphere in Adam Smith's Theory of Moral Sentiments." In *Shifting the Boundaries: Transformation of the Languages of Public and Private in the Eighteenth Century,* edited by Dario Castiglione and Lesley Sharpe, 172–95. Exeter: University of Exeter Press, 1995.

Peter, Jean-Pierre. "Disease and the Sick at the End of the Eighteenth Century." In *Biology of Man in History,* edited by Robert Forster and Orest Ranum, translated by Elborg Forster and Patricia Ranum, 81–124. Baltimore: Johns Hopkins University Press, 1975.

Philipp, Wolfgang. *Das Werden der Aufklärung in theologiegeschichtlicher Sicht.* Forschungen zur Systematischen Theologie und Religionsphilosophie, no. 3. Göttingen: Vandenhoeck und Ruprecht, 1957.

Philippi, Hans. *Landgraf Karl von Hessen-Kassel: Ein deutscher Fürst der Barockzeit.* Marburg: Trautvetter und Fischer, 1976.

Phillipson, Nicholas. "Towards a Definition of the Scottish Enlightenment." In *City and Society in the Eighteenth Century,* edited by Paul Fritz and David Williams, 125–47. Toronto: Hakkert, 1973.

———. *Hume.* London: Weidenfeld and Nicolson, 1989.

Pickerodt, Gerhart, ed. *Georg Forster in seiner Epoche.* Literatur im historischen Prozess, new ser., no. 4. Berlin: Argument, 1982.

Picon, Antoine. "Gestes ouvriers, opérations, et processus techniques: La vision du travail des encyclopédistes." *Recherches sur Diderot et sur l'Encyclopédie* 13 (1992): 131–47.

———. *L'invention de l'ingénieur moderne: L'École des Ponts et Chaussées, 1747–1851.* Paris: Presses de l'École Nationale des Ponts et Chaussées, 1992.

Pietropoli, Giuseppe. *L'Accademia dei Concordi nella vita rodigina.* Padua: Signum, 1986.

Pilkington, A. E. "'Nature' as Ethical Norm in the Enlightenment." In *Languages of Nature: Critical Essays on Science and Literature,* edited by Ludmilla Jordanova, 51–85. London: Free Association Books, 1986.

Polonoff, Irving. *Force, Cosmos, Monads, and Other Themes of Kant's Early Thought.* Kantstudien. Ergänzungshefte, no. 107. Bonn: Bouvier, 1973.

Pols, K. van der. "De introductie van de stoommachine in Nederland." In *Monumenten van Bedrijf en Techniek,* edited by P. Nijhof, 51–61. Zutphen: Walburg Pers, 1978.

Pomian, Krzysztof. *Collectionneurs, amateurs, et curieux: Paris, Venise, XVIe–XVIIIe siécle.* Paris: Gallimard, 1987.

———. *Collectors and Curiosities: Paris and Venice, 1500–1800.* Cambridge: Polity Press, 1990.

Pompa, Leon. *Human Nature and Historical Knowledge.* Cambridge: Cambridge University Press, 1991.

Poni, Carlo. "The Craftsman and the Good Engineer: Technical Practice and Theoretical Mechanics in J. T. Desaguliers." *History and Technology* 19 (1993): 215–32.

Porter, Roy. "Provincial Culture and Public Opinion in Enlightenment England." *British Journal for Eighteenth-Century Studies* 3 (1980): 20–46.

———. "The Enlightenment in England." In *The Enlightenment in National Context,* edited by Roy Porter and Mikuláš Teich, 1–18. Cambridge: Cambridge University Press, 1981.

———. "The Sexual Politics of James Graham." *British Journal for Eighteenth-Century Studies* 5 (1982): 199–205.

———. "The Scientific Revolution: A Spoke in the Wheel?" In *Revolution in History,* edited by Roy Porter and Mikuláš Teich, 290–316. Cambridge: Cambridge University Press, 1986.

———. "Barely Touching: A Social Perspective on Mind and Body." In *Languages of Psyche: Mind and Body in Enlightenment Thought,* edited by George Rousseau, 45–80. Berkeley and Los Angeles: University of California Press, 1990.

———. *The Enlightenment.* London: Macmillan, 1990.

Porter, Roy, and Mikuláš Teich, eds. *The Enlightenment in National Context.* Cambridge: Cambridge University Press, 1981.

Porter, Roy, Simon Schaffer, Jim Bennett, and Olivia Brown. *Science and Profit in Eighteenth-Century London.* Cambridge: Whipple Museum for the History of Science, 1985.

Porter, Theodore. *The Rise of Statistical Thinking, 1820–1900.* Princeton: Princeton University Press, 1986.

———. *Trust in Numbers: The Pursuit of Objectivity in Science and Public Life.* Princeton: Princeton University Press, 1995.

Preti, Giulio. *Alle origini dell'etica contemporanea: Adamo Smith.* Florence: La Nuova Italia, 1957.

Price, Derek J. de Solla. "Automata in History." *Technology and Culture* 5 (1964): 9–23.

Price, John Vladimir. "The Reading of Philosophical Literature." In *Books and Their Readers in Eighteenth-Century England,* edited by Isabel Rivers, 165–96. Leicester: Leicester University Press, 1982.

Probyn, C. T. *English Fiction of the Eighteenth Century, 1700–1789.* London: Longman, 1987.

Proust, Jacques. "Diderot, L'Académie de Pétersbourg, et le projet d'une Encyclopédie Russe." *Diderot Studies,* edited by Otis Fellows and Diana Guiragossian, 12:103–31. Geneva: Libraire Droz, 1969.

———. "L'image du peuple au travail dans les planches de l'Encyclopédie." In *Images du peuple au 18e siècle,* edited by Centre Aixois d'Études et de Recherches sur le Dix-Huitième Siècle, 65–85. Paris: Colin, 1973.

Pulte, Helmut. *Das Prinzip der kleinsten Wirkung und die Kraftkonzeptionen der rationalen Mechanik: Studia Leibnitiana,* Sonderheft no. 19. Stuttgart: Franz Steiner Verlag, 1989.

Puymèges, Daniel. "Les machines dans l'*Encyclopédie.*" *Milieux* 3/4 (1980): 88–96.

———. "Les anatomies mouvantes." *Milieux* 7/8 (1981–82): 62–69.

Raikov, Boris E. "Caspar Friedrich Wolff." *Zoologisches Jahrbuch (Systematik)* 91 (1964): 555–626.

Rann, Tovio V. "The Development of Estonian Literacy in the Eighteenth and Nineteenth Centuries." *Journal of Baltic Studies* 10 (1979): 115–26.

Reddy, William M. *Money and Liberty in Modern Europe: A Critique of Historical Understanding.* Cambridge: Cambridge University Press, 1987.

Reed, Arden. *Romantic Weather: The Climates of Coleridge and Baudelaire.* Hanover: University Press of New England, 1983.

Rehbock, Philip F. *The Philosophical Naturalists.* Madison: University of Wisconsin Press, 1993.

Reill, Peter Hanns. *The German Enlightenment and the Rise of Historicism.* Berkeley and Los Angeles: University of California Press, 1975.

Reiss, Timothy J. *The Meaning of Literature.* Ithaca: Cornell University Press, 1992.

Revel, Jacques. "Knowledge of the Territory." *Science in Context* 4 (1991): 133–61.

Reynolds, Terry. *Stronger than a Hundred Men: A History of the Vertical Water Wheel.* Baltimore: Johns Hopkins University Press, 1983.

Richards, Graham. "The Absence of Psychology in the Eighteenth Century: A Linguistic Perspective." *Studies in History and Philosophy of Science* 23 (1992): 195–211.

———. *Mental Machinery: The Origins and Consequences of Psychological Ideas.* London: Athlone Press, 1992.

Richardson, Ruth, and Brian Hurwitz. "Bentham's Self Image." *British Medical Journal* 295 (1987): 195–98.

Richetti, John J. *Philosophical Writing: Locke, Berkeley, Hume.* Cambridge: Harvard University Press, 1983.

Ricuperati, Giuseppe, and Dino Carpanetto. *Italy in the Age of Reason.* New York: Longman, 1987.

Rider, Robin. "Measure of Ideas, Rules of Language: Mathematics and Language in the Eighteenth Century." In *The Quantifying Spirit in the Eighteenth Century,* edited by Tore Frängsmyr, J. L. Heilbron, and Robin E. Rider, 113–40. Berkeley and Los Angeles: University of California Press, 1990.

Riedel, Manfred. "Historizismus und Kritizismus: Kants Streit mit G. Forster und J. G. Herder." In *Deutschlands kulturelle Entfaltung: Die Neubestimmung des Menschen,* edited by

Bernard Fabian, Wilhelm Schmidt-Biggemann, and Rudolf Vierhaus, 31–48. Studien zum Achtzehnten Jahrhundert, nos. 2/3. Munich: Kraus, 1980.

Riley, James C. *The Eighteenth-Century Campaign to Avoid Disease.* New York: Macmillan, 1987.

Risse, G. B. "Historicism in Medical History: Heinrich Damerow's 'Philosophical' Historiography in Romantic Germany." *Bulletin of the History of Medicine* 43 (1969): 201–11.

Roach, Joseph R. *The Player's Passion: Studies in the Science of Acting.* Newark: University of Delaware Press, 1985.

Robel, Gert. "Die Sibirien Expeditionen und das deutsche Russlandbild im 18. Jahrhundert." In *Wissenschaftspolitik in Mittel- und Ost-Europa: Wissenschaftliche Gesellschaften, Akademien, und Hochschulen im 18. und beginnenden 19. Jahrhundert,* edited by Erik Amburger, Michal Cieslaj, and Laszlo Sziklay, 271–94. Berlin: Verlag Ulrich Camen, 1976.

Roberts, Julian. *Walter Benjamin.* London: Macmillan, 1982.

Roberts, Lissa. "The Death of the Sensuous Chemist." *Studies in History and Philosophy of Science* 26 (1995): 503–29.

———. "Science Dynamics: The Dutch Meet the 'New' Chemistry." In *Lavoisier in European Perspective: Negotiating a New Language for Chemistry,* edited by Bernadette Bensaude-Vincent and Fernando Abbri, 87–112. Canton, Mass.: Science History Publications, 1995.

———. "Science Becomes Electric: The Static Electric Generator and Some Aspects of Dutch Culture in the Eighteenth Century." *Isis,* forthcoming.

Robinet, André. "Leibniz, l'automate, et la pensée." *Studia Leibnitiana* 4 (1972): 284–90.

Roche, Daniel. *Les républicains des lettres: Gens de culture et lumières au XVIIIe siècle.* Paris: Fayard, 1988.

Rodgers, James. "Sensibility, Sympathy, Benevolence: Physiology and Moral Philosophy in *Tristram Shandy.*" In *Languages of Nature: Critical Essays on Science and Literature,* edited by Ludmilla Jordanova, 90–116. London: Free Association Books, 1986.

Roe, Shirley A. *Matter, Life, and Generation: Eighteenth-Century Embryology and the Haller-Wolff Debate.* Cambridge: Cambridge University Press, 1981.

Roger, Jacques. *Les sciences de la vie dans la pensée française du XVIIIe siècle.* Paris: Armand Colin, 1963.

———. *Buffon: Un philosophe au Jardin du roi.* Paris: Fayard, 1989.

Rogers, Pat. *Johnson.* Oxford: Oxford University Press, 1993.

Rosenberg, Hans. *Bureaucracy, Aristocracy, and Autocracy: The Prussian Experience, 1660–1815.* Boston: Beacon Press, 1958.

Rossaro, Antonio. "Cristina Roccati e il suo tempo," *Atti dell'Imperiale Reale Accademia degli Agiati in Rovereto,* ser. 4, 1 (1914): 1–44.

Rousseau, George, and Roy Porter, eds. *The Ferment of Knowledge: Studies in the Historiography of Eighteenth-Century Science.* Cambridge: Cambridge University Press, 1980.

———, eds. *Exoticism in the Enlightenment.* Manchester: Manchester University Press, 1990.

Rudwick, Martin. "Cuvier and Brongniart, William Smith, and the Reconstruction of Geohistory." *Earth Sciences History* 15 (1996): 25–36.

Ruestow, E. G. *Physics in Seventeenth and Eighteenth Century Leiden: Philosophy and the New Science in the University.* The Hague: Nijhoff, 1973.

Rupp, Jan C. C. "Matters of Life and Death: The Social and Cultural Conditions of the Rise of Anatomical Theatres, with Special Reference to Seventeenth Century Holland." *History of Science* 28 (1990): 263–87.

Rusnock, Andrea. "Quantification, Precision, and Accuracy: Determinations of Population in the Ancien Régime." In *The Values of Precision,* edited by M. Norton Wise, 17–38. Princeton: Princeton University Press, 1995.

———. "The Weight of Evidence and the Burden of Authority: Case Histories, Medical Statistics, and Smallpox Inoculation." In *Medicine in the Enlightenment,* edited by Roy Porter, 289–315. Wellcome Institute Series in the History of Medicine. Amsterdam and Atlanta: Rodopi, 1995.

———, ed. *The Correspondence of James Jurin, 1684–1750, Physician and Secretary to the Royal Society.* The Wellcome Institute Series in the History of Medicine. Amsterdam and Atlanta: Rodopi, 1996.

Russell, E. S. *Form and Function: A Contribution to the History of Animal Morphology.* 1916. Reprint, Chicago: University of Chicago Press, 1982.

Russow, Fr. "Die lebenden Missgeburten (Monstra) bei der Kunstkammer." *Publications du Musée d'Anthropologie et d'Ethnographie de l'Académie Impériale des Sciences de St. Petersbourg* 1 (1900): 145–49.

Sahlgren, Jöran. "Linné som predikant." *Svenska Linnésällskapets Årsskrift* 5 (1922): 40–55.

Said, Edward. *Orientalism.* New York: Pantheon, 1978.

Saine, Thomas P. "Who's Afraid of Christian Wolff?" In *Anticipations of the Enlightenment in England, France, and Germany,* edited by Alan C. Kors and Paul J. Korshin, 102–33. Philadelphia: University of Pennsylvania Press, 1987.

Saisselin, Rémy G. *Taste in Eighteenth Century France: Critical Reflections on the Origins of Aesthetics; or, An Apology for Amateurs.* Syracuse: Syracuse University Press, 1965.

————. "The French Garden in the Eighteenth Century: From *Belle Nature* to the Landscape of Time." *Journal of Garden History* 5 (1985): 284–97.

Salmon, C. V. "The Central Problem of David Hume's Philosophy." *Jahrbuch für Philosophie und Phänomenologische Forschung* 10: 299–49.

Salomon-Bayet, Claire. *L'Institution de la science et l'expérience du vivant: Méthode et expérience à l'Académie Royale des Sciences, 1666–1793.* Paris: Flammarion, 1978.

Sarton, George. *Introduction to the History of Science,* 3 vols. in 5 parts. Washington, D.C.: Carnegie, 1927–48.

Saskolskij, I. P. "New Phenomena in the Baltic Trade of Russia in the Seventeenth Century," *Scandinavian Economic History Review* 34 (1986): 41–53.

Sauer, Lieselotte. *Marionetten, Maschinen, Automaten: der künstliche Mensch in der deutschen und englischen Romantik.* Bonn: Bouvier, 1983.

Savaris, Paola. *Cristina Roccati: Una rodigina del '700 tra scienza e poesia.* Tesi di laurea, Facoltà di Magistero, Università degli Studi di Ferrara, 1990–91.

Schaffer, Simon. "The Phoenix of Nature: Fire and Evolutionary Cosmology in Wright and Kant," *Journal for the History of Astronomy* 9 (1978): 180–200.

————. "Natural Philosophy." In *The Ferment of Knowledge: Studies in the Historiography of Eighteenth-Century Science,* edited by George Rousseau and Roy Porter, 55–91. Cambridge: Cambridge University Press, 1980.

————. "Natural Philosophy and Public Spectacle in the Eighteenth Century." *History of Science* 21 (1983): 1–43.

————. "Authorized Prophets: Comets and Astronomers after 1759." *Studies in Eighteenth-Century Culture* 17 (1987): 45–74.

————. "States of Mind: Enlightenment and Natural Philosophy." In *Languages of Psyche: Mind and Body in Enlightenment Thought,* edited by George Rousseau, 233–90. Berkeley and Los Angeles: University of California Press, 1990.

————. "Self-Evidence," *Critical Inquiry* 18 (1992): 327–62.

————. "The Consuming Flame: Electrical Showmen and Tory Mystics in the World of Goods." In *Consumption and the World of Goods,* edited by John Brewer and Roy Porter, 489–526. London: Routledge, 1993.

————. "The Show That Never Ends: Perpetual Motion in the Early Eighteenth Century," *British Journal for the History of Science* 28 (1995): 157–89.

Schama, Simon. "The Enlightenment in the Netherlands." In *The Enlightenment in National Context,* edited by Roy Porter and Mikuláš Teich, 54–71. Cambridge: Cambridge University Press, 1981.

Scheicher, Elisabeth. *Die Kunstkammer.* Innsbruck: Kunsthistorisches Museum, 1977.

Scheugl, Hans. *Show Freaks und Monster.* Cologne: DuMont Schauberg, 1974.

Schiebinger, Londa. *The Mind Has No Sex? Women in the Origins of Modern Science.* Cambridge: Harvard University Press, 1989.

————. *Nature's Body: Gender in the Making of Modern Science.* Boston: Beacon Press, 1993.

Schilling, Heinz. "Innovation through Migration: The Settlements of Calvinistic Netherlanders in Sixteenth- and Seventeenth-Century Central and Western Europe." *Social History* 21 (1983): 7–33.

Schmidt, James. "The Question of Enlightenment: Kant, Mendelssohn, and the Mittwochs-gesellschaft." *Journal of the History of Ideas* 50 (1989): 269–91.

———. "What is Enlightenment? A Question, Its Context, and Some Consequences." In *What is Enlightenment? Eighteenth-Century Answers and Twentieth-Century Questions,* edited by James Schmidt, 1–44. Berkeley and Los Angeles: University of California Press, 1996.

Schmidt, Jochen. *Die Geschichte des Genie-Gedankens in der deutschen Literatur, Philosophie, und Politik.* 2 vols. Darmstadt: Wissenschaftliche Buchgesellschaft, 1985.

Schmölders, Claudia, 1995. *Das Vorurteil im Leibe: Eine Einführung in die Physiognomik.* Berlin: Akademie Verlag, 1985.

Schneider, Manfred, and Rüdiger Campe, eds. *Geschichten der Physiognomik: Text, Bild, Wissen.* Freiburg: Rombach, 1996.

Schneiders, Werner. *Die wahre Aufklärung: Zum Selbstverständnis der deutschen Aufklärung.* Freiburg: Alber, 1974.

———. *Christian Thomasius, 1655–1728.* Hamburg: Meiner, 1989.

———, ed. *Christian Wolff, 1679–1754.* Studien zum Achtzehnten Jahrhundert, no. 4. Hamburg: Meiner, 1983.

Schöffler, Herbert. *Deutscher Osten im deutschen Geist: Vom Martin Optiz zu Christian Wolff.* Das Abendland, no. 3. Frankfurt am Main: Klostermann, 1940.

Schofield, Robert E. *Mechanism and Materialism: British Natural Philosophy in an Age of Reason.* Princeton: Princeton University Press, 1969.

Schrader, Wilhelm. *Geschichte der Friedrich-Universität zu Halle.* 2 vols. Berlin: F. Dummler, 1894.

Schultz, Uwe. *Immanuel Kant.* Hamburg: Rowohlt, 1965.

Schulze, Ludmilla. "The Russification of the St. Petersburg Academy of Sciences and Arts in the Eighteenth Century." *British Journal for the History of Science* 18 (1985): 305–35.

Schumacher, I. "Karl Friedrich Kielmeyer, eine Wegbereiter neuer Ideen." *Medizinhistorisches Journal* 14 (1979): 81–99.

Schwab, Raymond. *The Oriental Renaissance: Europe's Rediscovery of India and the East, 1680–1880.* Translated by Gene Patterson-Black and Victor Reinking. New York: Columbia University Press, 1984.

Schwartz, Richard B. "Boswell and Hume: The Deathbed Interview." In *New Light on Boswell: Critical and Historical Essays on the Occasion of the Bicentenary of "The Life of Johnson,"* edited by Greg Clingham, 116–25. Cambridge: Cambridge University Press, 1991.

Scott, Joan. "Gender: A Useful Category of Historical Analysis," In *Gender and the Politics of History,* 28–50. New York: Columbia University Press, 1988.

Scott, William R. *Adam Smith as Student and Professor.* Glasgow: Jackson, Son, and Co, 1937.

Seemen, H. von. *Zur Kenntnis der Medizinhistorie in der deutschen Romantik.* Leipzig: Füssli, 1926.

Séris, Jean-Pierre. *Machine et communication: Du théâtre des machines à la mécanique industrielle.* Paris: Vrin, 1987.

Sewell, William H. "Visions of Labor: Illustrations of the Mechanical Arts before, in, and after Diderot's *Encyclopédie.*" In *Work in France: Representations, Meaning, Organization, and Practice,* edited by Steven Laurence Kaplan and Cynthia J. Koepp, 258–86. Ithaca: Cornell University Press, 1986.

Shapin, Steven. "Discipline and Bounding: The History and Sociology of Science as Seen through the Externalism-Internalism Debate." *History of Science* 90 (1992): 333–69.

———. *A Social History of Truth: Civility and Science in Seventeenth-Century England.* Chicago: University of Chicago Press, 1994.

———. *The Scientific Revolution.* Chicago: University of Chicago Press, 1996.

Shapin, Steven, and Simon Schaffer. *Leviathan and the Air-Pump: Hobbes, Boyle, and the Experimental Life.* Princeton: Princeton University Press, 1985.

Sher, Richard B. *Church and University in the Scottish Enlightenment: The Moderate Literati of Edinburgh.* Princeton: Princeton University Press, 1985.

Shinn, Terry. *Savoir technique et pouvoir politique: L'École Polytechnique, 1794–1914.* Paris: Presses de la Fondation Nationale des Sciences Politiques, 1980.

Siebert, Donald. *The Moral Animus of David Hume.* Newark: University of Delaware Press, 1990.

Sitter, John. *Literary Loneliness in Eighteenth-Century England.* Ithaca: Cornell University Press, 1982.

Skinner, G. "'The Price of a Tear': Economic Sense and Sensibility in Sarah Fielding's *David Simple.*" *Literature and History,* 3d ser., 1 (1992): 16–28.

Skinner, Quentin. *Reason and Rhetoric in the Philosophy of Hobbes.* Cambridge: Cambridge University Press, 1996.

Slezkine, Yuri. "Naturalists versus Nations: Eighteenth-Century Russian Scholars Confront Ethnic Diversity," *Representations* no. 47 (1994): 170–95.

Sliggers, C. C. "Honderd jaar natuurkundige amateurs te Haarlem," In *Een elektriserend geleerde: Martinus van Marum, 1750–1837,* edited by A. Wiechmann and L. C. Palm, 67–101. Haarlem: Joh. Enschede en Zonen, 1987.

Sloan, Phillip R. "The Buffon-Linnaeus Controversy." *Isis* 67 (1976): 356–75.

———. "From Logical Universals to Historical Individuals: Buffon's Idea of Biological Species." In *Histoire du concept d'espèce dans les sciences de la vie,* edited by Scott Atran et al., 101–40. Paris: Fondation Singer-Polignac, 1985.

Sloan, Phillip R., and John Lyon, eds. *From Natural History to the History of Nature: Readings from Buffon and His Critics.* Notre Dame: University of Notre Dame Press, 1981.

Smith, Jay Michael. *The Culture of Merit: Nobility, Royal Service, and the Making of Absolute Monarchy in France, 1600–1789.* Ann Arbor: University of Michigan Press, 1996.

Smith, Pamela. *The Business of Alchemy: Science and Culture in the Holy Roman Empire.* Princeton: Princeton University Press, 1994.

Smith, Preserved. *The Enlightenment, 1687–1776.* 1934. Reprint, New York: Collier, 1962.

Smith, Roger. "The Language of Human Nature." In *Inventing Human Science: Eighteenth-Century Domains,* edited by Christopher Fox, Roy Porter, and Robert Wokler, 88–111. Berkeley and Los Angeles: University of California Press, 1995.

Smith, Wesley D. *The Hippocratic Tradition.* Ithaca: Cornell University Press, 1979.

Snelders, H. A. M. "J. S. C. Schweigger, His Romanticism, and His Crystal Electrical Theory." *Isis* 62 (1971): 328–38.

———. *Het Gezelschap der Hollandsche Scheikundigen.* Amsterdam: Rodopi, 1980.

———. "The New Chemistry in the Netherlands." *Osiris,* n.s. 4 (1988): 121–45.

———. "Professors, Amateurs, and Learned Societies: The Organization of the Natural Sciences." In *The Dutch Republic in the Eighteenth Century: Decline, Enlightenment, and Revolution,* edited by Margaret C. Jacob and Wijnand W. Mijnhardt, 308–23. Ithaca: Cornell University Press, 1992.

Spacks, Patricia Meyer. *Desire and Truth: Functions of Plot in Eighteenth-Century English Novels.* Chicago: University of Chicago Press, 1990.

Span, Walter, 1989. "Auf dem Wege zur theologischen Aufklärung in Halle: Von Johann Franz Budde zu Siegmund Jakob Baumgarten." *Halle: Aufklärung und Pietismus,* edited by Norbert Hinske, 71–89. Wolfenbütteler Studien zur Aufklärung, no. 15 (Zentren der Aufklärung, no. 1). Heidelberg: Schneider, 1989.

Spary, E. C. "Making the Natural Order: The Paris Jardin du Roi, 1750–1795." Ph.D. diss., Cambridge University, 1993.

Spekke, Aronds. *History of Latvia: An Outline.* Stockholm: M. Goppers, 1951.

Spencer, Samia I., ed. *French Women and the Age of Enlightenment.* Bloomington: Indiana University Press, 1984.

Spierenburg, Pieter. *De Sociale Functie van Openbare Strafvoltrekkingen.* Rotterdam: Centrum voor Maatschappijgeschiedenis, Erasmus Universiteit, 1979.

Spitzer, Leo. "Milieu and Ambiance." In *Essays in Historical Semantics,* 179–316. New York: S. F. Vanni, 1948.

Stafford, Barbara Maria. *Voyage into Substance: Art, Science, Nature, and the Illustrated Travel Account.* Cambridge: MIT Press, 1984.

———. *Artful Science: Enlightenment Entertainment and the Eclipse of Visual Education.* Cambridge: MIT Press, 1994.

Stafleu, Frans A. *Linnaeus and the Linneans: The Spreading of Their Ideas in Systematic Botany, 1735–1789.* Utrecht: Oosthoek, 1971.

Stallybrass, Peter, and Allon White. *The Poetics and Politics of Transgression.* London: Methuen, 1986.

Stangeland, Charles Emil. *Pre-Malthusian Doctrines of Population.* New York: Columbia University Press, 1904.

Star, Pieter van der, ed. *Fahrenheit's Letters to Leibniz and Boerhaave.* Amsterdam: Rodopi, 1983.

Starobinski, Jean. *The Invention of Liberty, 1700–1789.* Geneva: Skira, 1987.

Stavenow-Hidemark, Elisbeth, ed. *1700-tals textil. Anders Berchs samling i Nordiska Museet.* Stockholm: Nordiska Museets Förlag, 1990.

Steele, Brett D. "Muskets and Pendulums: Benjamin Robins, Leonard Euler, and the Ballistics Revolution." *Technology and Culture* 35 (1994): 348–82.

Steinbrügge, Liselotte. *The Moral Sex: Women's Nature in the French Enlightenment.* Oxford: Oxford University Press, 1995.

Stewart, Larry. *The Rise of Public Science: Rhetoric, Technology, and Natural Philosophy in Newtonian Britain, 1660–1750.* Cambridge: Cambridge University Press, 1992.

Stewart, Larry, and Paul Weindling. "Philosophical Threads: Natural Philosophy and Public Experiment among the Weavers of Spitalfields." *British Journal for the History of Science* 28 (1995): 37–62.

Stollberg-Rilinger, Barbara. *Der Staat als Maschine: Zur politischen Metaphorik des absoluten Fürstenstaat.* Historische Forschungen, no. 30. Berlin: Duncker, 1986.

Stone, Bailey. *The Genesis of the French Revolution: A Global-Historical Interpretation.* Cambridge: Cambridge University Press, 1994.

Stroud, Barry. *Hume.* London: Routledge and Kegan Paul, 1977.

Suleiman, Ezra N. *Elites in French Society: The Politics of Survival.* Princeton: Princeton University Press, 1978.

Sutter, Alex. *Göttliche Maschinen: Die Automaten für Lebendiges bei Descartes, Leibniz, La Mettrie, und Kant.* Frankfurt am Main: Athenäum, 1988.

Sutton, Geoffrey. "Electric Medicine and Mesmerism." *Isis* 72 (1981): 375–92.

———. *Science for a Polite Society: Gender, Cuture, and the Demonstration of Enlightenment.* Boulder: Westview Press, 1995.

Swianiewicz, Stanislaw. "The University of Wilno in Historical Perspective." *Polish Review* 27 (1982): 29–46.

Sydow, Carl-Otto von. "Vetenskapssocieteten och Henric Benzelius' Lapplandsresa, 1711." *Lychnos,* 1962, 138–61.

Tadmor, Naomi. "'In the Even My Wife Read to Me': Women, Reading, and Household Life in the Eighteenth Century." In *The Practice and Representation of Reading in England,* edited by James Raven, Helen Small, and Naomi Tadmor, 162–74. Cambridge: Cambridge University Press, 1996.

Tang, Frank, and Margriet Wigard. *Amsterdamse gasthuizen vanaf de Middeleeuwen.* Amsterdam: University of Amsterdam Press, 1994.

Taton, René, ed. *Enseignement et diffusion des sciences en France au XVIII siècle.* Paris: Hermann, 1964.

Taylor, E. G. R. *The Mathematical Practitioners of Tudor and Stuart England.* Cambridge: for the Institute of Navigation at the University Press, 1954.

Taylor, F. Sherwood. "The Origin of the Thermometer." *Annals of Science* 5 (1942): 129–56.

Terrall, Mary. "Maupertuis and Eighteenth-Century Scientific Culture." Ph.D. diss., University of California, Los Angeles, 1987.

———. "The Culture of Science in Frederick the Great's Berlin." *History of Science* 28 (1990): 333–64.

———. "Representing the Earth's Shape: The Polemics Surrounding Maupertuis's Expedition to Lapland." *Isis* 83 (1992): 218–37.

———. "Émilie du Châtelet and the Gendering of Science." *History of Science* 33 (1995): 283–310.

———. "Gendered Spaces, Gendered Audiences: Inside and Outside the Paris Academy of Sciences." *Configurations* 2 (1995): 207–32.

Thackray, Arnold, and Robert K. Merton. "On Discipline-Building: The Paradoxes of George Sarton." *Isis* 63 (1972): 673–95.

Thode, Britta. *Die Göttinger Anatomie 1733 bis 1828.* M.D. diss., University of Göttingen, 1979.

Thomas, Nicholas. *Colonialism's Culture: Anthropology, Travel, and Government.* Cambridge: Polity Press, 1994.

Thompson, E. P. *Customs in Common.* London: Merlin, 1991.

Thomson, Ann. *Materialism and Society in the Mid–Eighteenth Century: La Mettrie's Discours Préliminaire.* Geneva: Droz, 1981.

Tilly, Charles. *Coercion, Capital, and European States, AD 990–1992,* revised ed. Cambridge, Mass.: Basil Blackwell, 1992.

Tocqueville, Alexis de. *The Old Régime and the French Revolution.* Translated by Stuart Gilbert. Garden City: Doubleday, 1955. Original edition published 1856.

Todd, Janet. *Sensibility: An Introduction.* London: Methuen, 1986.

Tomaselli, Sylvana. "Moral Philosophy and Population Questions in Eighteenth Century Europe." In *Population and Resources in Western Intellectual Traditions,* edited by Michael S. Teitelbaum and Jay M. Winter, 7–29. Supplement to *Population and Development Review,* vol. 14. Cambridge: Cambridge University Press, 1989.

Tonelli, Giorgio. "La nécessité des lois de la nature au XVIIe siècle et chez Kant en 1762," *Revue d'Histoire des Sciences et de Leurs Applications* 12, no. 3 (1959): 225–41.

Tönneson, Kåre. "Tenancy, Freehold, and Enclosure in Scandinavia from the Seventeenth to the Nineteenth Century." *Scandinavian Journal of History* 6 (1981): 191–206.

Torlais, Jean. *Un ésprit encyclopédique en dehors de 'L'Encyclopédie': Réaumur: D'après des documents inédits.* Rev. ed. Paris: Albert Blanchard, 1961.

Tort, Patrick. *L'ordre et les monstres: Le débat sur l'origine des déviations anatomiques au XVIIIe siècle.* Paris: Le Sycomore, 1980.

Tröhler, Ulrich. "Quantification in British Medicine and Surgery, 1750–1830, with Special Reference to Its Introduction into Therapeutics." Ph.D. diss., University of London, 1978.

Truesdell, Clifford. *Essays in the History of Mechanics.* Berlin: Springer-Verlag, 1968.

Trumpler, Maria. "Questioning Nature: Experimental Investigation of Animal Electricity in Germany, 1790–1810." Ph.D. diss., Yale University, 1992.

Tullberg, Tycho. *Linnéporträtt: Vid Uppsala universitets minnesfest på tvåhundraårsdagen af Carl von Linnés födelse.* Stockholm: Aktiebolaget Ljus, 1907.

Turnbull, Gordon. "Boswell and Sympathy: The Trial and Execution of John Reid." In *New Light on Boswell: Critical and Historical Essays on the Occasion of the Bicentenary of "The Life of Johnson,"* edited by Greg Clingham, 104–15. Cambridge: Cambridge University Press, 1991.

Utterström, Gustaf. "Potatisodlingen i Sverige under frihetstiden: Med en översikt över odlingens utveckling intill omkring 1820." *Historisk Tidskrift* 63 (1943): 145–85.

———. "Labour Policy and Population Thought in Eighteenth-Century Sweden." *Scandinavian Economic History Review* 10 (1962): 262–79.

Van Sant, Ann Jessie. *Eighteenth-Century Sensibility and the Novel: The Senses in Social Context.* Cambridge: Cambridge University Press, 1993.

Van Tieghem, Paul. *Le sentiment de la nature dans le préromantisme européen.* Paris: A. G. Nizet, 1960.

Varep, E. "The Maps of Estonia Published by the Academy of Sciences, St. Petersburg, in the Eighteenth Century." *Imago Mundi* 31 (1979): 88–93.

Vartanian, Aram. "Trembley's Polyp, La Mettrie, and Eighteenth-Century French Materialism." *Journal of the History of Ideas* 11 (1950): 259–86.

———. *Diderot and Descartes: Scientific Naturalism in the Enlightenment.* Princeton: Princeton University Press, 1953.

———. *La Mettrie's L'Homme Machine.* Princeton: Princeton University Press, 1960.

Vatin, François. *Le travail: Économie et physique, 1780–1830.* Paris: Presses Universitaires de France, 1993.

Vaupaemel, Geert. "Experimental Physics and the Natural Science Curriculum in Eighteenth Century Louvain." *History of Universities* 7 (1988): 175–96.

Venturi, Franco. *Utopia and Reform in the Enlightenment.* Cambridge: Cambridge University Press, 1971.

———. *Italy and the Enlightenment.* Edited by Stuart Woolf, translated by Susan Corsi. New York: New York University Press, 1972.

Vercruysse, Jeroom. *Voltaire et la Hollande.* Vol. 46 of *Studies on Voltaire and the Eighteenth Century.* Geneva: Institut et Musée Voltaire, 1966.

Vermij, Rienk H. *Secularisering en Natuurwetenschap in the Zeventiende and Achttiende Eeuw: Bernard Nieuwentijt.* Amsterdam: Rodopi, 1991.

Vidler, Anthony. *The Writing of the Walls: Architectural Theory in the Late Enlightenment.* London: Butterworth, 1989.

Vincent-Buffault, Anne. *The History of Tears: Sensibility and Sentimentality in France.* Translated by Teresa Bridgeman. Basingstoke: Macmillan, 1991.

Vincenti, Walter G. *What Engineers Know and How They Know It: Analytical Studies from Aeronautical History.* Baltimore: Johns Hopkins University Press, 1990.

Visser, R. P. W. *The Zoological Work of Petrus Camper, 1722–1789.* Amsterdam: Rodopi, 1985.

———. "Camper en de Natuurlijke Historie der Dieren." In *Petrus Camper, 1722–1789: Onderzoeker van Nature,* edited by J. Schuller tot Peursum-Meijer and W. R. H. Koops, 59–64. Groningen: Universiteitsmuseum, 1989.

Völker, Karl. *Die Kirchengeschichtsschreibung der Aufklärung.* Tübingen: Mohr, 1921.

Völker, Klaus, ed. *Künstliche Menschen.* 1971. Reprint, Munich: Suhrkamp, 1994.

Vorländer, Karl. *Immanuel Kant: Der Mann und das Werk.* 3d ed. Hamburg: Meiner, 1992. First edition published 1924.

Voss, Irmela. *Das pathologisch-anatomische Werk Albrecht v. Hallers in Göttingen.* Göttingen: Vandenhoeck und Ruprecht, 1937.

Wachenfelt, Miles von. "Ett återfunnet Lapplandsporträtt av Carl von Linné." *Svenska Linnésällskapets Årsskrift* 30 (1947): 21–35.

Walling Howard, M. *The Influence of Plutarch in the Major European Literatures of the Eighteenth Century.* Chapel Hill: University of North Carolina Press, 1970.

Walters, Alice Nell. "Tools of Enlightenment: The Material Culture of Science in Eighteenth-Century England." Ph.D. diss., University of California, Berkeley, 1992.

Watt, Ian. *The Rise of the Novel: Studies in Defoe, Richardson and Fielding.* 1957. Reprint, London: Hogarth Press, 1987.

Weber, Max. "Die 'Objectivität' sozialwissenschaftlicher und socialpolitischer Erkenntnis." *Archiv für Sozialwissenschaften und Sozialpolitik* 19, no. 1 (1904): 22–87.

———. *From Max Weber.* Edited by and translated by H. H. Gerth and C. Wright Mills. New York: Oxford University Press, 1946.

———. *The Protestant Ethic and the Spirit of Capitalism.* Edited by Talcott Parsons. New York: Scribners', 1958. Original edition published 1904.

Weingarten, Michael. "Menschenarten oder Menschenrassen: Die Kontroverse zwischen Georg Forster und Immanuel Kant." In *Georg Forster in seiner Epoche,* edited by Gerhardt Pickerodt, 117–48. Literatur im historischen Prozess, new ser., no. 4. Berlin: Argument, 1982.

Weinrich, Harald. "Warum will Kant seinen Diener Lampe vergessen?" *Merkur* 51, no. 1 (1997): 41–51.

Weiss, John Hubbel. *The Making of Technological Man: The Social Origins of French Engineering Education.* Cambridge: MIT Press, 1982.

———. "Bridges and Barriers: Narrowing Access and Changing Structure in the French Engineering Profession, 1800–1850." In *Professions and the French State,* edited by Gerald Geison. Philadelphia: University of Pennsylvania Press, 1984.

Wellman, Kathleen. *La Mettrie: Medicine, Philosophy, and Enlightenment.* Durham: Duke University Press, 1992.

Wendorf, Richard. *The Elements of Life: Biography and Portrait Painting in Stuart and Georgian England.* Oxford: Clarendon Press, 1990.

Wenzel, Manfred. "Die Anthropologie Johann Gottfried Herders und das klassische Humanitätsideal." In *Die Natur des Menschen: Probleme der Physischen Anthropologie und Rassenkunde, 1750–1850*, edited by Gunter Mann and Franz Dumont, 137–67. Soemmerring-Forschungen, vol. 6. Stuttgart and New York: Fischer, 1990.

————, ed. *Samuel Thomas Soemmerring in Kassel, 1779–1784: Beiträge zur Wissenschaftsgeschichte der Goethezeit*. Soemmerring-Forschungen, vol. 9. Stuttgart and New York: Fischer, 1995.

Wetzels, W. D. *Johann Wilhelm Ritter: Physik im Wirkungsfeld der deutschen Romantik*. Berlin: de Gruyter, 1973.

Wiebenson, Dora. *The Picturesque Garden in France*. Princeton: Princeton University Press, 1978.

Wiechmann, A., and L. C. Palm, eds. *Een elektriserend geleerde: Martinus van Marum, 1750–1837*. Haarlem: Joh. Enschede en Zonen, 1987.

Wiklund, K. B. "Linné och lapparna." *Svenska Linnésällskapets Årsskrift* 8 (1925): 59–93.

Williams, Raymond. *Culture and Society, 1870–1950*. London: Chatto and Windus, 1958.

————. *Keywords: A Vocabulary of Culture and Society*. London: Fontana, 1976.

Wise, M. Norton. "German Concepts of Force, Energy, and the Electromagnetic Ether, 1845–1880." In *Conceptions of Ether: Studies in the History of Ether Theories, 1740–1900*, edited by G. N. Cantor and M. J. S. Hodge, 269–307. Cambridge: Cambridge University Press, 1981.

————. "Mediations: Enlightenment Balancing Acts or the Technologies of Rationalism." In *World Changes: Thomas Kuhn and the Nature of Science*, edited by Paul Horwich, 207–56. Cambridge: MIT Press, 1993.

Wittram, Reinhard. *Baltische Geschichte*. Munich: R. Oldenbourg, 1954.

————. *Das Interesse an der Geschichte*. Göttingen: Vandenhoeck und Ruprecht, 1958.

Wokler, Robert. "Apes and Races in the Scottish Enlightenment: Monboddo and Kames on the Nature of Man." In *Philosophy and Science in the Scottish Enlightenment*, edited by Peter Jones, 145–68. Edinburgh: Edinburgh University Press, 1989.

————. "From *l'Homme Physique* to *l'Homme Morale* and Back: Towards a History of Enlightenment Anthropology." *History of the Human Sciences* 6 (1993): 121–38.

Wolf, Abraham. *A History of Science, Technology, and Philosophy in the Eighteenth Century*. 2d ed. 2 vols. Edited by D. McKie. New York: Harper, 1961. First edition published 1938.

Wood, Paul B. "The Natural History of Man in the Scottish Enlightenment." *History of Science* 27 (1989): 89–123.

————. "Science and the Aberdeen Enlightenment." In *Philosophy and Science in the Scottish Enlightenment*, edited by Peter Jones, 39–66. Edinburgh: Edinburgh University Press, 1989.

————. "Hume, Reid, and the Science of the Mind." In *Hume and Hume's Connexions*, edited by M. A. Stewart and John P. Wright, 119–39. Edinburgh: Edinburgh University Press, 1994.

————. "The Science of Man." In *Cultures of Natural History*, edited by Nicholas Jardine, J. A. Secord, and E. C. Spary, 197–210. Cambridge: Cambridge University Press, 1996.

Woodbridge, Kenneth. *Princely Gardens: The Origins and Development of the French Formal Style*. New York: Rizzoli, 1986.

Woodmansee, M. *The Author, Art, and the Market: Rereading the History of Aesthetics*. New York: Columbia University Press, 1994.

Wundt, Max. *Die deutsche Schulphilosophie im Zeitalter der Aufklärung*. 1945. Reprint, Hildesheim: Olms, 1992.

Yolton, John W. *John Locke and the Way of Ideas*. Oxford: Clarendon Press, 1956.

————. *Thinking Matter: Materialism in Eighteenth Century Britain*. Oxford: Basil Blackwell, 1983.

Zabeeh, Farhang. *Hume, Precursor of Modern Empiricism: An Analysis of His Opinions on Meaning, Metaphysics, Logic, and Mathematics*. 2d ed. The Hague: Martinus Nihjoff, 1973.

Zapperi, Roberto. "Ein Haarmensch auf einem Gemälde von Agostino Carracci." In *Der "falsche" Körper: Beiträge zu einer Geschichte der Monstrositäten*, edited by Michael Hagner, 45–55. Göttingen: Wallstein, 1995.

CONTRIBUTORS

Ken Alder is associate professor of history at Northwestern University and the author of *Engineering the Revolution: Arms and Enlightenment in France, 1763–1815* (1997), which was awarded the 1998 Dexter Prize from the Society for the History of Technology.

William Clark currently teaches in the Department of History and Philosophy of Science, Cambridge University. He is coeditor with Peter Becker of *Little Tools of Knowledge* (forthcoming).

Lorraine Daston is director at the Max-Planck-Institut für Wissenschaftsgeschichte in Berlin and teaches at the University of Chicago. She is the author of *Classical Probability in the Enlightenment* (1988) and, with Katharine Park, *Wonders and the Order of Nature* (1997). Her current work concerns the history of the ideals and practices of scientific objectivity.

Paula Findlen teaches in the history department at Stanford University. She is the author of *Possessing Nature: Museums, Collecting, and Scientific Culture in Early Modern Italy* (1994) and many essays on science and culture in early modern Europe. Currently she is working on a book entitled *The Daughters of Galileo.*

Marina Frasca-Spada is a historian of philosophy. A former research fellow of Newnham College, Cambridge, she is an affiliated lecturer at the Department of History and Philosophy of Science, Cambridge University, and the associate and book reviews editor of *Studies in History and Philosophy of Science* and of *Studies in History and Philosophy of Biological and Biomedical Sciences.* She is the author of *Space and Self in Hume's "Treatise"* (1998). Her current work is on eighteenth-century eclectic writings and miscellanies as vehicles for the teaching and dissemination of logic, metaphysics, and moral philosophy.

Jan Golinski is associate professor of history and humanities at the University of New Hampshire. He is the author of *Science as Public Culture: Chemistry and Enlightenment in Britain, 1760–1820* (1992), and of *Making Natural Knowledge: Constructivism and the History of Science* (1998).

Michael Hagner is senior fellow at the Max-Planck-Institut für Wissenschafts-geschichte in Berlin. He is the author of *Homo cerebralis: Der Wandel vom See-lenorgan zum Gehirn* (1997) and has edited *Der "falsche" Koerper: Beiträge zu einer Geschichte der Monstrositäten* (1995). With Hans-Jörg Rheinberger and Bettina Wharig-Schmidt, he has coedited *Räume des Wissens: Repräsentation, Codierung, Spur* (1996).

Nicholas Jardine is professor of history and philosophy of the sciences at Cam-bridge University. His recent books include *Romanticism and the Sciences,* edited with A. R. Cunningham (1990), *The Scenes of Inquiry: On the Reality of Questions in the Sciences* (1991), and *Cultures of Natural History,* edited with James Secord and E. C. Spary (1996). He is editor of *Studies in History and Philosophy of Science* and *Studies in History and Philosophy of Biological and Biomedical Sciences.* His current research projects are on historiography of the sciences and, in collaboration with Alain Segonds, on priority disputes in early modern cosmology.

Lisbet Koerner studied history at the University of California, Berkeley, and at Harvard University. Her doctoral work addressed problems in eighteenth-century botany. She is currently an associate of the Department of the History of Science at Harvard University.

Dorinda Outram is professor of history at the University of Rochester. She is the author of *The Body and the French Revolution* (1989), and *The Enlighten-ment* (1996). After a research fellowship at Girton College, Cambridge, she has held visiting positions at Harvard University, the Dibner Institute for the His-tory of Science and Technology at MIT, Griffith University, Australia, and the Max-Planck-Institut für Wissenschaftsgeschichte, Berlin. She is currently at work on a book on exploration during the Enlightenment.

Lissa Roberts is associate professor of history at San Diego State University. Her research interests include the history of instrumentation, the chemical revolution, and the cultural history of eighteenth-century science in France and the Netherlands. She is currently working on a book entitled *Utility and Virtue: A Cultural History of Dutch Science in the Eighteenth Century.*

Andrea Rusnock is assistant professor of history in the Department of Science and Technology Studies, Rensselaer Polytechnic Institute. She is the editor of *The Correspondence of James Jurin, 1684–1750, Physician and Secretary to the Royal Society* (1996), and is currently completing a book on medical and polit-ical arithmetic in the eighteenth century.

Simon Schaffer is reader in history and philosophy of science at Cambridge University. He has recently published a number of articles on machinery and technique in eighteenth-century natural philosophy and is currently collaborating in a joint research project on science, travel, and instrumentation in the eighteenth and nineteenth centuries.

E. C. Spary is a research fellow at the Max-Planck-Institut für Wissenschaftsgeschichte, and an affiliated scholar of the Department of History and Philosophy of Science, Cambridge University. She has jointly edited *Cultures of Natural History* (1996) with Nicholas Jardine and James Secord. Her research to date, including a forthcoming book on the Jardin des Plantes in the French Revolution, *Naturalizing the Tree of Liberty,* has explored the settings and practices of natural history in late-eighteenth- and early-nineteenth-century Paris. At present she is preparing a book on the development of scientific accounts of food in Paris.

Mary Terrall is an assistant professor in the Department of History at the University of California, Los Angeles. Recent publications include articles on Émilie du Châtelet, on Maupertuis's theory of generation, and on issues of gender in Enlightenment science. She is working on a biographical study of Maupertuis.

Index